Handbook of Research on the Role of Tourism in Achieving Sustainable Development Goals

Filipa Brandão
GOVCOPP, University of Aveiro, Portugal

Zélia Breda
GOVCOPP, University of Aveiro, Portugal

Rui Costa
GOVCOPP, University of Aveiro, Portugal

Carlos Costa
GOVCOPP, University of Aveiro, Portugal

A volume in the Advances in Hospitality, Tourism, and the Services Industry (AHTSI) Book Series

IGI Global
PUBLISHER of TIMELY KNOWLEDGE

Published in the United States of America by
IGI Global
Business Science Reference (an imprint of IGI Global)
701 E. Chocolate Avenue
Hershey PA, USA 17033
Tel: 717-533-8845
Fax: 717-533-8661
E-mail: cust@igi-global.com
Web site: http://www.igi-global.com

Library of Congress Cataloging-in-Publication Data

Names: Brandão, Filipa, 1982- editor. | Breda, Zélia, 1975- editor. | Costa, Rui, 1975- editor. | Costa, Carlos, 1964- editor.
Title: Handbook of research on the role of tourism in achieving the sustainable development goals / Filipa Brandão, Zélia Breda, Rui Costa, and Carlos Costa, editors.
Description: Hershey, PA : Business Science Reference, 2021. | Includes bibliographical references and index. | Summary: "This book examines theoretical and practical approaches on the role of tourism in achieving sustainable development goals"-- Provided by publisher.
Identifiers: LCCN 2020013252 (print) | LCCN 2020013253 (ebook) | ISBN 9781799856917 (hardcover) | ISBN 9781799856931 (ebook)
Subjects: LCSH: Sustainable tourism. | Tourism--Social aspects. | Sustainable Development Goals.
Classification: LCC G156.5.S87 R65 2021 (print) | LCC G156.5.S87 (ebook) | DDC 338.4/791--dc23
LC record available at https://lccn.loc.gov/2020013252
LC ebook record available at https://lccn.loc.gov/2020013253

This book is published in the IGI Global book series Advances in Hospitality, Tourism, and the Services Industry (AHTSI) (ISSN: 2475-6547; eISSN: 2475-6555)

British Cataloguing in Publication Data
A Cataloguing in Publication record for this book is available from the British Library.

All work contributed to this book is new, previously-unpublished material. The views expressed in this book are those of the authors, but not necessarily of the publisher.

For electronic access to this publication, please contact: eresources@igi-global.com.

Advances in Hospitality, Tourism, and the Services Industry (AHTSI) Book Series

Maximiliano Korstanje
University of Palermo, Argentina

ISSN:2475-6547
EISSN:2475-6555

Mission

Globally, the hospitality, travel, tourism, and services industries generate a significant percentage of revenue and represent a large portion of the business world. Even in tough economic times, these industries thrive as individuals continue to spend on leisure and recreation activities as well as services.

The Advances in Hospitality, Tourism, and the Services Industry (AHTSI) book series offers diverse publications relating to the management, promotion, and profitability of the leisure, recreation, and services industries. Highlighting current research pertaining to various topics within the realm of hospitality, travel, tourism, and services management, the titles found within the AHTSI book series are pertinent to the research and professional needs of managers, business practitioners, researchers, and upper-level students studying in the field.

Coverage

- Service Training
- Leisure & Business Travel
- International Tourism
- Customer Service Issues
- Hotel Management
- Service Design
- Health and Wellness Tourism
- Sustainable Tourism
- Casino Management
- Cruise Marketing and Sales

IGI Global is currently accepting manuscripts for publication within this series. To submit a proposal for a volume in this series, please contact our Acquisition Editors at Acquisitions@igi-global.com or visit: http://www.igi-global.com/publish/.

The Advances in Hospitality, Tourism, and the Services Industry (AHTSI) Book Series (ISSN 2475-6547) is published by IGI Global, 701 E. Chocolate Avenue, Hershey, PA 17033-1240, USA, www.igi-global.com. This series is composed of titles available for purchase individually; each title is edited to be contextually exclusive from any other title within the series. For pricing and ordering information please visit http://www.igi-global.com/book-series/advances-hospitality-tourism-services-industry/121014. Postmaster: Send all address changes to above address.

Titles in this Series

For a list of additional titles in this series, please visit:
https://www.igi-global.com/book-series/advances-hospitality-tourism-services-industry/121014

Role and Impact of Tourism in Peacebuilding and Conflict Transformation
Jorge Tavares da Silva (GOVCOPP, University of Aveiro, Portugal) Zélia Breda (GOVCOPP, University of Aveiro, Portugal) and Fabio Carbone (Centre for Trust, Peace, and Social Relations, Coventry University, UK & GOV-COPP, University of Aveiro, Portugal)
Business Science Reference • © 2021 • 402pp • H/C (ISBN: 9781799850533) • US $195.00

Strategies for Promoting Sustainable Hospitality and Tourism Services
Maximiliano Emanuel Korstanje (University of Palermo, Argentina) Babu George (Fort Hays State University, USA) and Alexandru-Mircea Nedelea (Stefan cel Mare University of Suceava, Romania)
Business Science Reference • © 2020 • 243pp • H/C (ISBN: 9781799843306) • US $195.00

Handbook of Research on Resident and Tourist Perspectives on Travel Destinations
Patrícia Pinto (Faculty of Economics and Research Centre for Tourism, Sustainability, and Well-Being (CinTurs), University of Algarve, Portugal) and Manuela Guerreiro (Faculty of Economics and Research Centre for Tourism, Sustainability, and Well-Being (CinTurs), University of Algarve, Portugal)
Business Science Reference • © 2020 • 529pp • H/C (ISBN: 9781799831563) • US $285.00

Global Niche Tourism and Opportunities for the Medical Tourism Industry
Manjeet Singh (Central University of Jammu, India) and Subbaraman Kumaran (Alliance University Bangalore, India)
Business Science Reference • © 2020 • 300pp • H/C (ISBN: 9781799834274) • US $215.00

Cases on Tour Guide Practices for Alternative Tourism
Gulsun Yildirim (Recep Tayyip Erdogan University, Turkey) Ozlem Ozbek (Bandirma Onyedi Eylul University, Turkey) Ceyhun Caglar Kilinc (Akdeniz University, Turkey) and Abdullah Tarinc (Akdeniz University, Turkey)
Business Science Reference • © 2020 • 351pp • H/C (ISBN: 9781799837251) • US $245.00

Accelerating Knowledge Sharing, Creativity, and Innovation Through Business Tourism
Muhammad Waseem Bari (Government College University, Faisalabad, Pakistan) Sadia Shaheen (Government College University, Faisalabad, Pakistan) and Meng Fanchen (Beijing Institute of Technology, China)
Business Science Reference • © 2020 • 351pp • H/C (ISBN: 9781799831426) • US $225.00

Contemporary Management Approaches to the Global Hospitality and Tourism Industry
Abraham Pius (Arden University, UK) Husam H. Alharahsheh (University of Wales Trinity Saint David, UK) and Adenike A. Adesanmi (University of East London, UK)
Business Science Reference • © 2020 • 290pp • H/C (ISBN: 9781799822042) • US $225.00

IGI Global
PUBLISHER of TIMELY KNOWLEDGE

701 East Chocolate Avenue, Hershey, PA 17033, USA
Tel: 717-533-8845 x100 • Fax: 717-533-8661
E-Mail: cust@igi-global.com • www.igi-global.com

Editorial Advisory Board

List of Contributors

Table of Contents

Detailed Table of Contents

biodiversity loss—is discussed within the framework of the tourism sector. The study aims to propose a sustainable tourism vision and policy in line with the SDG 15. The study suggests the development of ecological tourism, which is a sustainable tourism vision, in line with SDGs.

Ecotourism is an emerging approach to attain sustainable development. The Morn Hills region of Haryana in India is emerging as a famous ecotourism destination, as it is endowed with rich and varied vegetation, flora and fauna, and forest area. The chapter empirically investigates the local community attitude towards ecotourism as a mean for sustainable development. The chapter used 200 data responses conveniently collected from respondents through a survey form and used suitable statistical tools to analyse the result. The study indicates huge prospects of ecotourism in the area and variation in the attitude of the local community towards the role of ecotourism in sustainable development. A highly positive attitude of locals as ecotourism brings sustainable and overall development in the area in the form of enhanced economic benefits, enrichment of socio-cultural values, increased concerns, and preservation of environmental and physical resources. The chapter concludes different stakeholders to come up with appropriate strategies for sustainable development of the area.

The mountains represent territories of exceptional environmental and heritage value, holding vital resources and functions for humanity. The conservation of its geo and biodiversity is decisive, given the growth in its demand, in particular for tourism and leisure activities. In addition to their immediate physical perception, they are spaces of science and knowledge and a resource for geoeducation to promote the natural and cultural values held. The creation of geoparks constituted a new paradigm in which geosciences, territories, their heritage (material and immaterial) and the ways of life of their communities are disseminated. In this context, geoparks are based on integrated territorial approaches to the enhancement, preservation and promotion of education, biodiversity, geological heritage, tourism and scientific research, among others. Tourism, as a strategy for geoeducation and territorial valorisation, is in the mission of Estrela Geopark, promoting educational programs with initiatives for the conservation of heritage and promotion of eco-cultural resources.

Peripheral Mediterranean regions, such as Calabria, have an urgent need to revitalise the economy through better use of natural resources. The increase in tourism can bring the sought after foreign exchange, resulting in the development of small and medium local enterprises. The chapter aims at evaluating the

Section 5
Sustainability in Tourism Service Providers

complementary services associated with their core offer, the partnership relationships they have with local stakeholders and what type, as well as the motivations for networking and development perspectives of the thermal spa sector in Portugal. The results presented are incorporated in this guideline to establish a competitive offer.

Foreword

The 2030 United Nations (UN) Sustainable Development Goals (SDGs) Agenda with the 169 targets represents a vision to transform the world. The Agenda comprises four sections: the first is a political declaration; the second is a set of 17 SDGs; the third states the means of implementation; and the final section represents the framework for follow up and review of the Agenda. It integrates the three dimensions of 'sustainable development', i.e. economic, social, and environmental. SDGs, such as ending poverty, improving health and education, gender equality, tackling climate change, and economic growth, reveal the ambitious commitment adopted by UN-member states on 25 September 2015. It is based on the assumption that each country is capable of achieving its full potential.

The role of tourism in sustainable development is mentioned several times in the 2030 Agenda, which is an acknowledgement of the global significance of the sector. The term 'sustainability', which came to prominence in the 1980s, evokes lots of discussions. Despite being an imprecise concept, according to the Brundtland Report, its utility lies in the notion of a development that meets the needs of the present without compromising the ability of future generations to meet their own needs. Tourism in the 2030 Agenda for Sustainable Development implies positive impacts on society, economy, and environment. From this optimistic perspective, responsible consumption of travel would bring positive and beneficial social changes. Alleviating poverty is number one in the 2030 Agenda: "We pledge that no one will be left behind" (UN). This pledge has been the guiding principle in tourism research on the so-called 'trickle-down effects' of tourism benefits. However, it is debatable that tourism can deliver on the 'trickle-down' promise. Commentators point to increase poverty and social inequality, and now with Covid-19 pandemic bringing the industry to stand still, the UN's vision to transform the world is said to be doubtful.

The pandemic has thrown international and domestic tourism out of gear, introducing new doubts on economic growth, and creating new global challenges. Not surprisingly, appreciation of this change of the world has important intellectual consequences that undermine old certainties about tourism and sustainable development. Things appear to have changed so much and seem to continue to change at such disconcerting rate that it seems impossible to fix explanation in any stable set of terms. Commentators are clear about the accelerated pace of change that seems set to continue. While readers will find guidance and direction in this collection of readings, they will not be provided with a route map towards a place where certainty will be found. The rapid spread of Covid-19 around the world has and will continue to have dramatic impacts on tourism and sustainability. Tragic consequences will have a major impact on the 2030 Agenda for Sustainable Development and the pledge to 'leave no-one behind'.

Such a ‘revolutionary’ and far-reaching set of goals clearly needs to be underpinned by academic studies, particularly those researchers working in the field of tourism and sustainable development. Tourism has a responsibility to ensure that people, the environment, and the climate are not negatively affected. SDGs and the 169 targets provide a valuable incentive for tourism to play a proactive role in achieving the 2030 Agenda goals. The international tourism community relies on the best available evidence to inform the decision making necessary to achieve the SDGs by 2030. This book brings together new research, diverse theoretical perspectives, and case studies. ‘The Role of Tourism in Achieving the Sustainable Development Goals’ is an edited collection of 19 chapters providing interesting perspectives on current debates concerning: poverty alleviation, environmental protection and conservation, gender equality and minorities, local communities’ participation, sustainability in tourism service providers, and education, leadership, and partnership. It covers current issues and challenges which will be of interest and relevance to scholars, policymakers, planners, managers of the industry and other stakeholders. The importance of this book lies in its diversity and its relevance to the current debates on the role of tourism in achieving the SDGs.

Eugenia Wickens
Buckinghamshire New University, UK

Preface

The concept of sustainable development emerged in the 1980s due to the increasing awareness of the deterioration of the environment and natural resources, and to its consequences for economic and social development. Bearing this in mind, the 1987 report "Our Common Future", prepared by the Brundtland Commission, launched the most cited definition of sustainable development as the "development that meets the needs of the present without compromising the ability of future generations to meet their own needs" (World Commission on Environment and Development [WCED], 1987, p. 43). The United Nations' 2005 World Summit acknowledges that the concept encompasses three interdependent and mutually reinforcing pillars: economic development, social development, and environmental protection (United Nations [UN], 2005).

The relation between tourism and sustainable development is long and has been studied from multiple perspectives. Governments and tourism authorities worldwide have been developing diverse strategies and measures to assure the sustainable development of tourism destinations and businesses. This is clear in the Agenda 21 for the Travel and Tourism Industry, which aims "to establish systems and procedures to incorporate sustainable development considerations at the core of the decision-making process and to identify actions necessary to bring sustainable tourism development into being" (WTTC, WTO, & The Earth Council, 1996, p. 38). According to the World Tourism Organisation (UNWTO, 1998), "sustainable tourism development meets the needs of the present tourists and hosts regions while protecting and enhancing opportunity for the future. It is envisaged as leading to management of all resources in such a way that economic, social, and aesthetic needs can be fulfilled while maintaining cultural integrity, essential ecological process, biological diversity, and life support systems". More than protecting them, tourism can enhance the natural and cultural resources of a given territory. According to Ritchie and Crouch (2003, p. 34), "sustainable tourism seeks to encourage the development and promotion of appropriate forms of tourism – many of which can enhance the environmental, social and cultural well-being of a destination in addition to increasing its economic prosperity".

While its traditional approach is based on the three above-mentioned pillars (economic, social and environmental), more recent approaches include additional dimensions, namely the technological, cultural and political (Choi, & Sirakaya, 2006; Eusébio, Kastenholz, & Breda, 2014a, 2014b, 2016), which provides an interesting theoretical and practical framework to embrace the new challenges posed by the Sustainable Development Goals (SDGs).

In 2015, with the end of the Millennium Development Goals, the UN concluded that, despite the applied efforts, there were still poverty, hunger and vulnerable people, gender inequalities, climate change, environmental degradation, a significant gap between the richer and the poorer and between rural and urban areas, and severe conflicts threatening human development. Within this context, the UN launched the Agenda for Sustainable Development, which gathered the agreement of 196 countries around 17 SDGs, which are:

1. No poverty
2. Zero hunger
3. Good health and well-being
4. Quality education
5. Gender equality
6. Clean water and sanitation
7. Affordable and clean energy
8. Decent work and economic growth
9. Industry, innovation, and infrastructure
10. Reduced inequalities
11. Sustainable cities and communities
12. Responsible consumption and production
13. Climate action
14. Life below water
15. Life on land
16. Peace, justice and strong institutions
17. Partnerships for the goals

Specifically, tourism is included as part of three SDGs, namely Goals 8 (decent work and economic growth), 12 (responsible and sustainable production and consumption), and 14 (life below water). However, it is acknowledged by the World Tourism Organisation (UNWTO) that, if properly planned and managed, tourism can have positive impacts within each of the 17 SDGs.

In fact, initiatives that sustain the cultural characteristics of the receiving community (Stević, & Breda, 2015); contribute to improving the quality of life of residents; stimulate the employment of local residents in the economic activities that integrate the tourism industry; sustain, enhance and conserve the landscape and habitats, considering the carrying capacity; stimulate the balance between the growth of tourism and the conservation of the environment; minimise negative environmental and socio-cultural impacts; optimise the satisfaction of visitors and residents; maximise long-term growth for the destination; sustain the tourist industry (Eusébio, Kastenholz, & Breda, 2014a, 2014b, 2016), promote regional-level innovation developed within a network of stakeholders including firms, government, education institutions and local communities (Brandão, Costa, & Buhalis, 2018), and networked planning towards sustainable destinations (Fernandes, Brandão, & Costa), public participation in tourism development (Costa & Brandão, 2011) are documented in literature. Despite it, the relationship between tourism and sustainable development needs to be deepened under the light of the recent challenges that societies, economies and the environment are facing, especially bridging the theoretical approach and the practical perspective, which is the aim of this book. It is fundamental to explore all tourism constructive opportunities and solutions for contemporary issues, aiming to offer new angles of understanding through case studies and best practices that could help students, researchers, mediators or entrepreneurs to deal in real case scenarios.

The book offers interesting perspectives on the ways that the tourism industry and related activities can promote to achieve UN SDGs, offering theoretical perspectives, case studies and best practices that can be of interest to academic faculty members, since the chapters are well suited for use in classes to consolidate knowledge; to students enrolled in any degree in tourism or a related area such as planning, management, economics, environmental and civil engineering (among others), sociology, anthropology; and researchers working on the topic of tourism and sustainability. Besides, it is a relevant tool for practitioners, as it creates awareness on the best practices implemented in different geographical settings around the world, in both developing and emerging economies, regarding the implementation of sustainable tourism development based on the SDGs.

The book is organised into six sections and 18 chapters. A brief description of each of the chapters follows:

Section 1: Poverty Alleviation

Chapter 1, "Tourism as a Tool for Poverty Eradication in Kenya," recognises tourism as a tool for poverty reduction, following its ability to create employment, entrepreneurial opportunities and impact societies' social infrastructure positively. Using case studies of the Maasai community and the coastal regional tourism-based of the Bomas and Safaris, this book chapter asserts the significance of tourism in poverty reduction among marginalised Kenyan communities. The concept, coupled with ecological sustainability, has opened up societies to share their cultures and tourist attractions for their own benefit. The opportunity cost of these advantages is a rise in local goods and neocolonialism tendencies characterised by dominant foreign control of the tourism sector and its gains, particularly at the coast. This situation is reversible by the adoption of community-based tourism, which ensures that tourism optimally benefits local communities.

Chapter 2, "Development-Tourism and Poverty Alleviation: Towards an Integrative Framework," is based on the assumption that the link between tourism and poverty alleviation has mainly been framed around the pro-poor tourism (PPT) approach, which aims to increase 'net' tourism benefits to the economically poor population. While the narrow focus on income-based poverty in PPT has received much criticism, alternative frameworks for research and practice that are capable of lifting the tourism-poverty link beyond the orthodox economic-centred approach have not received sufficient attention. In response, this chapter presents a conceptual framework amalgamates poverty alleviation and development-tourism genera, and the three key underpinning theoretical concepts of global citizenship, sustainable livelihoods and social entrepreneurship. This framework facilitates a more comprehensive understanding of the development-tourism-poverty alleviation nexus and offers new pathways to enhance tourism's contribution to Goal 1 of the SDGs.

Chapter 3, "Tourism and the Poverty Challenge: A Glance in Cape Verde," introduces a practical application of the first Sustainable Development Goal, which is "Tourism contribution to end poverty". It provides a recent literature review about the theme and presents the case study of Cape Verde, an archipelago with enormous tourist potential, but with severe poverty. As a destination, it needs to work on its sustainability, mainly to provide residents with better living conditions. It is urgent to study the main poverty challenges in Cape Verde in order to provide the right strategies. It is essential, also, to understand if tourism can help fighting local poverty. To achieve some conclusions, authors have done interviews with some local institutions as well as to several stakeholders to understand better the main issues concerning tourism challenges and its potential to deal with poverty in Cape Verde.

Chapter 4, "Poverty Alleviation, Ecotourism, and Biodiversity Protection in Príncipe Island," focuses on an insular territory where two-thirds of its seven thousand inhabitants are poor, and the productive fabric does not generate enough resources for poverty alleviation and biodiversity conservation. Besides, the archipelago belongs to the group of small islands developing states (SIDS), recognised by the United Nations Conference on Environment and Development as a particular case, and international funding for development and sustainability support is available. In addition, the autonomous government of Príncipe identified ecotourism as a niche market to develop, establishing goals for this activity. This chapter presents some results of the actions that have been promoted to reduce poverty through ecotourism and biodiversity protection initiatives.

Section 2: Environment Protection and Conservation

Chapter 5, "The Global Challenge and Re-Evaluation of Ecological Tourism in Line With Sustainable Development Goals," approaches SDG 15, which aims to protect, restore and promote sustainable use of terrestrial ecosystems, sustainably manage forests, combat desertification, and halt and reverse land degradation and halt biodiversity loss – within the framework of the tourism sector. The study aims to propose a sustainable tourism vision and policy in line with the SDG 15, suggesting the development of ecological tourism, which is a sustainable tourism product, in line with SDGs.

Chapter 6, "Ecotourism as a Tool for Sustainable Development in Morni Hills (Haryana), India," empirically investigates the local community attitude towards ecotourism as a mean for sustainable development. The chapter used 200 data responses conveniently collected from respondents through a survey form and used suitable statistical tools to analyse the result. The study indicates huge prospects of ecotourism in the area and variation in the attitude of the local community towards the role of ecotourism in sustainable development. A highly positive attitude of locals as ecotourism brings sustainable and overall development in the area in the form of enhanced economic benefits, enrichment of socio-cultural values, increased concerns and preservation of environmental and physical resources. The chapter concludes that different stakeholders advance with appropriate strategies for sustainable development of the area.

Chapter 7, "Tourism as a Strategy for Geo-Education and Cultural Valorization of Territories: Promoting Ecocultural Values and Geoconservation in Estrela Geopark, Portugal," asserts that the conservation of geo and biodiversity is decisive, given the growth in its demand, in particular for tourism and leisure activities. In addition to their immediate physical perception, they are spaces of science and knowledge and a resource for geoeducation to promote the natural and cultural values held. The creation of geoparks constitutes a new paradigm in which geosciences, territories, their heritage (material and immaterial) and the ways of life of their communities are disseminated. In this context, geoparks are based on integrated territorial approaches to the enhancement, preservation and promotion of education, biodiversity, geological heritage, tourism and scientific research, among others. Tourism, as a strategy for geoeducation and territorial valorisation, is in the mission of Estrela Geopark, promoting educational programs with initiatives for the conservation of heritage and promotion of ecocultural resources.

Chapter 8, "Eco-Tourism as a Trigger for Economic Development: A Resource-Based View of Calabria," is based on the idea that Peripheral Mediterranean regions, such as Calabria, have an urgent need to revitalise the economy through better use of natural resources. The increase in tourism can bring the sought-after foreign exchange, resulting in the development of small and medium local enterprises. The present chapter aims at evaluating the potential of the Calabria region. This was done by carrying out an analysis of a resource-based view of the aforementioned region. A list of potential nature-based tourism resource sites was identified and discussed. This study highlights key findings and recommendations to policy-makers.

Section 3: Gender Equality and Minorities

Chapter 9, "Promoting Gender Equality and Women Empowerment in the Tourism Sector," provide the readers with an interesting discussion on gender equality and women empowerment, which constitute a longstanding topic since the issues have never been fully addressed, although some progress has been made. The tourism sector is one area that seems to have more opportunities for women and, thus, fewer gender inequalities. However, cases of discrimination are still experienced. Hence, this chapter explores the situation of gender equality and women position in the field of tourism and establish the causes of the problem in the sector. The opportunities for females and avenues of achieving fairness and empowerment are also discussed.

Chapter 10, "In Pursuit of a Better Life? High-Skilled Migration in Portugal," alerts to an extremely important debate related to migration and labour market. Nowadays, people enjoy the sense of freedom by moving from one place to another, seeking personal and professional growth, and satisfaction with life. Borders have become fluid. In the global world, education and employment seem to be within easy reach. People are pursuing lifelong learning and training to adjust to a changing job environment and maximise career opportunities. However, migrating qualified individuals can face specific difficulties in their way. The present research aims to study high-skilled migration in Portugal. For this purpose, a qualitative approach, with a focus group methodology, was applied. Two focus groups were conducted with foreign-born individuals residing in Portugal. The chapter aims to add to the understanding of how highly qualified migrants make decisions on migration, and how they perceive the job market, professional development, and career growth opportunities in Portugal.

Section 4: Local Communities' Participation in Tourism Development

Chapter 11, "Community-Based Tourism as a Promoter of Sustainable Development in Rural Territories," is of great interest, not only because it reviews the concepts and models underlying 'community-based tourism', but it also applies to rural territories as a means to empower and improve the local economy and residents' quality of life. Nowadays, trends in the tourism sector are changing, and tourists are increasingly looking for an authentic experience. They wish to escape from mass tourism and urban spaces, as well as their awareness of sustainability, is increasing. In response to this trend, community-based tourism in rural settings has the potential to meet the demand requirements, while ensuring benefits to the territory and community of rural destinations. The present chapter aims to present the dynamics of community-based tourism and how it can promote regional development in rural, low-density regions, in accordance to the Sustainable Development Goals, and with high participation of residents in the several areas of the tourism industry.

Chapter 12, "Events as Sustainable Experiences for Local Communities of a Tourist Destination," proposes the analysis of the well-being of a destination's local community and its relationship with tourist events. There is an increasing concern for public stakeholders to develop sustainable tourism strategies, in order to satisfy the needs of today's society, without compromising future generations. This exploratory research used a qualitative approach by employing a semi-structured interview with the most relevant stakeholders of the destination. It is possible to conclude that local events can have different positive impacts on the quality of life of the residents. Greater support for local political strategies, better feedback and word-of-mouth, translating into a positive and sustainable promotion of the destination by the local community, are results that can be expected from organising events also designed for the residents of the destination, as a sustainable and inclusive experience.

Section 5: Sustainability in Tourism Service Providers

Chapter 13, "Communicating Sustainable Initiatives in the Hotel Industry: The Case of the Hotel Jakarta Amsterdam," points to the need of communicating sustainable initiatives in hotels, based on guests' perceptions, using the Hotel Jakarta Amsterdam as a case study. It was employed empirical research conducting semi-structured personal interviews with hotel guests by a non-probabilistic sample. Guests were randomly approached in the lobby of the hotel during August 2019. According to the data, 'A description on the digital distribution channels, such as Expedia, Priceline, Booking.com, etc.'; 'Signs inside the guestroom'; and an 'Informal chat by an employee when you check-in' were, by guests' perception, the favourite ways to communicating the sustainable initiatives developed by the hotel.

Chapter 14, "Accessible Tourism and Digital Platforms: A Study Applied to the City of Viana do Castelo," explores the digital accessibility in tourism, directed at tourists with special needs. The disclosure of accessibility conditions in tourism supply encourages the demand for destinations. The fact that a destination seeks solutions in order to satisfy the needs and preferences of tourists, regardless of the nature and degree of their disability, fosters a spirit of social and corporate responsibility, contributing to a differentiated, competitive and, at the same time, sustainable destination. Therefore, the present study aims to present the accessible resources in the city of Viana do Castelo, based on the Tur4ll platform. The objective is to analyse the tourism offer available and suitable for tourists with special needs (SN). Using a qualitative methodology, a diagnosis of the study context is presented, as well as the main constraints and limitations that are sometimes inherent to tourism with specific motivation.

The existence of several negative impacts in tourism motivates the analysis made Chapter 15, "Four Innovative Ways to Reduce Food Wasting in Hotel Restaurants," which addresses the problem of food wasting in hotel restaurants, especially when providing all-inclusive/all you can eat breakfasts and dinners. This study proposes activities which could lead to a food-wasting reduction. Through a questionnaire (with 132 respondents, from various countries), the authors wanted to find out respondents' attitudes towards suggested solutions and this problem in general. As part of the questionnaire, four possible activities were proposed to help to reduce food wasting and contribute to better food security in today's world. The survey results proved that respondents would appreciate such opportunities. Fridges (in which people could put their untouched leftovers), samples (before people take the whole portion of food), special cards (containing the ingredients used for the preparation of the meal) and daily announcements (of food wasting in kilograms) are four possible ways to reduce food wasting in hotel restaurants.

Section 6: Education, Leadership, and Partnerships

Chapter 16, "Leadership in Tourism: Authentic Leaders as Facilitators of Sustainable Development in Tasmanian Tourism," proposes to discuss the application of authentic leadership as the 'vehicle' forward for tourism. Specifically, how authentic leaders in the tourism industry can help achieve the United Nations Sustainable Development Goals (SDGs), and why they are important to the Tasmanian tourism industry. As such, authors propose the research question: How can authentic leadership enable the sustainable development of tourism in Tasmania? The chapter commences by exploring tourism in Tasmania and the related leadership gap found in the industry, followed by a brief explanation of our critical review method. The literature review then examines how tourism, a diverse industry, has the potential to contribute to the United Nation's SDGs. Authors aim to demonstrate how sound authentic leadership behaviours among tourist vendors facilitate ethical employment practices and economic growth in Tasmania. Finally, the chapter explores the possible implications of a synthesis of authentic leadership and sustainable development in the context of Tasmania.

Considering that international partnerships have a reinforcing effect and lasting benefits for the modern economy and social life, Chapter 17, "Intercultural Dialogue as a Tool for Maintaining Sustainable Partnerships in Tourism," argues that the principles of intercultural partnerships in tourism can be compared to other principles of sustainable development. This chapter recognises the ultimate role of sustainable development in the sphere of tourism while applying the theory of intercultural dialogue as an approach to the SDG 17 and tracing back partnerships to the bridge where different cultures meet. Such meetings very often need a bridge to cross over cultural gaps. Given the proposed research, one such bridge is presented by the International Tourism Fair Holiday & SPA EXPO 2020, and it is against the background of this event that the established partnerships mediated by intercultural dialogue are classified.

Chapter 18, "Inclusivity in Higher Tourism Studies: A Necessary Reality," embraces an interesting perspective on how the university as an institution plays a key role in its role in improving the quality of life of people. Education is shown as a fundamental pillar to achieving an ideal configuration of a heterogeneous, diversified, organised and inclusive society. Teamwork and respect among the people who are part of the academic community are reflected in the development of the teaching units that make up the Teaching Guides adapted to the European Education Area. With this premise, the objective of this research is to establish the interrelationship between inclusive education in the university and the tourism industry. To this end, it will be analysed to what extent the Tourism Degree curriculum contains subjects related to inclusiveness, particularly with aspects such as disability and mobility. A thorough literature review on this field of study and the analysis of real cases of a sample of Spanish universities is carried out.

Finally, Chapter 19, "Managing Partnerships Between Tourism and Hospitality Stakeholders: The Case of Health and Wellness Products in Portugal," presents a reflection on a new and resized (re) interpretation of the thermal spa sector in Portugal. Taking a holistic view of the strategic health and wellness tourism product, with the example of a case study (thermal spas in Portugal), it materialises a new proposal for a strategic management model. The study analyses the perspective of networking, which characterises a good part of the sector. Special attention is given to the training of organisations and their connection to integrated strategies to make the best use of resources, to the distinguishing orientation of preparation and placement of the tourism product on the market. The aim was to listen to thermal spa managers in Portugal regarding complementary services associated with their core offer, the partnership relationships they have with local stakeholders and what type, as well as the motivations for networking and development perspectives of the thermal spa sector in Portugal. The results presented are incorporated in this guideline to establish a competitive offer.

REFERENCES

Brandão, F., Costa, C., & Buhalis, D. (2018). Tourism innovation networks: A regional approach. *European Journal of Tourism Research*, *18*(1), 33–56.

Choi, H. C., & Sirakaya, E. (2006). Sustainability indicators for managing community tourism. *Tourism Management*, *27*(6), 1274–1289. doi:10.1016/j.tourman.2005.05.018

Costa, C., & Brandão, F. (2011). Tourism Innovation through effective public participation: A sustainable approach. In *Proceedings of the International Tourism Sustainability Conference 2011 'Embracing Social and Environmental Change: The Influence and Role of Tourism*. University of Technology of Mauritius.

Eusébio, C., Kastenholz, E., & Breda, Z. (2014a). Desenvolvimento sustentável de destinos rurais. In E. Kastenholz, C. Eusébio, E. Figueiredo, M. Carneiro, & J. Lima (Eds.), *Reinventar o turismo rural em Portugal. Cocriação de experiências turísticas sustentáveis* (pp. 51–58). UA Editora.

Eusébio, C., Kastenholz, E., & Breda, Z. (2014b). Tourism and sustainable development of rural destinations: A stakeholders' view. *Revista Portuguesa de Estudos Regionais*, *36*, 13–21.

Eusébio, C., Kastenholz, E., & Breda, Z. (2016). Tourism and sustainable development of rural destinations. In E. Kastenholz, M. J. Carneiro, C. Eusébio, & E. Figueiredo (Eds.), *Meeting challenges for rural tourism through co-creation of sustainable tourism experiences* (pp. 329–350). Cambridge Scholars Publishing.

Fernandes, S., Brandão, F., & Costa, C. (2017). Network-based planning for the sustainable development of tourism destinations: Conceptual approach. *Journal of Tourism & Development*, *27/28*(1), 1719–1731.

Ritchie, J. R. B., & Crouch, G. I. (2003). *The competitive destination: A sustainable tourism perspective*. CABI. doi:10.1079/9780851996646.0000

Stević, I., & Breda, Z. (2015). Networking for sustainable cultural tourism activities and dynamics: The case of Oporto. In Destination competitiveness, the environment and sustainability: Challenges and cases (pp. 182-191). Wallingford: CAB International.

United Nations (UN). (2005). *2005 world summit outcome*. Retrieved from https://www.who.int/hiv/universalaccess2010/worldsummit.pdf

UNWTO. (1998). *Guide for local authorities on developing sustainable tourism*. WTO.

WCED. (1987). *Our common future*. Oxford University Press.

WTTC, WTO, & The Earth Council. (1996). *Agenda 21 for the travel and tourism industry*. Retrieved from https://www.e-unwto.org/doi/book/10.18111/9789284403714

Acknowledgment

We would like to acknowledge the assistance of all academics and researchers who contributed to the successful completion of the book, in particular, the authors and the colleagues who took part in the review process. Without their support, this publication would not have been possible to bring to light.

We also would like to express our sincere gratitude to Eugenia Wickens for her insightful foreword.

Filipa Brandão
GOVCOPP, University of Aveiro, Portugal

Zélia Breda
GOVCOPP, University of Aveiro, Portugal

Rui Costa
GOVCOPP, University of Aveiro, Portugal

Carlos Costa
GOVCOPP, University of Aveiro, Portugal

Section 1
Poverty Alleviation

Chapter 1
Tourism as a Tool for Poverty Eradication in Kenya

Joan Mwihaki Nyika
https://orcid.org/0000-0001-8300-6990
Technical University of Kenya, Kenya

ABSTRACT

Tourism is widely recognised as a tool for poverty alleviation following its ability to create employment, entrepreneurial opportunities, and impact societies' social infrastructure positively. Countries, such as Kenya, have reaped from these benefits following evidence that the sector is the third-largest contributor to the economy following agricultural and manufacturing sectors. Using case studies of the Maasai community and the coastal regional tourism-based of the Bomas and Safaris, this chapter asserts the significance of tourism in poverty reduction among marginalised Kenyan communities. The concept, coupled with ecological sustainability, has opened up societies to share their cultures and tourist attractions for their own benefit. The opportunity cost of these advantages is a rise in local goods and neocolonialism tendencies characterised by dominant foreign control of the tourism sector and its gains, particularly at the coast. This situation is reversible by the adoption of community-based tourism, which ensures that tourism optimally benefits local communities.

INTRODUCTION

Sustainable tourism aims at reconciling intergenerational equity challenges, conserving the environment and promoting social justice and economic growth. This is because of its ability to recognise differences between guests and hosts as well as individuals and groups to facilitate fair treatment in the sectoral activities (Akama & Kieti, 2007). The implementation of this concept in practice could greatly contribute to sustainable socio-economic growth and resources management. The United Nations World Tourism Organisation (UNWTO) as noted by Manyara and Jones (2007) asserted that sustainable tourism is a roadmap to economic development and poverty eradication. Poverty eradication in the context of sustainable tourism is in line with the first sustainable development goal of "no poverty'. This is because it offers opportunities to trade goods and services, diversifies employment and economic opportunities for

DOI: 10.4018/978-1-7998-5691-7.ch001

locals of marginalised areas who are mainly poor; offers small-scale-labour-intensive opportunities just like the agricultural sector. Furthermore, it promotes gender equity by offering employment to women hence their empowerment; links local economies without their leakage and is based on landscape, wildlife and cultural assets, most of which belong to the poor (Manyara & Jones, 2007). According to Kibara et al. (2012), tourism results to 'dollar democratisation' since wealth and income are transferred from developed to developing countries leading to collaborative participation of economic sectors and the improvement of the quality of life among the hosts. Similarly, the World Tourism Organization (2002) asserted that tourism diversifies the livelihoods of the pro-poor in developing countries and reduces their vulnerability by offering a variety of economic and entrepreneurial opportunities. It is thus agreeable that tourism has a role to play in poverty eradication, economic development, and environmental conservation in line with the achievement of sustainable development goals.

In Kenya, the tourism sector is a great contributor to the economy. According to Manyara and Jones (2007), tourism is the leading foreign exchange earner and contributed at least 10% of the country's gross domestic profit (GDP). According to the Government of Kenya (GoK) (2011), the sector contributed Kenya shillings (Kshs.) 65.4 billion in 2007 which was an increment from Kshs 21.7 billion in 2002 and is a major economic contributor following the agriculture and manufacturing sectors. In recent statistics, tourism contributed to the creation of 500,000 jobs and had a share of 11% and 4% of the total GDP indirectly and directly, respectively in 2014 (Njoya & Seetaram, 2017). Additionally, in the same year, the income from the sector rose by 126% compared to the 2014 statistics amounting to United States dollar (US$) 2.1 billion while export earnings from international tourists amounted to 18% of the total exports (Njoya & Seetaram, 2017). Despite these achievements, the model of this sector in Kenya is, however, colonial-based, anachronistic and based on coastal and safari products, challenges that hinder its optimised potential as a tool for economic growth and poverty eradication. Coastal products involve the consumption of local goods produced at the coast while safari involves taking trips to see natural and man-made tourist attraction sites. For example, among the Maasai community and Kenyan coastal dwellers, the poverty level is very high despite the fact that these areas are endowed with numerous tourism attractions, according to Ondicho (2016). This contrast depicted in these two communities will be discussed as well as approaches to enforce affirmative action and ensure tourism benefits them towards their economic growth. In recognition of the economic importance of sustainable tourism, this book chapter explores how Kenya and other developing countries of Africa can leverage the sector for poverty eradication in line with the first sustainable development goal on zero poverty. The chapter also explores the opportunities Kenya has to better tourism for economic empowerment of marginalised communities, aspects that can be replicated by other developing countries endowed with tourism attractions.

Tourism and Poverty Nexus

Tourism occurs as a private sector business controlled by market forces and the need to serve the interests of investors and tourists. Among the poor, it is a tool used in diversifying from their conventional economic activities mainly agro-pastoralism hence it is known as pro-poor (Truong, 2014). Additionally, areas endowed with tourist attractions mainly have poor residents as Manwa and Manwa (2014) noted in an evaluation of the role of tourism at Botswana forest reserves in eradicating poverty among surrounding residents. The advocacy of tourism as a tool for development emerged in the 1970s and has grown to incorporate aspects of ecotourism and community-based tourism (CBT). Ecotourism seeks to preserve pristine and fragile environs and at the same time improve human welfare of primary stakeholders while

CBT focuses on localised control of tourism where host communities get a greater portion of benefits resulting from tourism activities (Juma & Khademi-Vidra, 2019).

Research focusing on the link between poverty alleviation due to tourism activities are limited, according to Vanegas et al. (2015). The empirical relationship between the two elements remains less understood and a subject of debate. For example, a number of studies support tourism-led poverty reduction (Akinboade & Braimoh, 2010; Dritsakis, 2004; Kim et al., 2006) while other establish a negative correlation of the two elements (Kim & Lee, 2012; Oh, 2005). Manwa and Manwa (2014) however disagree with this viewpoint claiming that tourism in poverty alleviation has received research attention in recent years following reported case studies from different parts of the world including South Africa (Ashley & Roe, 2002; Saayman et al., 2012), Ecuador (Erskine & Meyer, 2012), Ghana (Holden et al. 2011) and Mauritius (Sharpley & Naidoo, 2010) interlinking poverty and tourism. This is a more credible perspective considering that the growth of tourism research is rooted in the increased demand for sustainable tourism as part of sustainable development.

To understand the link between tourism and poverty reduction, it is essential to link trade liberalisation resulting from tourism activities to poverty alleviation, as Njoya and Seetaram (2017) suggested. In this context, increased tourism activities influence the poor in four aspects, which include 1) risk channels, 2) price, 3) tax and 4) income. Through tourism, poor households get revenue by engaging in resultant activities. In Costa Rica and Nicaragua, among other Central American countries, tourist arrival had positive effects on increasing household income as established using the unidirectional Granger causality effect (Vanegas et al., 2015). By contributing to the tax base at local and national governments, tourism increases income-generating opportunities used in improving social infrastructure. These include promoting gender equality, growing local institutions and positive effects in the educational sector. Tax is usually from businesses that offer goods and services to tourists. In Tanzania, tourism at the micro-level has led to the empowerment of women through small income enterprises to make goods for tourists, which is a step towards gender equality (Muganda et al., 2010). Additionally, it has contributed to income generation through levied taxes. Yu et al. (2019) and Gascon (2015) agreed that tourism results to an expanded tax base for national and local governments and the practice are stimulatory to natural, social, and cultural environment improvement through increased revenue.

Thirdly, tourist hosts incur opportunity costs by purchasing goods to enable the comfort of their guests (Truong, 2014). Consequently, the consumption of local goods and products including land, housing and food increases, which could lead to increased valuing and pricing of these commodities. This trend was reported in Thailand, which is tourism-intensive though these activities have aggravated household income distribution because of hiked prices of goods in the area (Wattanakuljarus & Coxhead, 2008). Consequently, the practice is less beneficial to the poor. The impact of price modifications depends on how well the tourist hosts consume their locally produced goods or otherwise. The aspect of risk focuses on the long-term dynamics of tourism to a region's economic development. Negative risks of tourism include environmental pollution and depletion of natural resources. Positive results of tourism include enhanced efforts towards conservation of biodiversity and increased finances to improve historical, natural and cultural resources towards expanded tourism. The inter-relationship between tourism and poverty is summarised in Figure 1.

Pro-poor tourism is also a developing research interest due to its ability to instil inclusivity for the underprivileged. This is because through equitable tourism distribution of land and water resources and their associated benefits can be enhanced according to Manwa and Manwa (2014). The interest of this concept has attracted even donor and international organisations. A case example is the development

Figure 1. The effects of tourism activities in poverty alleviation

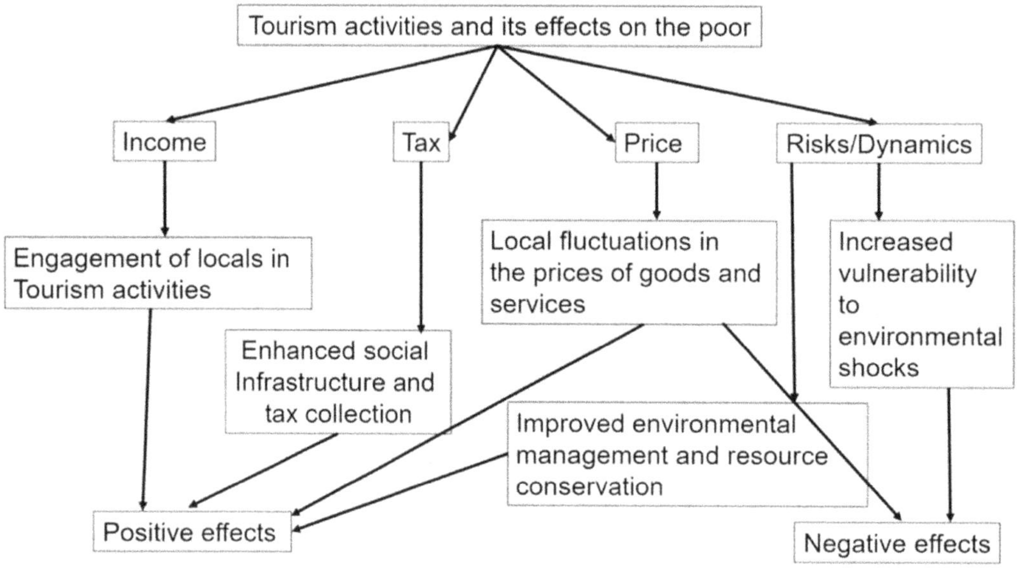

of sustainable tourism eliminating poverty (STEP) initiative under the United Nations world tourism organisation that came into place in 2002. STEP is aimed at promoting economic, environmental and social activities that create employment opportunities to the poor earning less than a dollar a day towards poverty reduction (UNWTO, 2014). Through this initiative, pro-poor tourism is not a product rather it is a development and management approach through which the poor are linked to development through its resultant business activities (Beeton, 2006).

Authors such as Mitchell and Ashley (2010) correlate tourism and poverty alleviation positively. They classify the benefits of tourism to the poor in three, direct, secondary and dynamic effects. Direct effects occur in the form of employment in the tourism sector, the establishment of enterprises such as craft markets and curio shops during cultural tourism. These effects promote the economic welfare of the poor (Scheyvens & Russell, 2012) but also cause negative externalities such as the loss of livelihoods to other community members as Mitchell and Ashley (2010) cautioned. Secondary effects often result through the creation of linkages in collaborations of CBT. Such linkages include the use of food and beverage made by local communities to entertain tourists (Torres & Momsem, 2004) and employing locals as housekeeping staff, chefs and salespersons to run businesses such as hotels and shops that are dependent on tourism activities (Trejos & Chiang, 2009). Dynamic effects are benefits resulting from structural changes associated with tourism. For instance, through tourism, roads, schools, clinics, access to water and other in-house facilities develop, which makes involved areas self-sufficient (Burns, 2004). Similarly, Yu et al. (2019) claimed that tourism in developing countries of sub-Saharan Africa has resulted in 6.6% and 4.7% employment and economic increment in the past decade. This is by increasing income rates among poor households directly and through participation in the activities indirectly.

Apart from the positive gains, pro-poor tourism has negative effects. Explored resources in a landscape, its social capital, people, heritage and local culture influence the aspects of tourism meant to promote poverty eradication (Yu et al., 2019). Different challenges of pro-poor tourism rely on these resources to promote economic development. However, the scope of tourism is not only dependent on the tangible resources (coastal, township and agricultural tourism) but also the intangible ones such as heritage tourism, tribal tourism and ethnic tourism (Pratt et al., 2013; Phommavong & Sorensson, 2014). Integrating the tangible and intangible aspects of tourism has proved difficult hence, the lack of inclusivity and sustainability in its dominant processing and benefits, as Yu et al. (2019) noted. Tourism is therefore known to enhance inequities among the poor and rich propagating poverty. Holden et al. (2011) argued that there are impediments such as lack of micro-finances for poor tourism developers and illiteracy/semi-illiteracy that prevent optimal reaping of tourism benefits by the poor in favour of empowered society members. In this case, the latter take advantage of the former to reap the optimal benefits of tourism activities. These suppositions have been proven using theoretical models, as Sahli and Nowak (2007) highlighted. The models assume that tourism will appreciate local currency and prices of local commodities, which leads to poorer households. Increased value of local currency eats away international competitiveness to non-tourist exports and hinders the economic capacity to generate income. This phenomenon is known as the Dutch disease effect, according to Njoya and Seetaram (2017). Wattanakuljarus and Coxhead (2008) who showed that an increase in tourism arrivals in Thailand by 10% worsened income distribution despite having an overall household income increase proved this phenomenon. Similarly, Blake et al. (2008), who applied the computable general equilibrium (CGE) model, showed an increase in prices of goods and services as a result of 10% increment in tourism spending would result to the poor only reaping a third of the resultant benefits at the expense of the rich. Other authors such as Mahadevan et al. (2016) reported of increased inequality in the income distribution of the poor due to tourism in Indonesia using the CGE model while Croes and Rivera (2017) pointed out a disproportional effect of the poor due to tourism development using the co-integration analysis. It is therefore conclusive to mention that tourism has both positive and negative effects on poverty and hence optimising the former is recommendable for sustainable development.

BRIEF PRESENTATION OF KENYA

Kenya is a country located in East Africa and has 47 counties. It is the world's 48th largest nation covering an area of about 580, 367 square kilometres. The country has a population of about 48 million people, making it the 29th most populous nation in the world (Kenya National Bureau of Statistics, KNBS, 2019). The capital city of Kenya is Nairobi while Mombasa is the oldest city in the country located on the coast. In sub-Saharan Africa, it is ranked the third leading economy after South Africa and Nigeria (Kimenyi et al., 2019). It borders Ethiopia, Somalia, Uganda, Tanzania and South Sudan to the north, east, west, south and northwest, respectively. Agriculture, industrialisation and manufacturing are the key economic contributors to the country's economy. Tourism is the second-largest economic contributor after agriculture. The main attractions are national parks, photo safaris, game reserves, historical forts, mosques, and mountains.

Tourism Situation in Kenya

Kenya is one of the favourite tourist destinations in Africa and is commonly known as the home of '*safari*', which refers to a tour at various wildlife habitats to experience the beauty of nature, its indigenous communities with their interesting culture, enchanting beaches and the open wilderness (Juma & Khademi-Vidra, 2019). According to Kihima (2015), domestic tourists, as well as residents from East

African countries such as Tanzania, Uganda, Rwanda and Burundi, visit major parks of the country. The country has a variety of tourist attractions including abundant natural habitats of wildlife, geographically diverse landscape, scenic beauty, unique communities with exceptional culture and many tropical beaches. The tourism sector is price sensitive and competitive whose niche products and premium parks are unique such that their attraction to tourists does not change even at peak periods. As such, tourism in Kenya has a multiplier effect with a capacity to steer up regional development through the growth of novel enterprises, increased demand for locally produced goods and services particularly from the agricultural sector.

Figure 2. Fluctuations in the number of international tourist arrivals to Kenya (TRI, 2018)

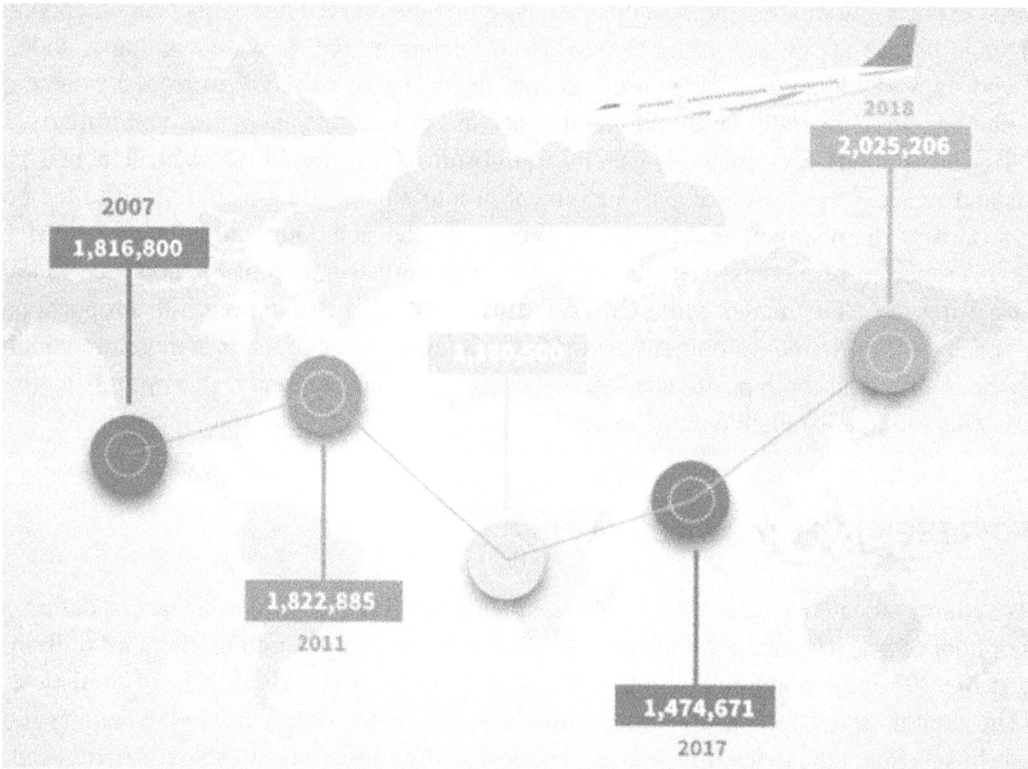

Tourism currently accounts for at least 10% of Kenya's GDP and is the third-largest contributor to the economy apart from the manufacturing and agricultural sectors. The sector is also a top foreign exchange earner generating about Kshs 65 billion in 2007, which is an increase from Kshs 22 billion in 2002 (International Monetary Fund, IMF, 2010). The employment rate in the sector is on the rise with a 3% annual raise and an 18% individual earning increment every five years (IMF, 2010). Through entry fees, license fees, duties and taxes, the sector is a major source of revenue for the national and county governments. In recent statistics of 2018, Kenya's tourism sector is on a growing trend. According to the Tourism Research Institute, TRI (2018), international tourist arrivals of 2018 were estimated at 2,025, 206 using data captured at the immigration services department. This was a 37.3% growth compared to the 2017 arrivals that were estimated at 1,474, 671 (TRI, 2018). Additionally, it was an all-time high compared to previous years since 2011, as shown in Figure 2. This trend could be attributable to political stability, open border policy with East African countries and increased investor confidence, unlike previous years where fluctuations are attributable to political uncertainties due to election cycles mulled with violence and the terrorism threat posed by neighbouring countries such as Somalia (TRI, 2018).

Figure 3. Distribution of international tourists' arrivals based on their country of origin (TRI, 2018)

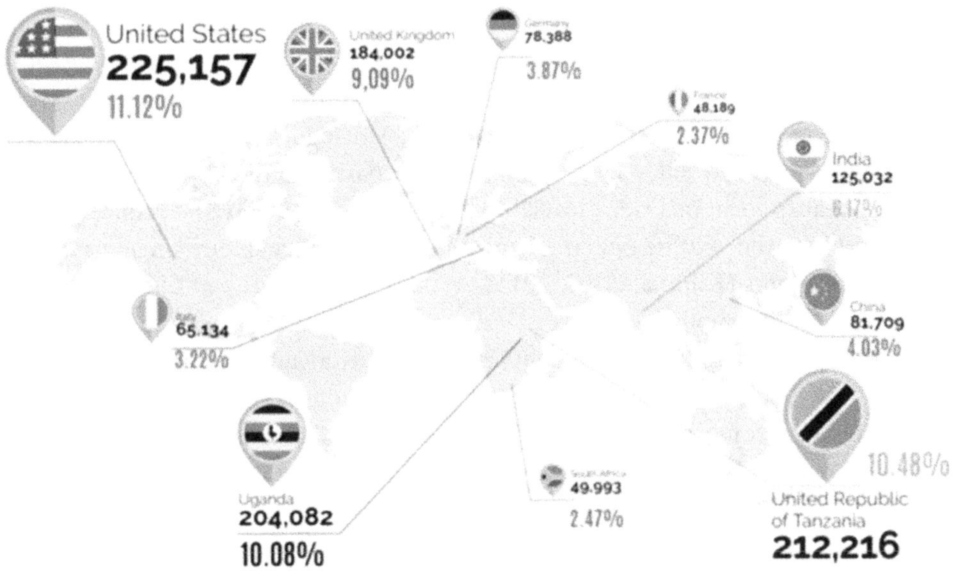

Majority of the tourists visiting Kenya were from the USA and the UK, while neighbouring countries, including Uganda and Tanzania, also had a significant share of the visitors. Figure 3 shows the share distribution of international tourists visiting the country in 2018. The purpose of their visits was mainly holidaying while other reasons such as conference and business activities and visiting friends and relatives took a minor share, as shown in Figure 4.

Figure 4. Distribution of international tourist arrivals based on their purposes of visiting (TRI, 2018)

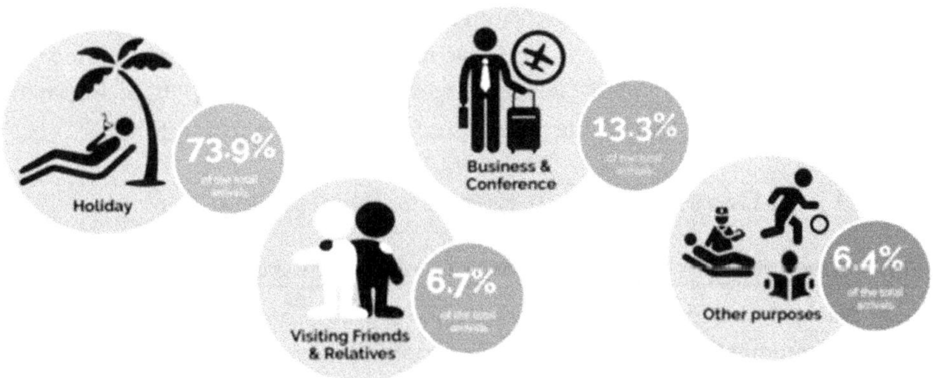

In addition to international tourist arrivals, there has been a surge in preference to domestic tourism in the country. According to the government of Kenya (2014), through the Kenya national bureau of statistics (KNBS), hotel bed nights by Kenyan locals were 1.57 million in 2008 and increased to 2.8 million in 2012. These statistics show a rising trend compared to the 2001-2005 period that has approximately 757,000 overnight stays by locals, as Kihima (2015) noted. In 2017 and 2018, the statistics were even higher as 3, 695,144 and 3,974,243 million local tourists, respectively, were quantified based on domestic bed-night estimates (TR1, 2018). This exponential growth is attributable to Kenya's ability to host international conferences, improved security, growth of the aviation sector to accommodate locals and revitalised marketing efforts. Kihima (2015) differs with this opinion attributing increased domestic tourism in Kenya to a number of initiatives. These include the *'Tembea Kenya'* meaning let us explore Kenya, and the *'Twende tujivinjari'* meaning let us go and have fun initiatives by the government to where local tourists can enjoy the country's attractions at subsidised prices. Other initiatives promoting domestic tourism include the low entry fees at conservation parks charged by Kenya Wildlife Service (KWS), provision of transport in bus rides to visit parks and lakes by KWS at reduced prices and the encouraged membership to the wildlife club of Kenya that promotes local tourism and subsidises prices for such activities to members (Kihima, 2010, 2015).

Tourism contributes significantly to Kenya's GDP in a number of ways. First, for every 11 tourists visiting whether local or international, one job opportunity to a Kenyan is created (Ministry of Tourism and Wildlife, 2016). The total contribution of tourism to job creation and the indirect effects from related investments and resultant income resulted to the creation of more than one million jobs in 2016, which accounted for 9% of total employment in Kenya. In addition to the figures, the ministry of tourism and wildlife (2016) projected that employment would increase by 8.9% to amount to 1,475,000 jobs by 2027 leading to an average 2.9% employment rise annually. Tourism also contributes to the economy from direct pay by tourists to airlines, coaches, trains, rental cars, hotels, conventions, entertainment centres and eateries. The consumed goods and services are supplied from other manufacturing sectors leading to their expansion and the ultimate creation of jobs. In 2018, for instance, the total revenue from tourism amounted to Kshs. 157,386,151,000, which was a 31% rise from the previous year, 2017 whose earnings were 119,900,000,000 (TRI, 2018). The 2018 revenue accounted for 12% contribution to the GDP, unlike previous statistics on the GDP shares at 11% documented by Njoya and Seetaram (2017) and 10% by GOK (2011). Therefore, tourism is a significant contributor to the economy and hence has a role to play in poverty alleviation. Although this is the perceived trend, a significant portion of the revenue leaks out of the economy. This occurs through the payment of salaries to employees of the sector, importation of items such as fruits and drinks consumed by tourists, payments to franchised companies, repayment of foreign loans and expatriation of proceeds to international companies among other expenses (Kihima, 2015). As such, only small portions of these earnings benefit the host communities, especially in a tourism sector like that of Kenya characterised by a seasonal nature and a high preference to international compared to domestic tourism. Weather changes largely influence the inflow of international tourists since they prefer visiting during their winter months and when it is warm in Kenya.

Policy Context and Development in Kenya's Tourism Sector

Before the 1990s, the tourism sector was on a growing trend and hence, a policy framework to ensure its development was not precedence (SNV Netherlands Development Organisation, 2007). This trend could have been attributable to the creation of a ministry of tourism to oversee the sector's development a few years post-independence. The ministry was rooted under the sessional paper No. 8 on tourism development in the country that set targets on the growth of the tourism sector under the government and private sector collaborations. The paper outlines the types of tourism to be promoted, approaches to conserve, and manage tourism attractions and develop infrastructure and the human capacity for the sector in addition to its marketing. However, the dynamics in the tourism market forced the ministry of tourism in collaboration with the Japan International Cooperation Agency (JICA) to formulate a master plan for the sector in 1995 after the government's realisation that it was taking a back seat in the sector's growth unlike other regional neighbours (SNV Netherlands Development Organisation, 2007). The master plan outlined the role of CBT in the sustainability of the tourism sector and was implemented piecewise, though not formally institutionalised by the ministry. Furthermore, the government launched the poverty reduction strategy paper (PRSP) in 2001 to reinforce the pro-poor role of tourism in the country's development agenda. PRSP implementation was further strengthened by the initiation of the economic recovery strategy of 2003 that aimed at promoting employment opportunities and wealth creation using tourism.

To strengthen and formalise these government efforts was the formulation of several policies aimed at optimising the economic gains of the tourism sector. These include the enactment of the tourism policy that outline the roles of local communities in promoting and benefiting from tourism. The costs borne by the tourism sector stakeholders were also reduced by the formulation of environmental impact assessment guidelines under the country's tourism trust fund. This was followed by the introduction of the national tourism policy of 2005, wildlife policy and bill and national museums and heritage act of 2008 and 2006, respectively, which reinforce the economic role of these sections as sub-sets of the tourism sector in the country (GoK, 2013). The implementation of the Kenyan constitution in 2010 devolved tourism to county governments, which gave locals an opportunity to optimise the development of their tourist attractions for their own gains. Furthermore, it empowered counties endowed with tourist attractions to focus more on developing them at a more decentralised approach, which would optimise their resultant economic gains for the benefit of their poor residents. This development was strengthened by the tourism act of 2011 and the sessional paper of 2012, that define the mandate of the tourism sector stakeholders in ensuring it leads to the economic growth of the country and poverty alleviation (GoK, 2013). The policy developments are also geared to attracting many local, and international arrivals in the country, as Ashley (2006) suggested that this approach is key in promoting pro-poor tourism.

Tourism and Poverty Alleviation in Kenya

In Kenya, tourism offers substantive opportunities to the poor living in areas endowed with tourist attractions. Some of these opportunities, based on the category of beneficiaries, are summarised in Table 1. In another research by Ashley (2006), the benefits of tourism to the poor are classified based on the businesses affected and the role of individuals in the sector. Affected sectors include the accommodation, food, crafts, transport and excursions while the individuals driving the sectors include business owners, employees, direct and indirect suppliers. These sectors and individuals are as summarised in Figure 5. A number of factors enhance these benefits in Kenya. These include the diversified nature of tourist attractions most of which are situated in marginalised areas, support through CBT projects and by non-governmental organisations, the multiplier effects of tourism to other sectors such as agriculture and its low capital and labour intensive nature (Kihima, 2010, 2015).

Table 1. Gains of pro-poor tourism in Kenya (SNV Netherlands Development Organisation, 2007)

Beneficiary	Gains
Local communities	- Development of tourist-led infrastructure. - Income from CBT initiatives. - Participatory management of tourist resources leading to community empowerment. - Donations from tourists to fund community projects.
Entrepreneurs	- Proceeds from establishments that are not capital intensive. - Diversification of livelihoods. - Psychological satisfaction from improved quality of life.
Workers	- Capacity to support dependants through earned wages and salaries. - On-job trainings for locals hence acquisition of additional skills. - Learning foreign languages and making new friends. - Improved living standards for locals.

Through CBT, small tourism enterprises could be initiated to facilitate a wider development route and support network, according to Manyara and Jones (2007). The impact of these initiatives would be great because of the participation of locals. The Kenyan government that perceives community-based enterprises as an avenue to develop tourism (GOK, 2013, 2014) has endorsed this idea and declared it a measure to promote indigenous ownership of tourist attractions by local communities. The initiation of the enterprises however should have specific objectives, enough support for their implementation and adequate information to make decisions concerning the marketing and product valorisation, as Kiss (2004) advised. In support of CBT and enterprises, the government of Kenya developed a legal and institutional framework to support these initiatives for economic development under the ministry of tourism and wildlife (Republic of Kenya, 2019). The decision was informed by the role of tourism in delivery of exponential growth in addition to being the third-best foreign exchange earner (Republic of Kenya, 2018). Through the private sector, organisations such as the Kenya community-based tourism (KECOBAT) and the federation of CBT organisation (FETCO) are supervising and shielding the owners of tourism-based enterprises from exploitation towards optimising their economic gains (Lapeyre, 2010). These NGOs noted that poor Kenyans living in tourist attraction-endowed areas are working together with other stakeholders of the sector to conserve involved resources, operate enterprises communally or individually and participate in planning on resource use and in tourist-related decision-making.

Although the ministry of tourism and wildlife provides limited information regarding the economic gains of tourism-based enterprises, Manyara et al. (2006) used a purposive sampling technique in a study focusing on Kenya's southern and coastal regions to identify these enterprises. The study concluded that the tourism-led enterprises offering opportunities to poverty alleviation are categorised into two: community-led (CLEs) and individually owned enterprises (IOEs). The latter is further classified as formal and informal where the formal are legally licensed businesses operating in established facilities while the latter operates on an ad-hoc basis in makeshift structures (Manyara et al., 2006).

Local communities who possess some tourism resources own CLEs. Membership to these enterprises is limited to a specific community who share proceeds once they are realised. Through contributions from individual members, CLEs do not have problems raising capital for their ventures. Examples of these ventures include Wasini women's group, Mwaluganje elephant sanctuary, Lumo wildlife sanctuary, Wilderness camp, Tofino Tali tours company and Kiteghe tours company (Manyara et al., 2006). Through collaboration with national and regional entities such as the KWS and the East Africa Wildlife

Figure 5. Benefits of tourism to the poor based on business ventures and individuals involved (Ashley, 2006)

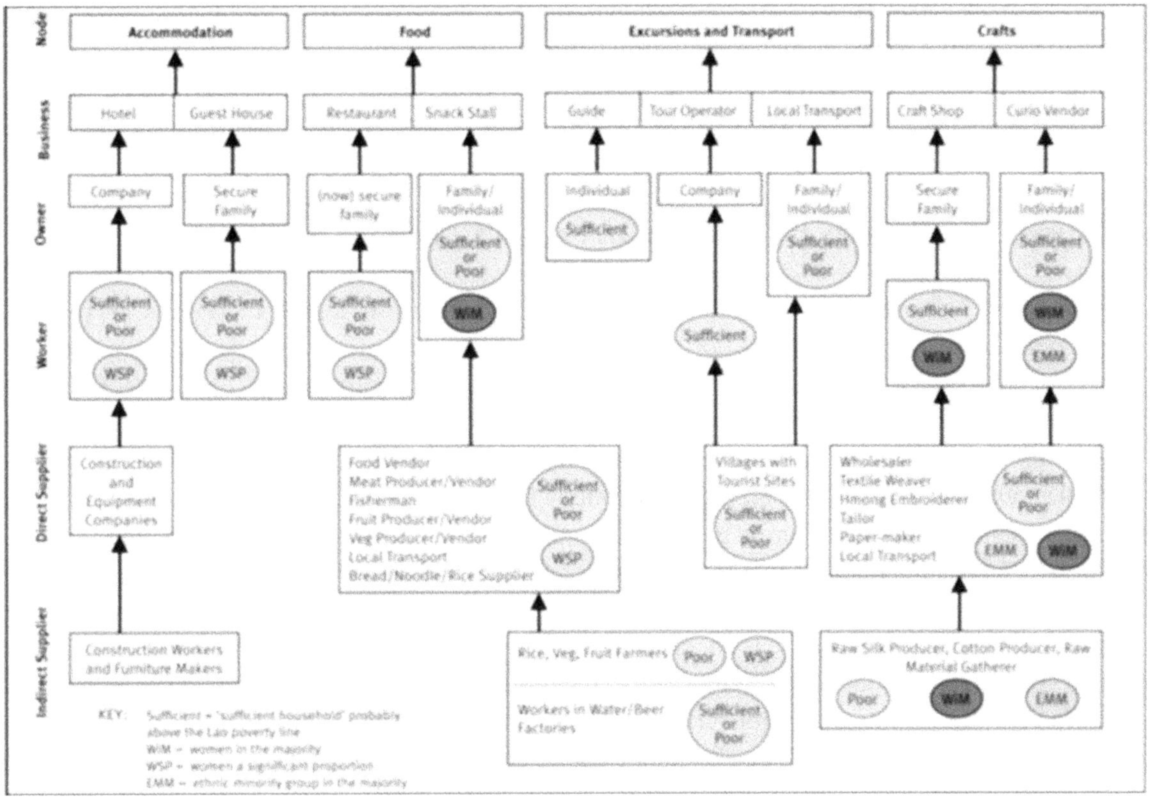

Service (EAWS), these ventures are managed via leasehold or fully owned and operated businesses of locals. Financial and capital limitations, as well as an inadequate human capacity, leads to the preference of leaseholds compared to full-ownership. Although these CLEs were conservation projects exclusively, they have evolved to offer a variety of services such as accommodation, catering and safari travels. These developments are enhanced by external support mobilising communities to work together and the drive towards environmental conservation while reaping optimal financial gains at the same time (Sofield, 2003). By diversifying the livelihoods of communities from pastoralism and subsistence farming, these ventures have improved the quality of life through extra income paid as dividends. Additionally, revenues from CLEs have been used to support social infrastructure by providing educational bursaries, equipping health facilities and schools as well as facilitating the access to clean water.

IOEs are solely maintained shops operated by owners and hardly employ locals. Most of these ventures do not have definitive objectives like CLEs and mainly venture in the sale of cultural artefacts of Kenyan indigenous communities such as souvenirs, beadwork, sculptures and carvings. The products are made locally and sold by informal enterprises. Formal businesses, on the other hand, offer accommodation and restaurant services. The informal IOEs offer self-employment and diversification from conventional livelihood activities such as fishing in the case of coastal dwellers. Examples include the Old town and Ewaso Nyiro curio shops of Mombasa and southern Kenya, respectively (Manyara et al., 2006). According to Ioannides and Petersen (2003), innovation to create unique tourism products and services increases the chances of realising proceeds. Formal IOEs provide opportunities for the novelty to offer unique yet competitive services and products. Examples include the Royal palms resort of Voi and the Mpunguti lodge of Wasini Island (Manyara et al., 2006). Both IOEs and CLEs provide opportunities to alleviate poverty in Kenya through this trend could be enhanced by increased support from the government unlike the current situation, where these ventures are NGO-based, which infuses neo-colonialism tendencies (Kiss, 2004; Manyara & Jones, 2007). Specific case studies in Kenya showing the progress made by tourism in poverty alleviation in Kenya are discussed in the following sections.

Ecotourism for Poverty Eradication in Kenya's Maasai Land

Although Kenya has made great strides in growing its economy post-colonialism, poverty persists in the arid and semi-arid regions, mainly the northern and southern regions of the country. The Maasai community, whose main source of livelihood is pastoralism occupy the southern regions (Ondicho, 2010, 2016). The indigenous culture of the community and a number of wildlife conservancies such as the Serengeti and Amboseli national parks are the main tourist attractions of the area. Although the country has established ecotourism as a poverty eradication strategy by offering employment and revenue-generating opportunities, the Maasai community has been left out in reaping these benefits. Wishitemi et al. (2015) expressed these sentiments claiming that the great potential affiliated with tourism in Kenya's wealth creation has not transferred these positive effects to the poor people of Maasai Mara and Amboseli national parks of Kenya. This is the case because the poverty rates in these areas are 50 to 60%. Similarly, Ondicho (2018) pointed out that communities living around the Amboseli national park get the least of its benefits despite the fact that it generates revenue from tourism activities. Consequently, the Maasai community has been alienated in their own native land despite external use of their resources aggravating pre-existent marginalisation and poverty.

In response to these challenges, the Maasai have turned to indigenous ecotourism to generate revenue and further diversify their livelihoods from predominantly agro-pastoralism. These tourism activities are operated in and by locals in their tribal lands for their own benefit (Zeppel, 2006). Using their own Maasai cultural Bomas, which is a form of indigenous ecotourism; the community has increased control of its tourism investments and reduced the leakage of their culture and its resultant income (Ondicho, 2016).

A cultural Boma is a Maasai village, constructed like a homestead for local entrepreneurs to sell locally made artefacts for revenue generation, unlike previous trends where the community used middlemen. These villages are set up by community enterprise groups and offer tourists an opportunity to interact with the Maasai, learn about their culture, enjoy their dances and songs, buy their designed arts and crafts and take photographs with them in their traditional attires. In this context, the Maasai community is able to alleviate poverty through ecotourism in two ways. First, they are 'objects of tourist gaze' (Ondicho, 2018) and, secondly, they are managers and entrepreneurs of these ventures. In six of the operational cultural Bomas in Amboseli national park vicinity, most of which became operational in the last two decades, it was reported that locals had greater control of the cultural heritage and used it to generate income and ultimately, alleviate poverty (Ondicho, 2010, 2016). Wishitemi et al. (2015) recommended the transfer of ecotourism resources to the Maasai to give them the power to use them for the benefit of the poor locals. This is with the precognition that local communities are more conversant with their resources and can best conserve them to reap their long-term benefits in line with poverty alleviation.

A number of indicators of this indigenous ecotourism as a poverty eradication tool have been documented. First, each Boma received a mean of five vans a day with each van occupying at least six tourists.

These values can fluctuate during on-peak and off-peak seasons of July-September and August-June, respectively. Tourists coming to visit the Bomas reside in nearby game lodges, require a tour guide and lodge vehicles. Villagers dressed in traditional costumes entertain them with ritual songs. These services are payable in addition to the purchase of artefacts. At the entrance of a Boma, a fee of US$20 is charged, and each individual can earn a range of US$ 50-100 weekly from the sales. Entry fees for each Boma range at a mean of US$ 300 a day. These financial benefits help entrepreneurs to support their dependants and improve the quality of their lives (Mbaiwa, 2015). Other resultant benefits of tourism, which are pro-poor, include increased employment opportunities. According to Ondicho (2018), Bomas at the Amboseli national part employ 400 to 500 locals. In addition to direct employment, spinoff activities such as transport services using bicycles and motorbikes, food supply, tour guiding, hawking foodstuff and currency exchange activities all leading to revenue generation for locals are on the rise. These activities enhance tourism's multiplier effects on economic growth and poverty eradication.

Other benefits accruing from their indigenous ecotourism is the help offered by voluntary and charitable organisations mainly comprising of tourists. These organisations have supported educational efforts in the area by offering scholarships to needy learners and funded the construction of boreholes, health facilities and schools. Infrastructural and social services have resulted in these regions, which were previously remote. A case example is the construction of the Namanga – Loitokitok road enabling access to the Bomas and the Amboseli national park by the government of Kenya in addition to the provision of network boosters to promote mobile phone access (Ondicho, 2010, 2016). Through the sale of locally made goods, the community's market has grown towards their improved livelihoods. Overall, these ecotourism gains have improved medical care, access to infrastructure, expanded educational opportunities and enhanced household food security among the Maasai (Wishitemi et al., 2015).

Tourism for Poverty Eradication in Mombasa, Kenya

In East Africa, Mombasa located at Kenya's coast is the leading tourist centre. It has the best tourist infrastructure in addition to a high concentration of tourists' hospitality facilities. According to Akama and Kieti (2007), Mombasa and its vicinities host a third of all tourists received in the country. Despite this phenomenon, the region, which includes counties of Mombasa, Kilifi, Kwale and Lamu has a high incidence of poverty by the locals living in deplorable conditions. More than 50% of the population is below the poverty line and have a revenue of below US$ 1 daily. The residents cannot meet their basic needs, and educational levels among other social amenities are below par. Such social deprivation and abject poverty are attributable to the high prevalence of insecurity, drug trafficking and abuse as well as prostitution in the area (Akama & Kieti, 2007). Tourism is, however, a growing tool for poverty eradication for the coastal region as expressed in a number of studies.

In a qualitative study in Changamwa, Likoni and Kisauni areas of Mombasa, 91% of residents expressed the positive benefits of tourism. Some of the mentioned benefits include the ability of tourism to provide employment, open up business opportunities, promote infrastructural development and enhance cross-cultural integration as expressed by 57, 34 5 and 4% of interviewees (Akama & Kieti, 2007). Similarly, Juma and Khademi-Vidra (2019) conducted a survey in Kenyan rural areas, including those of the coast, such as Mombasa and reported a number of tourism aspects that lead to poverty alleviation. In their study, tourism resulted in new markets for locally produced goods, created employment opportunities for the youth, supported local businesses and external investments to tourist-endowed areas. Additionally, the maintenance of identity for coastal communities and the sharing of their local traditions improved the quality of life in addition to providing revenue. A study by Kubo (2004) agreed that tourism has positive economic impacts on the coast of Kenya. It employs 13% of the population directly in addition to accounting to 24% of the total earnings in these regions. This trend is realised in the hotel industry with tourist hotspots being at Malindi, Diani and Mombasa North. However, Kubo (2004) pointed out a high preference to employ foreigners in the hotels whose management and ownership is mainly foreign-based. Consequently, locals benefit from seasonal and low-grade jobs despite being the owners of the resources. A study by the centre on ecotourism and sustainable development and documented by Nelson (2007) agreed that coastal tourism in Kenya and the larger part of East Africa is an economic booster through entrepreneurial, employment and economic growth opportunities. However, stronger policies to enhance its sustainability and benefit to the pro-poor are needed for optimal benefits to coastal communities (Nelson, 2007).

CONCLUSIONS AND RECOMMENDATIONS

This book chapter explores the relationship between tourism and poverty eradication using the case study of Kenya. Using the case of the coastal communities and the Maasai in Kenya, it is evident that tourism is a source of livelihood, provides employment to many individuals in addition to promoting secondary effects such as infrastructural development, education and healthcare benefits in marginalised areas endowed with tourism attractions. In the chapter, mixed views on the role of tourism in poverty eradication are highlighted. On the one hand, tourism creates employment and entrepreneurial opportunities for the poor. It also results in multiplied effects in other sectors, particularly the agricultural sector, opens up marginalised regions and creates avenues to improve social infrastructure. These positive effects in future must be optimised using CBTs to reap their best economic to the communities involved in tourism which are the primary stakeholders. On the other hand, tourism leads to the increased cost of goods and services and hence the high cost of living in regions endowed with related attractions in addition to neo-colonialist tendencies to control investments and gains from the sector at the cost of the poor who are custodians of such resources. With the best economic policies in Kenya and community empowerment, these negative effects can be controlled to ensure individuals involved in tourism are the ultimate beneficiaries of its economic outcomes in future. In Kenya, both the benefits and disadvantages of tourism are evident from the provided case studies. However, with recommendations provided in this chapter pegged on better tourism support by the government using stronger economic policies and the private sector in the provision of finances to the community, tourism in the country can be optimised to benefit the poor, conserve the environment and at the same time, improve social welfare for holistic sustainability.

REFERENCES

Akama, J., & Kieti, D. (2007). Tourism and socio-economic development in developing countries: A case study of Mombasa resort in Kenya. *Journal of Sustainable Tourism*, *15*(6), 733–748. doi:10.2167/jost543.0

Akinboade, O., & Braimoh, L. (2010). International tourism and economic development in South Africa: A granger causality test. *International Journal of Tourism Research, 12*(2), 149–163. doi:10.1002/jtr.743

Ashley, C. (2006). *How Can Governments Boost the Local Economic Impacts of Tourism?* Options and Tools. SNV and ODI.

Ashley, C., & Roe, D. (2002). Making tourism work for the poor: Strategies and challenges in South Africa. *Development Southern Africa, 19*(1), 61–82. doi:10.1080/03768350220123855

Beeton, S. (2006). *Community Development through Tourism*. Landlink Press. doi:10.1071/9780643093881

Blake, A., Arbache, J. S., Sinclair, M. T., & Teles, V. (2008). Tourism and Poverty Relief. *Annals of Tourism Research, 35*(1), 107–126. doi:10.1016/j.annals.2007.06.013

Burns, P. M. (2004). Tourism planning: A third way? *Annals of Tourism Research, 31*(1), 24–43. doi:10.1016/j.annals.2003.08.001

Croes, R., & Vanegas, M. (2008). Tourism and Poverty Alleviation: A Co-integration analysis. *Journal of Travel Research, 47*(1), 94–103. doi:10.1177/0047287507312429

Dritsakis, N. (2004). Tourism as a long-run economic growth factor: An empirical investigation for Greece using causality analysis. *Tourism Economics, 10*(3), 305–316. doi:10.5367/0000000041895094

Erskine, L. M., & Meyer, D. (2012). Influenced and influential: The role of tour operators and development organisations in tourism and poverty reduction in Ecuador. *Journal of Sustainable Tourism, 20*(3), 339–357. doi:10.1080/09669582.2011.630470

Gascon, J. (2015). Pro-poor tourism as a strategy to fight rural poverty: A critique. *Journal of Agrarian Change, 15*(4), 499–518. doi:10.1111/joac.12087

Government of Kenya. (2011). *Tourism Act 2010*. Government Printer.

Government of Kenya. (2013). Sector plan for tourism 2013-2017. Nairobi, Kenya: Author.

Government of Kenya. (2014). Kenya National Bureau of Statistics. Economic Survey. Nairobi, Kenya: Author.

Holden, A., Sonne, J., & Novelli, M. (2011). Tourism and poverty reduction: An interpretation by the poor of Elmina, Ghana. *Tourism Planning & Development, 8*(3), 317–334. doi:10.1080/21568316.2011.591160

International Monetary Fund (IMF). (2010). *Kenya: poverty reduction strategy paper*. IMF Country Report No. 10/224.

Ioannides, D., & Petersen, T. (2003). Tourism "non-entrepreneurship" in peripheral destinations: A case study of small and medium tourism enterprises in Bornholm. *Tourism Geographies, 5*(4), 408–435. doi:10.1080/1461668032000129146

Juma, L., & Khademi-Vidra, A. (2019). Community-based tourism and sustainable development of rural regions in Kenya; perceptions of the citizenry. *Sustainability, 11*(17), 4733. doi:10.3390u11174733

Kenya National Bureau of Statistics (KNBS). (2019). *2019 Kenya population and housing census results.* Available Online: https://www.knbs.or.ke/?p=5621

Kibara, O., Odhiambo, N., & Njuguna, J. (2012). Tourism and economic growth in Kenya: An empirical investigation. *International Business and Economics Research Journal, 11*(5), 512–528. doi:10.19030/iber.v11i5.6970

Kihima, B. (2015). Domestic tourism in Kenya: Trends, initiatives and practices. *The East African Review, 50*, 22–39.

Kihima, B. O. (2010). *Towards Unlocking Tourism Potential in Less Visited Destinations in Kenya: The Case of Western Kenya Region* (PhD Dissertation). University of Lyon, France.

Kim, H. J., Chen, M. H., & Jang, S. S. (2006). Tourism expansion and economic development: The case of Taiwan. *Tourism Management, 27*(5), 925–933. doi:10.1016/j.tourman.2005.05.011 PMID:32287716

Kim, N., & Lee, C. K. (2012). Granger causality among tourism development, foreign direct investment and economic growth: The case of Korea. *Korean Journal of Tourism Research, 27*(3), 43–57.

Kimenyi, M., Mwega, F., & Ndung'u, N. (2019). *The African lions: Kenya country case study*. Available Online: https://www.brookings.edu/wp-content/uploads/2016/07/kenya-country-case.pdf

Kiss, A. (2004). Is community-based ecotourism a good use of biodiversity conservation funds? *Trends in Ecology & Evolution, 19*(5), 231–237. doi:10.1016/j.tree.2004.03.010 PMID:16701261

Kubo, A. (2004). *The sustainability of coastal tourism. Research into the Kenya portfolio of TUI and its environmental performance.* Breda.

Lapeyre, R. (2010). Community-based tourism as a sustainable solution to maximise impacts locally? The Tsiseb Conservancy case, Namibia. *Development Southern Africa, 27*(5), 757–772. doi:10.1080/0376835X.2010.522837

Mahadevan, R., Amir, H., & Nugroho, A. (2016). Regional Impacts of Tourism-Led Growth on Poverty and Income: Inequality: A Dynamic General Equilibrium Analysis for Indonesia. *Tourism Economics*, 1–18.

Manwa, H., & Manwa, F. (2014). Poverty alleviation through pro-poor tourism: The role of Botswana forest Reserves. *Sustainability, 6*(9), 5697–5713. doi:10.3390u6095697

Manyara, G., & Jones, E. (2007). Community-based tourism enterprises development in Kenya: An exploration of their potential as avenues of poverty reduction. *Journal of Sustainable Tourism, 15*(6), 628–644. doi:10.2167/jost723.0

Manyara, G., Jones, E., & Botterill, D. (2006). Tourism and poverty alleviation: The case for indigenous enterprise development in Kenya. *Tourism, Culture & Communication, 7*(1), 19–37. doi:10.3727/109830406778493579

Mbaiwa, J. (2015). Ecotourism in Botswana: 30 years later. *Journal of Ecotourism, 14*(2–3), 204–222. doi:10.1080/14724049.2015.1071378

Ministry of Tourism and Wildlife. (2016). Kenya's tourism agenda 2018-2022. Nairobi, Kenya: Author.

Mitchell, J., & Ashley, C. (2010). *Tourism and Poverty Reduction: Pathways and Prosperity*. Earthscan.

Muganda, M., Sahli, M., & Smith, K. (2010). Tourism's contribution to poverty alleviation: A community perspective from Tanzania. *Development Southern Africa, 27*(5), 629–646. doi:10.1080/0376835X.2010.522826

Nelson, F. (2007). *East Africa: Trends in coastal tourism and strategies for promoting sustainable development. Centre on Ecotourism and Sustainable Development.* Stanford University.

Njoya, T., & Seetaram, N. (2017). Tourism contribution to poverty alleviation in Kenya: A dynamic computable general equilibrium analysis. *Journal of Travel Research, 57*(4), 513–524. doi:10.1177/0047287517700317 PMID:29595836

Oh, C. O. (2005). The contribution of tourism development to economic growth in the Korean economy. *Tourism Management, 26*(1), 39–44. doi:10.1016/j.tourman.2003.09.014

Ondicho, T. (2010). *Tourism, Power and Politics: The Challenges of Maasai Involvement in Tourism Development* (Doctor's Thesis). Massey University, Palmerston North, New Zealand.

Ondicho, T. (2016). Poverty alleviation in Kenya: A review of why anti-poverty strategies have failed. In A. A. Bard, J. B. Ndohvu, & B. Tiberius (Eds.), *Poverty and Human Rights: East African Experiences* (pp. 82–94). Focus Publishers.

Ondicho, T. (2018). Indigenous ecotourism as a poverty eradication strategy: A case study of the Maasai people in the Amboseli region of Kenya. *African Study Monographs, 56*, 87–109.

Phommavong, S., & Sorensson, E. (2014). Ethnic tourism in Lao PDR: Gendered divisions of labour in community-based tourism for poverty reduction. *Current Issues in Tourism, 17*(4), 350–362. doi:10.1080/13683500.2012.721758

Pratt, S., Gibson, D., & Mavono, A. (2013). Tribal tourism in Fiji: An application and extension of Smith's 4Hs of indigenous tourism. *Asia Pacific Journal of Tourism Research, 18*(8), 894–912. doi:10.1080/10941665.2012.717957

Republic of Kenya. (2018). *Kenya Tourism Agenda 2018–2022*. Government of Kenya Ministry of Tourism & Wildlife.

Republic of Kenya. (2019). Draft Community Based Tourism Framework. Republic of Kenya.

Saayman, M., Rossouw, R., & Krugell, W. (2012). The impact of tourism on poverty in South Africa. *Development Southern Africa, 29*(3), 462–487. doi:10.1080/0376835X.2012.706041

Sahli, M., & Nowak, J. J. (2007). Does Inbound Tourism Benefit Developing Countries? A Trade Theoretic Approach. *Journal of Travel Research, 45*(4), 426–434. doi:10.1177/0047287506295948

Scheyvens, R., & Russell, M. (2012). Tourism, land tenure and poverty alleviation in Fiji. *Tourism Geographies, 14*(1), 1–25. doi:10.1080/14616688.2011.593188

Sharpley, R., & Naidoo, P. (2010). Tourism and Poverty Reduction: The case of Mauritius. *Tourism and Hospitality Planning and Development, 7*(2), 145–162. doi:10.1080/14790531003737169

SNV Netherlands Development Organisation. (2007). Tourism and development: Agendas for action. Nairobi, Kenya: Author.

Sofield, T. (2003). *Empowerment for sustainable tourism development*. Pergamon.

Torres, R., & Momsen, J. M. (2004). Challenges and potential for linking tourism and agriculture to achieve pro-poor tourism objectives. *Progress in Development Studies*, *4*(4), 294–318. doi:10.1191/1464993404ps092oa

Tourism Research Institute (TRI). (2018). Tourism sector performance report, 2018. Nairobi, Kenya: Author.

Trejos, B., & Chiang, L. N. (2009). Local economic linkages to community-based tourism in rural Costa Rica. *Singapore Journal of Tropical Geography*, *30*(3), 373–387. doi:10.1111/j.1467-9493.2009.00375.x

Truong, V. (2014). Pro-poor tourism: Looking backward as we move forward. *Tourism Planning & Development*, *11*(2), 228–242. doi:10.1080/21568316.2013.864996

United Nations World Tourism Organisation (UNWTO). (2014). Available online: http://step.unwto.org/en/content/background-and-objectives

Vanegas, M. (2015). The triangle of poverty, economic growth, and inequality in Central America: Does tourism matter? *Worldwide Hospitality and Tourism Themes*, *6*(3), 277–292. doi:10.1108/WHATT-03-2014-0014

Wattanakuljarus, A., & Coxhead, I. (2008). Is tourism-based development good for the poor? A general equilibrium analysis for Thailand. *Journal of Policy Modeling*, *30*(6), 929–955. doi:10.1016/j.jpolmod.2008.02.006

Wishitemi, B., Momanyi, S., Ombati, B., & Okello, M. (2015). The link between poverty, environment and ecotourism development in areas adjacent to Maasai Mara and Amboseli protected areas, Kenya. *Tourism Management Perspectives*, *16*, 306–317. doi:10.1016/j.tmp.2015.07.003

World Trade Organization. (2002). *Tourism and poverty alleviation*. Author.

Yu, L., Wang, G., & Marcouiller, D. (2019). A scientometric review of pro-poor tourism research: Visualisation and analysis. *Tourism Management Perspectives*, *30*, 75–88. doi:10.1016/j.tmp.2019.02.005

Zeppel, H. (2006). Indigenous Ecotourism: Sustainable Development and Management. Ecotourism Series No.3. CABI. doi:10.1079/9781845931247.0000

KEY TERMS AND DEFINITIONS

Co-Integration Analysis: A statistical procedure that evaluates processes that have non-stationary time series to quantify changes over time.

Computable General Equilibrium (CGE) Model: A model that uses economic data to assess the effects of an economy following changes in technology, policy and/ or other external factors.

Dutch Disease: The term describes obvious causal relationships in economics. For example, the rise of economic development at a particular sector results to a decline in another. For instance, the development of the tourism sector leads to low agricultural practices and subsequent rise in prices of agricultural produce.

Intergenerational Equity: The concept that encourages justice and fairness from one generation to another especially during the use and sharing of environmental assets and natural resources.

Neo-Colonialism: The use of cultural, economic, and political pressure to influence and control decisions in independent countries that were former colonies. The phenomenon is predominant among developed nations to developing countries.

Sustainable Tourism: The concept of travelling places with a motive of influencing their economic, social and environmental sectors positively for the current and future generations.

Chapter 2
Development-Tourism and Poverty Alleviation:
Towards an Integrative Framework

Giang Thi Phi
Aalborg University, Denmark

ABSTRACT

The Sustainable Development Goals (SDGs) highlight the alleviation of multi-dimensional poverty in the global development agenda. Yet, for nearly two decades, the link between tourism and poverty alleviation has mainly been framed around the pro-poor tourism (PPT) approach, which aims to increase 'net' tourism benefits to the economically poor population. While the narrow focus on income-based poverty in PPT has received much criticism, alternative frameworks for research and practice that are capable of lifting the tourism-poverty link beyond the orthodox economic-centred approach have not received sufficient attention. In response, this chapter presents a conceptual framework amalgamates poverty alleviation and development-tourism genera, and the three key underpinning theoretical concepts of global citizenship, sustainable livelihoods, and social entrepreneurship. This framework facilitates a more comprehensive understanding of the development-tourism-poverty alleviation nexus, and offers new pathways to enhance tourism's contribution to Goal 1 of the SDGs.

INTRODUCTION

There is little argument that over time, poverty has been one of the most significant and persistent concerns in the global development agenda. Unsurprisingly then, "reducing extreme poverty and hunger" was identified as one of the priorities in the "Millennium Development Goals" (MDGs) - a document signed by 189 nations in 2000 with the aim of freeing human beings from multiple deprivations by 2015 (United Nations, 2003). As the MDGs come to an end in 2015, its successor - the Sustainable Development Goals (SDGs) were developed via a more transparent and participative manner, which resulted in a much broader and inclusive agenda to tackle the world's key issues (Ruhil, 2015). In particular, the

DOI: 10.4018/978-1-7998-5691-7.ch002

SDGs seek to "end poverty in all its forms anywhere and to ensure that all human beings can fulfil their potential in dignity and equality and in a healthy environment" (United Nations 2016, p.1). The SDGs thus highlight the importance of multi-dimensional poverty and an expanded vision of well-being in the global development agenda.

Yet to date, there have been limited attempts in developing a broaden agenda for the planning and development of tourism for poverty alleviation. The nexus of tourism and poverty alleviation is still largely guided by the Pro-poor Tourism (PPT) approach, which aims to increase 'net' tourism benefits to the economically poor population. Arguably, the focus on income poverty in the PPT discourse has resulted in the frequent absence of other poverty dimensions (e.g., social and cultural marginalisation, powerlessness) as well as local perceptions and conditions within the PPT action framework, leading PPT projects to largely concentrate on alleviating monetary poverty via a simplified economic-centred approach to poverty alleviation (Pleumarom, 2012).

This chapter explores in-depth the concept of development-tourism, which covers a plethora of alternative tourism genera (e.g., community-based tourism, exposure tourism, microfinance tourism; volunteer tourism, slum tourism) and which features poverty and/or diverse prospects of poverty alleviation via development as one of its key focuses. Currently, there is a lack of literature tying together these diverse development-tourism niches. Existing reviews (e.g., Frenzel, 2013; Scheyvens, 2012; Phi, 2017) suggest that the theoretical perspectives underpinning the link between development-tourism and poverty alleviation are also not clearly articulated. Not surprisingly then, the debate on development-tourism's role in poverty alleviation is highly fragmented and often leads to dialogue at cross-purposes, where different commentators advocate or criticise a development-tourism niche in isolation, without giving consideration to the broader development-tourism context or its overall theoretical concepts (Ausland, 2010).

The purpose of this chapter is to build an integrative framework in order to better understand and analyse the links between development-tourism and poverty alleviation. In doing so, the chapter seeks to pave the way for more coherent efforts among the diverse stakeholders and better inform decision-makers with regard to development-tourism and its contribution to the first goal of the SDGs – poverty alleviation. To achieve this, the chapter first defines the development-tourism and poverty alleviation space, before framing the development-tourism and poverty alleviation debate within the three key strands of literature: Global citizenship; sustainable livelihoods; social entrepreneurship. A vignette of practical case is presented under each key concept to facilitate student and practitioners' application. Finally, the implications of integrating these concepts into the discourse of development-tourism - poverty alleviation are discussed, and potential research agendas are proposed as a way forward for the future.

BACKGROUND: OVERVIEW OF DEVELOPMENT-TOURISM

Origins

Following a decade of strong criticism towards the conventional top-down approaches to development in general and mass tourism development in particular (e.g. de Kadt, 1979), the 1980s witnessed the rise of an alternative development paradigm and alternative tourism approaches that supported greater local control, grass-roots decision making and a livelihood-based approach for the local community (Gartner, 2008). Besides the heavy promotion of environmental-focus types of tourism (e.g., ecotourism and 'green' tourism), a variety of tourism approaches also emerged which aimed to establish direct links with the poor. These types of tourism approaches share the common characteristic of featuring poverty and/or the prospects of poverty alleviation as the main tourist 'attraction' (Scheyvens, 2012). As such, they are grouped under the banner of 'poverty tourism' or alternatively 'development-tourism'.

Categorising and Defining Development-tourism

In the 1960s and early 1970s, poverty was mainly seen as 'the condition where individuals lack the financial resources to satisfy their basic needs and/or a minimum standard of living' (Misturelli & Heffernan, 2008, p. 667). Subsequently, poverty alleviation was often seen only as 'a sub-goal or a natural outcome of regional/national economic growth' (Zhao & Ritchie, 2007, p. 120), while the development agenda mainly focused on economic development. Since the 2000s, the increasing recognition of multi-dimensional poverty (e.g., social-cultural deprivations, discrimination, powerlessness) shifted the global development agenda to focus on the alleviation of multi-dimensional poverty. This shift is clearly reflected in the first goal of the SDGs, which seek to 'end poverty in all its forms anywhere'.

Hence, for the purpose of this chapter, the term development-tourism is adopted for its capacity to highlight the potential of tourism to alleviate multi-dimensional poverty alleviation in the global development agenda (Ausland, 2010a, 2010b; Baptista, 2011). In addition, development-tourism is also deemed to be a more appropriate term than 'poverty tourism' as the term 'poverty tourism' is currently often used interchangeably with 'slum tourism', which is but one of the diverse tourism approaches incorporated under the banner of development-tourism. In short, the term 'development-tourism' better encapsulates the context of, and approach to, poverty alleviation.

Ausland (2010) observed that while 'development-tourism' is frequently discussed, a wide range of niche tourism activities are currently associated with, and often used interchangeably with this term. Each type of development-tourism approach has a different focus, influenced by diverse motives, values and backgrounds of the parties involved such as those of the tour providers, tourists, development agencies and local communities. In an attempt to steer the debate, Ausland (2010a) proposed a taxonomy of development-tourism, which categorises development-tourism based on tourism approaches as well as the foci/purposes of the associated tourism activity (i.e., learning, leisure & labour). Ausland's (2010b) taxonomy of development-tourism incorporates six tourism approaches that can be associated with poverty tourism: slum tours, voluntourism, exposure trips, study abroad, short-term volunteer trips and community tours.

In this chapter, Ausland's (2010b) taxonomy is condensed into four main approaches of development-tourism, which include volunteer tourism, community-based tourism, exposure tourism, and slum tourism (see Table 1). 'Short-term volunteer trips' such as church mission or missionary safaris are incorporated as part of volunteer tourism, and 'study abroad' (for development students) is categorised as one of the forms of exposure tourism. In addition, as Ausland's taxonomy was proposed almost decade ago, a more recent review of literature on tourism and poverty alleviation nexus was conducted by the author and revealed a new niche of development-tourism: microfinance tourism (e.g., Phi, 2017; Phi, Whitford, Reid & Dredge, 2017; Phi, Whitford & Reid, 2016).

Each type of development-tourism approach has a different focus, influenced by diverse motives, values and backgrounds of the parties involved such as those of the tour providers, tourists, development agencies and local communities. For instance, volunteer tourism tends to be fuelled by the demand of tourists to engage in some form of development (through tourists' contribution of economic and/or labour exchange) (Wearing, 2001). Community-based tourism (CBT) centres on the question of who initiates,

Table 1. Definitions of development-tourism approaches

Development-tourism approaches	Definition
Volunteer tourism	'Volunteer tourism encompasses those tourists who, for various reasons, volunteer in an organised way to undertake holidays that might involve aiding or alleviating the material poverty of some groups in society, the restoration of certain environments or research into aspects of society or environment' (Wearing 2001, p.1).
Community-based tourism (e.g., community tours and in some cases overlaps with cultural tours, indigenous tours, ethnic tours)	Community-based tourism is tourism projects that 'involves the host community in the planning, developing and the maintenance of those projects' (Blackstock, 2005, p. 181). Timothy (1999, cited in Sharpley & Telfer, 2008, p.152-153) defines CBT based on two key perspectives: public participation in decision-making and residents' benefits from tourism activities.
Exposure tourism (e.g., justice tours, donor tours, vision tours)	Exposure tourism allows travellers 'with an actual connection or interest in ongoing activities in the community (e.g., donors, activists) to gain greater understanding and appreciation of what is actually happening by seeing it first-hand' (Ausland 2010, p.1)
Slum tourism (e.g., township tours, favela tours)	Slum tourism or poverty tourism involves bringing tourists to visit impoverished urban areas, mainly in developing countries, to observe people living in poverty (Mekawy, 2012).
Microfinance tourism (e.g., microfinance tours, invest tours)	Microfinance tourism facilitates an opportunity for tourists to experience microfinance (i.e., the provision of microloans and other financial services to the poor population) in action by paying a visit to a small group of poor local micro-entrepreneurs operating in their daily environment. The tour profits are channelled into free or low-interest micro-loans for the hosts (Phi, 2017)

owns and manages the tourism activities, which, by definition, should rest with the local community (Blackstock, 2005; Higgins-Desbiolles, 2006). CBT is also frequently linked with the diversified livelihoods approach, especially in the context of rural tourism and rural development (Steinicke & Neuburger, 2012). Exposure tourism focuses on education and the delivery of a certain developmental message from providers to tourists (e.g., justice, charity, poverty alleviation) (Higgins-Desbiolles, 2008; Scheyvens, 2007; Wrelton, 2006). In contrast, lum tourism aims to satisfy the desire of tourists to witness or experience the lives of people who live in impoverished urban areas, which in most cases are in much worse living conditions than the rest of the people from the same city (Frenzel & Koens, 2012; Ma, 2010). Combining two major poverty alleviation approaches (i.e., microfinance and tourism), microfinance tourism has a dual focus on educating tourists of local/global poverty and microfinance as a means for poverty alleviation, whilst concurrently channelling tour profits into micro-loans for income-generating activities to support the poor local population (Phi et al., 2016).

These foci, however, are not mutually exclusive and the boundaries between the approaches of development-tourism are also not clear-cut, rendering the development-tourism space complex and messy (Ausland, 2010b; Baptista, 2011). For instance, besides exposure tourism and microfinance tourism, depending on the motives of the tour providers and of the tour participants, volunteer tourism and slum tourism also often claimed to be educational-oriented (Holdnak & Holland, 1996; Wearing, 2001). Moreover, Frenzel (2013, p.118) argued that a large part of modern slum tourism emanated from exposure tourism which has become 'a broader consumer leisure practice' with a market made up of 'political tourists interested in pursuing global justice and equity'. Concomitantly, Salazar (2004) and Ramchander (2003) argued that the approaches constituting development-tourism could be situated somewhere between 'cultural tourism' or 'ethnic tourism' where the focus of activities often centre on the notion of experiencing the 'other' and/or 'authentic poverty' (Baptista, 2011; Crossley, 2012; Rolfes, 2010). For instance, Crossley (2012) examined the perceptions of volunteer tourists to Kenya and found that many valued and indeed desired the opportunity to experience unique and authentic local poverty conditions.

Therefore, for the purpose of this chapter, development-tourism is defined by the author as:

the practice of travelling to impoverished areas for the purpose of experiencing lives of the poor in their everyday environment, and which may be associated with learning about aspects of development and/or contributing financial resources or labour to assist the individual poor/host community in improving their lives.

The next section discusses the literature on the development-tourism-poverty alleviation nexus.

DEVELOPMENT-TOURISM AND POVERTY ALLEVIATION

In the broader literature, many approaches and perspectives exist regarding what poverty is, who the poor are and how poverty should be alleviated (e.g., Alcock, 1993; Cobbinah, Black, & Thwaites, 2013; Dini & Lippit, 2009; Laderchi, Saith, & Stewart, 2003; Sen, 2000). For instance, on the one hand, the income-based approach looks at poverty as purely a lack of financial resources to afford basic material things for survival such as food, shelter and clothing (Cobbinah et al., 2013). This lack of income is often seen as being caused by economic inefficiency, which can be solved by macro-economic development (Haughton & Khandker, 2009, p. 2). On the other hand, both the capability-based approach & structural-based approach view poverty as multi-dimensional. The capability-based approach, however, focuses more on human development and the provision of basic services such as health and education (Sen, 2000). The structural-based approach focuses on the power relations embedded within the economic, social and political structures at the micro, meso and macro levels that allow some groups to have greater entitlement and access to resources and security, while at the same time restricting and/or excluding others (McCaston & Rewald, 2005). Additionally, the structural-based approach to poverty alleviation argues that poverty does not only reside with the poor or in developing countries but arises from a complex web of interaction which integrates the poor and non-poor as well as developing and developed countries (Sen, 2000; UNDP, 2000). Consequently, this approach often advocates empowering global citizens, especially those from developed countries, to help address the unequal power relations between the poor and non-poor as well as between developed and developing countries (e.g., Ferguson, 1990; Matthews, 2006; Veltmeyer & Petras, 2011).

The current debate of development-tourism and poverty alleviation centres around three key points. First, within the diverse and complex context of poverty alleviation, advocates of development-tourism claim that its various approaches can assist the poor to enhance and improve their livelihoods. This is achieved by creating direct sources of income for poor individuals/communities through providing tourism product and services and/or improve their capabilities to secure livelihoods through tourists' philanthropic activities (e.g., donations/loans to develop business, volunteer training of English/basic skills) (Medina-Muñoz et al., 2016; Scheyvens, 2012; Sharpley & Harrison, 2019). Critiques of development-tourism however, argue that the rhetoric of development-tourism tends to ignore the possibility that poor people, especially the poorest of the poor, still face significant constraints participating in, and recruiting benefits from, tourism activities. This may be due to a lack of capabilities (e.g., no skills/capital) and other structural constraints (e.g., the right to participate in and influence decision-making pertaining to development-tourism projects) (Harrison, 2008, Sharpley & Harrison 2019). This debate directly relates to sustainable livelihoods and poverty alleviation literature.

Second, development-tourism is seen as offering tourists a chance to better understand the lives of the poor and provide education pertaining to poverty issues, which in turn may inspire actions and advocacy in the quest to combat deep-seated causes of global poverty (Devereux, 2008; McGehee, 2006; Mekawy, 2012). Yet, as the concept of 'development' has become fashionable in the tourism industry, both poverty and poverty alleviation are increasingly objectified and commoditised with 'the private sector, governments, NGOs, media, the tourists and the hosts, all assisting in the creation and promotion of tourism as a 'developmental' act' (Baptista, 2011, p. 663). The educational impacts of development-tourism on visitors as such are still questionable (Lyons, Hanley, Wearing, & Neil, 2012). This debate is grounded within the literature of global citizenship and poverty alleviation.

Third, development-tourism is often criticised as 'a case of charity replacing the market', where development agencies and (I)NGOs inject significant funding into local tourism projects that seem successful on the surface, yet the actual returned economic benefits would only make up a small fraction of the initial investment (Butcher, 2006, Sharpley & Harrison, 2019). Indeed, it has been argued that mass tourism, in many cases, is less intrusive, easier to manage and able to generate more economic benefits than small-scale tourism projects (Thomlinson & Getz, 1996). Scheyvens (2012) thus maintained that tourism projects need to be commercially viable and financially sustainable (at the very least), to ensure the long-term delivery of the positive impacts on poverty alleviation. This argument strongly links with the emerging literature on social entrepreneurship.

These key strands of the debate focusing on the various development-tourism approaches to poverty alleviation are tightly linked to the broader literature of global citizenship, sustainable livelihoods and social entrepreneurship, yet no study thus far ties together these three concepts when discussing the tourism and poverty alleviation nexus. Thus, the following section aims to move the debate forward from analysing and/or criticising development-tourism as simply a niche sector of the tourism industry. To achieve this, the author draws on and incorporates the broader literature, in order to gain more enriched perspectives pertaining to development-tourism which in turn, can be articulated into more practical and appropriate applications of the development-tourism to support SDGs goal one.

KEY CONCEPTS OF DEVELOPMENT-TOURISM AND POVERTY ALLEVIATION

Sustainable Livelihoods

Sustainable livelihoods (SL) concept has increasingly been adapted and used within tourism-poverty alleviation context (e.g., Anderson, 2015; Laeis & Lemke, 2016; Snyman, 2014). Emerged in the 1980s, the SL concept was first officially mentioned in "Our Common Future" report by the Brundtland Commission in 1987 and the UN's first Human Development Report in 1990. These two reports set the key foundational concepts for SL, which were people-centred (i.e., prioritise the poor and communities' needs), participatory (i.e., allow and empower the poor to identify and prioritise livelihood strategies and goals), and sustainable (i.e., consider long-term impacts). Chambers and Conway (1992, p. 7) defined sustainable livelihoods as followed:

A livelihood comprises the capabilities, assets (including both material and social resources) and activities required for a means of living. A livelihood is sustainable when it can cope with and recover from stresses and shocks, maintain or enhance its capabilities and assets, while not undermining the natural resource base.

As indicated in the definition, the term 'livelihood' is used to emphasise the complex realities that poor people encounter in their daily lives, of which vulnerability occupies a central concern (Chambers, 2013). Hence, among other things, SL focuses on the development of preventative and protection mechanisms (e.g., diversify sources of income, insurance, social safety net) to help poor individuals and communities to reduce, mitigate and cope with the shocks and stresses that are specific to their living environments (Shaffer, 2008). Concurrently, SL helps to reveal the poor's strengths in dealing with shocks and stresses through the utilisation of their limited assets and sources of support (e.g., natural capital, social capital, financial capital, human capital) in diverse livelihood activities (e.g., trade, transport, casual wage work, farming, fishing). It is argued that such an appreciation and understanding of the poor's livelihoods and their adaptive mechanisms provide the best guide to develop poverty alleviation strategies that catalyse and build on strengths and potentials of the poor (Chambers, 2013).

Bennett et al. (1999) were among the pioneers who proposed ways for tourism to improve the poor's livelihoods: (1) Tourism brings tourists to the site of consumption, including the peripheral and isolated areas where the majority of the poor life; (2) Tourism encourages the creation of inter-sectoral linkages such as agriculture, artisan production and service sectors to support tourism activities and in turn, assists livelihood diversification of the poor; (3) Tourism has rather low barriers to entry; and (4) Tourism is generally labour intensive, employing a high level of females, young people and unskilled or less-skilled individuals who comprise the categories that make up a high percentage of the poor.

Notwithstanding the importance of income that tourism may bring to some individuals or communities, in many other communities, perceptions of SL or well-being may not at all be related to a monetary economy. Ethnic communities in Sapa, for instance, view rice sufficiency and undisrupted agricultural yield as the key to well-being (or lack thereof) (Truong, Hall & Garry, 2014). Alternatively, Eversole, McNeish and Cimadamore (2013) suggested that many indigenous groups across the globe view the protection of ancestral title to land and natural resources as the most important factor to achieve SL. Hence in a broaden framework to tourism and poverty alleviation, tourists can be encouraged to improve local poor's livelihoods in a variety of ways beyond economic exchange such as appropriate knowledge transfer (Novelli et al., 2016). For example, the Aurora Native Title Internship Program has been sending student volunteers to support various indigenous communities around Australia with baseline information and legal resources needed to secure their rights in the context of land struggles due to rapid economic development (Aurora Project, 2018). Thanks to the advent of technological innovations and global sharing platforms, there are also strong prospects of engaging an ever-larger tourist population in volunteering activities, and better matching volunteers' skills with the specific needs of local communities (Ong et al., 2014).

It is also important to note that depending on the context, tourism may not be the most suitable livelihood strategy to address poverty. For instance, the use of tourism as a livelihood strategy may further marginalise those who cannot cope with the principles of participating in tourism (e.g., the poorest of the poor with little to no resources). Many studies (e.g., Truong et al., 2014; Mgonja, 2015; Koens & Thomas, 2016) have suggested that even when opportunities are available through tourism development, there are many barriers that prevent the poor from gaining meaningful participation in, or being able to receive any long-term benefits from tourism activities. Yet the current practices – which favour the expansion of the industry – often present tourism as a compulsory strategy, leaving the presumed 'beneficiaries' little to no choice but to get involved. Indeed, Soumaré, (2015)'s study finds that tourism attracts a significant proportion of FDI inflows in North African countries, directing investment away from other sectors that are also important to local poverty alleviation such as agriculture, forestry and manufacturing.

For tourism to reflect the poor's real needs, its implications as a livelihood strategy need to be considered equally among a range of other potential strategies (Zhao & Richie 2007). Along the same line of thought, Shen et al. (2008) highlighted the need to distinguish the difference between the poor's participation in tourism and in the SL approach. While tourism is often prescribed as the 'magic pill' against poverty, the poor's perceived strengths and opportunities are captured in ways not predefined by any sector or discipline in SL. Ellis (1999, p.2) further argued that policies which facilitate the poor to "gain better access to opportunities (or to create their own opportunities) might turn out to be substantially more cost-effective for poverty alleviation than attempting, artificially, to support particular sectors or sub-sectors (such as tourism)".

Overall, SL should be incorporated in the tourism-poverty alleviation framework in a way that enables the poor's agency to be maximised to the point where they can reduce their involvement in tourism to a certain level, or refuse inclusion in tourism all together, if it is seen as diverting important resources from, or creating conflicts with, other livelihood activities that are deemed by the individual/community to be more suitable in a given situation.

Vignette 1 Diversify local livelihoods through community tourism in Vietnam

In most impoverished communities in Vietnam, local people tend to rely primarily on subsistence agriculture for their livelihoods, which expose them to adverse effects from unpredictable weather conditions and agricultural market fluctuations. In addition, many are also isolated geographically, and are from minority ethnic groups that experience discrimination and cultural assimilation. In this context, CBT Travel & Consulting social enterprise has re-defined community tourism development and poverty alleviation in Vietnam with 80 projects in the Northwest, Northeast, Central and Southern regions of Viet Nam between 2012–2019. In 2016, these community projects welcomed over 60,000 visitors, contributing over USD 1.2 million to the local communities (Phi & Whitford, 2017).

CBT Travel & Consulting was established in 2013 by Mr Binh Duong, a retired tour director and resort designer with the aim of supporting locally-owned tourism projects to diversify income and preserve local cultures, especially for impoverished ethnic minorities. CBT Travel focuses on providing hand-on coaching activities for tourism capacity development, along with homestay and destination activity designs and connect the local CBTs with tour operators. It is estimated that an average of 79 tourism jobs are created with each community tourism project, including 21 full-time positions in three homestays and 56 casual positions involved in the other tourism value-added services (e.g., traditional performance, laundry, tour guides, rafting, weaving etc.) (Phi & Whitford, 2017). Indirect jobs, such as supplying local products and handicrafts to tourists, also helped to distribute tourism profits more widely among community members.

Not only do these diverse value-added activities assist locals in diversifying their income sources, but they also contribute to increase local pride and reduce discrimination through recovering local cultural practices and performances that are in danger of disappearing.

Global Citizenship

As many causes of poverty are deeply rooted within inequitable structures, it is becoming clear that economic and skill transfers via tourism often provide symptom reliefs instead of addressing global structural poverty issues in the long run. Rather, there is growing evidence that transformative tourism experiences which include direct interactions with the poor can help to increase tourists' knowledge of poverty issues, poverty causes as well as empowering the participants to assist with poverty alleviation in more comprehensive ways (Scheyvens, 2012; Wrelton, 2006; Wearing, 2001). Tourism as such is often positioned as an important tool to promote and foster an ethic of global citizenship (GC) to tourists, who are no longer seen as merely global consumers but rather global citizens (e.g., Phi et al., 2017).

The concept of GC has a long history, tracing back to the Socratic notion of "citizen of the world", and to the cosmopolitanism of Enlightenment philosophers (Nussbaum, 2002). In the last few decades, GC has received renewed philosophical and practical interest, with the development of a strand of literature specifically linking global citizenship to poverty alleviation (see e.g., Edwards, & Gaventa, 2014; Carter, 2013). This link is embedded within the transformational approach to globalisation, which realises that "in both the North and South, there exist concentrations of wealth and power along with increasing poverty and exclusion" (Shultz, 2007, p. 255). There is thus, a growing global consensus for citizens, regardless of their states and societies, to share the responsibility of actively creating and enhancing global social justice through poverty alleviation and to improve the life capabilities of the marginalised (Edwards, & Gaventa, 2014).

Various tourism approaches may contribute to spread the ethics of GC. Zarandian et al., (2016) for example found that the experience of living and volunteering in developing countries often resulted in the voluntourists' heightened awareness of social inequalities and injustices, which could lead to their increased engagement with social movements. Similarly, Phi et al., (2017) and Frenzel (2013) suggest that with proper development, microfinance tourism and slum tourism does not only help to better inform tourists of the various aspects and causes of poverty, but also enable global linkages and solidarity between developing and developed countries. This in turn "leading to the formation of alliances" that help to address the social-political root causes of global poverty (Frenzel 2013, p.118). Given the right conditions, the targeted poor population can also act as powerful educators of GC. By directly engaging with tourists and sharing their struggles and aspirations for a better life, the tourist-host encounter could lead to a better understanding and a reduction of poverty stereotypes (Phi et al., 2017).

Despite GC's potential, a number of authors have questioned tourism's capacity to actually foster GC (e.g., Butcher, 2017; Jakubiak & Iordache-Bryant, 2017). Phi et al., (2017) find that while a continuum of 'thin' versus 'thick' GC exists, many tourism providers tend to promote only 'thin' GC for practical and commercial reasons (i.e., depoliticised view that poverty can be solved through charitable or tourism consumption). It is thus important for all key stakeholders involved in GC education through tourism (e.g. tourism providers, tour guides, tourists, funding organisations, local communities, and governments) to be informed of the different perspectives regarding GC and make conscious decisions on the goals and practices of fostering GC. There is also a need to better train tour guides and provide pre-trip (e.g., discussion prior to getting to the destination) and post-trip support, (e.g., online forum) to engage tourists throughout the experiential learning process and deepen their reflections. Similarly, Taplin, Dredge and Scherrer (2014) call for the investigation into more transparent and accountable approaches to monitoring and evaluation of commercial volunteer tourism. This should also include measures that better giving voices to the host community/host organisations such as a two-way review system on global online platforms.

Beyond the education of tourists, a body of literature which explores GC education to the diverse range of stakeholders involved in tourism has also slowly emerged, as Caton, Schott & Danielle (2014, p.125) put it "A global practice requires a global ethic from those who plan and manage it". Education effect can partly come from the tourists who continue to reflect and engage in discussions on issues they witnessed on the ground after returning home. Through voicing their opinions and ideas in traditional forums as well as via emerging technology such as online petition and social media, these tourists can become GC educators themselves, facilitating the networking of new social movements and providing constant feedback for NGOs, government and corporations regarding practices and policies in poverty alleviation (Mayo, 2005). Second, there is also increasing call for higher education sector to embrace their roles in effectively educating tourism students of GC and the ethics of care for everyone, especially those disadvantaged and marginalised (Canziani, 2014). Tourism Education Futures Initiative (TEFI) for instance stresses the importance of moving away from the idea that tourism is a pure industry governed by market values and extraction of resources, towards teaching students to develop respect and understanding to individuals, communities and places to which we are all a part of (Caton et al., 2014).

Vignette 2: The Tripadvisor – Kiva partnership to educate travellers of global citizenships

TripAdvisor Inc. manages a number of branded travel websites, with more than 390 million monthly unique visitors per month (Tripadvisor, 2017). As the largest online travel community, a common issue that the Tripadvisor travellers often encounter is that they enjoy the travel experience but do not know how to give back to the local communities in a meaningful way. In response, a unique partnership between TripAdvisor and Kiva.org - the world's largest online microlending platform was announced in December 2012. The partnership aims to combine the power of TripAdvisor's crowdsourcing and Kiva's crowdfunding to make meaningful change to the lives of the poor on a global scale. Premal Shah, the President of Kiva stated:

As global citizens, none of us are visitors, we are all connected. Nothing makes this clearer than when we travel. Through this partnership, the affinity you develop for the people and places you visit can continue when you return home.

Under the partnership, a traveller, who has been to a KIVA-supported country and written a review about that trip on TripAdvisor, would be given US$25 to support a KIVA borrower of their choice from the same country. With the initial fund of US$250,000 from TripAdvisor, the first phase of the partnership aims to educate at least 10,000 travellers about how they can directly contribute to the fight against global poverty. (Vignette adapted from Newton, 2012; Tripadvisor, 2017 and Phi, Dredge & Whitford 2013)

So far, the discussion on tourism and poverty alleviation has focused on supportive and transformational aspects of tourism from a social perspective, driven by social needs and concerns. Many of the given examples are drawn from the non-profit sector, led by NGOs. This is unsurprising as most NGO operations are guided by un-commodified agendas and social values, which are in stark contrast to the way global tourism practices is increasingly shaped into a private-led industry (Wearing, McDonald & Pointing, 2005). However, the non-profit sector lacks the capacity to direct tourism and poverty alleviation activities on a more significant scale and for longer time-frame than project funding would usually allow. Zhao and Ritchie (2007, p.127) stressed that tourism for poverty alleviation 'cannot or should not be a form of charity'. The next section thus discusses the concept of social entrepreneurship as another key emerging concept that could contribute to realise the potential of tourism as a means for poverty alleviation.

Social Entrepreneurship

Social Entrepreneurship (SE) describes a process involving innovative ways of organising "resources and networks to pursue opportunities that can lead to social change and/or address social needs" (Mair & Martí, 2006, p. 37). SE has become a global phenomenon in the last few decades as a response to the failure of the traditional approaches in dealing with many social issues, including poverty.

Arguably, despite the notion that "very few of our actions could be said to be motivated by anything so simple as untrammelled greed or utterly selfless generosity" (Graeber, 2007, p. 2), the current system of non-profit and for-profit sectors divides human nature into two separate dimensions, effectively compartmentalising and separating social values and goals from economic values and goals (Zebrowski, 2009). While the non-profit sector is burdened with the role of addressing social needs, ideological constraints (e.g., maintain low overheads and low marketing costs, pursue only social goals) lead them to have no economic freedom, coupled with the constant struggle to capture resources from donors (Zebrowski, 2009). NGOs are thus seen as frequently moving away from the poor's real needs to favour the views of donors (van Beers, 2011). In contrast, Polak (2009) argued that the current conceptualisation of the private sector only allows room for economic motives, despite increasing expectations on businesses to integrate social goals and involve poverty alleviation. This dichotomy is clearly demonstrated in Burns' (2004) continuum of tourism planning in developing countries. On one end of the spectrum is the "tourism first" approach with an emphasis on market-led, industry expansion and economic growth. At the other end of the spectrum is the "development first" approach, led by the non-profit sector with a focus on developmental goals. In this context, SE offers a pragmatic structure of prioritising social values while striving to also achieve financial self-sufficiency and financial sustainability (Volkmann, Tokarski & Ernst, 2012). As such, SE supports both the incorporation of business practices into the non-profit sector and the incorporation of social values into the private sector. Either way, by integrating social and economic elements, SE challenges the conventional structures and gives way to new thinking that has the ability to transcend traditional boundaries. This enables SE to overcome the limitations of each sector in dealing with social issues (Zebrowski, 2009). Within the development and poverty alleviation context, social entrepreneurs are creating and showing new pathways and solutions to achieve global and /or local development goals through their ability to 'resist being trapped by the constraints of ideology or disciplines' (Mair, Robinson & Hockerts 2006, p. 6).

Unsurprisingly, SE is gaining more attention in the recent tourism and poverty alleviation literature (e.g., Marko & Jelena, 2014; Mottiar, Boluk & Kline, 2018; Sloan, Legrand & Simons-Kaufmann, 2014; Sheldon & Daniele, 2017). Marko and Jelena (2014) consider SE as an important alternative business model for a responsible tourist destination, especially to better distribute tourism benefits within the context of community-based tourism. Phi, Whitford and Dredge (2017) investigate the knowledge dynamics in tourism SE and highlight the capacity of tourism social entrepreneurs as "knowledge boundary spanners" that can bring together diverse knowledge reside within the public, private and third sectors to create innovative solutions for poverty alleviation. Laeis and Lemke (2016)'s study on tourism SE in Cape Town, South Africa found that tourism social enterprises strongly links to, and supports the sustainability of local livelihoods. They further highlight that SE does not simply operate as a conduit to channel monetary gain from businesses into the community, but also has the power to influence changes in local structures and institutions. Novelli et al. (2016, p.829) view SE as part of 'modern philanthropy', which has potential to enhance global (corporate) citizenship and foster long-term solutions to address the diverse social, economic and institutional infrastructure that perpetuates poverty, while Biddulph (2018) explores the contribution of social enterprises in Cambodia to create inclusive tourism opportunities. The link between SE, tourism and poverty alleviation, however, is still in its infancy, with increasing calls for more integration of the concept into various tourism types such as indigenous tourism (Rebutin, 2009), ecotourism (Situmorang & Mirzanti, 2012), community-based tourism (Phi et al., 2017) and hospitality (Johnson, 2013).

Beyond the integration of SE into current practices of leisure tourism, the ultimate promise of SE perhaps lies in the context of inclusive tourism and mobility: unlocking travel freedom for the majority of the world. The last few decades have witnessed the rapid rise of globalisation a 'new mobilities paradigm' (Sheller & Urry, 2006), yet there remain billions of people living in poverty who are socially, politically and economically excluded from travel (Hall, 2010). For this so-called 'immobile' population, the ability to travel for any period of time and for any length of distance (even just from their home village to the nearest city) means much more than a leisure experience. It may open up, among other things, access to proper medical and legal services, education and economic opportunities and new livelihood ideas that are instrumental for a better life when returning home (Phi & Dredge, 2019). Rogerson (2014) for instance, called for more attention and support to the 'migrant entrepreneurs' or the 'necessity tourists' who travel almost daily across borders in sub-Saharan Africa in order to make a living through subsistent trading. Inclusive tourism and fulfilling human being's mobility rights as such is significant in realising the global vision of well-being as set out by the SDGs, yet the "obligation to fulfil the precepts of social tourism" has so far only been given to the governments – especially governments of developed countries to subsidy their disadvantaged citizens to participate in leisure tourism (Higgins-Desbiolles, 2006, p.1201). SE – with the innovative organisations of resources at its core, could potentially give rise to the creation of affordable travel products and services that specifically cater to the people at "the bottom of the pyramid" all over the world.

Despite its potential, SE has been criticised for aligning with and contributing to the dominant neoliberal agenda of social services' marketisation and expansion of free-market principles (Kim, 2016). This is due to SE's efforts to integrate and balance social missions/values with market logics and financial sustainability. In reality, the social and commercial forces are constantly clashing, especially when the initial SE initiative becomes widely adopted by the private sector (Phi, 2016). Thus, SE may run the risk of mission-drift or in worst-case scenarios, expose the impoverished and marginalised to be taken advantage of by powerful commercial interests. The utilisation of SE in tourism, therefore, should be taken with caution, and with a more transparent debate, to develop frameworks and measures that contribute to regulate and minimise cases of mission-drift and misconduct.

Vignette 3: Fundación En Vía – empowering local female entrepreneurs through social entrepreneurship

Fundación En Vía is a non-profit organisation that works 'to promote the empowerment of women, the well-being of families and the strengthening of communities' (Envia, 2018, p.1). The organisation was found by Carlos Topete (Mexico) and Emily Berens (United States) in 2010 and use tourism revenue to provide interest-free loans and educational programs to women in Mexico.

Since its modest start in 2010, En Vía's operations in Mexico have expanded considerably, from servicing only one community in the beginning to covering six communities in the Tlacolula Valley of Oaxaca, Mexico. By 2018, nearly 1,500 tourists went on microfinance tours, generating revenue of around 1.7 million USD. The tour profit supported 254 entrepreneurial women who participated directly in the program, and subsequently benefited a population of over 1000 local people (Envia, 2018). As these micro-loans are repaid, the capital is accumulated and continue to be recycled in the community to support more female entrepreneurs. The locals also benefit from educational business courses, English courses and other relevant courses for community development such as health and nutrition, design and marketing of textile accessories. En Vía's efforts in alleviating local poverty through microfinance tourism were recognised by a high-profile organisation, namely the World Travel Tourism Council (WTTC, 2016).

Unlike normal tourism poverty alleviation approaches which attempt to integrate the poor permanently into the tourism system, poor individuals/households, in this case, tend to have only one-off interaction with tourists who come to visit their houses. Microfinance tourism thus gives people living in poverty greater flexibility in choosing which livelihood opportunities they would like to pursue, in turn, more effectively addressing their needs, while utilising local resources and knowledge (Phi, 2017). Through combining microfinance and tourism, En Vía's case also highlights how the stakeholders from diverse sectors/disciplines within a destination can work together to reduce overlapping efforts and create better synergy to support the local community.

SOLUTIONS AND RECOMMENDATIONS

The development-tourism – poverty nexus is tightly linked to the three key strands of literature on SL, GC and SE, yet to date, no study has tied together these three concepts to scaffold our understanding of tourism for poverty alleviation. The above discussion contributes essential elements for the development of an integrative, conceptual tourism-poverty alleviation framework that contributes to a broaden agenda for research and practices of development-tourism for poverty alleviation (Figure 1).

Figure 1. Integrative framework for development-tourism and poverty alleviation

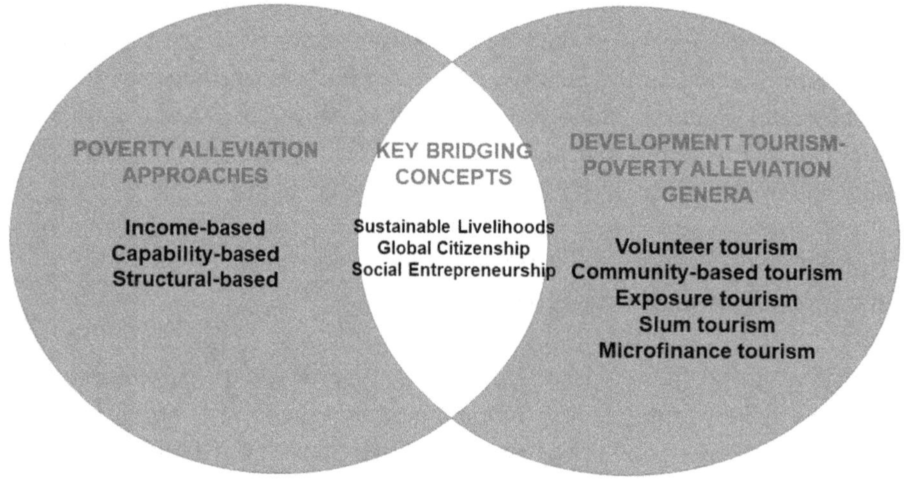

The framework includes three key poverty alleviation approaches (i.e., income-based, capability-based and structural-based), and five primary tourism-poverty alleviation genera. Each genus elucidates a different focus on the alleviation of multi-dimensional poverty through the utilisation of tourism. Three key bridging concepts (i.e., sustainable livelihoods, global citizenship, and social entrepreneurship) situate in the middle, which bring together the current, fragmented literature pertaining to various research niches related to tourism and poverty alleviation. The framework thus can potentially facilitate future collaboration and knowledge exchange among researchers, practitioners and policymakers with diverse perspectives to poverty alleviation, as well as those who tend to focus only on one of the six tourism-poverty alleviation genera. The bridging concepts could also be further utilised in diverse and innovative ways to assist in improving the poor's well-being, as demonstrated in the above discussion.

FUTURE RESEARCH DIRECTIONS

The chapter also identified a number of gaps and proposals for future studies. First, through the analysis of poor peoples' strengths and weaknesses, studies may wish to explore more flexible ways that tourism can be used to better assist poor peoples' needs. This may include integrating tourism with other poverty alleviation strategies in a destination to deliver more significant impacts on people's lives. More attention should also be exerted into exploring the different ways tourism can be applied to improve the livelihoods of poor people in urban versus rural areas.

Second, there is a need for studies that seek to understand how tourists, tour providers, funding organisations and the local communities construct their perceptions of the poverty issue, and how this construction can be influenced to better educate GC. Future studies may investigate further into specific tourism genus of the framework and explore the factors that constrain or help to foster global citizenship within each genus. Furthermore, an area with potential but has received little to no attention in the nexus of tourism and GC is the extent to which the ethics of GC is promoted to the poor community through tourism activities (Moufakkir & Kelly, 2015). Tourism, in many ways, acts as an opening channel for a community to connect with people from various parts of the world, which may have profound psychological and spiritual impacts on the way each member of the community functions. Moufakkir and Kelly (2015, p. 193) hence posed further questions for research "Does tourism open the poor host community's minds, meet the needs for their search for the self and change their relationship with the world?".

Third, SE provides an important opportunity for researchers and practitioners to move beyond the duality of the "tourism first" or "development first" agenda, leading to a new wave of innovative tourism models that are not only financially sustainable over the long run but also assist in resolving systematic local/global issues. In order to facilitate the development of SE in tourism, new studies can explore various examples of SE innovation in relation to the various elements of the integrative framework; map out the typology of tourism social entrepreneurs as well as the opportunities and constraints they face in initiating and scaling their tourism SE models. Importantly, more critical studies on SE are needed to enhance the long-term social/economic balance of tourism SE models and prevent potential 'hijack' from those with pure for-profit motives.

CONCLUSION

As long as there are people living in poverty across the globe, our goal, and arguably our duty, should be the continual quest to realise poverty alleviation. The aim of this chapter, therefore, has been to identify and synthesise the various perspectives and approaches on tourism-poverty alleviation (beyond the rhetoric of PPT) in the leading up to, and after the release of the SDGs. This chapter has demonstrated that while development-tourism is utilised as a vehicle to address poverty and contribute to the achievement of the first goal of the SDGs, overall, knowledge and understanding of development-tourism is still underdeveloped and fragmented. Moreover, this chapter has revealed that the development-tourism debate is consistently peppered with disparate dialogue which affords little to no consideration of the broader development-tourism context or its overall theoretical concepts. Subsequently, this chapter has proposed an integrative framework to help better understand and analyse the development-tourism - poverty alleviation nexus. During the development of this framework, the chapter has effectively synthesised cross-purpose dialogue pertaining to the development-tourism and poverty-alleviation nexus by first defining the elements of development-tourism and poverty alleviation and then reframing the development-tourism and poverty alleviation debate within the three key strands of literature: GC, SL and SE. By integrating these three concepts into the discourse of development-tourism - poverty alleviation, this chapter provides a clear theoretical scaffolding for both practitioners and researchers to underpin their goals in relation to poverty alleviation through development-tourism activities.

REFERENCES

Alcock, P. (1993). *Understanding Poverty*. Macmillan.

Anderson, W. (2015). Cultural tourism and poverty alleviation in rural Kilimanjaro, Tanzania. *Journal of Tourism and Cultural Change*, *13*(3), 208–224. doi:10.1080/14766825.2014.935387

Aurora Project. (2018). *Overview| The Aurora Project*. Retrieved from https://auroraproject.com.au/overview

Ausland, A. (2010a, August 27). Poverty taxonomy 2.0. *Huffington Post*. Retrieved from https://www.huffingtonpost.com/aaron-ausland/poverty-tourism-taxonomy-_b_697399.html

Ausland, A. (2010b). *Poverty tourism: A debate in need of typological nuance. Staying for tea*. Retrieved from http://stayingfortea.org/2010/08/17/poverty-tourism-a-debate-in-need-of-typological-nuance/

Baptista, J. (2011). The tourists of developmentourism– representations 'from below'. *Current Issues in Tourism*, *14*(7), 651–667. doi:10.1080/13683500.2010.540314

Bennett, O., Ashley, C., Roe, D., & Britain, G. (1999). *Sustainable tourism and poverty elimination study: a report to the department for international development*. Deloitte & Touche.

Biddulph, R. (2018). Social enterprise and inclusive tourism. Five cases in Siem Reap, Cambodia. *Tourism Geographies*, *20*(4), 610–629. doi:10.1080/14616688.2017.1417471

Blackstock, K. (2005). A critical look at community based tourism. *Community Development Journal: An International Forum*, *40*(1), 39–49. doi:10.1093/cdj/bsi005

Burns, P. (2004). Tourism planning: A third way? *Annals of Tourism Research*, *31*(1), 24–43. doi:10.1016/j.annals.2003.08.001

Butcher, J. (2017). Citizenship, global citizenship and volunteer tourism: A critical analysis. *Tourism Recreation Research*, *42*(2), 129–138. doi:10.1080/02508281.2017.1295172

Canziani, B. F. (2014). Values-based Internships: Combining TEFI Values, Career Internships, and Community Engagement. *Journal of Teaching in Travel & Tourism*, *14*(2), 129–148. doi:10.1080/15313220.2014.907957

Carter, A. (2013). *The political theory of global citizenship* (2nd ed.). Routledge. doi:10.4324/9780203183397

Caton, K., Schott, C., & Daniele, R. (2014). Tourism's Imperative for Global Citizenship. *Journal of Teaching in Travel & Tourism*, *14*(2), 123–128. doi:10.1080/15313220.2014.907955

Chambers, R. (2013). *Ideas for development* (2nd ed.). Earthscan. doi:10.4324/9781849771665

Chambers, R., & Conway, G. (1992). *Sustainable rural livelihoods: practical concepts for the 21st century*. Institute of Development Studies.

Cobbinah, P., Black, R., & Thwaites, R. (2013). Dynamics of Poverty in Developing Countries: Review of Poverty Reduction Approaches. *Journal of Sustainable Development*, *6*(9), 25–35. doi:10.5539/jsd.v6n9p25

Crossley, É. (2012). Poor but happy: Volunteer tourists' encounters with poverty. *Tourism Geographies*, *14*(2), 235–253. doi:10.1080/14616688.2011.611165

de Kadt, E. (1979). Social planning for tourism in the developing countries. *Annals of Tourism Research*, *6*(1), 36–48. doi:10.1016/0160-7383(79)90093-8

Devereux, P. (2008). International Volunteering for Development and Sustainability: Outdated paternalism or a radical response to globalisation? *Development in Practice*, *18*(3), 357–370. doi:10.1080/09614520802030409

Dini, A., & Lippit, V. (2009). *Poverty, from Orthodox to Heterodox Approaches: a Methodological Comparison Survey*. University of California- Riverside.

Edwards, M., & Gaventa, J. (2014). *Global citizen action*. Routledge. doi:10.4324/9781315071916

Ellis, F. (1999). *Rural livelihood diversity in developing countries: evidence and policy implications*. Overseas Development Institute London.

Envia. (2018). *Fundacion En Via*. Retrieved from https://www.envia.org/

Ferguson, J. (1990). *The anti-politics machine: " development, " depoliticisation, and bureaucratic power in Lesotho*. CUP Archive.

Frenzel, F. (2013). Slum tourism in the context of the tourism and poverty (relief) debate. *Journal of the Geographical Society of Berlin*, *144*(2).

Frenzel, F., & Koens, K. (2012). Slum tourism: Developments in a young field of interdisciplinary tourism research. *Tourism Geographies*, *14*(2), 195–212. doi:10.1080/14616688.2012.633222

Gartner, C. M. (2008). Tourism, Development, and Poverty Reduction: A Case Study from Nkhata Bay, Malawi (Master of Environmental Studies in Geography - Tourism Policy and Planning). University of Waterloo.

Graeber, D. (2007). Army of Altruists: on the alienated right to do good; Essay. *Harper's Magazine, 314*(1880), 31.

Hall, C. M. (2010). Changing paradigms and global change: From sustainable to steady-state tourism. *Tourism Recreation Research, 35*(2), 131–145. doi:10.1080/02508281.2010.11081629

Harrison, D. (2008). Pro-poor tourism: A critique. *Third World Quarterly, 29*(5), 851–868. doi:10.1080/01436590802105983

Haughton, J. H., & Khandker, S. R. (2009). *Handbook on poverty and inequality*. World Bank Publications.

Higgins-Desbiolles, F. (2006). More than an "industry": The forgotten power of tourism as a social force. *Tourism Management, 27*(6), 1192–1208. doi:10.1016/j.tourman.2005.05.020

Higgins-Desbiolles, F. (2008). Justice tourism and alternative globalisation. *Journal of Sustainable Tourism, 16*(3), 345–364. doi:10.1080/09669580802154132

Holdnak, A., & Holland, S. M. (1996). Edu-tourism: Vacationing to learn. *Parks & Recreation, 31*(9), 72–75.

Jakubiak, C., & Iordache-Bryant, I. (2017). Volunteer tourism in Romania as/for global citizenship. *Tourism Recreation Research, 42*(2), 212–222. doi:10.1080/02508281.2017.1299344

Johnson, C. (2013). *Social Entrepreneurship - Call for papers*. https://www.emeraldgrouppublishing.com/products/journals/call_for_papers.htm?id=4978

Kim, J. (2016). Neoliberalism and the Politics of Social Enterprises in South Korea: The Dynamics of Neoliberal Governmentality and Hegemony (PhD). Georgia State University.

Koens, K., & Thomas, R. (2016). "You know that's a rip-off": Policies and practices surrounding micro-enterprises and poverty alleviation in South African township tourism. *Journal of Sustainable Tourism, 24*(12), 1641–1654. doi:10.1080/09669582.2016.1145230

Laderchi, C. R., Saith, R., & Stewart, F. (2003). *Does it matter that we don't agree on the definition of poverty? A comparison of four approaches*. University of Oxford.

Laeis, G. C. M., & Lemke, S. (2016). Social entrepreneurship in tourism: Applying sustainable livelihoods approaches. *International Journal of Contemporary Hospitality Management, 28*(6), 1076–1093. doi:10.1108/IJCHM-05-2014-0235

Lyons, K., Hanley, J., Wearing, S., & Neil, J. (2012). Gap year volunteer tourism: Myths of Global Citizenship? *Annals of Tourism Research, 39*(1), 361–378. doi:10.1016/j.annals.2011.04.016

Ma, B. (2010). *A Trip into the Controversy: A Study of Slum Tourism Travel Motivations*. Paper presented at the 2009-2010 Penn Humanities Forum on Connections.

Mair, J., & Martí, I. (2006). Social entrepreneurship research: A source of explanation, prediction, and delight. *Journal of World Business*, *41*(1), 36–44. doi:10.1016/j.jwb.2005.09.002

Mair, J., Robinson, J., & Hockerts, K. (Eds.). (2006). *Social entrepreneurship*. Palgrave Macmillan. doi:10.1057/9780230625655

Marko, P., & Jelena, D. (2014). Systems thinking and alternative business model for responsible tourist destination. *Kybernetes*, *43*(3/4), 480–496. doi:10.1108/K-07-2013-0132

Matthews, S. (2006). Responding to poverty in the light of the post-development debate: Some insights from the NGO Enda Graf Sahel. *Africa Development. Afrique et Developpement*, *31*(4), 52–72.

Mayo, M. (2005). *Global citizens: Social movements and the challenge of globalisation*. Zed Books.

McCaston, M. K., & Rewald, M. (2005). Unifying Framework for Poverty Eradication & Social Justice: The Evolution of CARE's Development Approach: Technical report. Atlanta, GA: Cooperative for Assistance and Relief Everywhere (CARE).

McGehee, N. (2006). *Volunteer Tourism: Sustainable Innovation in Tourism, or just "Pettin' the Critters"?* Best Education Network Think Tank VII: Innovations for Sustainable Tourism.

Medina-Muñoz, D. R., Medina-Muñoz, R. D., & Gutiérrez-Pérez, F. J. (2016). The impacts of tourism on poverty alleviation: An integrated research framework. *Journal of Sustainable Tourism*, *24*(2), 270–298. doi:10.1080/09669582.2015.1049611

Mekawy, M. A. (2012). Responsible slum tourism: Egyptian experience. *Annals of Tourism Research*, *39*(4), 2092–2113. doi:10.1016/j.annals.2012.07.006

Mgonja, J. T., Sirima, A., Backman, K. F., & Backman, S. J. (2015). Cultural community-based tourism in Tanzania: Lessons learned and way forward. *Development Southern Africa*, *32*(3), 377–391. doi:10.1080/0376835X.2015.1010710

Misturelli, F., & Heffernan, C. (2008). What is poverty? A diachronic exploration of the discourse on poverty from the 1970s to the 2000s. *European Journal of Development Research*, *20*(4), 666–684. doi:10.1080/09578810802464888

Mottiar, Z., Boluk, K., & Kline, C. (2018). The roles of social entrepreneurs in rural destination development. *Annals of Tourism Research*, *68*, 77–88. doi:10.1016/j.annals.2017.12.001

Moufakkir, O., & Kelly, I. (2015). Developing a tourism poverty reduction strategy. In Y. Reisinger (Ed.), *Transformational Tourism: Host Perspectives*. CABI. doi:10.1079/9781780643922.0180

Newton, M. (2012). *TripAdvisor and Kiva Launch Partnership*. Retrieved from http://ir.tripadvisor.com/news-releases/news-release-details/tripadvisor-and-kiva-launch-partnership

Novelli, M., Morgan, N., Mitchell, G., & Ivanov, K. (2016). Travel philanthropy and sustainable development: The case of the Plymouth–Banjul Challenge. *Journal of Sustainable Tourism*, *24*(6), 824–845. doi:10.1080/09669582.2015.1088858

Nussbaum, M. (2002). Education for citizenship in an era of global connection. *Studies in Philosophy and Education*, *21*(4-5), 289–303. doi:10.1023/A:1019837105053

Phi, G. (2017). *A Critical Analysis of Microfinance Tourism and Poverty Alleviation: Characteristics, Opportunities and Constraints* (PhD thesis). Griffith University, Brisbane, Australia.

Phi, G., Dredge, D., & Whitford, M. (2013). *Fostering Global Citizenship through the Microfinance-tourism Nexus*. Paper presented at the 7th Annual Tourism Education Futures Initiative Conference 2013 (TEFI7): Tourism Education for Global Citizenship: Educating for Lives of Consequence, Oxford, UK.

Phi, G. T., & Whitford, M. (2017). Social entrepreneurship in Vietnam. In *Managing Growth and Sustainable Tourism Governance in Asia and the Pacific* (pp. 132–139). UNWTO.

Phi, G. T., Whitford, M., & Dredge, D. (2017). Knowledge dynamics in the tourism-social entrepreneurship nexus. In *Social Entrepreneurship and Tourism* (pp. 155–172). Springer. doi:10.1007/978-3-319-46518-0_9

Phi, G. T., Whitford, M., Dredge, D., & Reid, S. (2017). Educating tourists for global citizenship: A microfinance tourism providers' perspective. *Tourism Recreation Research*, *42*(2), 235–247. doi:10.1080/02508281.2017.1290738

Phi, G. T., Whitford, M., & Reid, S. (2016). What's in the black box? Evaluating anti-poverty tourism interventions utilising theory of change. *Current Issues in Tourism*, 1–16.

Pleumarom, A. (2012). *The Politics of Tourism, Poverty Reduction and Sustainable Development*. Third World Network.

Polak, P. (2009). *Out of Poverty: What Works When Traditional Approaches Fail*. Berrett-Koehler Publishers.

Ramchander, P. (2003). Towards the responsible management of the socio-cultural impact of township tourism. *Tourism and politics: Global frameworks and local realities*, 149-173.

Rebutin, J. (2009). *Social Enterprise & Tourism: The key to a better Integration of Indigenous populations*. Université de Toulon et du Var.

Rolfes, M. (2010). Poverty tourism: Theoretical reflections and empirical findings regarding an extraordinary form of tourism. *GeoJournal*, *75*(5), 421–442. doi:10.100710708-009-9311-8

Salazar, N. B. (2004). Developmental tourists vs. development tourism: A case study. In A. Raj (Ed.), *Tourist behaviour: A psychological perspective* (pp. 85–107). Kanishka Publishers.

Scheyvens, R. (2007). Exploring the Tourism-Poverty Nexus. *Current Issues in Tourism*, *10*(2-3), 231–254. doi:10.2167/cit318.0

Scheyvens, R. (2012). *Tourism and Poverty*. Routledge. doi:10.4324/9780203831595

Sen, A. (2000). *Social exclusion: Concept, application, and scrutiny. Social Development Papers No. 1*. Office of Environment and Social Development: Asian Development Bank.

Shaffer, P. (2008). *New Thinking on Poverty: Implications for Globalisation and Poverty Reduction Strategies*. DESA Working Paper No. 65.

Sharpley, R., & Harrison, D. (2019). *A research agenda for tourism and development*. Edward Elgar. doi:10.4337/9781788112413

Sharpley, R., & Telfer, D. (Eds.). (2008). *Tourism and Development: Concepts and Issues*. Channel View Publications.

Sheldon, P. J., & Daniele, R. (Eds.). (2017). *Social Entrepreneurship and Tourism*. Springer. doi:10.1007/978-3-319-46518-0

Shen, F., Hughey, K. F., & Simmons, D. G. (2008). Connecting the sustainable livelihoods approach and tourism: A review of the literature. *Journal of Hospitality and Tourism Management, 15*(01), 19–31. doi:10.1375/jhtm.15.1.19

Shultz, L. (2007). Educating for Global Citizenship: Conflicting Agendas and Understandings. *The Alberta Journal of Educational Research, 53*(3), 248–258.

Situmorang, D. B. M., & Mirzanti, I. R. (2012). Social Entrepreneurship to Develop Ecotourism. *Procedia Economics and Finance, 4*(0), 398–405. doi:10.1016/S2212-5671(12)00354-1

Sloan, P., Legrand, W., & Simons-Kaufmann, C. (2014). A survey of social entrepreneurial community-based hospitality and tourism initiatives in developing economies: A new business approach for industry. *Worldwide Hospitality and Tourism Themes, 6*(1), 51–61. doi:10.1108/WHATT-11-2013-0045

Snyman, S. (2014). The impact of ecotourism employment on rural household incomes and social welfare in six southern African countries. *Tourism and Hospitality Research, 14*(1-2), 37–52. doi:10.1177/1467358414529435

Soumaré, I. (2015). Does FDI improve economic development in North African countries? *Applied Economics, 47*(51), 5510–5533. doi:10.1080/00036846.2015.1051655

Steinicke, E., & Neuburger, M. (2012). The impact of community-based afro-alpine tourism on regional development. *Mountain Research and Development, 32*(4), 420–430. doi:10.1659/MRD-JOURNAL-D-11-00102.1

Taplin, J., Dredge, D., & Scherrer, P. (2014). Monitoring and evaluating volunteer tourism: A review and analytical framework. *Journal of Sustainable Tourism, 22*(6), 874–897. doi:10.1080/09669582.2013.871022

Thomlinson, E., & Getz, D. (1996). The Question of Scale in Ecotourism: Case Study of Two Small Ecotour Operators in the Mundo Maya Region of Central America. *Journal of Sustainable Tourism, 4*(4), 183–188. doi:10.1080/09669589608667267

Tripadvisor. (2017). *TripAdvisor Network Effect and the Benefits of Total Engagement*. Retrieved from https://www.tripadvisor.com/TripAdvisorInsights/w828

United Nations. (2003). *Human Development Report, Millennium Development Goals: A Compact among Nations to End Human Poverty*. Oxford University Press.

United Nations. (2016). *Transforming our world: the 2030 Agenda for Sustainable Development*. Retrieved from https://sustainabledevelopment.un.org/post2015/transformingourworld

van Beers, T. (2011). *Discourse in Tourism Development: A Narrative of Inclusive Business* (MSc Thesis Leisure, Tourism and Environment). Wageningen University.

Veltmeyer, H., & Petras, J. (2011). *Beyond Pragmatic Neoliberalism: From Social Inclusion and Poverty Reduction to Equality and Social Change.* Paper presented at the UNESCO "Rethinking Development: Ethics and Social Inclusion", Mexico City.

Volkmann, C. K., Tokarski, K. O., & Ernst, K. (2012). *Background, Characteristics and Context of Social Entrepreneurship Social entrepreneurship and social business*. Springer. doi:10.1007/978-3-8349-7093-0

Wrelton, E. (2006). *'Reality Tours' to Chiapas, Mexico: The Role of Justice Tourism in Development.* Academic Press.

Zarandian, N., Shalbafian, A., Ryan, C., & Bidokhti, A. A. (2016). Islamic pro-poor and volunteer tourism—The impacts on tourists: A case study of Shabake Talayedaran Jihad, Teheran—A research note. *Tourism Management Perspectives, 19*, 165–169. doi:10.1016/j.tmp.2015.12.005

Zebrowski, R.L. (2009). *Social Entrepreneurship as a New Paradigm: Rethinking Development Intervention Ideologies.* Academic Press.

Zhao, W., & Ritchie, B. (2007). Tourism and poverty alleviation: An integrative research framework. *Current Issues in Tourism, 10*(2-3), 119–143. doi:10.2167/cit296.0

ADDITIONAL READING

Daniele, R., & Quezada, I. (2017). Business Models for Social Entrepreneurship in Tourism. In *Social Entrepreneurship and Tourism* (pp. 81–100). Springer International Publishing. doi:10.1007/978-3-319-46518-0_5

Frenzel, F., Koens, K., & Steinbrink, M. (Eds.). (2012). *Slum tourism: Poverty, power and ethics* (Vol. 32). Routledge. doi:10.4324/9780203136751

Scheyvens, R. (2007). Exploring the tourism-poverty nexus. *Current Issues in Tourism, 10*(2-3), 231–254. doi:10.2167/cit318.0

Sharpley, R., & Naidoo, P. (2010). Tourism and poverty reduction: The case of Mauritius. *Tourism and Hospitality Planning & Development, 7*(2), 145–162. doi:10.1080/14790531003737169

Vodopivec, B., & Jaffe, R. (2011). Save the world in a week: Volunteer tourism, development and difference. *European Journal of Development Research, 23*(1), 111–128. doi:10.1057/ejdr.2010.55

Westley, F., Zimmerman, B., & Patton, M. (2009). *Getting to maybe: How the world is changed.* Vintage Canada.

Zapata, M. J., Hall, C. M., Lindo, P., & Vanderschaeghe, M. (2011). Can community-based tourism contribute to development and poverty alleviation? Lessons from Nicaragua. *Current Issues in Tourism, 14*(8), 725–749. doi:10.1080/13683500.2011.559200

KEY TERMS AND DEFINITIONS

Capacity-Based Poverty: A significant lack of important social-psychological assets such as health, literacy, as well as a lack of income/material assets.

Income Poverty: A significant lack of income/material assets.

Development-Tourism: The practice of travelling to impoverished areas for the purpose of experiencing lives of the poor in their everyday environment, and which may be associated with learning about aspects of development and/or contributing financial resources or labours to assist the individual poor/host community to improve their lives.

Global Citizenship: A call for all citizens to build awareness of their own and others' roles within an interconnected global context; to take responsibility for their actions; and to take actions towards development outcomes that are socially, economically, environmentally and politically just and sustainable.

Social Entrepreneurship: A process of innovatively organising resources and networks to pursue opportunities that can lead to long-lasting social change and/or address social needs.

Structural-Based Poverty: A significant lack of social, psychological and economic assets due to factors inherent in social, economic and institutional structures that favour some groups over others.

Sustainable Livelihoods: The development of preventative and protection mechanisms to help poor individuals and communities to reduce, mitigate and cope with the shocks and stresses that are specific to their living environments.

Chapter 3
Tourism and the Poverty Challenge: A Glance in Cape Verde

Mariana Marques
Lusophone University of Humanities and Technologies, Portugal

Eduardo Moraes Sarmento
Lusophone University of Humanities and Technologies, Portugal

ABSTRACT

This chapter introduces a practical application of the first Sustainable Development Goal, which is "Tourism Contribution to End Poverty." It provides a recent literature review about the theme and presents the case study of Cape Verde. Cape Verde is an archipelago with enormous tourist potential, but, on the other hand, it is a country that must deal with the poverty problem. So, it is a destination that needs to work on its sustainability, mainly to provide residents with better living conditions. It is urgent to study the main poverty challenges in Cape Verde in order to provide the right strategies. It is essential, also, to understand if tourism can help fighting local poverty. To achieve some conclusions, authors have done interviews with some local institutions as well as to several stakeholders to understand better the main issues concerning tourism challenges and its potential to deal with poverty in Cape Verde.

INTRODUCTION

Cape Verde is a small archipelago located in Africa that has been facing since its independence numerous vulnerabilities which may arise from different aspects, such as reduced dimension, limited domestic resource base, distance to major markets and geographic isolation, limited institutional capacity, openness to international markets and instable external flows. All these constraints, which are typical to small island development state (SIDS), have imposed several challenges to various national governments over the years.

DOI: 10.4018/978-1-7998-5691-7.ch003

Cape Verde is a country with an economic system that has never been based on its richness in natural resources. Cape Verde has unequivocally been able enhancing the human resources at its disposal, exploring its privileged geographical location, deepening its regional and world economic integration, capitalising on infrastructure investments (mainly tourism) benefiting from political stability and security.

To achieve new development and growth paths and to try to control population poverty (a major problem in this country), tourism has been elected as the main economic instrument. As a matter of fact, tourism has consolidated its importance since its independence and has become a resilient economic activity with huge importance. So, it is easy to understand why tourism has been elected as a strategic instrument not only as a strategic activity but also as a valuable instrument to fight poverty.

Despite the fact that, in the last years, Cape Verde achieved some remarkable economic as well as social improvements, several islands still face one considerable and undesirable social problem due to the existence of poverty focus among its population. This is a calamity that the Millennium Report has been trying to eradicate with SDG1. Under this context, this chapter aims at answering the following questions:

- How important is sustainability in the resolution of poverty in Cape Verde?
- What are the main poverty challenges in Cape Verde?
- How is Cape Verde dealing with this reality and what is the importance of Tourism in this strategy?

To answer these questions, this chapter is divided into four sections. In the first one, we present a theoretical discussion about sustainability and poverty. In the second section, we analyse the country performance. In section three, we present the methodology adopted to achieve the conclusions, and in the final section we reflect on the main results and future challenges for key players.

THEORETICAL FRAMEWORK

The SDG chosen in this chapter was the first one that is related to tourism contribution to ending poverty. This SDG focus on complete eradication of poverty by 2030, so, it represents a huge challenge. We must be aware that poverty is a multidimensional concept, and because of that it is extremely difficult to manage. The big question on this topic is if tourism can help to end poverty in all its forms and everywhere. Although tourism has been considered for some years as a way of alleviating poverty, it is necessary to understand whether tourism is indeed strong enough to end poverty and improve the quality of life of over one billion people living in poverty around the world (Scheyvens & Hughes, 2019). Within the context of theoretical context, authors will present topics such as the Sustainable Development Goals, the relationship between tourism, poverty alleviation and the concept of community-based tourism.

Sustainable Development Goals and Millennium Report

On March 2000, the United Nations presented the Millennium Report. This report offers a plan to improve globalisation effects for people all over the world. The values reflected in this report are freedom, equity and solidarity, tolerance, non-violence, respect for nature and shared responsibility. Its vision was to fight poverty in all dimensions. So, the report created eight Millennium Development Goals (MDGs), which were oriented to the proposed vision. According to The Millennium Development Goals Report (2015), in 1990, the number of people living in extreme poverty was 1.9 billion, although that number decreased to 1 billion in 2011. According to recent estimates, in 2015, 10% of the world's population or 734 million people lived on less than \$1.90 a day. Unfortunately, due to the COVID-19 crisis along with the oil price drop, this trend will probably reverse in 2020. This COVID-19 crisis is expected to have a disproportionate impact on the poor, through job loss, loss of remittances, rising prices, and disruptions in services such as education and health care. Recent research estimates that, by 2030, up to two-thirds of the global extreme poor may be living in fragile and conflict-affected economies, reinforcing that without an intensified action, the global poverty goals will not be met (World Bank, 2020).

In the context of current trends in tourism development, it is impossible not to mention the Sustainable Development Goals. In fact, tourism cannot grow beyond measure without taking certain aspects into account, namely regarding sustainability. This is because it has reached a point where it is no longer viable, in the long term, to keep many things as they were until now, with the possibility of running out of resources, in the very near future. In addition, there are key areas where it is essential to focus, and which are intricately linked to human rights. These areas are the people, planet, prosperity, peace, and

Figure 1. Central areas for SDG (personal elaboration)

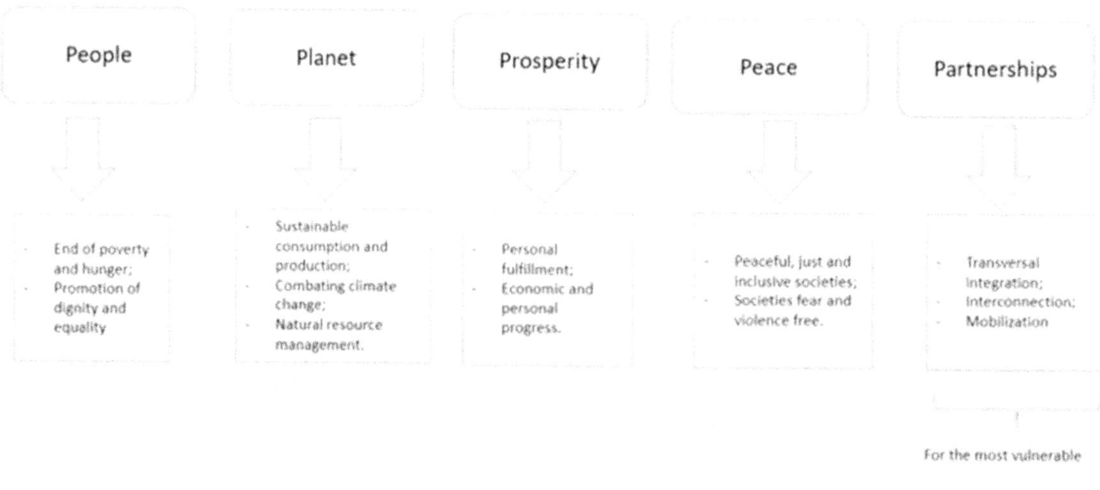

partnerships and can be seen in the next scheme where we list the main focus on each of the central areas.

The United Nations 2030 Agenda for Sustainable Development defines some sustainable development goals that represent major guides to help tourism policymakers such as industry and DMOs - Destination marketing organisations. As a matter of fact, these Sustainable Development Goals (SDGs) associated with the MDGs have become extremely important to the study of tourism's role to sustainable development (Hall, 2019). According to the United Nations World Tourism Organization (UNWTO), we can find 17 Sustainable Development Goals, represented in figure 2.

If we properly analyse these seventeen SDGs, they are, in some way, related to the five central themes identified in figure one. In fact, each of them intends to focus and plan measures with the objective of developing policies related to people, planet, prosperity, peace, or partnership. There is no doubt that tourism is a sector that has the potential to contribute, directly or indirectly to each one of these SDGs.

Figure 2. The 17 Sustainable Development Goals (UNTWO)

So, the sustainability concept is deeply important to tourism development. In some of the poorest regions of the planet, tourism is almost the only way residents may survive. In fact, at these places, tourism is one of the activities that employ the most people, and many of them live just with the outcome of their local small businesses like hotels and restaurants, small local markets, handicrafts and tour guide services, among others.

Tourism and Poverty Alleviation

Traditionally, there are several approaches to poverty reduction using tourism. Without intending to be too exhaustive, we can highlight the following (Spenceley, Ashley & Kock, 2009):

- Inclusive tourism. It can promote the creation of bonds and interactions between the different actors of tourism, generate partnerships with private agents and, above all, create incentives for the local economy, integration among women and participation of the communities themselves (Scheyvens & Biddulph, 2018).
- Sustainable tourism. According to its definition, it must commit itself to the promotion of local prosperity by seeking to maximise its contribution to economic growth in the places of destination. In this way, it should generate income and decent employment for workers without affecting the environment and the culture of the tourist destination places and ensure the viability and competitiveness of these same destination places and companies so that they can continue to prosper and generate benefits from the long term.
- Responsible tourism. It aims to minimise the negative economic, environmental and social effects and it is expected that it will lead to greater economic benefits for the local population, in addition to promoting the well-being of the host communities and better working conditions and access to the sector. It should also desirably involve local people in decisions that affect them, in addition to making positive contributions to the conservation of natural and cultural heritage and the maintenance of diversity in the world. Finally, if successful, this experience should strengthen the links between tourists and local populations (Shen, Sotiriadis, & Zhou, 2020).
- Favourable tourism for people affected by poverty is the one that will generate the most net benefits for people who are in a situation of poverty. Tourism favourable to people in poverty should not be a specific product or niche, but as an approach towards development and better tourism management. In other words, it must strengthen the links between tourism companies and people in situations of poverty so that the contribution of tourism to poverty reduction is more intense and people affected by poverty are able to participate more actively and effectively in the development process of different products.
- Fair trade-based tourism. It aims to ensure that people whose land, natural resources, knowledge, work, and culture are used in tourism activities can effectively benefit from them.
- Community tourism. It is considered that in which local populations (many of whom are rural, affected by poverty and economically excluded) invite tourists to visit their communities, providing facilities and developing activities for this purpose (Robinson & Wiltshier, 2011).
- Ecotourism. It combines conservation, community participation and sustainable travel. It implies responsible travel to natural areas that preserve the environment and improve the well-being of local people.

It should be noted that tourism governance should focus on three key issues: sustainability, responsibility, and pro-poor (Musavengane, Siakwab & Leonard, 2019). Another major issue is the fact that tourism has been considered crucial for economic development, for countries development and growth, and in this case, specifically for its potential to reduce or fight poverty (Mahadevan & Suardi, 2019).

However, while tourism may be considered a strategic activity due to its capacity of job creation, country development and money income, some recent perspectives stressed what they considered as the darker side of tourism development. According to this perspective, tourism may be seen as an industry led by large entities that exploit employees, create bad working conditions, cause environmental degradation, accentuate the inequality gap and turn aspects of traditional culture into commodities (Scheyvens & Hughes, 2019).

According to Luo and Bao (2019), poverty has three main characteristics, such as:

- Lack of opportunities to participate in economic activities.
- Lack of discourse power in making important decisions related to livelihoods.
- High likelihood of being impacted by economic and related issues.

According to these authors, poverty may be caused both by individual and institutional factors. As far as the individual factors, they include mostly the lack of motivation to accomplished big things, the satisfaction with a low-income from the employment and a low level of independence. The institutional factors include poor distribution of resources, harming the poor and favouring the richest, unequal power relationships and bad policies of governance (Scheyvens & Hughes, 2019). Despite the economic impact of inbound tourism to reduce poverty, we must also consider the domestic perspective since both domestic and inbound tourism have a significant effect in reducing poverty (Llorca-Rodríguez, García-Fernández, & Casas-Jurado, 2018).

It is important to remember that tourism is a multifaceted industry that has a high impact, both positive and negative in diverse areas. Therefore, it is necessary to think about how these Sustainable Development Goals can also be adapted to the tourism industry for a more sustainable, fair, and equitable future. In addition, there is an urgent need to analyse aspects related to biodiversity, climate change, conservation of cultural and environmental diversity, as well as the development of well-being and living conditions, aspects that are fundamental for tourism. Sustainability represents the balance between socio-economic interests and the ecological system, and we must remember that it is essential the conservation of cultural and environmental diversity. So, the SDGs arise to help the tourism industry to be more sustainable, equitable and create a fair future around the world (Boluk, Cavaliere, & Higgins-Desbiolles, 2019).

Poverty has been a problem discussed avidly in recent years and represents a persistent concern of humanity. In fact, reducing extreme poverty and hunger was identified as one of the priorities in the context of the Millennium Development Goals. This document was signed by 189 nations in 2000, and poverty alleviation has become a focal point in tourism policies. As a matter of fact, has been started some anti-poverty tourism interventions (ATIs) that represent a set of policies and strategies that relate tourism to poverty alleviation thru regional or national economic development and defined some projects or approaches to use tourism as a way to improve the lives of poor population (Phi & Reid, 2018).

In fact, there are two principal issues in the country development that are environmental conservation and poverty alleviation. Some authors defend that poverty is a result of non-sustainable use of resources, especially the natural ones (Yergeau, 2020). Tourism is a sector that can help in the development of less developed countries and can be important in reducing or eliminating poverty in some other countries. However, tourism development may also be a cause of poverty. So, it is important that countries can define some strategies in terms of local government to be able to take advantage of tourism potential (Luo & Bao, 2019).

Cooper and Hall (2008), consider that a tourist destination should be considered much more than a sum of producers related to the tourism area, since it is the physical space where tourism takes place, but it is also the place where communities they live and work, a place surrounded by cultural and historical symbols. So, the residents are also important and not just tourists.

As previously described, tourism has emerged as a resilient activity in many countries with a huge capacity to overcome several economic, political, and environmental imbalances in a sustainable way. However, despite its capacity to overcome many of the previous constraints and problems, as well as achieving a sustainable growth strategy in the long run, one social problem has not yet been completely eradicated in many countries since we still find high levels of unemployment, short family incomes, job instability among other social issues.

The management and planning of tourism must be cautious, so as avoid the appearance of conflicts between tourists and local inhabitants as well as to make tourism compatible with the conservation of the main ecosystems and with the preservation and the good use of the historical-cultural heritage. Historically, socio-economic, and socio-environmental effects, especially in developing countries, have shown their potential for resilience and capacity to become one of the main drivers of development (Szymańska, 2018).

Community-Based Tourism and Poverty Alleviation

It is important that Community-based Tourism (CBT) must be seen as one of the strategies that can help in poverty alleviation. This concept was developed by Murphy in 1985 in a study that analyses tourism issues and impacts on local communities in developing countries. CBT defends that is the community itself that has control over tourism management and gain with the benefits generated by tourism (Juma & Khademi-Vidra, 2019). According to Giampiccoli and Saayman (2017), it can be seen as a source of local growth and also as a way to poverty alleviation. However, this concept of CBT must be related to responsible tourism.

In fact, community-based and responsible tourism are focal points to some pro-poor tourism strategies. Also, it is extremely important to involve local people from poor communities in tourism development (Medina-Muñoz, Medina-Muñoz, & Gutiérrez-Pérez, 2016). Community participation can create a higher level of tourism sustainability because it will allow the creation of better job opportunities for the community and by increasing the benefits through participation in tourist activities (Sebele, 2010). If this strategy is successful, the level of tourism development in a region and its degree of competitiveness will have a direct positive impact on the level of development and competitiveness of the tourist region (Kambur, Petryshchenko, & Sehorina, 2018).

So, even though the United Nations 2030 Agenda for Sustainable Development stressed the importance of SDGs associated with the MDGs, we still find of outmost importance that tourism may be conducted in order to reduce the population levels of poverty as well as to create higher levels of employability and incomes so that tourism may become an effective instrument against social exclusion or in other words to become an instrument of social inclusion.

METHODOLOGY

To achieve the present results, the authors took advantage of a World Bank international project in Cape Verde in order to do several local in-depth interviews. In this context, both authors have interviewed the following personalities/entities in April/May/September 2019 about the main issues concerning tourism challenges and its potential to deal with poverty in Cape Verde. Authors used in-depth interviews to collect major information. Interviews varied from fifteen minutes up to one hour. This methodology is a qualitative research technique that involves conducting intensive individual interviews with a small number of respondents to explore their perspectives on an idea, program, or situation (Boyce & Neale, 2006). This methodology allowed authors to extract more detailed information from the participants and promoted a more flexible way to obtain extra information. In-depth interviews are often seen as one of the most efficient methods of collecting primary data (Rivas & Gibson-Light, 2016). Since these interviews allow relatively large freedom of the interviewer in terms of content and design, this situation often increases the willingness to provide information and the spontaneity of the respondents (Morris, 2015).

The main purpose of semi-structured interviews for data collection is to gather information from key informants who must have personal experiences, attitudes, perceptions, and beliefs related to the topic of interest (Dejonckheere & Vuaghn, 2019). This method also has the advantage of collecting new, exploratory data related to a research topic and to triangulate with other data sources or validate findings through member checking. After all, when we use a mixed-method approach, the semi structured interviews can also be used in a qualitative phase to explore new concepts (Dejonckheere & Vuaghn, 2019).

Table 1. List of entities interviewed (personal elaboration)

Entities	Position
Ministro do Turismo e Transportes e Ministro da Economia Marítima	Minister José Gonçalves
Binter CV	Director
Enapor	Director
Hotel Pescador	Director
Hotel Praia Mar	Director
Direção Nacional do Ambiente	Director
ZDTI	Administrator
INE	Coordinator
Comunidade cabo-verdiana de cruzeiros	President
Câmara M. Ribeira Grande Santiago	President and Director
Câmara M. São Lourenço dos Órgãos	President
Câmara M. Santa Cruz	President
Câmara M. Tarrafal	Tourism councilor
Câmara M. Praia	Tourism councilor
Instituto Património Cultural - Museus	Director
Instituto Património Cultural – Património Imaterial	Director
Universidade de Cabo Verde	Dean
Polícia Nacional	Director
TACV	Marketing Director
Fundo do Turismo	Manager
DGTT (Tourism)	Tourism Director
DGTT (Tourism)	Ministry of Transport and Tourism Assessor
ASA	Airport Director
Electra	Directors
CV Trade Invest	Administrator
Associação Turismo Santiago	President

Source: Personal Elaboration

The collected data may be considered primary data as it was obtained only for the accomplishment of the current study, having not been previously collected (Malhotra & Birks, 2006). In addition, several main institutional documents about this issue were also analysed, and different places in Cape Verde were also visited to evaluate these problems.

CASE STUDY

Cape Verde Geography

Cape Verde is a democratic and parliamentary republic, recognised as one of the most stable governments in Africa and as a model in terms of governance. It is a small archipelago with ten islands, located in the Atlantic Ocean, located about 455 km from the African coast. The extension is about 4,033 sq. Km and were formed by the accumulation of rocks, resulting from eruptions on the underwater platforms. The islands are divided into two groups, Barlavento and Sotavento, from where the wind blows and where the wind flows, respectively. Thus, Barlavento includes Santo Antão, S. Vicente, Santa Luzia, S. Nicolau, Sal and Boa Vista while Maio, Santiago, Fogo and Brava are part of the Sotavento group. Except for the islands of Boa Vista, Sal and Maio, the relief of the archipelago is rugged, with altitudes exceeding 1000 meters on some islands reaching 2,882 meters on the island of Fogo, the highest point in the whole country.

The climate of the most rugged islands is varied and with some rainfall. It is tempered thanks to the moderating action that the ocean and trade winds have on temperature, with annual averages rarely rising above 25ºC, never falling below 20ºC. The seawater temperature varies between 21ºC in February and March and 25ºC in September and October. Cape Verde's climatic stability guarantees the possibility of tourism all year round. Cape Verde's has various nature reserves and parks. Among the main species of fauna, birds are especially important. In terms of fauna, there are 755 plant species, of which 83 are endemic and 224 are indigenous. The dragon tree is particularly important because it is a tree whose presence precedes the discovery of the archipelago itself.

Regarding tourism, Cape Verde is located at the crossroads of Europe, Africa and South America being only about 1 hour from Dakar, about 4 hours from Fortaleza in Brazil and Lisbon. The islands offer a stable climate, mostly with sunshine and some beautiful natural landscapes. Cape Verde has four international airports which are particularly important for tourism growth. However, they face some constraints that they need to resolve to develop the tourism sector, such as water supply, electricity distribution, internet connection and roads. To help with that, the government has recently dedicated significant funds and efforts and launched an investment incentive program to investors. To develop the branding of destination, Cape Verde plans on promoting the archipelago as a global destination, with diverse offerings under the "one country, ten islands" umbrella brand (Market Report Cape Verde, 2018).

Cape Verde's Economy

Economic Cape Verdean growth has been improving in the last four years (+5% in Gross Domestic Product [GDP] in 2019), mainly due to the good performance of exports of goods and services and tourism receipts, as reported in figure 3 (Banco Portugal [BP], 2019).

This situation has been complemented by the external accounts situation that also improved due to a good dynamic of goods exports, the solidity of tourism revenues, emigrant remittances, and lower outflows of Foreign Direct Investment (FDI) incomes (Figure 4) (BP, 2019).

To achieve the prior results, Cape Verdean Government has invested heavily over the past 15 years at the expense of huge public debt, as depicted in figure 5 (Governo de Cabo Verde [GCV], 2016).

Even though several reforms have been undertaken and several advances were achieved in different areas, Cape Verde still faces an economic development challenge, high unemployment and youth

Figure 3. Real DGP, annual variation rates - % (BP, 2019)

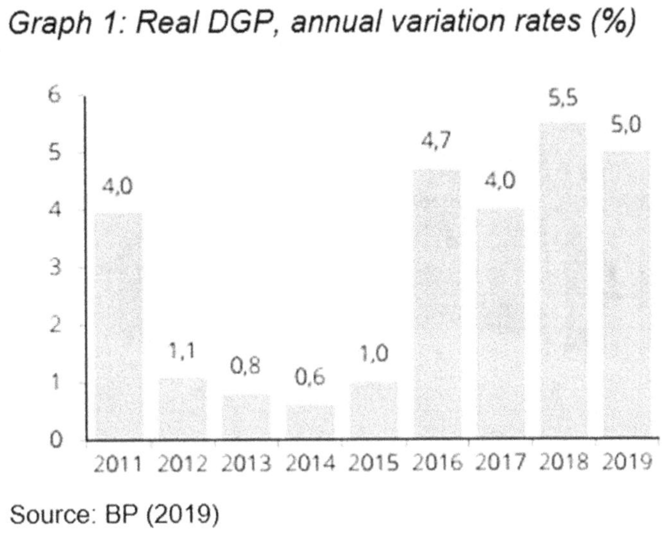

Figure 4. External accounts – in % of the DGP (BP, 2019)

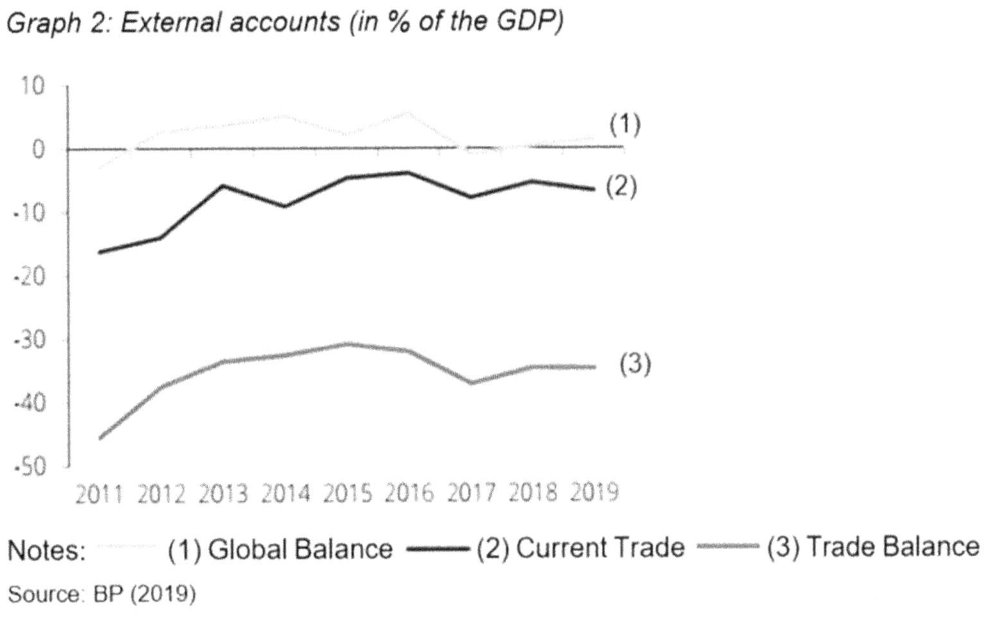

unemployment rates, the business environment is bad, and insecurity has been increasing to worrying levels. There are also severe air and maritime transport problems (GCV, 2016). On the other hand, the data on national accounts shows that each island's contribution to shaping the economy presented an unequal distribution in recent years, especially the island of Santiago, which represents 52% of GDP, followed by São Vicente with 16% and Sal with 12%. Such fact reflects the regional inequalities within

Figure 5. Public debt – in % of the DGP (BP, 2019)

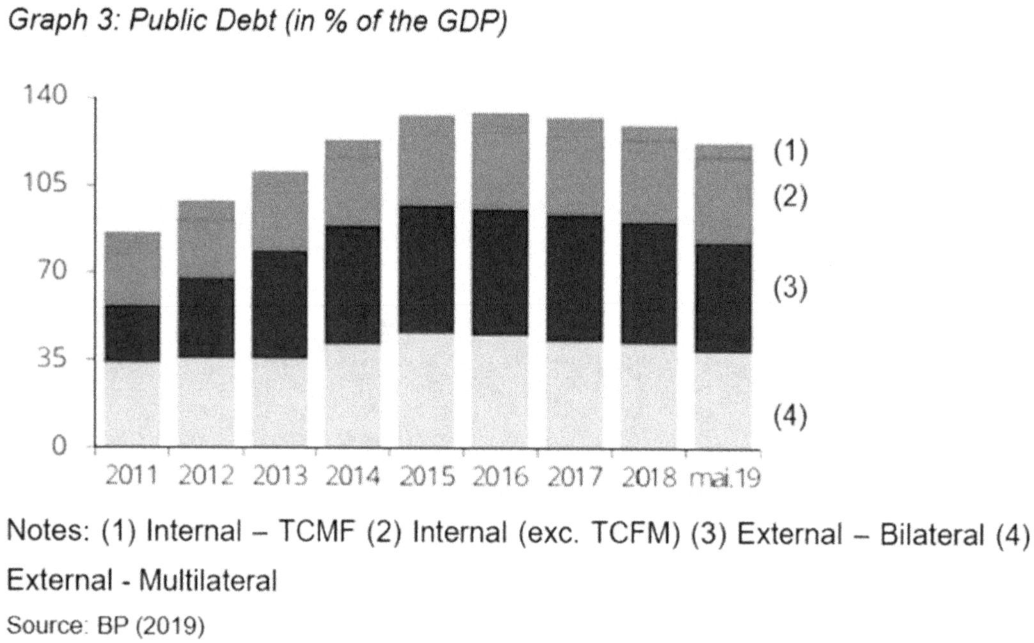

the country, with the remaining six inhabited islands representing 20% of GDP. One of the direct consequences of these asymmetries is the manifestation and distribution of poverty throughout the country. In 2015, 35% of the population lived in poverty, while 10.6% in extreme poverty (Governo de Cabo Verde [GCV], 2018). Therefore, population face an increasingly uncertain in a world of greater complexity, which means that the government must create a strategy in order to better deal with the vulnerabilities and the external dependency that the country still faces (GCV, 2016, 2018).

Nowadays, everyone is looking for happiness and better living conditions. Under this goal, the Cape Verdean government defined as the main goal to ensure sustainable development of the country, aimed at bringing happiness to Cape Verdeans on the basis of full employment, increasing the average income of the country, promoting greater and better security and providing a better quality of life for all. In other words, they are committed to ensure: (i) reduction of poverty and eradication of hunger and extreme poverty in the country promoting inclusive economic growth; (ii) full employment and decent work for all; (iii) put Cape Verde on the list of the top ten best-rated small island countries when assessed by the Human Development Index; (iv) place care for dependents – children, older people and people with disabilities, traditionally seen as an exclusive social mandate for families (GCV, 2016).

In order to fulfil this goal, among other options, tourism has been elected as a key driver for some years. From the selected strategy, several measures in different sectors have been taken that may positively to reduce poverty as we will see in the next section.

The total contribution of tourism to GDP was CVE 76.0bn in 2017 (44.9% of GDP) and CVE 78.4bn (44.4% of GDP) in 2018 (Figure 6). It is forecast to rise by 5.0% pa to CVE 127.8bn by 2028 (47.9% of GDP) (World Travel & Tourism Council [WTTC], 2018).

The total contribution of Travel & Tourism to employment (Figure 7) achieved a total of 93,500 jobs in 2017 (39.3% of total employment), a total of 95,500 jobs (38.9% of total employment) and it is

Figure 6. Total contribution of Travel & Tourism to GDP – Cape Verde (WTTC, 2018)

Graph 4: Total contribution of Travel & Tourism to GDP (Cape Verde)

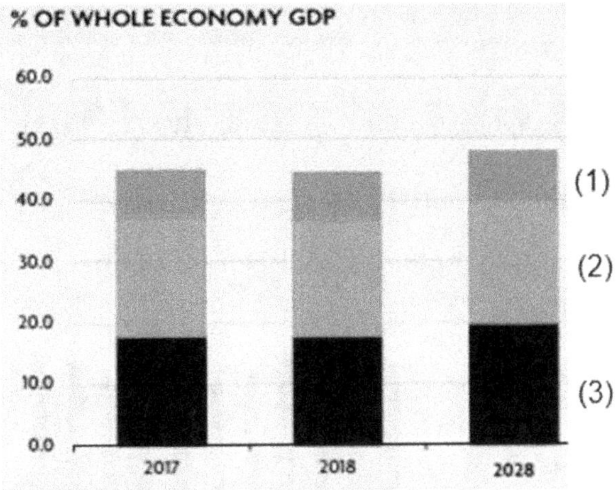

(1) Induced (2) Indirect (3) Direct

Source: WTTC (2018)

Figure 7. Total contribution of Travel & Tourism to employment – Cape Verde (WTTC, 2018)

Graph 5: Total contribution of Travel & Tourism to employment (Cape Verde)

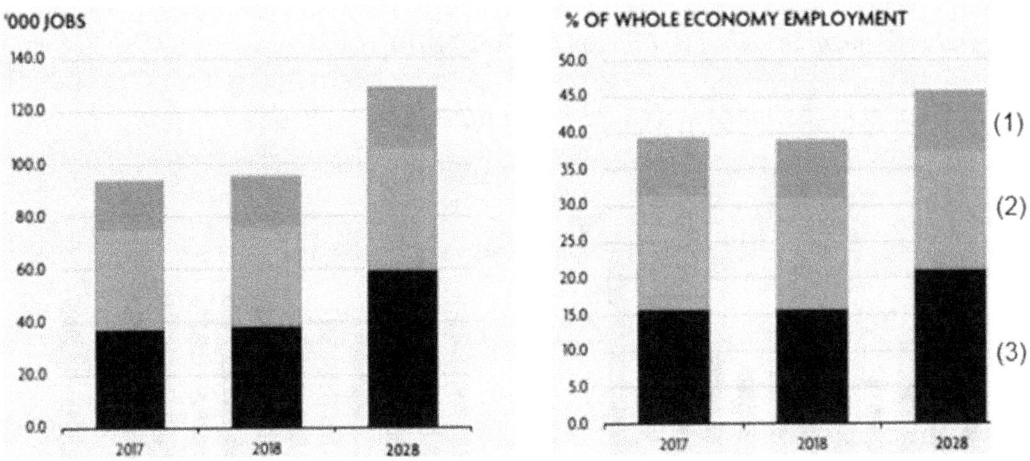

(1) Induced (2) Indirect (3) Direct

Source: WTTC (2018)

expected to support 129,000 jobs in 2018 (45.7% of total employment), an increase of 3.1% pa over the period (WTTC, 2018).

Visitor exports play a decisive role in the total contribution of Travel & Tourism. In 2018 Cape Verde attracted 634,000 international tourist arrivals, as demonstrated in figure 8. By 2028, international tourist arrivals are forecast to achieve 801,000, generating expenditure of CVE 78.5bn (an increase of 5.5% pa) (WTTC, 2018).

Figure 8. Visitor exports and international tourist arrivals – Cape Verde (WTTC, 2018)

Graph 6: Visitor exports and international tourist arrivals (Cape Verde)

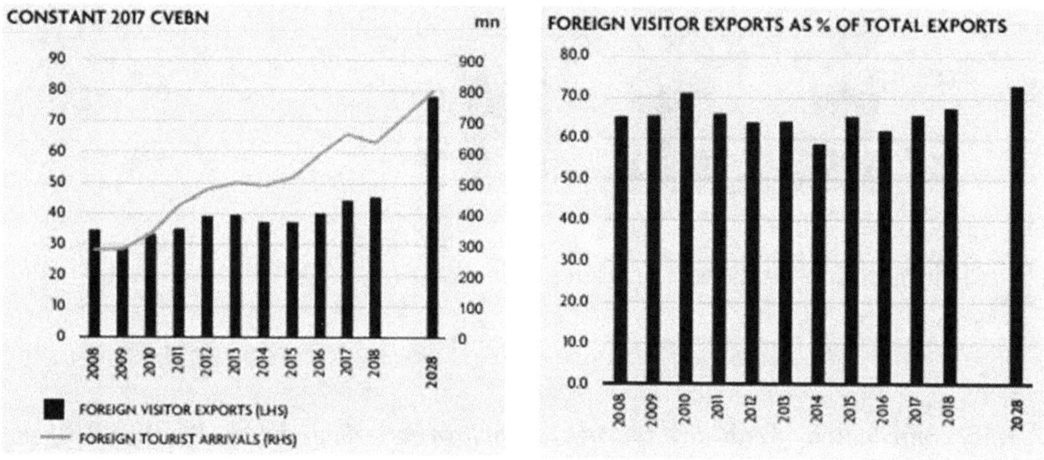

Source: WTTC (2018)

Figure 9. Capital investment in Travel & Tourism (WTTC, 2018)

Graph 7: Capital investment in Travel & Tourism (Cape Verde)

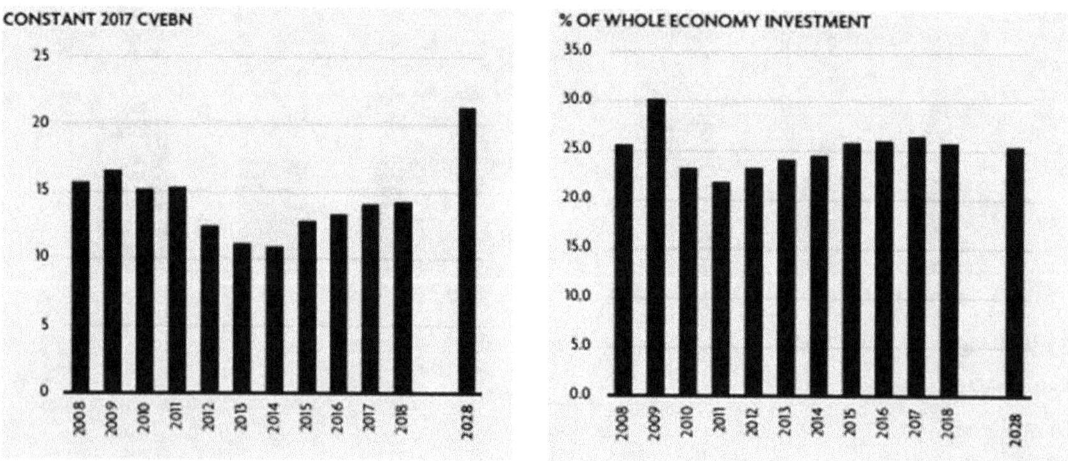

Source: WTTC (2018)

Finally, travel & tourism attracted capital investment of CVE 14.0bn in 2017 (Figure 9) and is expected to rise by 4.1% pa over the next ten years to achieve CVE 21.2bn in 2028 (25.4% of the whole economy investment) (WTTC, 2018).

So, it can be concluded that tourism has been playing a major role in Cape Verde and it is expected to grow its importance soon.

CASE STUDY DISCUSSION

Cape Verde has been historically confronted with two major challenges: (i) the challenge of ensuring the survival of the poor resident community from a perspective of reduction of Official Development Assistance (ODA), and the challenge of self-sustained and accelerated development to respond to legitimate aspirations of its population to achieve higher living standards (GCV, 2018).

From the prior interviews as well as from official documents, we can see that Cape Verde's tourism currently faces several challenges: (i) competitiveness, (ii) sustainability, (iii) concentration and (iv) the challenge of maximising the impact on wealth and welfare of the Cape Verdeans poor people.

As stated earlier, the tertiary sector has been the main generator of employment in Cape Verde with about 61 out of 100 employees, and the phenomenon is predominantly urban, with 72.6% of employment. In the countryside, the primary sector generates 54 out of 100, and this is the main reason why there is such a high incidence and concentration of poverty in this sector. The imbalance in accessing economic opportunities and employment fosters the imbalance in access to income. Santiago and São Vicente comprised in 2017 about 70% of the employment generated in the country. Adding the island of Sal, it goes up to 79% of the employment opportunities in the country (GCV, 2018).

Simultaneously, many rural municipalities have significantly higher poverty rates than urban areas. According to the latest Poverty Profile Report, there were 179.909 poor people in Cape Verde (global absolute poverty) in 2015, of which 54.395 extremely poor (extreme absolute poverty). The proportion of the rural poor is significantly higher than in the urban area, 48.5% against 27.8 (Instituto Nacional de Estatística [INE], 2015).

Extreme poverty has a much higher incidence in rural areas, with 37,036 people living in extremely poor rural areas (68.1%), compared to 17,360 in urban areas (31.9%). The total poverty incidence rate is 35.2% in Cape Verde. In regional terms, 59% of the poor live in Santiago, 22% in Praia. Extreme poverty is most concentrated in the municipalities of Santa Cruz, with 14.6% of the population in extreme poverty, Praia, with 11.9% of the population, São Filipe, and Santa Catarina with 10.7% and 10.9% respectively (INE, 2015). Another trend is that poverty in Cape Verde has a female (53%) and young face (60% were under 25 years old) and essentially affects populations with less access to education (15% were intern migrants, and 44% had a basic level of education (5 to 6 years of schooling) (GCV, 2018; INE, 2015).

Thus, taking into consideration that poverty is a multidimensional phenomenon that simultaneously includes economic, social, and cultural factors, its combat presupposes the adoption of new, innovative solutions that directly address poverty constraints such as tourism (GCV, 2019).

According to various interviews, tourism was referred as an important driver of development and an instrument to fight poverty due to (i) its capacity to value various resources such as warm climates, cultural heritage, intact landscapes and biodiversity that may be particularly found in rural areas; (ii) its interconnection. Since the tourist product consists of different stakeholders, it has a large and varied value chain and tourism expenses may benefit a different range of sectors such as agriculture, crafts, transport and services; (iii) the capacity of bringing together consumers and producers. Tourism is an activity that brings consumers and producers closer together and is consumed at the production site.

Under this strategy, we can stress one new program entitled “One family, one tourist”. This innovative government program aims to encourage the development of housing tourism in villages and to boost a set of economic activities of tourism at a community level in rural areas in Cape Verde. So, in addition to valuing and qualifying the potential of tourism in localities in all municipalities of the country, this program intends to bring the benefits of tourism at the level of families in rural villages, aiming at improving the tourist experience in these spaces and contributing to the creation of wealth and employment in local communities, thus reducing poverty.

So, to conclude, this program is directed to rural localities with strong tourist potential. It will be anchored in the creation of local accommodation of standard dimensions, conditions, and basic quality, specifically supporting families to qualify their homes, rehabilitating, or building 1 to 3 rooms/suites to receive tourists.

Complementary, these families will also be encouraged to create a set of business initiatives in terms of related and complementary tourist services that allow the tourist to provide a differentiated local experience, such as restaurants or typical pasture house, housekeeping services, guide, tourist transport, handicrafts, tourist animation, trekking and hiking, agritourism, fisheries, promotion of historical-cultural heritage and promotion of endogenous products, among others.

The development challenge appears, therefore, as a main challenge to the country. Tourism is expected to become the main engine of economic activity, acting as a catalyst, to which the other sectors should be linked, on the path of building a sustainable economy.

According to the Government program, “tourism is elected as one of the central pillars Cape Verde’s economy, a key to the relaunch of private investment, employment and economic growth. The policy for tourism will be refocused on a new size and quality, which exceeds the concept and achieve the multi-specialisation of the Cape Verdean economy” (GCV, 2018, p. 8).

So, we have seen that Cape Verde Government has been committed to (i) promote positive externalities of tourism through agriculture, fisheries, culture, and sports; (ii) strengthen and consolidate the model Sun, Beach and Sea Tourism. Under this goal, social economy has been selected as a key role in the expansion of the employment, equal opportunities and promotion of social, environmental, and historical assets that support local and regional development and therefore as a tool against poverty. Thus, government´s action has been trying to stimulate employment through several different instruments: (i) elimination of Social Security on companies that recruit young people; (ii) launching and co-financing of professionals internships, job creators and facilitators, in partnership with the private business sector; (iii) effective promotion of entrepreneurship; (iv) launch of a program of support for the social economy and solidarity and the promotion of social innovation (GCV, 2018 and interviews).

Another important sustainable issue is concerned with the valorisation of the historical, cultural and tourist heritage. This is a way of better understanding and approach heritage as a product of marketing, job creation, differentiating territories, and promoting regional and national identities. Tourism, being associated with heritage, may enhance local cultures and, most importantly, their people, as a development instrument to attract visitors and share genuine cultural exchanges, providing social inclusion and generating equity of opportunities.

CONCLUSION

Cape Verde is a small island developing country facing several challenges and vulnerabilities. One of them is the poverty that is found in many of the islands of this archipelago. Throughout the years, several economic strategies have been taken, but this problem persists and is still found.

Cape Verde has seen a huge rise in the tourism activity. If responsibly managed, this activity may become a vital instrument to contribute to reducing and if possible, to eradicate poverty.

Under this context, we found that local municipalities, key players as well as the central government are all aware of this issue. However, as stated by the local government, it is still important to take several initiatives to achieve this sustainable development goal. It is important to promote multiple instruments like the elaboration and implementation of a true Plan to Combat Poverty and Social Exclusion, focus on family-oriented social policies, vulnerable groups (e.g. elderly and disabled and the most depressed territorial spaces with the highest incidence of poverty) and the definition of Priority Intervention Zones (ZIPs) for spatial targeting of social policies (housing nutrition, access to water, sanitation, energy, basic education, health and protection etc.) and economic (income-generating activities, microfinance schemes, market access, vocational training, etc.) (GCV, 2016).

So, to see the success rate of Tourism as an instrument to achieve better poverty fight, we will briefly stress the main results that have been achieved.

1. Ocean economy. Several fishermen have been qualified, thus improving the skills of several population segments as well as survival modes. Fisheries still have a huge potential, and it is crucial that artisanal fisheries may be improved so that poor population may answer the tourism demands.
2. Agriculture. Knowing that agriculture is an income-generating sector, which brings prosperity and social recognition, several measures have been adopted to achieve more competitive agriculture and agro-industry and, thus, to guarantee not only the food and nutritional security of the population but also raise the export capacity of the country. Agriculture has a huge potential due to the expanding tourism sector and has a market that can rapidly reach 1.5 million consumers, of which about 1 million with high power purchase. Simultaneously, Cape Verdean agriculture has been changing its vocation because it has been evolving from family farming to modernised and competitive farming, as well as the development of the rural world. Even though these achievements, there are still some commitments to fulfil: (i) get more quality and well-being to low-income families living in the rural area; (ii) be able to transform agriculture into an export sector that generates income and social recognition; (iii) integrate agricultural policy with environmental protection, particularly in the fight against sand extraction and protection of streams; (iv) provide environmentally friendly practices that effectively reduce poverty.
3. Economic performance. Small entrepreneurs must still benefit from investment opportunities to get new transactions that come with tourism. However, the real challenge is with PME´s that need several institutional supports.
4. Trade, services, and catering serving tourism are all activities that create more jobs, help to stabilise domestic demand, and generate higher incomes. So, their success, as well as their competitiveness, are crucial to fight poverty.
5. Transports and infrastructures. The unification of the country is vital, and Cape Verde is facing various challenges that not only may affect tourism but also affect the population. We can stress the problems that arise from the lack of modernisation of the port infrastructure, including Ro-Ro ramps and port equipment in almost all the islands. Also, the relaunch ship maintenance activity

and shipbuilding activity can also be positive for future poverty fight. The aerial transportation is also limited and may limit the tourism activity, thus reducing the opportunities for local companies. Concerning the terrestrial transport system and road infrastructure, they are unanimously referred as being fundamental and decisive in reducing the economic distance between the municipalities, reducing the pressure for urban concentration cities and to better distribute productive activity, public services and incomes across different locations, thus creating an opportunity to reduce poverty form populations that are far from the main urban centres. In Cape Verde, it is important to not only to promote and ensure the conservation, operation and development planning of the road network but also provide municipalities as well as citizens and tourists with a road transport infrastructure network meeting the technical conditions for convenience and safety. The public transportations also appear to be a weak point, and it is important to stress the opportunity to create more jobs to help tourism.

6. In a country facing poverty, a qualified and flexible labour market with a proper wage policy, a skilled and flexible labour market is an essential component of a favourable investment climate. Even though there have been improvements, a study on the labour market and wage policy is still an important challenge to capture investment and create more employment through the promotion of high productivity and economic efficiency. The labour market still must be regulated through an effective commitment to certification as well as a training system in partnership with the private sector, even at management of training promotion institutions capable of qualifying the Cape Verdean workforce in accordance with the tourism market needs.
7. Water and sanitation. In a country with a huge shortage of water in many islands, this is a huge challenge for tourism and agriculture activities. Thus, Cape Verde Government must assume water efficiency as a priority vector looking for universal and balanced access to safe drinking water for all in order to achieve a better improvement of the socio-economic conditions of the population along with the welfare of citizens and tourists.
8. Adoption of circular economy by using waste as a renewable source through the development of water reuse projects irrigation and urban cleaning, along with a soil protection strategy that obliges companies engaged in hazardous activities to assess the quality of their soil and take responsibility for decontamination.

RECOMMENDATIONS

The proper implementation of a sustainable tourism strategy presupposes that governments (in this case Cape Verde government) recognise tourism as a fundamental and multifaceted activity and that, if properly regulated, it can contribute to achieving a higher social inclusion of the most deprived population. In this context, governments should provide for the provision of structures that may facilitate the proper planning of this activity and therefore ensure the effective inclusion of the resident and most deprived community in the tourism expansion process (Delgado, 2018). In other words, the search for greater social sustainability of tourism means that if it this activity is successful, this will not only lead to the expansion of the tourism market, but will also provide a more satisfactory experience for visitors and increase the social sustainability of these experiences, maximising the full potential of tourism for visitors and natives (Wearing & Darcy, 2011).

Community inclusion in tourism is crucial, as it will contribute to the internationalisation of the country as well as to obtain better organisational levels. These are key aspects for continuing to grow and expand business (Delgado, 2018). So, in this context, it is imperative to opt for the effective goal of guaranteeing more and better jobs for the population, especially the most disadvantaged fringes, which as we have seen may have an extra opportunity with the tourism development. However, we must not forget that employment in tourism activity faces, in many cases, some specificities that are mainly linked to seasonality, with a direct connection between the number of jobs and the tourism season. The strategic and sustainable development of tourism presupposes that maximum employment is guaranteed throughout the year to reduce the economic risks associated with seasonality and provide greater economic stability, which includes necessary stable employment (Diakonidze, 2018).

On the other hand, taking into account the pressure that urban centres generally suffer from tourism, it is important to boost regions outside these large urban centres as viable alternatives for attracting tourism with a clear objective: sustainable development (Daghri, Jorio & Kasmi, 2018). This strategy seeks to provide long-term social, economic and environmental welfare activities for a vulnerable community, strengthening participation in local governance processes, promoting entrepreneurship, building more resilient communities and ensuring adequate sanitation facilities, better levels of education, better infrastructure, drinking water, health facilities, environmental protection, resource efficiency, waste management, renewable energy with the ultimate goal of providing employment and taking advantage of female cooperatives as a source of production (Daghri, Jorio, & Kasmi, 2018).

In other words, it allows the implementation of new policies and strategies to achieve sustainable development and, ultimately, economic growth. If tourism keeps the behaviour and trend of recent years, it is important to emphasise the need to pay more attention to avoid the negative side effects of tourism in order to properly determine the final effects since it requires an appropriate comparison between the benefits received and the costs incurred by the host communities.

Traditionally, this balance has been difficult to achieve, mainly because of the negative and positive externalities, which are difficult to quantify (Concu & Atzeni, 2012).

FUTURE RESEARCH DIRECTIONS

This chapter is a first attempt to analyse the importance of tourism and its effect on poverty alleviation under the SDGs. As described earlier, one must be aware that poverty is a multidimensional concept, and, hence, it is extremely difficult to manage without a proper model. This chapter offers various options to deal with poverty and economic vulnerabilities in a small developing country through a tourism option based on a sustainable path. However, it would be interesting and important for stakeholders or tourism companies to have a quantitative analysis of this goal both in Cape Verde as well as in other islands or destinations to statistically be able to understand the importance of several key variables. Moreover, it would also be important to measure the direct, indirect, and induced effect of tourism in the quality of job creation based on a statistical model so that stakeholders and the government could know and understand which strategies should be created in each island in order to overcome the prior vulnerabilities successfully.

REFERENCES

Banco Portugal (BP). (2019). *Evolução das Economias dos PALOPS e de Timor-Leste 2018-2019*. Banco Portugal.

Boluk, K., Cavaliere, C., & Higgins-Desbiolles, F. (2019). A critical framework for interrogating the United Nations Sustainable Development Goals 2030 Agenda in tourism. *Journal of Sustainable Tourism*, *27*(7), 847–864. doi:10.1080/09669582.2019.1619748

Boyce, C., & Neale, P. (2006). *Conducting in-depth interviews: a guide for Designing and Conducting In-Depth Interviews for Evaluation Input*. Pathfinder International.

Cape Verde Government Website. (n.d.). Available at https://www.governo.cv/

Concu, N., & Atzeni, G. (2012). Conflicting preferences among tourists and residents. *Tourism Management*, *33*(6), 1293–1300. doi:10.1016/j.tourman.2011.12.009

Cooper, C., & Hall, M. (2008). *Contemporary tourism: An international approach*. Elsevier.

Daghri, T., Jorio, H., & Kasmi, S. (2018). Communicating a new vision of tourism: The case of Volubilis, Meknès, Morocco. *International Scientific Conference on Economic and Social Development Proceedings*, 297-306.

Dejonckheere, M., & Vuaghn, L. (2019). Semistructured interviewing in primary care research: A balance of relationship and rigour. *BMJ Journals*, *7*(2), 1–8. doi:10.1136/fmch-2018-000057 PMID:32148704

Delgado, R. (2018). Community inclusiveness for tourism governance and globalisation: The case of Kanagawa prefecture, Japan. *International Scientific Conference on Economic and Social Development Book of Proceedings*, 642-647.

Diakonidze, M. (2018). Tourism industry and prospects of development of the Georgian labor market. *Managerial Issues in Modern Business,* 98-103.

Giampiccoli, A., & Saayman, M. (2017). Community-based tourism, responsible tourism, and infrastructures development and poverty. *African Journal of Hospitality, Tourism and Leisure*, *6*(2), 1–28.

Governo Cabo Verde (GCV). (2016). *Programa do Governo – IX Legislatura*. Cabo Verde.

Governo Cabo Verde (GCV). (2018). *PEDS: Plano Estratégico de Desenvolvimento Sustentável 2017/2021*. Cabo Verde: Ministério Finanças – Direção Nacional Planeamento.

Governo Cabo Verde (GCV). (2019). Grandes Opções do Plano *Estratégico de Desenvolvimento Sustentável do Turismo para 2018-2030*. Cabo Verde: Boletim Oficial I Série, N.º 2.

Hall, C. (2019). Constructing sustainable tourism development: The 2030 Agenda and the managerial ecology of sustainable tourism. *Journal of Sustainable Tourism*, *27*(7), 1044–1060. doi:10.1080/09669582.2018.1560456

Instituto Nacional Estatística (INE). (2015). *III Inquérito às Despesas e Receitas Familiares (IDRF)*. Cabo Verde: INE.

Juma, L., & Khamedi-Vidra, A. (2019). Community-Based Tourism and Sustainable Development of Rural Regions in Kenya; Perceptions of the Citizenry. *Sustainability*, *11*(4733), 1–23. doi:10.3390u11174733

Kambur, O., Petryshchenko, N., & Sehorina, N. (2018). Factors affecting the development of tourism industry. *Conference on Economic and Social Development Proceedings*, 125-133.

Llorca-Rodríguez, C., García-Fernández, R., & Casas-Jurado, C. (2018). Domestic versus inbound tourism in poverty reduction: Evidence from panel data. *Current Issues in Tourism*, 1–20. doi:10.1080/13683500.2018.1494701

Luo, X., & Bao, J. (2019). Exploring the impacts of tourism on the livelihoods of local poor: The role of local government and major investors. *Journal of Sustainable Tourism*, *27*(3), 344–359. doi:10.1080/09669582.2019.1578362

Lv, Z. (2019). Deepening or lessening? The effects of tourism on regional inequality. *Tourism Management*, *72*, 23–26. doi:10.1016/j.tourman.2018.11.009

Mahadevan, R., & Suardi, S. (2019). Panel evidence on the impact of tourism growth on poverty, poverty gap and income inequality. *Current Issues in Tourism*, *22*(3), 253–264. doi:10.1080/13683500.2017.1375901

Malhotra, N., & Birks, D. (2006). *Marketing Research: An Applied Approach*. Pearson. doi:10.1108/S1548-6435(2006)2

Market Report Cape Verde. (2018). Available at https://www.hospitalitynet.org/file/152008345.pdf

Medina-Muñoz, D., Medina-Muñoz, R., & Gutiérrez-Pérez, F. (2016). The impacts of tourism on poverty alleviation: An integrated research framework. *Journal of Sustainable Tourism*, *24*(2), 270–298. doi:10.1080/09669582.2015.1049611

Mitchell, J., & Li, S. (2017). Autonomy found: estimating the local benefit from tourism in SIDS: The case of Cape Verde. *Journal of Policy Research in Tourism, Leisure & Events*, *9*(2), 182–200. doi:10.1080/19407963.2016.1261145

Morris, A. (2015). *A practical introduction to in-depth interviewing*. SAGE Productions., doi:10.4135/9781473921344

Musavengane, R., Siakwah, P., & Leonard, L. (2019). The nexus between tourism and urban risk: Towards inclusive, safe, resilient and sustainable outdoor tourism in African cities. *Journal of Outdoor Recreation and Tourism*, *29*, 100254. doi:10.1016/j.jort.2019.100254

Phi, G., & Reid, M. (2018). What's in the black box? Evaluating anti-poverty tourism interventions utilising theory of change. *Current Issues in Tourism*, *21*(17), 1930–1945. doi:10.1080/13683500.2016.1232703

Rivas, R., & Gibson-Light, M. (2016). Exploring culture through in-depth interviews: Is it useful to ask people about what they think, mean, and do? *Cinta de Moebio*, *57*(57), 316–329. doi:10.4067/S0717-554X2016000300007

Scheyvens, R., & Biddulph, R. (2018). Inclusive tourism development. *Tourism Geographies*, *20*(4), 589–609. doi:10.1080/14616688.2017.1381985

Scheyvens, R., & Hughes, E. (2019). Can tourism help to "end poverty in all its forms everywhere"? The challenge of tourism addressing SDG1. *Journal of Sustainable Tourism, 27*(7), 1061–1079. doi:10.1080/09669582.2018.1551404

Sebele, S. (2010). Community-based tourism ventures, benefits and challenges: Khama rhino sanctuary trust, central district, Botswana. *Tourism Management, 31*(1), 136–146. doi:10.1016/j.tourman.2009.01.005

Shen, S., Sotiriadis, M., & Zhou, Q. (2020). We must be aware that poverty is a multidimensional concept and because of that, it is extremely difficult to manage. *Sustainability, 12*, 1–21. doi:10.3390u12041470

Spenceley, A., Ashley, C., & Kock, M. (2009). *Tourism-led poverty reduction programme: Core training module*. Centro Comércio Internacional.

Szymanska, E. (2018). Sustainable tourism: Theory versus practice. *International Scientific Conference on Economic and Social Development Book of Proceedings*, 17-25.

World Travel & Tourism Council (WTTC). (2018). *Travel & Tourism economic impact 2018 in Cape Verde*. WTTC.

United Nations (UN). (2015). *The Millennium Development Goals Report*. New York: UN.

Wearing, S., & Darcy, S. (2011). Inclusiveness of the 'Othered' in Tourism. *Cosmopolitan Civil Societies: An Interdisciplinary Journal, 3*(2), 18–34.

Wiltshier, P., & Robinson, P. (2011). Community Tourism. *Research Themes for Tourism,* 97-99.

World Bank. (2020). *Fragility and conflict: On the front lines of the fight against poverty*. World Bank.

Yergeau, M. (2020). Tourism and local welfare: A multilevel analysis in Nepal's protected areas. *World Development, 127*, 1–19. doi:10.1016/j.worlddev.2019.104744

KEY TERMS AND DEFINITIONS

Community-Based Tourism (CBT): Is tourism in which local residents (often rural, poor and economically marginalised) invite tourists to visit their communities with the provision of overnight accommodation.

Destination Marketing Organisation (DMO): They are not-for-profit organisations that help to promote and market local attractions, accommodation options, tourism services and related sectors.

Gross Domestic Product (GDP): Is the most commonly used measure of economic activity to track the health of a nation's economy. It represents the value of all goods and services produced over a specific time period within a country's borders.

Millennium Development Goals (MDGs): Some Development Goals that United Nations defined and that the 191 UN member states have agreed to try to achieve by the year 2015. The United Nations Millennium Declaration, signed in September 2000, commits world leaders to combat poverty, hunger, disease, illiteracy, environmental degradation, and discrimination against women.

Small Island Developing States (SIDS): Distinct group of developing countries facing specific social, economic and environmental vulnerabilities.

Sustainable Development Goals (SDGs): They are 17 goals that were adopted by all United Nations Member States in 2015, and they are expected to be achieved by 2030.

Chapter 4
Poverty Alleviation, Ecotourism, and Biodiversity Protection in Príncipe Island

Alda Matos
Escola Superior Náutica Infante D. Henrique, Portugal

ABSTRACT

Príncipe is the smallest of the two São Tomé and Príncipe (STP) islands and site of a UNESCO Biosphere Reserve. Two-thirds of its seven thousand inhabitants are poor, and the productive fabric does not generate enough resources for poverty alleviation and biodiversity conservation. Moreover, STP archipelago belongs to the group of small island developing states (SIDS), recognised by the United Nations Conference on Environment and Development as a particular case, and international funding for development and sustainability support is available. In addition, the autonomous government of Príncipe identified ecotourism as a niche market to develop, establishing goals for this activity. This chapter presents some results of the actions that have been promoted. In future work, the author studies the case of Cape Verde, another Portuguese speaking SIDS, aiming to understand what has been done in terms of poverty alleviation and its impacts.

INTRODUCTION

Príncipe Island, henceforth Príncipe, is an autonomous region of the state of São Tomé and Príncipe (STP), located in the Gulf of Guinea, 220 km off the African west coast. The STP archipelago has a total area of 1001 sq. km, homeland of 178,739 inhabitants (estimated in 2018), of whom 7,000 are from Principe, and two-thirds of the country's population is poor. In addition, the insularity condition affects the economic development and business environment thereby impeding competitiveness and poverty alleviation (Dallimer & King, 2008). The World Bank (2005) highlighted that STP is not a poor country; it has become poor as the result of poor national resource management and inadequate policy options.

In 1900, STP was the world`s biggest producer of cacao, but the plantations collapsed in the 1970 decade. From 1986 to 1990, and after a decade of abandon of the country´s colonial legacy of the cocoa

DOI: 10.4018/978-1-7998-5691-7.ch004

monoculture plantation economy, the country`s authorities have undergone a rehabilitation program of some state-owned estates, conceding them to private foreign management, financed by international institutions, in order to increase cocoa output and achieve economic recovery. Due to the poor condition of the cacao trees and the international cocoa prices decrease´, the higher annual production was insufficient to increase exports´ revenues (Seibert, 2016). In the following years, the failure of the agricultural development increased rural migration, urban poverty and emigration mainly towards Portugal, Gabon and Angola (Poças, 2016).

A key event in the archipelago´s recent history is the discovery of offshore oil reserves in 1997. In the following years, oil companies paid sizable "signing bonuses" to start exploratory oil drills, and an agreement was signed with Nigeria, in 2001, on the creation of the Joint Development Zone (JDZ) in the maritime area. Meanwhile, the exploratory oil drillings yielded unsatisfactory results and the exit from the JDZ of Chevron, Addax, Sinopec, and Total companies diminished oil prospects for the foreseeable future (IMF, 2013). Present in STP since 2015, Galp acquired a 45% stake in Block 6, in which it is an operator. In the following year, Galp acquired a 20% stake in blocks 5, 11 and 12 (Galp, 2019). In 2017, a 16,000 sq. km 3D seismic acquisition campaign was launched across all blocks, but the impacts of the eventual financial return to the local economy are still awaited.

Meanwhile, since 2000 little improvement has been achieved in poverty mitigation in the archipelago: roughly 67,4% of the local population has an annual *per capita* expenditure of less than STP "*dobra*" 10.9 million, equivalent to an amount under the threshold of 445 € annual *per capita* revenue estimated as the poverty line for the country (Gama & Sanoussi, 2017). The lack of up-to-date poverty data undermines efforts to reduce poverty in the country. Still, recent World Bank estimates, based on growth assumptions, indicate that around one-third of the country's population lives on less than US$1.9 PPP per day and more than two-thirds of the population is poor, at a poverty line of US $ 3.2 in 2019, with an increase of roughly 2 percentage points in comparison with 2010. This small change in poverty has been mainly attributed to increases in the value of household income rather than the redistribution of income across the population. Moreover, inequality in STP remains high for international standards (Gini index of 56.3 in 2017), and additional welfare indicators such as the Human Development Index (HDI) indicate 0.59 for STP under the average for selected peers which is 0.62 (The World Bank, 2018, 2019). Ever since 2010, the African Development Bank considers STP a '*Fragile State*' due to persisting poverty, economic vulnerability to external shocks, weak government capacity and inadequate provision of basic social services to the population (ADB, 2012).

In the literature on the economics of small island states, STP is encompassed in the MIRAB (Migration, Remittances, Aid and Bureaucracy) economic model (Bertram, 2006; Tisdell, 2016). Indeed, the archipelago has strong external aid that has financed 90% of the 150 million USD country´s annual budget at least during the past four decades (ADB, 2012; Seibert, 2016). Around 21,9% of the country's population has emigrated overseas, but their remittances decreased from 5.7% of the Gross Domestic Product (GDP) in 2015 to 4.6% in 2016, as economic growth slowed in Portugal and stagnated in Angola (The World Bank, 2018).

Nonetheless, the official scarcity of remittances from emigrant is partly justified by the fact that informal channels are preferred in detriment of formal remittance service providers (such as banks) due to their excessive costs (UNCTAD, 2012) and so official statistics are incomplete. Public administration and defence are not quite relevant for employment and accounted 11,9% of the national wealth (Gama & Sanoussi, 2017); tourism still has a minor value of 5% of the country´s GDP, both data from 2015 (AICEP, 2017).

Príncipe has a land area of 136 sq. km and is part of the West Africa biodiversity hotspot of tropical forests, with a high level of ecological diversity including primary tropical and shadow forests, marine ecosystems and lowland riparian habitats of indigenous and endemic species (UNESCO, 2012). The island topography and difficult accessibilities reinforce the development of cohesive communities, with a strong sense of identity linked by a system of formal and informal relations. But, a wild and remote area may become a 'spatial poverty trap' that combines lowest human population density and lack of access, making it less attractive for delivery of development assistance or for work by development organisations (Redford et al., 2008). For the Príncipe`s 7.000 inhabitants, of whom 69% are poor, that would be the worst perspective, and it has been quite important for Príncipe's stakeholders to find out a way not to fall in such a spatial poverty trap' scenery.

To sum up, the smallness and fragility of Principe's ecosystems are not compatible with mass tourism (Cardoso, 2015) and ecotourism is a niche market more adaptable to the island's area. As the local economy is fragile and the national stakeholders don´t possess enough resources to protect inhabitants and the whole Biosphere Reserve, the solution is to find out external resources, public and private ones.

This chapter aims to analyse the challenge of poverty alleviation and biodiversity protection, and the opportunities of getting international public resources available, and ecotourism foreign investments, while maintaining the "Leve-Leve" life philosophy of local inhabitants who live at the rhythm of nature (Relego, 2017). The chapter is divided into the following parts: firstly, the problem formulation; secondly a literature review regarding the eventual linkage between poverty alleviation and diversity protection; thirdly some data are presented concerning international public mechanisms available to finance directly local biodiversity protection activities, and international private investments on ecotourism, maintaining nature protection and employment stimulus in Príncipe; finally some conclusions and future work.

PROBLEM FORMULATION

Príncipe is one of the 669 Biosphere Reserves of the World (UNESCO, 2018), part of the World Network of Islands and Coastal Zone Biosphere Reserves (Abreu, 2012). The Principe´s Biosphere Reserve, approved by UNESCO in 2012, is a kind of sustainable development umbrella and a major non-financial asset for local government. The reserve has three different zones (UNESCO, 2018):

- a *core area* of 17,242.37 ha (marine 11,198.55 ha; land 6,043.82 ha), a strictly protected ecosystem for the conservation of landscapes, ecosystems, species and genetic variation;
- a *buffer zone* of 11,769.92 ha (marine 10,323.18 ha; land 1,446.74 ha) that surrounds or adjoins the core areas;
- a *transition area* of 42,580.27 ha (marine 36,081.76 ha; land 6,498.51 ha) home of resident population and tourism facilities.

In fact, the reserve includes the island, some inhabited islets and their adjacent marine areas which represent 80% of its whole (Dias, 2017). Local people rely on the ocean for fish, the forest for everything else, and the combination of rich soil and tropical weather enables cultures of vegetables and fruits everywhere (Scott Dunn, 2018). At the local scale, the challenge is to gather different stakeholders in order to manage natural resources of biodiverse ecosystems and get better living conditions for the local 7,000 inhabitants (Kepe et al., 2004; FMI, 2014).

Resources may be obtained from international funding for eligible projects concerning development, climate, and ecosystem sustainability support, and some data is mentioned in a later section.

Both Principe and S. Tomé are part of the Small Island Developing States' group recognised by the United Nations as a special case both for their environment and development. In addition, according to UNWTO, tourism is the overarching challenge and opportunity for SIDS to support further economic growth while protecting and benefitting island environments and communities (UNWTO, 2020).

Most data presented below refer to the country and are published by the Instituto Nacional de Estatística, located in S. Tomé. Indeed, half of the 136 sq. km of Príncipe's land area is an uninhabited reserve and the local government does not have a statistics office of his one. Using aggregated data, between 2010 and 2016, 131,001 tourists visited the country, with an annual increase from 8 thousand to 29 thousand, but the exploration of the tourist potential of natural resources is still embryonic (INE-STP, 2019).

Regarding tourist revenues, data from the World Tourism Organisation indicate an increase in the number of tourism revenues in STP, reaching a record high value of US $ 56 million in 2013 (18,5% of GDP). In 2014, the amount reported was slightly lower ($ 52 million) and not in line with the increase of tourists visiting the country, so the real value is expected to be higher than the existing official data (STP, 2018).

In 2018 the government announced the Strategic and Marketing Plan for Tourism in STP, with an investment forecast of 10.28 million euros to be implemented in 5 years, with 2 million euros of annual budget and a *vision* for the tourism sector (2018):

In 2025, São Tomé and Príncipe is the most preserved island tourist destination in Africa Equatorial, with a unique nature and biodiversity, paradisiacal beaches, where the Santomeans' hospitality, based on its historical-cultural legacy of coffee fields and cocoa, share their way of life and the warm way of receiving (…).

According to the Strategic Plan, already referred, in 2017, the STP's hotel offer amounted to 54 hotel units, 723 rooms and 1,508 beds represented in figure 1.

Figure 1. Hotel units and rooms (Instituto Nacional de Estatística, STP, 2018)

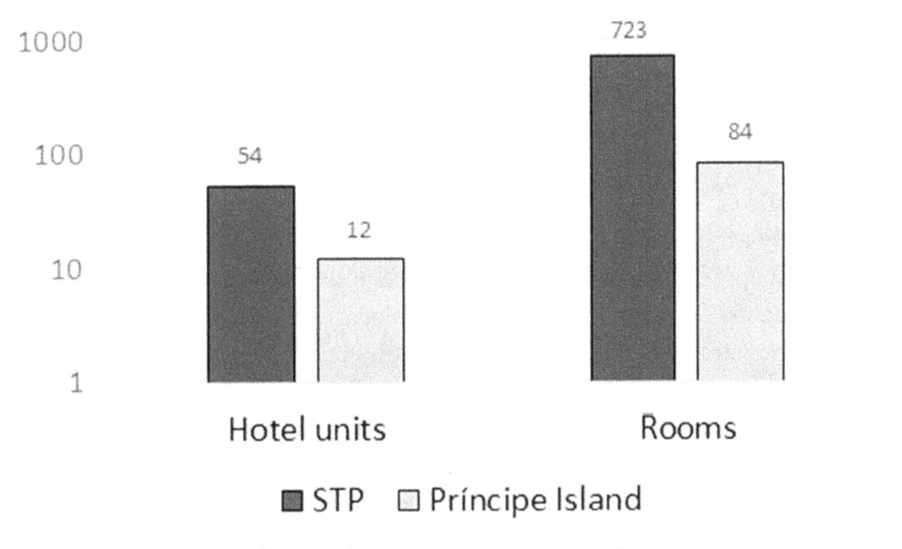

In addition, the hotel units in Príncipe are more exclusive and geared towards luxury tourism, highlighting the Here Be Dragon hotel group which has developed an important work in terms of training, the involvement of the local population and recovery of the built heritage of *Roças* (former colonial mansions). This innovative model results in a significant local social counterpart (STP, 2018) (more detail in section 4.2.).

Moreover, in 2017 there were 1,834 people working in the tourism sector, corresponding to 1% of the country's total population. Besides, in 2016, tourism contributed directly to 14.2% of the country´s jobs and, according to the World Travel and Tourism Council - Data Gateway, 56% of those jobs were in the capital's district (S. Tomé) and the next 11% in Príncipe (STP, 2018). The country government also intends to carry out and support training in specific functions of hotels, restaurants, tourist entertainment, tourist guides, tourist transport, among others, which will allow training human resources with quality for the sector.

Nonetheless, in Príncipe, the tourism development has already begun in 2011, when its regional government chose to work with Mark Shuttleworth, a South African entrepreneur from Here Be Dragons (HBD), who started its operation upon three main pillars: Tourism, Agriculture and Conservation. At the time, an opposite option could have been to accept the business advances of a major palm oil company that would jeopardise the Príncipe`s forest, like it has been done in S. Tomé with an enormous palm oil plantation around Angolares. The opposite option followed by Príncipe regional government lead to the "UNESCO Biosphere Reserve" classification in 2012, and the beginning of a program of long-term concessions rental to foreign investors: no more than 17% of the territory, to build up to 15 hotels with 50 rooms or less. HBD rent seven concessions across the island: *Bom Bom*, *Macaco*, *Boi*, *Paciência*, *Praia Grande*, *Uba*, and *Sundy*, and for each one the corporate created a sustainable hotel concept, except for the *Praia Grande* (the biggest beach of Príncipe) which is a protected area for turtle conservation. More detailed information is presented in section 4.2.

LITERATURE REVIEW

In the 20th century, economic growth was a dominant priority in detriment of environmental problems that were seen as solvable in a later future (Adams, 2004). For decades, a strong debate has taken place about the relationship between poverty alleviation and biodiversity conservation without a consensus on how to reconcile the two antagonistic goals (Kepe et al., 2004). The years after the 1980s marked some awareness to the topic of sustainable development, but scepticism remained about community-based approaches to conservation. Since then, poverty alleviation became first in the list of the UN Millennium Development Goals (2008), a priority of wealthier nations endorsed then by over 150 heads of state (Redford et al., 2008).

However, it is widely accepted that biodiversity loss and poverty are connected topics, and current debate reflects different approaches and analysis of the relationship between the two, their connections and disconnections (Adams, 2004). A first approach emphasises the existence of protected areas in poor developing countries to guarantee the maintenance of vulnerable species (such as forest primates) and considers poverty alleviation as a different matter to be treated separately. Incentives on conservation outcomes follow the basic principle of "you get what you pay", with two paths to achieve the goal: for example, pay to protect a rain forest or, indirectly, give some incentives to improve ecotourism which would finance conservation (Ferraro & Kiss, 2002). Indeed 'the relationship between tourism and poverty alleviation largely remains *terra incognita* among tourism academics (Zhou & Ritchie, 2007). Most popular investments encourage economic activities that indirectly protect ecosystems and species. Nevertheless, a direct approach has proved to be more effective and efficient than indirect ones with some ongoing pioneering experiences in developing nations (Bruner et al., 2001; Conrad et al., 2001; Gullison et al., 2000; Kiss, 2002). Briefly, priority goes to biodiversity criteria instead of poverty alleviation. Besides, conservation only benefits poverty reduction in a quite indirect manner and touristic arrangements in protected areas present risks of misallocating conservation resources and may compromise biodiversity preservation (Oates, 1999; Terborgh, 1999).

In a second approach, poverty is considered as a critical constraint on conservation with its alleviation being necessary for an effective strategy with priority to the involvement of local communities. As a consequence, the population would benefit from employment creation, participation in planning processes, sharing revenue from wildlife tourism in protected areas or other activities that might increase locals' income (Barrett & Arcese, 1995; Hulme & Murphree, 2001; Redford et al., 2008).

A third approach considers the specific needs of conservation and poverty alleviation, considering their conflicts and incompatibilities. The main focus is on conservation so long as it does not have a negative impact on poverty or on the livelihoods of the poor (Schmidt-Soltau & Brockington, 2004). Strategies might include the full payment of local opportunity costs of conservation in protected areas, but the few billion dollars funds allocated worldwide each year roughly cover 5% of the minimum requirements for effective conservation (Bruner et al., 2003).

A fourth approach considers poverty reduction dependent on living resource conservation and its sustainable use. Poor people depend on living species in biodiverse ecosystems, which means that they are entangled with the improvement of living conditions through appropriate conservation activities being translated into environmental protection (Adams, 2008). Conservation can and should be a powerful tool for wealth creation and poverty reduction due to the provision of ecosystem services, employment opportunities and preservation of cultures (Schmidt-Soltau & Brockington, 2004). Because activities of the poor threaten the environment, conservation organisations should focus on the 'pro-poor conservation agenda' (Roe, 2008).

A later approach has been introduced with the recognition of Small Island Developing States (SIDS) as a distinct group of (52) developing countries facing specific social, economic and environmental vulnerabilities, particularly vulnerable to global climate change, climate variability and sea-level rise. SIDS were recognised as a special case both for their environment and development at the United Nations Conference on Environment and Development (UNCED), also known as the Earth Summit, held in Rio de Janeiro, Brazil (3–14 June 1992). This recognition was made specifically in the context of Agenda 21 (Chapter 17 G) (UN, 1992).

The common challenges faced by SIDS also exist in STP, namely: narrow resource base, depriving the archipelago of the benefits of economies of scale; small domestic market and heavy dependence on a few external and remote markets; high costs for energy, infrastructure, transportation, communication and servicing; long-distance from export markets and import resources; low and irregular international traffic volumes; little resilience to natural disasters; limited opportunities for the private sector and a proportionately large reliance of the economy on the public sector; and fragile natural environment. By consequence, the archipelago is highly disadvantaged in its development process and requires special financial support from the international community. Literature in the 1990s and several United Nations conferences confirmed the configuration of the sustainable development principle based on environmental protection, economic growth and social development (Nocca, 2017). Indeed, SIDS host most of the world's biodiversity hotspots and during the Conference in Samoa it was agreed that those states could not bear the full costs of protecting biodiversity by themselves, considering that many of the benefits of those biodiversity hotspots are global (Mercer et al., 2014).

In addition, the Small Islands Big(ger) Stakes Basis for Action from UN states tourism as the catalyst for SIDS development and its Action 11 foresees that local authorities should identify and develop facilities to meet specific niche markets, particularly in ecotourism, nature and cultural tourism, and involve local populations in the identification and management of natural protected areas (UN-OHRLLS, 2017). The World Tourism Organization from the United Nations (2014) also added that sustainable tourism could play a key role in stronger support to the Príncipe's development agenda and be a path to advance socio-economic welfare while protecting its fragile ecosystem. In fact, recalling sustainable tourism development as meeting the needs of present tourists and host regions while protecting and enhancing opportunities for the future (WTO, 1998, p.21), in broader terms, sustainable tourism would transfer the principles of sustainable development into the context of tourism needs (Hardy & Beeton, 2002; Lu & Nepal, 2009).

Nowadays, discussions concerning the importance of biodiversity conservation in the development agenda are included in the focus on climate change. The solution is to invest both in biodiversity conservation and poverty reduction, because climate change affects both agendas. After more than 50 years of debate, climate change may, therefore, be the glue that binds nature conservation and poverty alleviation together and invigorates a time-pressured search for sustainable solutions (Roe, 2008).

Nature-based tourism (ecotourism) (Ceballos-Lescuráin, 1987) is believed to be a fast-growing sector and could help in providing solutions for lack of funding for wildlife and biodiversity protection through targeted efforts to increase revenue-sharing at the local level (The World Bank, 2018; Walpole et al., 2001).

Prospects for adaptative co-managing protected areas for sustainable tourism have already been treated in the literature (Plummer & Fennell, 2009). However, the extent to which ecotourism offsets costs to preserve a protected area is not clear, and some examples show that it may not be the ultimate solution to finance biodiversity protection or alleviate poverty, as the following information indicates: The Komodo National Park experience (Indonesia, Islands Rinca, Komodo, and Padar) shows that only 69% of park management costs were recovered and only 0.2% of the total expenditure by tourists visiting this Park had accrued income to local villages adjacent to the Park (Walpole et al., 2001); The Royal Chitwan National Park in Nepal only affected the income of 6% of households living nearby, from the 50,000–100,000 people visiting the Park each year (Bookbinder et al., 1998).

Besides, tourism can also have marked impacts on whether the economic returns from development are retained in local economies in poorer regions or are substantially expropriated by interests in wealthy metropolitan centres (Bramwell & Lane, 2008). The World Bank estimates that 55% of tourism gross revenues leaks back to developed countries in the form of repatriated earnings, advertising costs, and payments for tourism-related imports (Frueh, 1988), that´s to say, only the minor part of 45% of tourism gross revenues stays in the visited place. There is no 'one size fits all' development strategy (Rodríguez-Pose, 2010). Let`s see if sustainable tourism may be an important key driver of the Príncipe's development strategy.

DISCUSSION

The north of Príncipe contains most of the population and the ultimate challenge is how to shape human interactions with nature, in landscapes where people live (Brockington et al., 2006). Somehow, ways must be found of increasing tropical conservation investments with fairly distribution of conservation benefits to local community in order to offset local opportunity costs of conservation and achieve development goals (Reynolds et al., 2001). Some international financing mechanisms are available as presented in next two sub-sections.

Financing Mechanisms

The Small Islands Big(ger) Stakes Basis for Action from UN mentions to provide SIDS with the necessary and adequate funding, including through innovative sources of financing, to assist SIDS in their mitigation and adaptation measures and efforts (UN-OHRLLS, 2017). The 'financing cycle' is presented by Büge et al (2015) as in Figure 2, shows the flow of capital from potential sources to their use in biodiversity protection (inner circle) and the accompanying policies to mobilize channel and use these funds (outer circle).

Figure 2. Financing cycle (Büge et al., 2015)

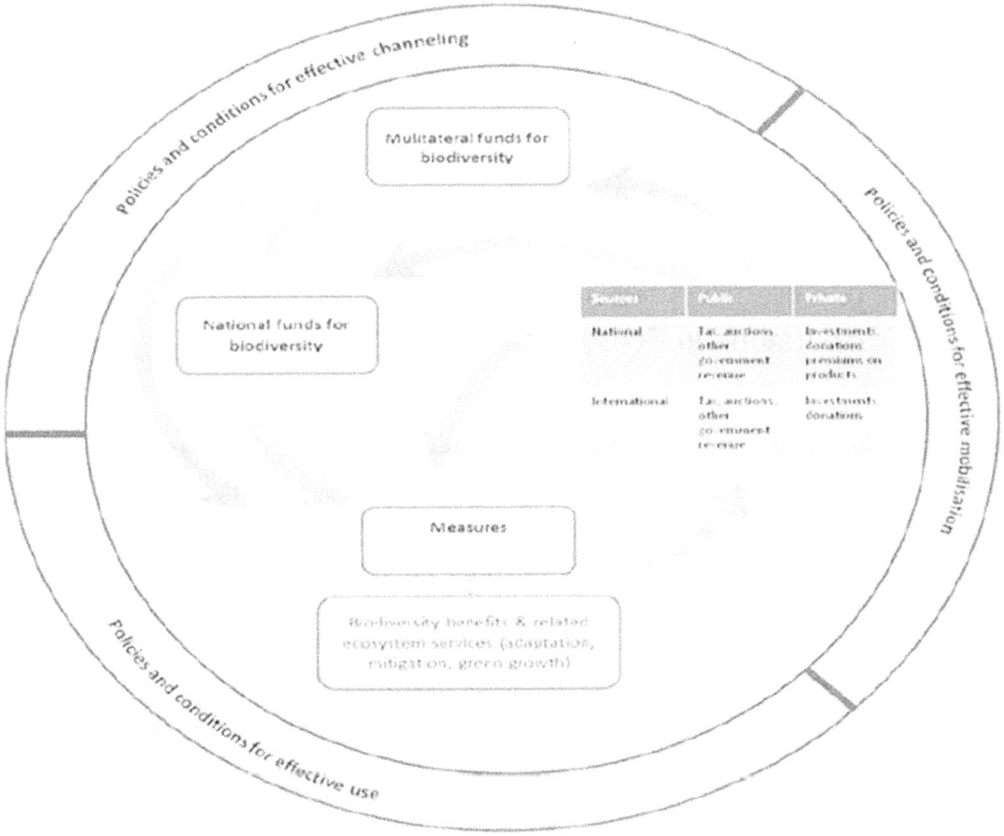

Funds from private/public sources may flow directly to local biodiversity protection activities, be transferred to the country's government, or be channelled through multilateral mechanisms (continuous arrows). In turn, biodiversity benefits can indirectly contribute to mobilising international resources (dashed arrow). In this cycle, biodiversity conservation has strong inter-linkages with other development goals such as poverty reduction and community development.

Some international funds are already being used in STP´s state from the Global Environment Facility (GEF) which is the agency of the United Nations Development Programme (UNDP) for development, climate, and ecosystem sustainability support. Data approved to STP´s country include 33 projects and an amount of USD 185.70 million of GEF Grant Funding, and an additional co-financing of USD 670.46 million (GEF, 2018) - there are no separate amounts for Príncipe.

The UNDP-GEF has a Small Grants Programme (SPG) of sustainable development by "thinking globally, acting locally" which supports eligible projects to conserve/restore the environment and enhance livelihoods of the population. SPG grants up to USD 50,000 per project, average around USD 25,000, to community-based organisations and civil society ones. SPG funds go to Biodiversity, Climate change and Land degradation (The Global Environment Facility (GEF), 2018) but no data was found for STP.

Official Development Assistance (ODA) is another source of finance that remains resilient, steadily increasing since the turn of the century and doubling since 2000 (OCDE, 2018). An ODA financing mechanism is the debt-for-nature-swaps that allow developing countries to convert their foreign debts into local currency funds for biodiversity conservation or other environmental and ecological projects. At end-2008, STP had reached the completion point and irrevocable Heavily Indebted Poor Countries Initiative and Multilateral Debt Relief Initiative but the debt-for-nature-swaps mechanism has not been very useful to STP due to the threat it represents on macroeconomic stability of such a small indebted country (Braga & Dömeland, 2009).

SIDS receive a share of 5.7% of total ODA and most aid flows are concentrated in the Pacific. On the whole, SIDS have not been able to leverage as much climate and environmental aid as may have been expected, due to their limited capacities to apply for and manage climate finance (typically administered via complex funds), as it is the case of STP, as well as unfulfilled donor commitments and a donor bias towards mitigation rather than adaptation finance (Hurley, 2015).

International Investments in Ecotourism

The Príncipe's autonomous government has been negotiating with international investors since 2009 and has established long term concession contracts for the recuperation of colonial plantation houses and other investments provided monetary fees payment that are not published.

The most relevant international investor is the HBD that has been working together with the government of Príncipe and the UNESCO Biosphere Reserve`s stakeholders to develop its projects in the island in a sustainable and responsible way as exemplified next: all the HBD hotels have soft illumination designed to leave turtles and other animals in the rainforest undisturbed; the style of the villas at the Sundy Praia was chosen to achieve a reversible design construction, natural light, low impact and minimal use of concrete; the whole structure of the restaurant made of bamboo, and an atmosphere inviting nature inside the room.

The Roça Sundy manor house[1] was completely restored by local workers who learned new and traditional construction techniques during an intensive three-year training course at nearby Roça Paciência. The original purpose for Roça Sundy was to become a Hotel School providing training to Sundy Praia's future staff - the training was so successful that the school was turned into a 4-star boutique hotel fully managed by locals and all staff working at Roça Sundy are from STP. HBD also manages its own carpentry team, where wooden furniture for the hotels is crafted, as well as housing for Príncipe's population.

The Príncipe Trust (FPT) is a Conservation NGO, created in 2015, to reorganise the environmental and social part of the tourism investment company HBD, with a complementary strategy to support the sustainable development of the island. Today, the FPT is working towards becoming an independent structure, operating alongside HBD and the local government, and supported closely by Fauna and Flora International an international conservation NGO, focuses on building and strengthening local NGO. The FPT´s activities are spread into different areas and we refer some of them:

- Training activities - in partnership with the Azores Tourism School, 100 students were trained for nine months in restaurant operation, housekeeping, gardening and reception, to work at the HBD´s hotels.
- Water & Recycling - a project started in 2013 with the Biosphere Reserve aimed to encourage people to collect plastic, exchanging it for reusable bottles. More than 300,000 plastic bottles were collected in 2 years (the second phase of the project is the transformation of recycled plastic into bricks). Simultaneously, ten water stations were installed in key points around the island (hospital, schools, airport, hotels and others) providing filling water points free of charge to the population.
- Waste management - with the support of the communities, they separate and recycle plastic, glass and organic matter and transform it into construction material and some jewellery to sell to the hotels.
- Agricultural project at the Paciência plantation (*Paciência Organics*) that provides organic food and cosmetics to the hotels – people from local community work in the gardens and laboratories, producing jams, muesli and other foodstuffs to the hotels, and organic products to the spas and hotels. Visitors come to visit the place and buy local handmade products.
- To expand knowledge of Príncipe's biodiversity in order to influence future decision-making processes in several areas of the Island: Botanical Sampling; Evaluation of the Seabird populations of the Tinhosas Islands; Evaluation of the coral habitats; Marine Turtle Conservation; Whale watching; New technologies.
- Reducing illegal and harmful practices to Príncipe's Biodiversity ("Omali Vida Nón") to promote better management of marine resources through the participation of fishers and fish traders.
- "Bumbu d'Iê Project" promotes beekeeping and agroforestry as an income alternative and better management of forests and restoration of degraded areas.
- Zero Capture to promote a change of attitude of the population regarding the protection and conservation of the biodiversity, mainly of the sea turtles.
- Biosphere Trails aims to create opportunities for social and economic development through offering initiatives for Ecological and Sustainable Tourism and the training and qualification of Local Guides.
- Increase of FPT sustainability – Seabirds: Identification, Biology and Field Methodology; – Diving Open Water course; – Managing and Leading Conservation Projects, – Organizational Management Skills for Protected Area Managers; – Photography: Terrestrial and Marine; – Reed Check: Methodology and Identification of species, Marine Turtle Conservation, among many others.

There are no official data about the investments amounts of the HBD group, but several articles in the press point to around 120 million euros in five years to remodel old manor houses, to build hotels, investments in the social sector, road construction, the runway of the Príncipe`s airport and in staff formation (Panapress, 2020). Nowadays, within hotels, agriculture and other activities, HBD employs 10% of the population of Príncipe (HBD, 2020).

Besides the HBD investments and development support activities, there is some local biological cocoa farming associated with the business of chocolate and coffee refinery called "Claudio Corallo Cacao and Coffee" (2018) whose manufactured chocolates and coffee flavours have already won international awards.

The meticulous work is done with the collaboration of Santomean workers that Cláudio Corallo himself takes charge of training in a whole process, from the cocoa plantation on Príncipe to the manufacture of chocolate in the capital; he employs about 250 people. This line of final production is sited in S. Tomé and if developed in Príncipe could attract the segment of tourism of distinction, of people of high cultural level who seek authentic and extraordinary experiences in their travels, with new flavours, exotic or gourmet preparations (Camacho-Gómez, 2016; Mitchell & Mitchell, 2001) as the different aromas and tastes of Príncipe Island's aromatic herbs and medicinal plants, viewed in the reserve's trails.

CONCLUSION

This paper deals with the fragile capacity of Príncipe Island to reduce poverty, which 69% of its 7.0000 inhabitants. An attempt to revert the poverty situation was made by the STP´s government, from 1986 to 1990, with a rehabilitation program of some state-owned estates, conceding them to private foreign management, financed by international institutions, in order to increase cocoa output and achieve economic recovery. The program failed its objectives due to the international price decrease of cocoa. In the first decade of this century, oil exploration was announced, but it has not yet produced the attended revenues. The Príncipe fragile productive tissue is based on agricultural and fishing activities, and local stakeholders do not possess viable tradable resources to sustain its population above the poverty line and to assure de conservation effort of the Biosphere Reserve classified by UNESCO in 2012. In this context, international resources are needed either public or private ones.

Some international funds are already being used in STP namely from the Global Environment Facility (GEF) which is the agency of the United Nations Development Programme (UNDP) for development, climate, and ecosystem sustainability support, to an amount of USD 185.70 million of GEF Grant Funding, and an additional co-financing of USD 670.46 million (GEF, 2018), but there are no separate data amount for Príncipe. An Official Development Assistance (ODA) financing mechanism is the debt-for-nature-swaps that allow developing countries to convert their foreign debts into local currency funds for biodiversity conservation or other environmental and ecological projects, but at end-2008 STP reached the completion point and irrevocable Heavily Indebted Poor Countries Initiative and Multilateral Debt Relief Initiative and the debt-for-nature-swaps mechanism has not been used due to the threat it represents to (monetary policy and) macroeconomic stability.

STP is part of the SIDS group recognised by the UN, and those island states receive a share of 5.7% of total ODA, but most aid flows are concentrated in the Pacific, and most islands states have limited capacities to apply for and manage finance typically administered via complex funds, as it is the case of STP.

In addition, the Small Islands Big(ger) Stakes Basis for Action from UN states tourism as the catalyst for SIDS development and its Action 11 foresees that local authorities should identify and develop facilities to meet specific niche markets, particularly in ecotourism, nature and cultural tourism, and involve local populations in the identification and management of natural protected areas (UN-OHRLLS, 2017).

Since 2009 the Príncipe's autonomous government has been negotiating with international investors and it has established long term concession contracts for the recuperation of colonial plantation houses and other investments provided monetary fees payment that are not published.

The most relevant international investor is the HBD, that has been working together with the government of Príncipe and the UNESCO Biosphere Reserve`s stakeholders to develop its tourism projects in the island in a sustainable and responsible way. Moreover, the restoration of the Roça Sundy manor house was completely done by local workers who learned new and traditional construction techniques during an intensive three-year training course at nearby Roça Paciência, financed by HBD. The original purpose for Roça Sundy was to become a Hotel School providing training to Sundy Praia's future staff, but the training was so successful that the school was turned into a 4-star boutique hotel fully managed by locals and all staff working at Roça Sundy are from STP. HBD also manages its own carpentry team, where wooden furniture for the hotels is crafted, as well as housing for Príncipe's population.

The Príncipe Trust (FPT) is a Conservation NGO, created in 2015, that reorganises the environmental and social part of the tourism investment company HBD, with a complementary strategy to support the sustainable development of the island. Today, the FPT is working towards becoming an independent structure, operating alongside HBD and the local government, and supported closely by Fauna and Flora International an international conservation NGO, focuses on building and strengthening local NGO. The FPT´s activities are spread into different areas, and we refer some of them: Training activities for local people in partnership with the Azores Tourism School in restaurant operation, housekeeping, gardening and reception, to work at the HBD´s hotels; Waste management and recycling; an agricultural project at the Paciência plantation (*Paciência Organics*) to provide organic food and cosmetics to the hotels, where people from local community work in the gardens and laboratories; multiple projects in different areas of the Island: Botanical Sampling; Evaluation of the Seabird populations of the Tinhosas Islands; Evaluation of the coral habitats; Marine Turtle Conservation; Whale watching; New conservation technologies. Biosphere Trails to create opportunities for social and economic development through offering initiatives for Ecological and Sustainable Tourism and the training and qualification of Local Guides. There are other activities linked with sustainability, such as Seabirds Identification, Biology and Field Methodology, Managing and Leading Conservation Projects, Identification of species, Marine Turtle Conservation, and others.

There are no official data about the investments amounts of the HBD group, but several articles in the press point to around 120 million euros in five years in the different projects: remodelling of old manor houses, new hotels facilities, social sector, road construction, the runway of the Príncipe`s airport and in staff formation (Panapress, 2020). Nowadays, within hotels, agriculture and other activities, HBD employs 10% of the population of Príncipe (HBD, 2020).

Besides the HBD investments and development support activities, there is some local biological coffee and cocoa farming associated with the business of chocolate manufacture of the Cláudio Corallo's corporate (2018). He takes charge of training in the whole process from the cocoa plantation on Príncipe to the manufacture of chocolate in S. Tomé and employs about 250 people. The line of hand-produced chocolates and coffee flavours could also be developed in Príncipe to be included in the niche of tourism of distinction, for people of high cultural level who seek authentic and extraordinary experiences in their travels, with new flavours, exotic or gourmet preparations.

REFERENCES

Abreu, A. (2012). Príncipe Island, UNESCO Biosphere Reserve, a tool for the sustainable development. *2nd Meeting of the World Network of Island and Coastal Biosphere Reserves.*

Adams, W. (2004). Biodiversity Conservation and the Eradication of Poverty. *Science, 306*(5599), 1146–1149. doi:10.1126cience.1097920 PMID:15539593

Adams, W. (2008). *Green Development, Environment and sustainability in the Third World.* Academic Press.

ADB. (2012). *Democratic Republic of São Tomé and Príncipe, Country Strategic Paper, 2012-2016.* Academic Press.

AICEP. (2017). Ficha de Mercado. Academic Press.

Barrett, C. B., & Arcese, P. (1995). Are Integrated Conservation-Development Projects (ICDPs) Sustainable? On the conservation of large mammals in sub-Saharan Africa. *World Development, 23*(7), 1073–1084. doi:10.1016/0305-750X(95)00031-7

Bertram, G. (2006). Introduction: The MIRAB model in the twenty-first century. *Asia Pacific Viewpoint, 47*(1), 1–13. doi:10.1111/j.1467-8373.2006.00296.x

Bookbinder, M. P., Dinerstein, E., Rijal, A., Cauley, H., & Rajouria, A. (1998). Ecotourism's Support of Biodiversity Conservation. *Conservation Biology, 12*(6), 1399–1404.

Braga, C. A. P., & Dömeland, D. (2009). *Debt Relief Beyond Lessons Learned and Challenges Ahead.* Academic Press.

Bramwell, B., & Lane, B. (2008). Priorities in Sustainable Tourism Research. *Journal of Sustainable Tourism, 16*(1), 1–4. doi:10.2167/09669580803489612

Brockington, D., Igoe, J., & Schmidt-Soltau, K. (2006). Conservation, Human Rights, and Poverty Reduction. *Conservation Biology, 20*(1), 250–252. doi:10.1111/j.1523-1739.2006.00335.x PMID:16909680

Bruner, A., Gullison, R., Rice, R., & da Fonseca, G. (2001). Effectiveness of Parks in Protecting Tropical Biodiversity. *Science, 291*(5501), 125–128. doi:10.1126cience.291.5501.125 PMID:11141563

Bruner, A., Hanks, J., & Hannah, L. (2003). *How much will effective protected area systems cost?(updated version).* http://www.citeulike.org/group/342/article/432012

Büge, M., Meijer, K., Wittmer, H., Bordon, I., Hein, J., Lindenberg, N., Rodriguez De Francisco, J. C., Richerzhagen, C., Willem Te Velde, D., & Wolff, P. (2015). *International financial instruments for biodiversity conservation in developing countries – financial mechanisms and enabling policies for forest biodiversity Background paper for the European Report on Development 2015.* Academic Press.

Camacho-Gómez, M. (2016). Chocolate Business in the Hospitality Industry. *ECORFAN Journal ECORFAN Journal-Mexico Journal*, *7717*(17), 7–17.

Cardoso, M. (2015). Migrações e Estratégias de Desenvolvimento dos Pequenos Estados Insulares em Desenvolvimento. Estudos de caso: Cabo Verde e São Tomé e Príncipe. *Cadernos de Estudos Africanos*, *29*(29), 103–128. doi:10.4000/cea.1804

Ceballos-Lescuráin H. (1987). The future of ecotourism. *Mexico Journal*, 13–14.

Cláudio Corallo. (2018). *As 13 melhores imagens em Chocolate from Claudio Corallo no Pinterest | Chocolates, Coisas engraçadas e Palavras*. https://www.pinterest.pt/culinarious/chocolate-from-claudio-corallo/

Conrad, J., Ferraro, P., Alm, J., Cummings, R., Morrison P., C. J., Taylor, L., Uchida, T., & Weitzman, M. (2001). *Habitat Conservation: the dynamics of direct and indirect payments The Dynamics of Direct and Indirect Payments*. Academic Press.

Dallimer, M., & King, T. (2008). Habitat preferences of the forest birds on the island of Príncipe, Gulf of Guinea. *African Journal of Ecology*, *46*(3), 258–266. doi:10.1111/j.1365-2028.2007.00891.x

Dias, C. (2017). São Tomé e Príncipe | Na ilha do Príncipe, como em muitas outras, já se ouve a orquestra do Titanic. *Público*.

Ferraro, J., & Kiss, A. (2002). Direct Payments to Conserve Biodiversity. *Science*, 298. PMID:12459569

Frueh, S. (1988). *Report to WWF on Tourism to Protected Areas*. Academic Press.

Gama, I., & Sanoussi, F. (2017). African Economic Outlook. Academic Press.

GEF. (2018). *Sao Tome and Principe | Global Environment Facility*. https://www.thegef.org/country/sao-tome-and-principe

Gullison, R., Rice, R., & Blundell, A. (2000). 'Marketing'species conservation. *Nature*, *404*(923). PMID:10801103

HBD. (2020). *The Island of Príncipe*. https://press.hbdprincipe.com/pdf/press_release/Beyond_Green-Principes_sustainability_story.pdf

Hulme, D., & Murphree, M. (2001). African wildlife and livelihoods: the promise and performance of community conservation. Academic Press.

Hurley, G. (2015). *Financing for development and small island developing states: A snapshot and ways forward. UNDP and UN-OHRLLS*. https://sustainabledevelopment.un.org/content/documents/2181(UNDP & OHRLLS 2015) Financing for development and SIDS A snapshot and ways forward.pdf

IMF. (2013). *IMF Country Report nº14/2, Democratic Republic of São Tomé and Príncipe*. Academic Press.

INE-STP. (2019). *Boletim Estátistico da Direção Geral do Turismo e Hotelaria*. https://www.ine.st/index.php/component/phocadownload/file/467-boletim-estatistico-da-direcao-geral-do-turismo-e-hotelaria

Kepe, T., Saruchera, M., & Whande, W. (2004). Poverty alleviation and biodiversity conservation: A South African perspective. *Oryx*, *38*(382), 143–145. doi:10.1017/S0030605304000262

Kiss, A. (2002). *"Making biodiversity conservation a land use priority" in Getting Biodiversity Projects to Work Towards More Effective Conservation and Development*. Academic Press.

Lu, J., & Nepal, S. (2009). Sustainable tourism research: An analysis of papers published in the Journal of Sustainable Tourism. *Journal of Sustainable Tourism*, *17*(1), 5–16. doi:10.1080/09669580802582480

Mercer, J., Kurvits, T., Kelman, I., & Mavrogenis, S. (2014). Ecosystem-Based Adaptation for Food Security in the AIMS SIDS: Integrating External and Local Knowledge. *Sustainability*, *6*(9), 5566–5597. doi:10.3390u6095566

Mitchell, M. A., & Mitchell, S. J. (2001). Consumer Experience Tourism. *Journal of Food Products Marketing*, *6*(3), 1–16. doi:10.1300/J038v06n03_01

Nocca, F. (2017). The Role of Cultural Heritage in Sustainable Development: Multidimensional Indicators as Decision-Making Tool. *Sustainability*, *9*(10), 1882. doi:10.3390u9101882

Oates, J. F. (1999). *Myth and reality in the rain forest: how conservation strategies are failing in West Africa*. University of California Press.

OCDE. (2018). *Geographical Distribution of Financial Flows to Developing Countries 2018*. OECD Publishing.

Panapress. (n.d.). *São Tomé e Príncipe: sul-africano investe 12 milhões em resort na ilha do Príncipe*. Retrieved March 30, 2020, from https://hotelcontinentalluanda.com/index.php/pt/noticias/541-sao-tome-e-principe-sul-africano-investe-12-milhoes-de-euros-em-resort-na-ilha-do-principe

Plummer, R., & Fennell, A. (2009). Managing protected areas for suatainable tourism:prospects for adaptative co-management. *Journal of Sustainable Tourism*, *17*(2), 149–168. doi:10.1080/09669580802359301

Poças, F. (2016). *Sustainable Development in Small Islands Developing States: Public/Private Cooperation in the case of Príncipe Island*. ISCTE.

Redford, H. K., Levy, M. A., Sanderson, E. W., & Sherbinin, A. (2008). What is the role for conservation organisations in poverty alleviation in the world's wild places? *Oryx*, *42*(04), 516. doi:10.1017/S0030605308001889

Relego, D. (2017). *São Tomé e Príncipe. Academic Press.*

Reynolds. (2001). *Conservation of Exploited Species*. Academic Press.

Rodríguez-Pose, A. (2010). *Do institutions matter for regional development?* http://personal.lse.ac.uk/RODRIGU1

Roe, D. (2008). The origins and evolution of the conservation-poverty debate: A review of key literature, events and policy processes. *Oryx*, *42*(04), 491. doi:10.1017/S0030605308002032

Schmidt-Soltau, K., & Brockington, D. (2004a). The social and environmental impacts of wilderness and development. *Oryx*, *38*(382), 140–142.

Scott Dunn. (2018). *Here Be Dragons – Bom Bom Island*. Academic Press.

Seibert, G. (2016). São Tomé and Príncipe 1975-2015: Politics and economy in a former plantation colony. *Estudos Ibero-Americanos*, *42*(3), 987–1012. doi:10.15448/1980-864X.2016.3.22842

STP. M. F. (2018). *Plano Estratégico e de Marketing para o Turismo de São Tomé e Príncipe*. https://interlusofona.info/wp-content/uploads/2018/10/Plano-Estratégico-e-de-Marketing-para-o-Turismo-de-São-Tomé-e-Príncipe.pdf

Terborgh, J. (1999). *Requiem for nature*. Island Press.

The Global Environment Facility (GEF). (2018). *Funding | Global Environment Facility*. Academic Press.

The World Bank. (2005). *São Tomé and Príncipe Joint IDA-IMF Staff and Advisory Note on the Poverty Reduction Strategy Paper*. https://documents.worldbank.org/curated/en/824571468299160559/pdf/31726.pdf

The World Bank. (2018). *São Tomé e Príncipe: aspectos gerais*. https://www.worldbank.org/pt/country/saotome/overview

The World Bank. (2019). *São Tomé and Príncipe Recent Developments*. http://pubdocs.worldbank.org/en/766771492188172104/mpo-stp.pdf

Tisdell, C. (2014). *The MIRAB Model of Small Island Economies in the Pacific and their Security Issues: Revised Version*. Social Economics, Policy and Development Working Papers.

Tisdell, C. A. (2016). Models of the international economic dependence of Pacific Microstates: A critical review with important implications for international policies and relations. *Journal of Self-Governance and Management Economics*, *4*(2), 7–27. doi:10.22381/JSME4220161

UN. (1992). *Earth Summit UN Conference on Environment and Development (1992)*. Academic Press.

UN-OHRLLS. (2017). *Small Island Developing States, Small Islands Big(ger) Stakes*. http://www.sustainablesids.org/wp-content/uploads/2017/08/UN_SIDS_booklet_5x6-5_062811_web.pdf

UNCTAD. (2012). *The Least Developed Countries Report 2012*. Academic Press.

UNESCO. (2012). *The Island of Príncipe*. Academic Press.

UNESCO. (2018). *Biosphere Reserves*. http://www.unesco.org/new/en/natural-sciences/environment/ecological-sciences/biosphere-reserves/

UNWTO. (2014). *Tourism in Small Island Developing States*. SIDS.

UNWTO. (2020). *Small Islands Developing States (SIDS)*. https://www.unwto.org/sustainable-development/small-islands-developing-states

Walpole, M., Goodwin, H., & Ward, K. (2001). Pricing Policy for Tourism in Protected Areas: Lessons from Komodo National Park, Indonesia. *Conservation Biology*, *15*(1), 218–227. doi:10.1111/j.1523-1739.2001.99231.x

WTO. (1998). *Guide for local authorities on developing sustainable tourism.* https://www.e-unwto.org/doi/pdf/10.18111/9789284402809

Zhou, W., & Ritchie, J. (2007). Tourism and poverty alleviation: An integrative research framework. *Current Issues in Tourism, 10*(2-3), 119–143. doi:10.2167/cit296.0

ENDNOTE

1 Roça Sundy, where on May 29, 1919 English astronomers led by Arthur Eddington observed and photographed a total eclipse of the Sun and confirmed for the first time experimentally Einstein's Theory of General Relativity, presented in November 1915 to the Prussian Academy of Sciences.

Section 2

Environment Protection and Conservation

Chapter 5
The Global Challenge and Re-Evaluation of Ecological Tourism in Line With Sustainable Development Goals

Gul Yilmaz
Istanbul Arel University, Turkey

ABSTRACT

In the 2030 Agenda for Sustainable Development adopted by the UN General Assembly on 5 September 2015, 17 targets were set in line with the Sustainable Development Targets. In this study, the Sustainable Development Goal 15 (SDG15)—protect, restore, and promote sustainable use of terrestrial ecosystems, sustainably manage forests, combat desertification, and halt and reverse land degradation and halt biodiversity loss—is discussed within the framework of the tourism sector. The study aims to propose a sustainable tourism vision and policy in line with the SDG 15. The study suggests the development of ecological tourism, which is a sustainable tourism vision, in line with SDGs.

INTRODUCTION

Special ecosystems and nature around the world come across dangers today. The harm caused by human beings has reached extreme dimensions with the increase in the human population and the development of technology. That is, the advanced user of technology destroys natural areas, especially about environmental pollution and unusual changes occur in the ecosystem due to climate change. On the other hand, Irregular agriculture and uncontrolled chemical drugs used in agriculture with ecosystem corruption, the opening of forest lands to zoning and deforestation, garbage piles threw into the oceans, biodiversity loss as a result of significant reduction of animal and plant species are examples of the negative effects of human beings on nature.

If we talk about biodiversity loss, we see that some countries take various precautions in this regard, but, clearly, some countries are seriously harming nature. When we examine the countries taking

DOI: 10.4018/978-1-7998-5691-7.ch005

various precautions, the harm of nature is prevented by both public awareness, state support, and legal procedures but if we give an example of the country of Brazil, we can say that Amazon Forests have hurt a great loss in the last ten years.

International organisations also carry out various activities about climate change, biodiversity conservation, and the future of the world. At this point, 17 targets were determined in the direction of 2030 Sustainable Development Goals accepted by the United Nations (UN) General Assembly on September 5, 2015. Sustainable Development Goal 15 is determined as the protecting sustainable use of terrestrial ecosystems, restore and promote, manage the forests as sustainably, struggling against desertification and stopping land corruption and loss of biodiversity and to reverse this situation. Based on this, the emergence of different studies in many different fields has become unavoidable, and the tourism sector is one of these areas.

Considering the human population that is expected to reach 9 billion in 2050, the sustainable development goals of the 2030 Sustainable Development Agenda become even more important in this global challenge (UNGA, 2015). Success in the global challenge that threatens the future of our species and other species necessitates the re-evaluation and regulation of all sectors based on sustainability (Gomiero, 2016). To achieve this objective, the development agenda on "Mainstreaming biodiversity for well-being" adopted by the UN Environmental Assembly in 2015 aimed to improve integration for the protection of biodiversity in various sectors and the development of sustainable policies (UNGA, 2015). Tourism also has an important role in these sectors. The importance of developing a sustainable tourism vision within the scope of sustainable policies has become more important than ever. In a sense, this means accepting the negative share of traditional tourism vision and policies on a critical basis in the global challenge, which has become an important threat today. Therefore, it is important to re-evaluate ecological tourism in all aspects as a sustainable tourism vision in line with sustainable development goals.

A great increase has occurred in production and consumption activities due to the development of technology after the industrial revolution and to meet the unlimited wishes of the human being. A rapid increase in production and consumption activities has led to a rapid decrease in natural resources. In order to prevent resources from being decreased and then run out of overtime, "Wasteful World View" was replaced with "Sustainable World View" (Ceylan, 2019).

This study is evaluated in the context of utilising sustainable development objectives within the framework of the tourism sector. The purposes of this study are to provide that 15. Sustainable Development Goal (SDG) becomes a project that can be applied not only in certain countries but also all over the world, providing examples to develop a sustainable tourism vision and policy and to make suggestions for the development of ecological tourism in the direction of sustainable tourism targets.

The goal of this study was to provide solution offers on the actions that may be carried out for further development of tourism in the world and Turkey in line with sustainable tourism. The main purpose of human beings should be to care and protect natural resources to leave a sustainable and liveable world to the next generations. In this regard, the term sustainability and sustainable tourism have been clarified in this study. Later on, sustainable tourism and ecotourism terms have been discussed. The environmental, economic, and socio-cultural effects of ecotourism have been addressed. It is drawn attention to creating environmental consciousness by implementing environmentally friendly tourism activities to overcome problems that may occur during tourism activities. A literature review on what are the sustainable tourism management and ecotourism practices in the world and Turkey has been carried out and significant examples from the world and Turkey have been given.

SUSTAINABLE DEVELOPMENT AND CONCEPTUAL FRAMEWORK

The importance of the term sustainability could not be understood until the conference on the Human Environment held by the United Nations in Stockholm in 1972. Following this, the studies on this term increased and the term sustainability was reshaped by the Brundtland Report published in 1987. The most common definition of the term of sustainability is the one included in the report called Our Common Future (which is also known as Brundtland Report) published by the World Commission on Environment and Development in 1987. In this report, sustainability has been defined as a process carried out to provide environmental, economic and, social development on both local and global scales or a case that could be maintained to a certain degree (Yenil & Akyazi, 2019).

The phenomenon of sustainable tourism is based on sustainable development. The goal of sustainable tourism is to maintain tourism by protecting local characteristics of touristic regions. In this respect, the environmental resources and values supported by economic resources can be passed down to future generations.

Due to its close relationship with the environment and natural resources, tourism, considered as an important industry in progress and regional development, necessitates that development should be carried out in compliance with the environment. At this point, the term of sustainable tourism emerges by urging the primary goals of developing tourism with environment and economy with increasing importance as part of each other, maintaining environmental quality, equalising development (Solmaz, 2019).

Special attention is drawn to the protection of the environment and the necessity of protecting the ecological values provided by nature for people in the Sustainable Development Goals (SDG) of the United Nations. It is drawn attention to preparing action plans for the protection of the ecosystem in development strategies (Wood et al., 2018). Although today's economic and technological developments brought along with globalisation seem to improve life standards, they may sometimes damage the nature in certain points. Therefore, there may occur problems in long term use of natural resources. Solutions to these problems lie in "sustainability management" enabling to use of resources for a longer time. For this reason, it is possible to implement sustainability and its management in tourism. Sustainable tourism is a type of tourism integrated into development and supporting development at the same time.

According to Kotob (2000), The conceptual framework of the concept as well as the history of sustainability, research areas, and how it has been obtained should be examined in depth to define the concept of sustainability. While Kotob defines the concept of sustainability, he has been made a sharp difference between profitable sustainability efforts and non-profit sustainability efforts because according to Kotob (2000), there should be no profit in the concept of sustainability. Besides, Kotob could not give a clear definition of sustainability because of the concept of sustainability is typically lacking and the current general understanding of what sustainability means is wrong. Institutions and organisations which carrying out sustainability studies should not prepare social responsibility projects only for profit (Kotob, 2000).

According to Giovannoni and Fabietti (2014), the concept of sustainability is accepted as multidimensional and various dimensions of the concept of sustainability have turned into separate concepts by labelling different discourses over time. At this point, many practices have remained only discourse qualification. Based on this, Giovannoni and Fabietti (2014) approached the concept of sustainability from a different perspective by considering the structures, systems, and governance models of this concept and featured the concept of "integrated sustainability" to the front. Based on an integrated approach, it has been claimed that the integrated structure of the concept of sustainability in environmental, social, and commercial contexts should be examined. Besides, arguments about how important an integrated approach has been put forward an idea by pointing out to the United Nations Sustainable Development Conference. There are other definitions at this point. Drexhage and Murphy (2010) explained the integrated approach between three dimensions (environmental, social, and commercial) as follows: "sustainable development out of the environment "box" and considering wider social, economic, and geopolitical agendas".

When we consider other definitions regarding the concept of sustainability, it is defined as a vision for the world where people are reasonably healthy; communities and nations are safe, peaceful and successful; there is an economic opportunity for everyone; and the integrity of the life-supporting biosphere, a level necessary to make these goals possible (Cortese & Rowe, 2017).

Ben (2015, p. 6) described this concept as follows in his special study on sustainability: "A dynamic equilibrium in the process of interaction between a population and the carrying capacity of its environment such that the population develops to express its full potential without producing irreversible, adverse effects on the carrying capacity of the environment upon which it depends."

Economic development theory and ecological studies constitute the theoretical framework of the concept of sustainable tourism. Within the framework of economic development theory, tourism is an important factor for national economies but focusing exclusively on mass tourism can pose serious risks for countries. Starting from this, within the framework of sustainable development objectives; tourism and ecological tourism have created a serious field of activity by coming into prominence.

Today, there are countries with serious ecological resources. Sustainable tourism also has huge areas of activity for some countries. Starting from this, different strategies and methods may occur both for national economies and for tourism innovations. On the other hand, within the framework of sustainable tourism, experimenting with different strategies and methods has also created international principles for sustainable tourism (Aall, 2014).

Many countries have made successful efforts for sustainable tourism based on innovations in tourism. When we examine the sustainable tourism industry, we see three main parameters in the form of natural resources, in particular, human product resources and socio-cultural natural resources. However, we have already stated that mass tourism can pose serious risks. From this perspective, the most important stages of sustainable tourism are natural resources and protection of these resources, prevention of human-made disasters, and the use of their with sustainable tourism.

Based on the definitions of sustainable development, there are different opinions on sustainable development, both theoretically and practically. When we evaluate the opinions, we see that economists say that the concepts of development and sustainability create reverse effects but neoclassical economists claim the opposite (Klarin, 2018). Within the framework of political power, economic policy and colonialism neoclassical economists claimed that the concepts of development and sustainability have different theoretical perspectives today and they suggested that it could be tried in many different fields in the practical field. The concept of sustainable development has gained different dimensions with these theoretical developments (Klarin, 2018).

The concept of development can be interpreted as a process aimed at continuous progress by improving the quality and capacity of life. At this point, based on modern liberal, neoliberal and modern development theories, countries can operate the opportunities which they have. At this point, the concepts of sustainability and development can be considered in parallel with the tourism sector and can be adapted to practical life. Based on the examples to be given later in the study, we see that there are very successful sustainable tourism projects.

The concept of ecological tourism emerging in recent years is directly related to sustainable development projects. The main thing we need to examine is what kind of effects ecotourism creates in terms of environmental, economic, and socio-cultural sustainability.

ENVIRONMENTAL, ECONOMIC AND SOCIO-CULTURAL EFFECTS OF ECOTOURISM WITHIN THE SCOPE OF SUSTAINABILITY

The concept of sustainable tourism started to be discussed in the 1990s, and serious changes occurred in the tourism sector after the 2000s. Tourism has had different qualities with the development of technology and increasing options in a continuously globalising world. At this point, the most important thing we need to discuss is the loss of influence of mass tourism with time and change in people's demands and tendencies. This is the main factor in the emergence and development of the concept of sustainable tourism.

The World Tourism Organisation (UNWTO) indicates that sustainable tourism models must meet the necessities of both the recipient region and current tourists. At the same time, they must also protect the resources on which said activities rely, given that doing so will secure future opportunities (UNEP & WTO, 1993). In this sense, the tourism industry is acknowledged as having the potential to act as a social and economic tool for development, while balancing social, economic, and environmental interests - even when the said balance is not automatic or assured (Martinez et al., 2019). The countries should organise their tourism activities along with sustainable tourism.

The ecotourism is a nature-based tourism type helping to protect nature. The difference between ecotourism and nature tourism is the fact that ecotourism is based on the creatures in natural systems and it covers the aspects of both environmental and cultural heritage. Ecotourism is a sustainable form of tourism based on the ecological principle and sustainable development theory (Bansal & Kumar, 2011).

Increasing world population and environmental pollution show the necessity of protecting sustainable areas. While protecting these areas, they should also be protected against intensive tourism use because tourists may damage tourism areas by mass tourism. Some special areas may be damaged due to overuse. For example, Salda, a crater lake in the Lakes Region of Turkey and called the Maldives in Turkey, has similar features to Mars.

Since these structures named after scientifically as "stromatolite rocks" are also seen in Mars, the red planet is believed to resemble Salda millions of years ago. The scientists urge that there is evidence related to the existence of rivers, lakes in Mars in previous periods. Since the atmosphere of Mars is 100 times thinner than Earth, the water on Mars has evaporated over time and these rock structures have remained. Salda, one of the two places in the world with such rock formations, is being destroyed every day due to the unconscious use of tourists since it is not under protection (Salda Gölü Rehberi-Türkiye'nin Maldivleri, 2020).

When the negative activities occurred within tourism are replaced with environmentally friendly and sustainable practices such as ecological tourism, the negative effects of tourism on the environment may be limited and positive effects could be achieved. These positive effects are classified under the basic titles, such as an increase in recreational areas that may be used by local people and tourists, protection of historical and cultural resources, development of infrastructure facilities, increase in environmental awareness, and its value (Solmaz, 2019). The areas used intensively due to mass tourism are damaged, and they are affected by environmental pollution, noise pollution, and air pollution caused by exhaust gas of vehicles. This leads to ecosystem degradation. When countries make plans on tourism activities, they should initiate a management plan on increasing the economic benefits of local people and establishments by protecting the natural environment and local culture.

Ecotourism can provide much-needed revenues to protect national parks and other natural area-based facilities that may not be available from other sources. Also, ecotourism can provide a viable economic alternative to development for local communities with few other income-generating options. In addition, ecotourism can increase education and activity among travellers, making them more enthusiastic and effective conservation agents (Shedenov et al., 2019).

This situation is also connected with political and economic developments in the world. The collapse of the Soviet Union, The fall of the Berlin Wall, and choosing the free market economy in which countries governed by the communist regime have increased the globalisation trend all over the world; communication between people without barriers have been increased and travelling between countries has become easy. The tourism sector supported the creation of many different original projects by mobilising with the development and spread of mass media. Although mass tourism first came into prominence, the consumption trend lost its effect at some point. The intense interest of people about consumption in countries that turn to a free market economy has led to the early consumption of many demands and trends. At this point, people began a quest for different searches.

Many authorities claim that ecotourism is not a form of tourism, but a philosophy. The realities about how mass tourism affects nature have also placed the position of ecotourism overtime on a solid basis. Campaigns against economic consumption, negative effects of people on nature, the belief that many people should be in harmony with nature, the emergence of alternative and renewable technologies, minimising the damage to nature, losing the power of political and economic authorities with actions and pressures allow of the spread of ecotourism, and many people and authorities commented that this was a philosophy (Nail & Wearing, 1999).

Protection of nature as an environmental impact is the first thing to be evaluated within the framework of ecotourism. People and communities have become self-aware of the negative effects of mass tourism with the proliferation of ecotourism. When we evaluate this issue as socio-culturally, people travelling as domestic and foreign tourists met people from different cultures. This situation led to different perspectives for individuals trapped in urban life. That's why ecotourism is interpreted by many as a philosophy.

The introduction of ecotourism in our lives is based on old histories and is not a new concept. In this regard, ecotourism is an old concept both in terms of its definition and its historical background. Ecotourism can be interpreted as an alternative tourism experiment and a philosophy because of the effects of the consumer society, the negative effects of capitalism, and changing world conditions. Anti-government actions in the American community and its spread to all world is one of the main reasons which lead to the spread of ecotourism. In addition, the transformations in literature have enabled many people around the world to develop different perspectives on nature and the environment. Individuals who have reached individual awareness in many respects have canalised alternative searches. At this point, we need to give the definitions of ecotourism.

By the 1990s, the components of ecotourism began to take shape. These components can be stated in the form of principles, features, criteria, themes, and dimensions. Based on this, according to Buckley (1994), ecotourism offers us a size-based model that includes conservation, sustainability, environmental education, and nature-based activities. It has defined many ecotourism principles in addition to the basic components. The formation of the principle of ecotourism concept in time points to this concept has a philosophical depth and has evolved with theoretical studies. In this respect, Wallace and Pierce

(1996) hypothesised that ecotourism has six principles. These principles are as follows: Providing infrastructure and minimising negative impact, raising awareness, carrying out works for the protection and financial and moral support, developing democratic infrastructure in decision-making processes, observing community benefit and educational support, disseminating ecotourism with education. There are different definitions regarding ecotourism principles and components. Bjork (2000) determined different ecotourism principles and components (Ryan & Bernard, 2000). The listing, creating educational opportunities, determining planning issues and applying them in practice, looking out for the principle of ethical responsibility and raising public awareness should be the basic principles of ecotourism and also if we compare with other approaches, Bjork (2000) is one of the researchers who brought a methodological quantitative approach to ecotourism (Ryan & Bernard, 2000).

Ecotourism has achieved a legitimate ground at world standards with ith the expansion of ecotourism definitions and the development of principles; this concept, which started to form a methodology in itself has reached a satisfactory frame with complementary projects, effective tourism management plans, increase of research projects and starting from this, examples that can be exemplified in the world, ethically ecotourism minimising harm and adoption of environmental standards by environmentalists and finally developing active activities against organisations such as industrial organisations (Ryan & Bernard, 2000).

Countries that support ecotourism from different perspectives and create awareness through training are the leading points about ecotourism. These countries are usually among the countries that send the most ecotourists. Economically underdeveloped or developing countries are the leading countries that attract the most ecotourists. Activities practised within the scope of ecotourism differ from country to country, region to region, and region to region within regions. These differences exchange according to the environmental, economic, and socio-cultural characteristics of each country, region, and country. While there are generally positive effects in countries visited by conscious ecotourists, the activities abused by the administrators create negative effects on the local community base, especially the environment in the places where mass tourism is in the foreground.

In recent years, countries such as the Republic of South Africa, Kenya, Nepal, Sri Lanka have made a lot of effort to attract ecotourists. Many countries are working to strengthen police agencies and minimise security issues in the country; they establish special tourism units and take many precautions for the safety of tourists. In table 1, the important natural areas of ecotourism, income, tourist numbers, and employment rates of the leading and prominent countries in ecotourism are presented in detail.

Based on all this, many countries have discovered the importance of ecotourism and started working in this field. Today, there are countries with serious ecological resources. Some countries prefer to save their resources and use ecotourism and the future of the world, however, some countries operate their resources due to many factors such as mines, mass tourism, and construction. One of the most important reasons for this situation is that countries are economically weak and they develop different strategies and methods to economically strengthen. As an example, it is one of the best examples that we can give Costa Rica's announcement of all cloud forests in its country as a private reserve area and open it to ecotourism.

Costa Rica, one of the important ecotourism destinations, is not only a leading country in ecotourism but also it is one of the few countries where ecotourism activities are performed sustainably following its definition. The ecotourism has the biggest share in the Gross Domestic Product (GDP) of this country. National Ecotourism Certification Program is implemented in Costa Rica (Kasalak, 2015).

Table 1. Ecotourism statistics of countries (BAKA, 2012; Hawkins & Lamoureux, 2001)

Countries	Employment Statistics	Average Number of Visits	Income	Natural Areas
Germany	The farm has received more than 1400 quality certificates, rural tourism accommodation with a capacity of at least 184 thousand beds, over 23,000 farms that accept tourists	1 billion people on average		
Australia	130,000 people in a 10-year period	Number of 3,000,000 visitors between 1983 and 1993		Great Barrier Reef Marine Park: 1.6 million tourists, $ 1 billion ($ 988 million)
The United States of America			Ecotourism volume: According to 2008 statistics, 27.1 billion USD volume	Yellowstone National Park 2010 3,640,184 visitors, Florida Keys Area: $ 75 million annually
Equator			$ 100 million per year	Galapagos Island
France	Until 2002, 94500 rural hotel beds, 50 thousand rural residences, 30 thousand guest houses, 37 thousand holiday villages, 28 thousand tourism houses, 308 thousand tent areas			
England				Devon city countryside
Spain	From 1993 year, 6534 active eco farms until 2003			
Italy			Income between 2000-2008 4.114.286 Euro	Apulia Area
Cambodia	(Cambodia, Laos, Myanmar, Thailand, Vietnam, People's Republic of China, Yunnan Province) GMS Economic Cooperation Program	1.3 million ecotourists until 2007, 2 million visitors in 2010	290 million USD according to 2010 data	
Kenya		865,300 people in 1994		
Costa Rica	Costa Rica National Ecotourism Certification Program	This number increased to 64,000 in 1964 and 792,000 in 1995.	From $ 10 million to $ 661 million from 1964 to 1995	
Nepal				
Tanzania		350,000 people in 1995		
Jordan			$ 2.1 billion with 2010 data	Regions other than Wadi Rum

When we examine the sustainable tourism in Costa Rica at full length, we see that Costa Rica is very systematic about sustainable tourism. At this point, We need to talk about The Mesoamerican Biological Corridor. This corridor aims that preservation of large areas called bio corridors that are relatively well preserved or connected by restored habitat areas. As part of the bio corridor, Sistema Nacional de Áreas de Conservación-MINAE / Conservation Areas System was chosen as the focal point in Costa Rica. Ecosystems are mapped at the regional level with the project that started in 1998. More than 80% of ecosystems in the country are protected by developing maps providing classification system. On the other hand, carbon pools have been created in many regions. The risk considerably reduced by taking endangered animals under preservation. At this point, the most important support is nationwide efforts to raise awareness of the education system with state support; also, one of the issues we have to mention was X The Paid Environmental Services Program (PES) system created by Costa Rica. This system is a compensation system for created to protect forest plantations and it provides financial support for the protection of forest plantations across the country as a highly constructive system (Zamora & Obando, 2001). Table 2 lists the most important sustainable tourism projects in Costa Rica.

Table 2. The most important sustainable tourism projects in Costa Rica (Zamora & Obando, 2001)

Project	Area	Administrator
INBioparque at INBio	Santo Domingo de Heredia	INBio
Ecotouristic Shelter La Catarata	La Fortuna de San Carlos	Private family association
Monteverde City Centre	Monteverde, Puntarenas	A community with ecotourism
Monte Alto Forest Reserve	Hojancha, Nicoya, Guanacaste	Monte Alto Foundation

Based on examples of Vietnam and Costa Rica; tourists who have visited Costa Rica, Mexico, Belize, Ecuador, and the Dominican Republic (about 46%) travel to see specially protected wildlife in these countries. At this point, some African countries will be as effective in ecotourism and sustainable tourism as other countries in the event of necessary international support is provided. Countries such as South Africa, Kenya, Tanzania have made significant gains from ecotourism in recent years because ecotourism, which has a dynamic structure due to its structure acts as an important market for countries. Underdeveloped countries such as Vietnam and Nepal are also developing policies for infrastructure requirements related to ecotourism. Sustainable tourism examples in Vietnam are as important as the example of Costa Rica's cloud forests. Tourists can swim with a hippo in Vietnam, can deal with fish especially and also; they can visit silk villages and can take private lessons on natural life. Tourists can also play golf as a modern sport if they want (MacNall & Sobieszcyk, 2016).

Vietnam is making efforts to increase efficiency in sustainable tourism day by day. Authorities and bureaucrats want to improve sustainable tourism in groups of three provinces in North (Ninh Binh, Phu Tho, Ha Giang), Central (Quang Nam, Thua Thien Hue, Quang Binh) and South (An Giang, Can Tho, Kien Giang) and do a feasibility analysis. These areas are predetermined by the authorities and ecotourism potential is considered to be high in these regions. Vietnamese authorities have identified three main

Table 3. Statistics about foreign tourists visiting Vietnam (Dien et al., 2019)

Areas	2005	2010	2015
Northern Coastal	2384.1	3376.0	4150.0
Ratio compared to the whole coastal area (%)	37.7	33.6	27.5
North Central Coastal	377.4	2154.0	4540.0
Ratio compared to the whole coastal area (%)	16.6	21.4	28.2
South Central Coastal	406.7	767.0	1200.0
Ratio compared to the whole coastal area (%)	7.6	7.6	7.4
Coastal Region	2793.3	4263.0	5930.0
Ratio compared to the whole coastal area (%)	28.5	37.4	35.0
Total marine tourism	5191.4	10056.0	22000.0
Whole country	6346.1	13390.0	31000.0

parameters: local people, government promotion, and scoring mechanism in this regard. These indications show that Vietnamese officials are sensitive to sustainable tourism and that they are serious about tourism (ILO, 2017). Statistics on foreign tourists who visited Vietnam and coastal tourism have been shared in table 3. Vietnam showed a serious development in sustainable tourism according to these statistics.

As seen in table 3, the northern coast of Vietnam is one of the most important attractions centres of the country. The Coastal Region and Northern Coastal region are popular with tourists increasingly every year. What is more, the Great Himalayan National Park, one of the best national parks in the world, is one of the other important beauties in the world with its rich biological diversity (Bansal & Kumar, 2011).

In addition, New Zealand, Peru, Kenya, and Tanzania are other countries attracting tourists in terms of ecotourism. Kenya, located in Africa, arouses curiosity with its untouched nature and wildlife, and it is estimated that the tourism incomes constitute one-third of foreign exchange earnings of the country (Kasalak, 2015). The developing countries try to attract tourists by opening their unique geographies, untouched natural areas, their culture to ecotourism. The tourism named after the smokeless industry plays a great role in the economies of some countries by generating revenue more than the industry sector.

After Vietnam and Costa Rica, one of the countries we need to mention about is Turkey. Turkey has significant diversity in the context of biodiversity and natural areas. As we mentioned before, Turkey is home to a different corridor project as similar to the corridors in Costa Rica. "GAP EcoTourism Corridor", which combines the GAP Corridor with the Winter Corridor, is one of these projects. This project is one of Turkey's leading projects; especially in terms of biodiversity and ecotourist attraction potential. Besides, The Lake District Eco-Tourism Development Region, which is a private region, is a similar project to the GAP EcoTourism Corridor.

There are seven regions in Turkey. There are nine provinces in the region called South-eastern Anatolia Region (GAP). This is the first tourism planning study evaluating this region as a whole on the regional scale in Turkey. The significance of this region comes from its being known as the cradle of civilisation in the history of humanity and which is also called Upper Mesopotamia and Fertile Crescent. This region has witnessed the world history with the works left by many civilisations since prehistoric ages, and there are many natural, historical, and archaeological values important for the tourism sector. Göbeklitepe, included in the UNESCO world heritage list and called "zero points in history" is located in this region.

There are many factors in drawing tourists, both Turkey's foreign tourists and ecotourists. Turkey is surrounded by the sea on three sides and has biodiversity and sharp climate differences between regions and historical richness; thus, Turkey is privileged in tourism. Turkey, however, faces many problems because it doesn't move systematically like Costa Rica. According to data received by Turkey Statistical Institute (TSI), there is no 'nature purpose' option in surveys which is applied to tourists arriving in Turkey (Turkish Statistical Institute, 2020). Therefore, tourism directorates do not act systematically in terms of data collection. This situation causes incomplete data on the evaluation of the areas suitable for ecotourism and whether they can be evaluated within the scope of sustainable tourism. However, Vietnam and Costa Rica are very meticulous about this issue. In table 4, biodiversity and natural areas and areas with special status in the status of Turkey and with the potential for tourism activities that can be carried out in these areas are indicated.

As shown in table 4, many regions of Turkey have very important natural areas compared to other countries. Many regions in the table have a potentially applicable ecotourism project status. Issues like government incentives, government policies and some investors in the private sector prefer mass tourism are the obstacles to about Turkey's sustainable tourism, but Turkey's biggest advantages are having undiscovered destinations, the private sector raising highly qualified employees in the tourism sector, increasing local awareness about preserving natural areas and finally rich local food culture. These advantages seriously increased the incentive for tourism in Turkey in the last ten years (Yalınkılıç & Arpa, 2005).

When considering Turkey according to a Swot analysis (Table 5) in terms of the natural beauties related to sustainable tourism, the strengths of the country are being a bridge linking Asia and Europe due to its geopolitical and geographical location, and being surrounded on three sides by the sea. The touristic advantages are its climatic conditions in which four seasons are seen, natural and untouched environmental resources. It also draws attention not only with its natural beauties but also its rich history and hosting ancient Anatolian civilisations. Since the rate of the young population is high, this makes it easier for these youth to be employed in the tourism sector as a workforce.

The weaknesses of Turkey are insufficient international promotion and the inability to get enough shares from the world tourism pie, although it has rich tourism potential. The country cannot promote its potential in terms of culture tourism, nature tourism, ecotourism, and other alternative tourism types, and the tourists in desired numbers cannot be drawn.

When considering the opportunities, it can be ranked that there are many undiscovered areas, and it has rare animal and plant species in the wildlife which may attract other world countries, and there are natural spring and healing waters.

When making a situation assessment related to threats, there are important cases such as political and economic instability, the high share of other countries in the Mediterranean basin from the tourism market, insufficient tourism image of Turkey. In addition to these, insufficient development of daily tourism routes for local tourists and underdeveloped establishments, global damages given to the environment in the world and Turkey, epidemics, rapid development in mass communication, and environmental pollution.

Increasing demand for tourism activities in global tourism markets will result in positive results for Turkey in the near future. Turkey will see the value it deserves in the global markets in the next periods.

Increasing transportation and communication opportunities due to technological development, increasing demands of individuals in terms of travelling, entertaining, and resting due to the increase in income levels are constantly developing tourism in global markets (Neto, 2003). An important indicator of the continuous development of the tourism sector in global markets is the increase in the number of tourists.

Table 4. Biodiversity and natural areas (Batı Akdeniz Kalkınma Ajansı, 2012)

Activity area	City or Area	Potential Tourism Activity Field	Status or Feature
St. Paul's Way	Western Mediterranean Region	Mountain and trekking	Turkey's second-longest walking route.
Lycian Way, Mount Olympus climbing	Western Mediterranean Region	Mountain and trekking	One of the 10 routes that must be followed has been chosen.
Lake District	Western Mediterranean Region	Bird watching	
Egirdir Lake	Isparta	Bird watching	Important Bird Area, Drinking, Potable Water Protection Area and Natural Protected Area
Lake Burdur	Burdur	Bird watching	
Santa Bird Paradise	Antalya Demre	Bird watching	
Belek Boğazkent Beach	Antalya		Caretta carettas in the Mediterranean the second (after the Greek island of Zakynthos) and Turkey's largest spawning area
Konyaaltı district Çakırlar location	Antalya	Rural Tourism	Rural Tourism and Agricultural Settlement Area
Barla and Dedegöl Mountains; Isparta, Egirdir and Kovada Lake	Aegean and Mediterranean Region	Plant and Wildlife Observation and Photo Safari	Endemic plant
Karacaören Dam Lake	Burdur	Plant and Wildlife Observation and Photo Safari	Sweetgum Forests
İnsuyu, Karain, Damlataş, Dungeon and Pınargözü Caves	Mediterranean Region	Cave Tourism	
Shipwreck off the arch marina	Antalya	Underwater Diving and Water Sports	Paris Shipwreck, sunken in World War I
Canyon Reef	Northeast of Picnic Island	Underwater Diving and Water Sports	Dive Point
Köprülü Canyon, Written Canyon, Dim Stream, Manavgat Stream and Alara Tea	Aegean and Mediterranean Region	Rafting and Adrenaline Sports	
Orphan Island	Antalya Manavgat	Underwater Diving and Water Sports	Cuttlefish from the shark family
Patara Canyon	Antalya Kaş	Nature and Adrenaline Sports	Coral and sponge types
Uzungöl	Trabzon	Tourism Center	Plateau Tourism Project
Altındere Valley	Rize	Nature and Rafting	Plateau Tourism Project
Gallipoli Peninsula	Marmara Region		Sensitive Area to be Certainly Protected

Accordingly, it is possible to say that the number of tourists increased significantly in 2019 at the global level. According to UNWTO data, the number of international tourists increased by 4% in the first nine months of 2019 compared to the same period of the previous year. Regionally, there was a 9% increase in the Middle East, 5% in Asia-Pacific and Africa, 3% in Europe and 2% in America (UNWTO, 2019a). As of 2018, the total revenue of the tourism sector at the global level has increased to 1.7 trillion dollars. In this way, the international tourism sector accounts for 29% of global services exports and 7% of total exports. It is also possible to say that the growth rate of the tourism sector in recent years is higher than the growth in total exports of goods and services (UNWTO, 2019b).

The tourism sector has many economic contributions such as an increase of public and private income, expansion of employment, diversification of the economy, and development of infrastructure and superstructure investments (Lerner & Haber, 2001). Given the role of the tourism sector in the global economy, the economic, social, cultural, and environmental impacts of the sector become important. Excessive use of tourism resources and depletion of resources, degradation of natural structure, pollution and wastes resulting from increased tourism activities, excessive use of potable water resources, destruction of forests, damage to coastal and beaches and cultural heritage are environmental problems caused by the tourism sector (Ruhanen, 2015). It is possible to say that the natural areas and the environment are seriously damaged especially as a result of traditional tourism. Consequently, to tackle these problems affecting the whole ecosystem, traditional approaches should be reviewed in the context of the tourism sector (Spenceley, 2005). The tourism sector is an important sector for achieving sustainable development goals due to both its environmental impacts and economic role. Therefore, it is necessary to develop solutions in the context of the tourism sector in line with the sustainable development goals against global environmental problems. Sustainable tourism projects and application areas implemented in the section below will be evaluated, and work will be supported with advice. Besides, it will examine the relationship between renewable energy and sustainable tourism practices in this section.

Table 5. SWOT analysis (own elaboration)

Strengths	**Weaknesses**
• Being a bridge linking Asia and Europe due to its geopolitical and geographical location and being surrounded on three sides by the sea • The climatic conditions in which four seasons are seen • Natural and untouched natural resources • The infrastructure suitable for the tourism industry • Appropriate legal regulators for legal regulations • High young population which makes easier for young people to be employed in the tourism sector as a workforce • Rich historical heritage and hosting ancient Anatolian civilisations	• Insufficient international promotion and the inability to get enough shares from the world tourism pie although it has rich tourism potential • Inability to promote its potential in terms of culture tourism, nature tourism, ecotourism, and other alternative tourism types and to draw tourists in desired numbers • Insufficient government support to ecotourism investors • Inadequacies are related to transportation and communication network for areas to be opened to ecotourism.
Opportunities	**Threats**
• There are many undiscovered areas, and it has rare animal and plant species in the wildlife which may attract other world countries • There are natural spring and healing waters • Contribution of new job opportunities in rural areas to employment • The positive contribution of Turkey's having a share in the international tourism market to the national economy	• Political and economic instability in the country • The high share of other countries in the Mediterranean basin from the tourism market • Insufficient development of daily tourism routes for local tourists and underdeveloped establishments • Global damages are given to the environment in the world and Turkey, epidemics, rapid development in mass communication and environmental pollution

SUSTAINABLE TOURISM MANAGEMENT AND SAMPLE ECOTOURISM PRACTICES

Resource efficiency, which is one of the 12 basic principles of sustainable tourism, is directly related to renewable energy (UNEP & WTO, 2005). According to this, it is necessary to minimise the use of scarce and non-renewable resources in the development of tourism activities and services and within tourism enterprises. One of the basic principles of sustainable tourism must be directly related to renewable energy because the type and amount of energy used in sustainable tourism activities and destinations is an important issue. The use of renewable energy sources in sustainable tourism destinations is one of the most important topics for both to achieve the resource efficiency target and to achieve other environmentally renewable tourism targets. Ecotourism investments, which also include renewable energy resource investments; except own value of returns; also affect other sectors economically in sustainable tourism destinations. For example, "Green Transformation Strategy" has been developed which includes energy efficiency, greenhouse gas emission reduction, lower water consumption, and better waste management practices in consequence of problems with waste management and their risk of extinction of important natural resources such as coral reefs and marine animals in Spain; in one of the world's leading beach and sun destinations; as in most tourist destinations. Over 100 per cent is recycled with the strategy which is developed in many issues such as operational efficiency of hotel buildings, operational efficiency of other buildings, design of green buildings, clean water transfer, wastewater decontamination and preservation of biodiversity (UNEP & WTO, 2012).

Today, it is possible to say that the transition to renewable energy use in tourism destinations is slow since the first investment cost of renewable energy investments, which directly and indirectly affect sustainable tourism development positively, is high. In this context, cooperatives stand out as an important alternative, especially to solve financing problems in the transition to renewable energy use. Thus, cooperatives enable the propagation of ownership to the monetary base and combining small savings into a source of finance (Akkaya, 2018). It is possible to say that local and renewable energy investments; whether on a small or large scale, in the city or the countryside; make important contributions in the context of renewable energy sources.

Within the framework of sustainable tourism management, many ecotourism practices can be both encouraging for many projects and exemplary to other people and institutions in the world. Within the frame of researches, this situation has become widespread and turned into volunteer projects. Foreign tourists who travel between countries voluntarily participate in projects in sustainable ecotourism practices in the countries they travel to; thus tourists are not only having difficulty with accommodation and food but also they become conscious by socialising with many people. An example of the platforms that bring close together project owners and volunteer tourists is the Workaway Platform.

Many of the ecotourism practices are not only limited to certain areas but also include cities and even countries. For example, the city of Gijon has a "Biosphere World Urban Destination" certificate and is a leading city in the field of ecotourism. Bio Hotel Organic Suites in Colombia, Sustainable hotel model in the Dana biosphere reserve in Jordan, Eco-Luxury Resort in Brazil are the applications we have come across in recent years (Biosphere Tourism, 2015). Only 15 districts have been declared *cittaslow* cities in Turkey and it is a very positive development in terms of ecotourism. Regions that offer an ordinary life perspective have experienced a great transformation in terms of environmental, cultural, economic and socio-cultural projects with *cittaslow* projects in studies which the support of the local population, ecotourism professional and mayors in Turkey and it has provided the recognition of many natural beauties known only by the local people in the world (Turkish National Network, 2020). Although some cities have never attracted tourists, it has caused the extreme interest of both domestic and foreign tourists. We can also show examples of the Şirince village in a beautiful ecotourism practice in Turkey except for *cittaslow*. Many features make Şirince village stand out in the field of sustainable tourism. Agriculture and village markets, gastronomy, winemaking, crafts, and the House of Virgin Mary near Sirince are the parameters that make the village of Sirince special. Many projects are being developed in Şirince within the scope of sustainable and creative tourism. Improving existing tourism awareness of local people, sustaining handicrafts with classical methods, a mixed kitchen riches, historical assets, rich resources suitable for rural tourism and having advanced winemaking according to many regions of Turkey create alternative feasible projects for the Sirince village (International Vision University, 2017). However, in Şirince, sustainable construction principles should be applied within the scope of sustainable tourism. However, in Şirince, sustainable construction principles should be applied within the scope of sustainable tourism. At this point, many road works have been neglected, there are no elevators, escalators and ropeway systems in the village, there is no tourist information centre and guidance services are insufficient. Since the village does not have a parking system, it becomes quite complicated during busy visits (Koca, 2019).

One of the countries that we can show as an example of ecotourism projects in Mexico. Mexico, located on the land where the Mayan civilisation, an important civilisation, was founded and destroyed has successfully included Mayan culture and civilisation into sustainable tourism management within the framework of ecotourism. Though Mexico is a politically and economically turbulent country, thousands of tourists travel to Mexico every year but applications of ecotourism which approaching mass tourism over time have become capable of harming large and important ecosystems such as evergreen forests, coniferous forests and cloud forests (Valle-García, 2014). At this point, many countries have similar problems, not only Mexico. Besides, serious damages can be caused directly to important ecosystems except for ecotourism. This issue is another point that sustainable tourism attracts attention. Many factors such as unconsciousness, government policies, seeking for economic power with resources other than tourism, mines, seconds, road projects, poaching, unplanned urbanisation prevent the formation of sustainable tourism projects. Destruction of rainforests in Malaysia and Indonesia for palm oil and endangering the orangutan generation contain serious dangers for both ecosystems and the future of the world.

The palm oil industry has many positive and negative effects as socio-economically and culturally. Although it is a strong industry, the damage which it creates on nature is quite high. The benefits of the palm oil industry for local communities and contribute to country economies an also, its cultivation and development of palm oil are effective in many areas to make this industry attractive. However, in countries such as Indonesia, Malaysia, Thailand, forests get harmed greatly, especially in uncontrolled and unauthorised works. For this reason, many regions that are very important for sustainable tourism are also damaged (Sunarminto, 2019).

Barrage which will be held on the historic town of Hasankeyf in Turkey, which has thousands of years of culture is one of the biggest losses in the framework of sustainable ecotourism projects. Although there was a reaction from many organisations around the world, not only Turkey Turkey entered into a construction project due to political power. This makes it mandatory that there is a certain authority in the world concerning sustainable tourism and that both cultural heritage sites and ecosystems are protected.

Ecotourism can provide much-needed revenues to protect national parks and other natural area-based facilities that may not be available from other sources. Also, ecotourism can provide a viable economic alternative to development for local communities with few other income-generating options. In addition, ecotourism can increase education and activity among travellers, making them more enthusiastic and effective conservation agents.

CONCLUSION

Within the framework of this study, different strategies and methods for sustainable tourism are evaluated with examples and alternative areas were evaluated for ecotourism and its applications in the context of international standard principles. We can say that sustainable tourism has many exemplary practices in the world and is still developing rapidly. We can list the main reasons for the rapid spread of sustainable tourism in the world as it has an economically strong effect and many countries both want to protect their nature and ecosystem and due to the rapidly increasing natural disasters want to convert it into an economic power. As mentioned in the background of the study, experimenting with different strategies and methods for sustainable tourism has established international standard principles. These standard principles that feature ecological tourism and nature protection were created to seek alternative solutions to the perception of classical tourism at the first stage and later have been designed with the support of the necessary institutions and people and projects have been successful in practice. Within the framework of this study, many practical sustainable tourism projects have been evaluated starting from the theoretical framework, and it has been decided that the necessity of producing alternative solutions. Developing flexible and idealistic policies and creating projects in a way not to harm ecosystems within the scope of ecotourism will be a continuous, inclusive, and permanent employment area for sustainable tourism. Mass tourism can be prevented in the world by promoting sustainable tourism in many regions. It can be advantageous in many ways, but an international authority organisation is required like UNESCO.

When we think about tourism in the context of sustainability, we can see that while Cittaslow city projects are designed technology is used effectively. Renewable energy sources are also used and the urban fabrics are protected contrary to classical tourism approaches. This perception of preservation includes ecological areas and nature.

Many countries in the world have worked hard to protect their ecological areas and natural resources due to natural disasters and human-made disasters. These countries have made their ecological areas and resources useful for tourism by constituting the necessary legal protection status, but from a legal protection status point of view, many aspects such as access to areas, access to clothing, materials, equipment, maximum durations have strict control. Getting support from local people in such regions is a very important issue. Besides, the protection of tourist diversity is just as important as local public support. Sustainable, inclusive, and sustainable economic growth and encouraging sustainable projects with permanent employment can be created for everyone living in the region through flexible and idealistic policies.

If we end our work with good examples from the world, we can show an example of unique projects from many parts of the world. Inkaterra Region in Peru would be a good example for this study. One of the first eco-friendly hotels in the world was built On the land purchased by José Koechlin von Stein in 1975. Inkaterra Hotel appealed to a wide audience thanks to its vision. The ecotourism movement, which has a structure integrated with the war and anti-government actions in the world is quite developed with such projects. The hotel, which has developed over time, has started to shape its surroundings. effective hotel and vision in the Amazon rainforest, an important tourism destination like Machu Picchu, cloud forests, in the Sacred Valley of the Incas, and the city of Cusco region led the opening of many different businesses (Responsible Travel to Peru, 2017).

Tunisia and Morocco have improved quite well in tourism in the past ten years. These two countries that present their culture and geography to the world in different frameworks are highly developed in favour of tourism. Especially Morocco hosts many projects within the framework of desert safari, different communities, and cultures, and interaction of life and unique culture in alternative cities with the West However, the fact that mass tourism enters these countries has affected many alternative sustainable development projects negatively. For example, Kasbah du Toubkal is a typical shack right in the middle of a local Berber community in the Atlas Mountains, located within the borders of Morocco. The shack, which started operating in 1995, carried out many activities to protect the Imlil valley, which is also within the borders of Morocco, from mass tourism. The foundation and its structure, which adopts the Berber culture and lifestyle, alternatively opposes mass tourism and struggles to keep its culture alive (Responsible travel to Morocco, 2017). Tunisia also hosts many projects such as Morocco. For example, Chikly Island, Ickheul National Park, Boukornine National Park are projects that lead to Tunisia and have very positive developments for ecotourism (UNESCO World Heritage Sustainable Tourism, 2020). Ickheul National Park has struggled with dams for years. Many studies have been carried out because the dams cut off the entire flow of freshwater and disrupt the natural flora. At this point, The Ichkeul National Park Management Committee and the Ichkeul Agricultural Development Group are one of the few commissions involved in the study. Over time, Unesco stepped in, and commissions and committees that received a certain amount of funds enabled local businesses and communities to participate in tourism activities, so they saved a considerable flora from both extinction and mass tourism.

One of the main reasons for the example of Ichkeul National Park in Tunisia at the end of the study is that can be a model to countries such as Turkey, Brazil, Indonesia, Mexico, and Malaysia and shows that very nice projects can be created by struggling over time.

To conclude, it can be urged that studies should be carried out by individuals, organisations, and governments to identify the components of tourism and to design assessment and implementation methods of these components for developing sustainable tourism management. In these studies, national and international strategies should be determined, and these strategies should be implemented in a more widespread extent in various regions of the world. The awareness of people on nature protection, showing respect to nature, and protecting nature socially should be raised. It should be the duty of all people to leave a clean world for future generations.

REFERENCES

Aall, C. (2014). Sustainable tourism in practice: Promoting or perverting the quest for a sustainable development? *Sustainability*, *6*(5), 2562–2583. doi:10.3390u6052562

Aldırmaz Akkaya, F. (2018). *Türkiye'de kadın kooperatiflerinin sürdürülebilir kırsal turizmdeki önemi ve finansal performanslarının analizi*. Akademisyen Kitabevi.

Bansal, S. P., & Kumar, J. (2011). Ecotourism for community development: A stakeholder's perspective in Great Himalayan National Park. *International Journal of Social Ecology and Sustainable Development*, *2*(2), 31–40. doi:10.4018/jsesd.2011040103

Batı Akdeniz Kalkınma Ajansı, B. A. K. A. (2012). Ekoturizm sektör raporu. Academic Press.

Ben, M. (2005). *Sustainability: Definition and five core principles a new framework.* Sustainability Laboratory Publication.

Biosphere Tourism. (2015). Retrieved from: https://www.biospheretourism.com/en/blog/17-sustainable-tourism-examples-for-this-2017/2

Buckley, R. (1994). A framework for ecotourism. *Annals of Tourism Research, 21*(3), 661–665. doi:10.1016/0160-7383(94)90126-0

Ceylan, Y. (2019). Sürdürülebilir Turizm Kapsamında Turizmde Eko Eko Etiketler. *Turizm Ekonomi ve İşletme Dergisi, 1*(1), 65–80.

Cortese, A. D., & Rowe, D. (2017). Sustainability strategic plan. In *Sustainable Haverford.* Haverford College, Council on Sustainability and Social Responsibility Final Draft.

Dien, N. X., Hoa, H. T., Huong, N. T. T., Thuy, L. T., Nga, N. T. Q., & Nguyen, P. T. (2019). Sustainable development of Vietnam's Marine Tourism. *Environmental Management and Sustainable Development, 8*(4).

Drexhage, J., & Murphy, D. (2010). Sustainable development: From Brundtland to Rio 2012. Background paper for the high-level panel on global sustainability. New York: United Nations.

Giovannoni, E., & Fabietti, G. (2010). *What is sustainability? A review of the concept and ıts applications. Springer International Publishing Switzerland.* Department of Business and Law, University of Siena.

Gomiero, T. (2016). Soil degradation, land scarcity and food security: Reviewing a complex challenge. *Sustainability, 8*(3), 281. doi:10.3390u8030281

Hawkins, D. E., & Lamoureux, K. (2001). Global growth and magnitude of ecotourism. In D. B. Weaver (Ed.), *The encyclopedia of ecotourism.* CABI International.

ILO. (2017). *Rapid market assessment of responsible tourism ın Vietnam.* International Labour Organization.

International Vision University. (2017). Yaratıcı Turizm Kapsamında Şirince Köyü İçin Bir Uygulama Modeli Önerisi ve Köyün Yaratıcı Turizm Kaynakları. In *Mediterranean International Congress on Social Sciences MECAS II.* International Vision University Ohrid.

Kasalak, M. A. (2015). Dünya'da Ekoturizm Pazarı ve Ekoturizm'in Ülke Gelirlerine Katkıları. *Journal of Recreation and Tourism Research, 2*(2), 20–26.

Klarin, T. (2018). The concept of sustainable development: From its beginning to the contemporary issues. *Zagreb International Review of Economics & Business, 21*(1), 67–94. doi:10.2478/zireb-2018-0005

Koca, G. (2019). Evaluation of traditional Şirince Houses According to sustainable construction principles. *International Journal of Architecture & Planning, 7*(1), 30–49. doi:10.15320/ICONARP.2019.65

Kotob, F. (2000). *What is sustainability?* University of Wollongong, Faculty of Business.

Lerner, M., & Haber, S. (2001). Performance factors of small tourism ventures: The interface of tourism entrepreneurship and the environment. *Journal of Business Venturing*, *16*(1), 77–100. doi:10.1016/S0883-9026(99)00038-5

MacNall, S. G., Lang, D. Q., & Sobieszcyk, T. (2016). Ecotourism in Costa Rica and Vietnam: Is ıt sustainable? *Sustainability: The Journal of Record*, *9*(3), 144–154. doi:10.1089/SUS.2016.29058.sgm

Martinez, J. M., Martin, J. M., Fernandez, J. A., & Mogorron-Guerrero, H. (2019). An analysis of the stability of rural tourism as a desired condition for sustainable tourism. *Journal of Business Research*, *100*, 165–174. doi:10.1016/j.jbusres.2019.03.033

Nail, J., & Wearing, S. (1999). *Reed Educational and Professional Publishing Ltd.* Linacre House.

Neto, F. (2003). A new approach to sustainable tourism development: Moving beyond environmental protection. *Natural Resources Forum*, *27*(3), 212–222. doi:10.1111/1477-8947.00056

Responsible travel to Morocco and Peru. (2017). Retrieved from: https://www.wanderlust.co.uk/content/sustainable-tourist-projects-maldives-morocco-thailand/

Ruhanen, L., Weiler, B., Moyle, B., & McLennan, C. (2015). Trends and patterns in sustainable tourism research: A 25-year bibliometric analysis. *Journal of Sustainable Tourism*, *23*(4), 517–535. doi:10.1080/09669582.2014.978790

Ryan, G. W., & Bernard, H. R. (2000). Data management and analysis methods. In *Handbook of Qualitative Research* (2nd ed., pp. 769–802). Sage Publications, Inc.

Salda Gölü Rehberi- Türkiye'nin Maldivleri. (2020). Retrieved from: http://www.bizevdeyokuz.com./SaldaGölüRehberi/Türkiye'ninMaldivleri

Shedenov, U., Litvishko, O., Kazbekov, B., Suyunchaliyeva, M., & Kazbekova, K. (2019). Improvement of ecological tourism on the principles of sustainable economic development. *E3S Web of Conferences*, *135*, 04047. Doi:10.1051/e3sconf/201913504047

Solmaz, S. A. (2019). Ekim). Sürdürülebilir Turizm ve Türkiye'de Çevre Politikaları. *Uluslararası Sosyal Araştırmalar Dergisi*, *12*(66), 1176–1186.

Spenceley, A. (2005). Nature-based tourism and environmental sustainability in South Africa. *Journal of Sustainable Tourism*, *13*(2), 136–168. doi:10.1080/09669580508668483

Sunarminto, T., Mijiarto, J., & Prabowo, E. D. (2019). Socioeconomic and cultural impacts of oil palm plantation development in Indonesia. *IOP Conf. Series: Earth and Environmental Science, 336*. 10.1088/1755-1315/336/1/012008

Turkish National Network. (2020). *Cittaslow Turkey*. Retrieved from: https://www.cittaslow.org/network/turkish-national-network

Turkish Statistical Institute. (2020). *Tourism Statistics*. Retrieved from: http://www.tuik.gov.tr/PreTablo.do?altid=1072

UNEP & WTO. (1993). *Sustainable Development Goals*. Retrieved from: https://www.unwto.org/sustainable-development

UNEP & WTO. (2005). *Making tourism more sustainable: A guide for policymakers*. Retrieved from: https://www.unwto.org/sustainable-development

UNEP&WTO. (2012). *Tourism in the green economy: Background report*. Retrieved from: https://www.unwto.org/sustainable-development

UNESCO. (2020). *World Heritage Sustainable Tourism Toolkit*. Retrieved from: http://whc.unesco.org/sustainabletourismtoolkit/guide-2---case-study-ichkeul-national-park-tunisia

United Nations. (2020). Retrieved from: https://www.un.org/wcm/webdav/site/climatechange/shared/gsp/docs/GSP1-6_Background%20on%20Sustainable%20Devt.pdf

United Nations General Assembly. (2015). Transforming our world: The 2030 Agenda for sustainable development. Resolution 15-16301. United Nations.

UNWTO. (2019a). *UNWTO world tourism barometer* (nov 2019). *UNWTO World Tourism Barometer, 17*(8), 1–44.

UNWTO. (2019b). *International tourism highlights* (2019 edition). World Tourism Organization.

Valle-García, S. E. (2014). Ecotourism: Sustainable indigenous policies and its effects in Mayan communities in Southern Mexico. *WIT Transactions on Ecology and The Environment*, 187.

Wallace, G. N., & Pierce, S. M. (1996). An evaluation of ecotourism in Amazonas, Brazil. *Annals of Tourism Research, 23*(4), 843–873. doi:10.1016/0160-7383(96)00009-6

Wood, S., Jones, K., & Brauman K. (2018). Distilling the role of ecosystem services in the Sustainable Development Goals. *Ecosystem Services, 29*(Part A), 70-82. Doi:10.1016/j.ecoser.2017.10.010

Yalınkılıç, M. K., & Arpa NY. (2005). *Türkiye'deki Korunan Alanlar ve Ekoturizm*. Korunan Doğal Alanlar Sempozyumu Sözlü Bildiriler Kitabı, 3-13, Isparta. Yeşil P. 2009. Kelkit Havzası Tokat Kesiminin Biyosfer Rezervi.

Yenil, Ü., & Akyazı, N. (2019). Sürdürülebilir Turizm Kapsamında Kültürel İz Koridoru Değerlendirmesi. *Niğde Ömer Halis Demir Üniversitesi Mühendislik Bilimleri Dergisi, 8*(2), 1179–1192.

Zamora, N., & Obando, V. (2001). *Biodiversity and tourism in Costa Rica*. Instituto Costarricense de Turismo, ICT; Sistema Nacional de Áreas de Conservación, SINAC; Ministerio del Ambiente y Energía, MINAE; Instituto Nacional de Biodiversidad, INBio.

Chapter 6
Ecotourism as a Tool for Sustainable Development in Morni Hills (Haryana), India

Lakhvinder Singh
Government College, Kaithal, India

ABSTRACT

Ecotourism is an emerging approach to attain sustainable development. The Morn Hills region of Haryana in India is emerging as a famous ecotourism destination, as it is endowed with rich and varied vegetation, flora and fauna, and forest area. The chapter empirically investigates the local community attitude towards ecotourism as a mean for sustainable development. The chapter used 200 data responses conveniently collected from respondents through a survey form and used suitable statistical tools to analyse the result. The study indicates huge prospects of ecotourism in the area and variation in the attitude of the local community towards the role of ecotourism in sustainable development. A highly positive attitude of locals as ecotourism brings sustainable and overall development in the area in the form of enhanced economic benefits, enrichment of socio-cultural values, increased concerns, and preservation of environmental and physical resources. The chapter concludes different stakeholders to come up with appropriate strategies for sustainable development of the area.

INTRODUCTION

Tourism is the emerging sector in the global marketplace, and ecotourism plays a crucial role in the development of tourism (Fennell & Dowling, 2003). Ecotourism is majorly based on natural tourist resources of the destination and involves a visit to uninterrupted natural areas with an aim to learn about the unique experience of local communities by providing environmental, economic and socio-cultural paybacks to local people of the destination without causing any negative impact (Lai & Nepal, 2005; Gheorge, 2013). It is also further elaborated that in addition to less impact on the natural environment of destination, ecotourism involves local people in tourism to extends its benefits among locals to a large extent and enhance the tourist experience.

DOI: 10.4018/978-1-7998-5691-7.ch006

Ecotourism is a sub-component of the field of sustainable tourism. Ecotourism's perceived potential as an effective tool for sustainable development is the main reason why developing countries are now embracing it and including it in their economic development and conservation strategies. Ecotourism, as alternative tourism, involves visiting natural areas in order to learn, to study, or to carry out activities environmentally friendly, that is, tourism based on the nature experience, which enables the economic and social development of the local community (Zhong, & Liu, 2017).

The characteristics of ecotourism suitably fit into the sustainable mechanism as it helps to bring overall development of the area including all aspects of development, such as the welfare of locals, promote local business firms, increase membership of locals in tourism planning and management, providing entrepreneurial prospects and facilitate the education and training to local communities (Theingthae, 2017). Hence, a number of emerging tourist destinations is considering this form of tourism as a major segment of their tourism market and placed a lot of emphases to tap its natural resources, improve the social and economic conditions of the residents of the tourist places.

Some studies supposed that although ecotourism helps to achieve sustainable development, it also contributed towards the degradation of the environment to a extend due to poor management of eco-tourism activities (Li, 2004). Ecotourism helps in community development by providing the alternate source of livelihood to the local community, which is more sustainable. Its aim is to conserve resources, especially biological diversity, and maintain sustainable use of resources, which can bring ecological experience to travellers, conserve the ecological environment, and gain economic benefit (Bashar, 2018). However, achieving the aims in ecotourism depends on whether they are environmentally and ecologically sustainable and economically applicable. In order for ecotourism to encourage patterns of sustainability, which can benefit local communities, protect the environment, and be economically viable, it must be comprehensive and account for the complexity of issues that have been mentioned in this chapter. Therefore, it is essential to carefully evaluate the dimensions that make ecotourism a sustainable development tool. The sustainable development approach also contributed towards the major global agenda (MDG) focuses on maintaining eco-balances in developmental work (Chandel & Mishra, 2016). Hence the present chapter considered this fact and aimed to study the attitude of locals towards the role of ecotourism in the sustainable development of Morni hills region in Haryana.

Morni Hills is a hill resort in the Panchkula District of Haryana. This hill is an offshoot of the Shivalik mountain ranges of the Himalayas. Morni derived its name from that of a queen who ruled the city. The village is on the mountainside at an altitude of 12,20 metres overlooking the panoramic view of the Himalayas. There are two lakes and the river Ghaggar meanders through the mountain ranges. These lakes seem to be interlinked since the level of water remains identical in both the lakes, though they are separated by a mountain. These lakes are considered sacred by the local populace.

Morni is a sleepy village with a general rhythm where people move about on their daily occupation like farming on the terraced fields or tending to their animals. It is believed that it is an ancient place as fossils of certain extinct animals were found during the period 1831 to 1843 while digging the Daob Canal. The fossils were found by the team led by Captain Cautley. It is stated that Morni was the bed of a sea and the soil was pushed up to make it a firmer ground when the sea receded. Going by the excavations and the artefacts dug up from the ancient Thakur Dwara Temple, it is said to have been in existence in the 10th Century AD. A lot of artefacts consisting of Brahminical sculptures from the Prathihara period which include Kartikkeya, Surya, Siva, Uma-Maheswara, Ganesha, Lakulisa, Nataraja, Trimurthi and Siva Lings have been excavated.

The Meers of Kotaha ruled over this place from Kotaha, existing a small fort in Morni. This old fort is in ruins, and the locals believe that it belonged to Queen Morni. The fort had been built sometime in the 17th century at the time when the Meers of Kotaha ruled from Kotaha. It came under the Raja of Sirmur in 1814 when the Gurkhas attacked his kingdom. Later the fort was taken over by the Gorkha commander, GoureeSah until its surrender in 1815 to the British.

There are three human-made lakes: Tikkar Tal, Bada Tikkar and ChhottaTikkar. There is also an old temple on the banks of the lake – Temple of the Holy Trinity –, considered to be built in the 12th century. It is believed that a Shiva temple was there at this location, previously. An important Gurudwara for the Sikhs also located in Morni hills region of Haryana (Gurudwara Nada Sahib). The impressive mud cliff, i.e. Mandhana, from where the forests and the occasional nullahs can also be seen in the study region. A human-made park made by the government is between the twin lakes of Tikkar (The Adventure Park). This park offers several amusements for the kids. Near the park is a waterfall at Thalapur.

Adventure tourism activities can also be enjoyed by tourists in Morni hills: trekking, hiking, hang gliding, and rock climbing are popular activities in the hills. February and March are best suited for trekking and climbing hills. Morni Hills are also game for wild nature enthusiasts. It is a place for the birdwatchers as the hills are abode to crested kingfisher, bar-tailed tree creepers, oriental turtle doves, Himalayan bulbul, red jungle fowl, black francolin, wallcreepers and quails. The Beruwala check post is one place for the bird enthusiasts to watch for migratory birds from Siberia, Afghanistan, and Europe. The best time for bird watching is in the months from September to December. Water sports like boating, canoeing, kayaking, and sailing are also present in the lakes.

The hill stations have accommodations provided by the Haryana Government, the Forest Department, and a rest house by the Public Works Department. There are dormitory accommodations provided by Haryana Tourism Corporation (HTC) at Tikkar Tal. The best season to visit Morni Hills is between September and March as the climate will be pleasant at that time. Summer is from April to June and is hot and humid. In July and August, there are intermittent rains and winters are cool and pleasant. There are roads to all major towns in Haryana. Government buses and taxies ply these roads. The distance between Chandigarh and Morni Hills is 42 km, Shimla is 152 km, and Delhi is 276 km by road. The nearest railway station is at Chandigarh. Chandigarh has the nearest airport.

THEORETICAL BACKGROUND

Tourism is a highly complex activity and thus requires tools to assist in effective decision making to come to terms with the competing economic, social, and environmental demands of sustainable development (Fadahunsi, 2011). The principle of sustainable tourism should include indicators of social and economic sustainability, adding up to indicators of environmental sustainability. In addition, Wall (1997) has argued that for tourism to contribute to sustainable development, it must be economically viable, environmentally sensitive, and culturally appropriate, and the forms that this might take are likely to vary with location.

The concept of sustainable development refers to meeting the present needs without comprising the future generation needs (Tosun, 2001) with an aim to develop in all spheres such as provides opportunities for job creation and economic development, enhances social progress and strengthens communities, encourages the protection of the environment and contributes actively to conservation. The global tourism agencies (UNWTO, WTTC) in association with different global level campaigns (UNEP) also working in this area to ensure the optimum utilisation of different resources to achieve sustainable development (Niñerola, Sánchez-Rebull, & Hernández-Lara, 2019).

As far as the Indian government (GoI) is concerned, it also focuses on attaining sustainable development. The Ministry of Tourism of the country (MoT) also farmed sustainable guidelines for the different tourism & hospitality enterprises in the country to achieve this major aim of sustainable development goal (SDG).

This concept in ecotourism is considered as a tool to focus on economic, socio-cultural, and ecological sustainability (Loach et al., 2017). Additionally, the sustainable development approach also focuses on safety and security measures at the destination, unique experience. The recent research appears that local communities form an integral part of the tourism development agenda (Jamal & Stronza, 2009) as they were closely related with tourism industry at destination such as indulging tourism services such as hotelier, drivers, waiters, escort & guides, porters, photographers etc. Hence, it was also noted that the involvement of residents in tourism planning and management nurturing the local residents and helps to attain sustainable development goal (Dyer et al., 2007). Furthermore, the ecotourism must support the development of educational and training opportunities, increasing awareness regarding maintenance of the ecosystem of the destination to sustainably develop the destination. In addition, Wall (1997) has argued that for tourism to contribute to sustainable development, it must be economically viable, environmentally sensitive, and culturally appropriate, and the forms that this might take are likely to vary with location. The following table 1 develops sustainability goals of ecotourism.

Table 1. Three system of sustainability in ecotourism development (Wall,1997)

Sustainability Dimensions of Ecotourism	Indicators
Environmental sustainability	ü Promotes environmental protection ü Impact assessment and environmental planning ü Waste minimisation and litter drainage control ü Water & energy conservation ü Minimisation of the transport effect ü Minimum impact on wildlife ü Provides environmental education ü Increases public environmental consciousness ü Fosters healthy attitudes/behaviours to nature ü Encourages donations to contribute to the ü Protection of local natural resources
Socio-cultural sustainability	ü Promotes local people's active participation ü Promotes local ownership ü Empowers local people -e.g. builds up local people's confidence/self-esteem ü Enhances local community's equilibrium ü Encourages intercultural appreciation ü communication between host communities and tourists
Economic sustainability	ü Contributes to local economic development ü Creates permanent jobs for local people ü Drives the development of related industries ü Upgrades local infrastructure ü Profits earned retained within local community ü Equal distribution of revenues ü Promotes consumption and production ü Finances the establishment and maintenance of protected areas ü Uses natural resources efficiently

Thomas et al. (2014) in a study on ecotourism noticed a variation in the opinion of locals towards the core components such as environmental, socio-cultural, and economic aspects are essential to consider while measuring the impact of ecotourism in sustainable development. Barzekar et al. (2011) discovered the significant role of ecotourism in sustainable development through different identified components and their indicators such as twenty-one indicators of environment and ecology, eight indicators of economic development, 27 indictors of socio and cultural component and five for the institutional component of ecotourism destination. Aziz et al. (2015) based on identified dimensions and indicators in the field of ecotourism noted variation in the attitude of respondents towards all the identified indicators. Theingthae (2017) applied seven dimensions including environment, socio-culture, heritage, economy, promotion, religion-based aspects of the tourist places and sub-indicators to measure the role of ecotourism in sustainable development and study noted significant variation in the attitude of locals towards these dimensions. However, the study noted the poor attitude of locals towards the role of ecotourism in economic sustainability.

The extensive review of the literature observed the understanding of the locals' attitudes toward ecotourism is valuable for formulating sustainable tourism planning, as well as it was also noted that no specific study had been conducted in the area covered by researchers which indicate a research gap in the research line. Therefore, the present researchers attempted to fill this research gap by conducting a meaningful study focusing on the measurement of the attitude of locals towards ecotourism as a tool for sustainable development in Morni Hills area of Haryana in India.

RESEARCH APPROACH

The present chapter is confined to Morni hills located in the Haryana in the northern part of the country. It is a famous ecotourism place in the area which spread over Shivalik Hills with scenic landscapes, high hills, deep valleys, beautiful lakes and forests, and long stretched green farmlands attract the attention of the visitors and make this land a must-visit site. The hills have an extensive forest area and variety of flora and vegetation. Ecotourism activities in Morni hills include Bird watching, Bird Nests, collection of information in species, wildlife and nature tourism and other non-forestry activities. Some ecotourism nature camps have also been built by the Forest Department of the state. The famous camps are nature camp at Thapli (Panchkula), nature camp at Morni fort, nature camp at Chuharpur (Yamunanagar) with reserve forest and herbal garden, nature camp at Ban Santaur (Yamunanagar), natural camp at Kalesar (Yamunanagar) and Kalesar National Park (Yamunanagar) are few of them. The entire region from last decade experiences the large tourist footfalls. But, despite, its huge potential for ecotourism and its importance from tourism business point of view, no study has been conducted so far to evaluate the role of ecotourism in sustainable development in Morni hills.

The author of present chapter majorly used primary data sources as well as, to some extent, previous research work, used to provide a theoretical framework for the study and design to a structured questionnaire based on five-point Likert scale. The questionnaire was distributed to locals residing near ecotourism sites in the Morni Hills region of Haryana. The questionnaire was distributed to respondents at ten major tourist destinations in Morni region such as Morni Fort Museum, GarhiKotaha Fort, Masoompur Fort, Tikkar Tal, ChotaTikkar Tal, Morni Waterfall, Thakurdwar Temple, Bhuri Singh God Temple, Herbal Forest, and Moutainquail Tourist Resort. The researcher contacted 350 respondents covering 35 at each selected tourists' sites. But out total contacted respondents, only 200 completely filled up the questionnaire. Moreover, the selection of 200 respondents also in line with recommendations of previous studies (Canadian Tourism Commission, 2007), which indicate that to generate inputs for a tourism impact assessment is recommended to survey a minimum of 200 respondents (Canadian Tourism Commission, 2007). The questionnaire was developed in two different sections. The first section used to collect background and demographic information of respondents. The section of the questionnaire was developed to identify the attitude of locals regarding the role played by ecotourism in achieving sustainability in the area. The section was based on the rating of the respondents measured through using a five-point Likert scale ranging from strongly disagree (1) to strongly agree (5). The author used to visit study region repeatedly and to contact the locals and other stakeholders to get more insight of ecotourism in the area and to frame the statements of the questionnaire.

Regarding data collection for the chapter, the author carried out filed survey from April 2017 to August 2018 with an aim to remove the seasonal bias from the survey responses of locals. The author adopted convenient sampling technique to contact the local people and distributed questionnaire forms to 350 locals, but out of them only 200 usable and completely filled in questionnaire returned and used for the empirical findings of the chapter. The collected data was properly coded and entered in the data entry sheet of statistical package for social science (SPSS). After coding and categorisation of collected data, a suitable statistical tool was carried by using SPSS. After entering values of different statements of the questionnaire into SPSS data entry sheets, the author applied factor analysis method to identify the important dimensions from raw data of questionnaire and remove the unimportant and insignificant variables with poor values.

The chapter also carried out the measurement of reliability of the questionnaire through the value of Cronbach's alpha analysis. The items with a value of more than 0.80 were used that were more than the threshold level of alpha value (0.70) as suggested by earlier research studies (Nunnally & Berstein,1994). Some variable items of the questionnaire were dropped due to less reliability score and eigenvalues. Therefore, the remaining variable items were considered for further identification of the Kaiser-Meyer-Olkin Measure of Sample Adequacy (KMO Measure) sample adequacy and Bartlett's test of result to ensure the appropriateness of factor analysis. The calculated KMO value (0.891) and Bartlett's test value (p-value =0.000) supported that all the variable items were suitable to conduct the factor analysis. Finally, the author used descriptive tools such as frequency distribution, percentage analysis, arithmetic mean, and standard deviation. The result of the study was presented through a table to make it more understandable.

MAIN FOCUS OF THE CHAPTER

Regarding the characteristics of the respondents, it was noticed that gender was almost balanced with males and female counterparts accounting almost equal number of respondents. One-third of the respondents belong to the senior age group, followed by more than 25% of the young age group (21-40 years). Only one-fifth of the respondents were in the teenage group (up to 20 years). The majority of the respondents (65.93%) have at least a graduate or above degree, and the remaining have concluded up to the senior secondary level. Most of the respondents (43%) were indulged in accommodation related business activities in the area. Almost one-fifth (24%) of the respondents indulged transportation-related business activities, and remaining were in retail, food stalls, restaurants, tea/coffee shops and other activities. This indicated that locals have sufficiently involved in tourism-related business activities to earn their livelihood. The majority of the respondents were working in tourism-related entrepreneurial activities from more than five years. Respondents on their income indicated only two-fifth (40%) earned high income due to tourism activities, whereas the remaining were indicated themselves in the middle- or low-income group. This brings to notice that the concerned stakeholders need to incorporate some measures to raise awareness and training to locals to ripe maximum economic leverages from ecotourism activities in Morni Hills.

For identification of different dimensions of sustainable ecotourism, an exploratory factor analysis (Table 1) was performed. The study conducted three factor analyses for identification of factors of different dimensions of sustainable development, including economic dimension, socio-culture dimension, and environmental dimension. Regarding its all the items of three dimensions, the measurement of reliability of the questionnaire items the value of Cronbach alpha indicated good reliability of the measurement scale (with p-value of more than 0.8) which is more than the threshold level of alpha value (0.70) by some researchers (such as Nunnally & Berstein, 1994). Some variable items of the questionnaire were dropped due to less reliability score and eigenvalues. Therefore, the remaining variable items were considered for further identification of the Kaiser-Meyer-Olkin Measure of Sample Adequacy (KMO Measure) sample adequacy and Bartlett's test of result to ensure the appropriateness of factor analysis. The calculated KMO value (0.891) and Bartlett's test value (p-value =0.000) supported that all the variable items were suitable to conduct the factor analysis.

Table 2. Dimensions of ecotourism as a tool for sustainable ecotourism (Author's Creation)

Dimension	No. of Items	Factor Loading	Cumulative Percent of Variance	α Value
Economic Dimension				
Infrastructure Development	7	0.732-0.550	20.816	0.881
Poverty Alleviation	6	0.718-0.510	26.577	0.870
Living Cost and Dependency	6	0.712-0.494	32.112	0.861
Employment and Entrepreneurship	6	0.698-0.455	48.271	0.839
Socio-Cultural Dimension				
Social Value and Lifestyle	8	0.790-0.603	8.273	0.905
Participation of Women in Society	8	0.759-0.453	14.960	0.890
Involvement of Locals in Ecotourism Planning	6	0.709-0.465	37.587	0.853
Cultural Awareness and Preservation	4	0.594-0.461	62.182	0.801
Environment Dimension				
Pollution and Disease	6	0.706-04376	43.010	0.846
Environment and Community	5	0.621-0.423	53.054	0.833
Natural Resources and Agriculture	5	0.617-0.403	57.632	0.828

For further analysis of local residents attitude towards ecotourism as a tool for sustainable development, the study used Best's (1977) translation level of opinion which rated the items of mean score .00-1.80 as strongly disagree, mean score 1.81-2.60 as disagree, mean score 2.61-3.40 as neither agree nor disagree, mean score 3.41-4.20 as agree and mean score 4.20- 5.00 as strongly agree with the items.

So, the respondents' level of expectation is considered as very low, low, moderate, high and very high in the same order.

The locals' attitude towards infrastructure development indicator and poverty alleviation indicator economic sustainability dimension were considered as a moderate role in sustainable development. Regarding living cost and dependency indicator of sustainable development, the attitude of locals indicated partial consideration of ecotourism role in sustainable development as three items considered ecotourism is important for sustainable development to a large extent as well as three indicated a moderate extent. The impact of ecotourism on employment and entrepreneurship indicated a high impact, and locals also considered this as important to a large extent to achieve sustainable development in the area.

Table 3. Locals attitudes towards ecotourism as a tool for economic sustainability (Author's Creation)

Dimensions	Items	Mean	SD.
Infrastructure Development	It provides more golf courses, herbal park for recreation of the locals in the area	3.41	.636
	It stimulates the local traffic and transport management	3.26	.726
	It stimulates the improvement of public utilities such as drinking water supply, electricity supply and communication services	3.27	.770
	It brings the development of medical and healthcare facilities in the area	3.21	.713
	It encourages the improvement in banking and ATM services in the area	3.06	.799
	It encourages the establishment of more educational institutes in the area	3.16	.811
	Safety and Security arrangements are improved due to ecotourism in the area	3.06	.715
Poverty Alleviation	Ecotourism contributes to local economic development	3.02	.788
	Ecotourism enhances the sale of handicrafts of the local area	3.08	.778
	It helps to development increases the personal income of locals	3.18	.723
	Support the sale of local agricultural products in the area	3.07	.772
	Encourage tourists to consume local products and services such as farmhouse, homestay, local guide etc.	3.09	.588
	The money spent by tourists at ecotourism sites remains in my community	3.12	.678
Living Cost and Dependency	Ecotourism is beneficial only to a few people in the local area	3.26	.674
	Ecotourism draws outside labour/worker who refrains locals from earning from tourism-related work	3.32	.632
	Ecotourism leads to the increased price of goods and services in the area	3.50	.658
	Ecotourism causes prices increase of land and housing in the area	4.05	.758
	Ecotourism brings only seasonal employment to locals which increased risk of unemployment	3.27	.778
	Ecotourism development brings additional infrastructure cost (water, sewer, power, medical, banking) development	3.47	.741
Employment and Entrepreneurship	Ecotourism created jobs for locals	3.99	.687
	Ecotourism employs local youth	3.61	.636
	Ecotourism increase the business of local industries	3.56	.726
	Ecotourism attracts more people towards small business	3.74	.770
	Ecotourism attracts investment and development projects	4.14	.713

Table 4. Locals attitudes towards ecotourism as a tool for socio-culture sustainability (Author's creation)

Dimensions	Items	Mean	SD.
Social Values and Lifestyle	It improves the quality of life of locals	3.30	.758
	It transforms the narrow thinking of the locals	3.51	.778
	It helps to develop politeness and calmness among locals	3.57	.741
	It enhances the image and popularity of the local area	3.44	.735
	It promotes cultural exchange among host and guest	3.44	.663
	It deteriorates the local traditions and customs	3.48	.695
	It increases the chances of demonstration effect on locals	3.30	.765
	It increases the chances of crimes such as drug abuse, child labour and prostitution etc. in the area	3.38	.700
Women Participation in Society	Ecotourism also gives employment opportunities to local women in the area	3.23	.721
	It increases the self-awareness and self-economic dependence among local women in the area	3.25	.694
	Its development managerial and entrepreneurial skills among local women	3.36	.619
	Social contact and self-confidence of women extend due to ecotourism in the area	3.21	.565
	It promotes the decision-making power in the women in the area	3.20	.674
	It helps to increase the level of education and health of women in the area	3.19	.704
	It increases the work pressure and stress among working women in the area	3.24	.653
	Women face a high risk of sexual harassment and exploitation in ecotourism related business in the area	3.32	.687
	Encourage tourists to consume mostly local products and services such as farmhouse, homestay, local guide etc.	3.09	.588
	The money spent by tourists at ecotourism sites remains in my community	3.12	.678
Involvement of Locals in Ecotourism Planning	It supports the involvement of locals in the service and management aspects of ecotourism related work in the area	3.26	.735
	It helps to force the government to provide education and training to locals on Ecotourism development	3.61	.663
	It encourages to consider locals point of view while the formulation of Ecotourism policies and schemes in the area	3.40	.695
	It supports the involvement of locals while making a decision about any related developmental work in the area	3.15	.765
	The concerned ecotourism authority increase awareness among locals to indulge in ecotourism business activities	2.95	.700
	Locals have no participation and control on ecotourism in the area	3.33	.721
Culture Awareness and Preservation	Ecotourism promotes the preservation of local historical and religious places in the area	3.67	.632
	It promotes conservation of local arts and crafts	3.42	.658
	It improves the locals understanding of culture and tradition	3.77	.758
	It enhances the awareness regarding culture-heritage of the area	3.39	.778

The locals' attitude towards social value and lifestyle dimension based on scores of six out of eight items, considered to a large extent that ecotourism plays a role in sustainable development in Morni hills. In response to women participation in society due to ecotourism, locals considered to a moderate extent that ecotourism is important for sustainable development in the area. The impact of ecotourism on the involvement of locals' dimension of sustainable development also considered important to a moderate extent, except for one item. Lastly, the locals consider the role of ecotourism on cultural awareness and preservation for sustainable development up to a large extent.

Table 5. Locals attitudes towards ecotourism as a tool for environmental sustainability (Author's Creation)

Dimensions	Items	Mean	SD.
Pollution and Diseases	It improves the cleanliness and hygienic condition in the area	3.25	.694
	Increased ecotourism traffic brings solid waste and pollution in the area	3.94	.619
	It deteriorates the environment by spreading noise pollution	3.62	.565
	It increases the risk of spreading diseases in the area	3.90	.674
	It increases the pollution to the local water resources in the area	3.61	.704
	It increases local traffic congestion and crowding	2.83	.653
Environment and Community	It provides extra income or incentive to protect the natural environment	2.73	.799
	It restrains the overexploitation of local water and forest reserves	3.40	.811
	It promotes awareness regarding environment among locals	2.85	.715
	It helps to improve the visual and aesthetic appearance of the local area	3.04	.788
	It enhances the protection of local monuments and buildings	3.96	.778
Natural Resources and Agriculture	The large tourist footfalls over-consume electricity and water resources in the area	2.82	.723
	Excess tourist traffic cause overexploitation of local natural resources	2.62	.772
	It causes degradation to landscapes, historic sites and monuments in the area	3.49	.588
	It increases the destruction of flora and fauna for infrastructure development	2.96	.678
	It increases the loss to the natural landscape and agricultural land for the development of tourist facilities	4.08	.674

The ecotourism contributed sufficiently towards the increase in pollution and disease in the area, as most of the item of pollution and disease dimension indicated its consideration to a large extent among locals. As far as the impact of ecotourism on natural resources and agriculture in the area, the locals, up to some extent, considered that ecotourism helps to impact this dimension of sustainability. The ecotourism towards the environment and community indicator believe that ecotourism has impacted the achievement of environmental sustainability in Morni hills to a large extent as four out five items supported this by giving high and positive mean rating towards the role of ecotourism in environmental sustainability.

Ecotourism is playing an important role in sustainable development in the study region by fulfilling the current requirement of residents as well as tourists without negotiating the requirement of the forthcoming generations. In the economic principle of sustainable development, it plays a crucial role in riping the economic gains from tourism activities in the area. The residents of the area got several new avenues for tourism-related employment and business opportunities. The local government also increases its earning through entry fee and taxes. Regarding the socio-cultural aspects of sustainable development, the study found ecotourism brings many advantages to the local community if they integrate themselves with it otherwise it can lead to communication resulting in erosion of culture and community cohesion. However, the study found a lack of consideration of residents by the government in tourism planning and development. Hence, it becomes essential for them to get involved in different phases of it for their own benefits. Also, the local communities are very familiar with the local resources, it is essential to involve them in the decision-making. Stakeholder participation and community involvement are very important in the decision-making process since they have the greatest stake in the outcome of any ecotourism project. The physical components of sustainable development are an important element of ecotourism. Although to some extent, the ecotourism brings some hazard such as deforestation, land erosion, overcrowding, increased solid waste, but the planned development of ecotourism helps to minimise these environmental costs by emphasising low-impact construction, controlling the number of visitors, taking care in interacting with local flora and fauna. Hence, a balance between development and conservation should be created to avoid the degradation of the natural environment.

SOLUTIONS AND RECOMMENDATIONS

The result of the empirical investigation in chapter offered several valuable observations and suggestions. The implication of these suggestions and recommended steps ensure the development and expansion of the tourism sector with maximum economic benefits in the study area. Few of the recommended actions and suggestions are advocated below. According to the findings of the study, the author suggested to maintain the neat and clean ambience at a tourist site by proper management of wastage should be prioritised to the satisfaction of the visitors. The findings of the chapter also highlight to apply a sustainable approach in ecotourism sites for conservation of important resources like water, electricity etc. Hence, it also recommended using a solar power plant and a rainwater harvesting system to achieve sustainable development.

The findings of the chapter also noticed a lack of professionalism among tourism and hospitality workforce in the study region. Therefore, emphasised improve this weakness to match service delivery system with rising expectations of visitors. After observing the poor image and marketing of the destination are on the global tourist map, the author portrayed to formulate strategic promotional policies to attract more tourists to the area. For converting the study region into a holistic destination, the regional tourism authorities should work towards promoting different types of tourism activities including cultural tourism, pro-poor tourism and increasing the involvement, participation and cooperation of the local community in decision-making, development activates is very necessary.

In order to remove the seasonality syndrome, the destination stakeholders must pay attention to better touristic facilities related with food, accommodation, drinking water, parking, sanitation, transportation, medical facilities, safety measures, and Wi-Fi and mobile network connectivity in the area. The suitable measure to transform the behaviour of the tourism personnel and professional should be taken into consideration by relevant agencies and departments to ensure a positive image of destination among tourists.

The planning and forecasting through government need to reframe in order development of tourism infrastructure and wayside amenities. The private sector should be encouraged to provide good quality facilities and services to please the tourists and ensure their repeat visits for the survival of the tourism sector in the state. For ensuring adequate and optimum utilisation of resources, the author from the findings of the study also suggested to conduct a proper audit and check by concerned authorities or agencies that bring transparency, accountability, and responsibility in the system.

FUTURE RESEARCH DIRECTIONS

Although the present chapter covered different aspects of ecotourism and sustainable development and described the role of ecotourism towards achieving overall development in the study region, some other avenue of relevant research paradigms still has room for consideration in future research agenda. Some of the future research can be conducted on a useful theme, such as the role of tourism in reducing poverty, participation of locals in tourism planning and management and other topics related to community-based tourism. The present chapter also provides a basis for future research such as a comparative study of covering Morni hills and other states in the northern region of the country as well as future studies can also contribute in the measurement of the role of tourism in socio-cultural and environmental impact to destination and measurement of the contribution of tourism in sustainable development.

CONCLUSION

Travel and tourism in the contemporary world have attained a prominent position due to financial and non-financial benefits arise from its development. Tourism is the fastest-growing sector across the globe. It is this sector which benefits not only the world economy, but it is beneficial to the destinations in the form of socio-cultural enrichment and, eventually, it benefits geopolitical milieu of the globe. In a nutshell, the study area is undoubtedly full of tourist resources which need to be tapped, and it is the need of the hour to avail the benefits of this most lucrative industry. Its magnificent opportunities for upliftment of various types of tourist activities coupled with the rich culture including the fort, palaces, handicrafts, folk music, folk dances etc. can go in a sustainable manner. The state should make integrated efforts with the neighbouring states to boost tourism. It is evident from the study that the need of the hour is to properly maintain and preserve the existing resources and provide more opportunities in the form of development of activity-based tourism like amusement parks, adventure sports, etc.

The Morni Hills region of Haryana state has the rich potential of ecotourism development. The study empirically identified eleven dimensions of sustainable development out which four dimensions were about economic sustainability, four dimensions for socio-cultural sustainability and three for environmental sustainability in Morni hills. Based on further investigation of the attitude of locals with descriptive statistics, it came to notice that ecotourism is very important for the achievement of sustainable and overall development in the area. The analysis of inferential statistic also supported this and noticed a positive and significant consideration among locals towards the role of ecotourism in economic, socio-cultural and environmentally sustainable development. On the basis of findings of it can be concluded that ecotourism is an important tool for sustainable development in Morni Hills.

The study from its findings, also provide implications for concerned stakeholders to come up with efficient strategies to focus local community involvement to accord the maximum benefits to them. The study further recommends the improvement in management and supervision of ecotourism activities, enhanced cooperation among different stakeholders, the introduction of a visitor management system, raising training and awareness among locals towards the importance of ecotourism to sustainable development, so that the overall development in are can be achieved easily.

This study has important implications for policy developers as they should be aware of the importance of tourism for achieving economic development. The adoption of offered suggestive and recommended activities ensures the attainment of more economic gains from this industry in Morni Hills.

REFERENCES

Anup, K. C., & Rijal, K., & Sapkota. (2014). Role of ecotourism in environmental conservation and socioeconomic development in Annapurna conservation area, Nepal. *International Journal of Sustainable Development and World Ecology*, *22*(3), 251–258.

Aziz, A., Barzeker, G., Ajuhar, Z., & Idris, N. H. (2015). Criterion and Indicators for Monitoring Ecotourism Sustainability in a Protected Watershed Sustainable Tourism: A Delphi Consensus. *IOSR Journal of Environment Science. Toxicology and Food Technology*, *9*(3), 1–9.

Barzeker, G., Aziz, A., Mariapan, M., Ismail, M. H., & Hoseni, S. M. (2011). Delphi Technique for Generating Criterion and Indicators in Monitoring Ecotourism Sustainability in Northern Forest of Iran: Case Study on Dohezar and SehezarWatersheds, Folia. *Forestalia Pronica*, *53*(2), 130-141.

Bashar, M. A. (2018). Vision on biodiversity: Ecotourism and biodiversity conservation in Bangladesh. *Journal of Biodiversity Conservation & Bio-resource Management*, *4*(1), 1–10. doi:10.3329/jbcbm.v4i1.37871

Best, J. W. (1977). *Research in education* (3rd ed.). Englewood Cliffs, NJ: Prentice Hall, Inc.

Bunruamkaew, K., & Murayama, Y. (2012). Land Use and Natural Resources Planning for Sustainable Ecotourism Using GIS in Surat Thani, Thailand. *Sustainability*, *4*(1), 412–429. doi:10.3390u4030412

Canadian Tourism Commission. (2007). *Guidelines: Survey Procedures for Tourism Economic Impact Assessments of Gated Events and Festivals*. Alberta Tourism, Parks, Recreation & Culture. Retrieved from https://www.destinationbc.ca/getattachment/Research/Resources-for-esearchers/How-To-Guides/Gated_Events_Festivals-economic-impact.pdf.aspx

Chandel, A., & Mishra, S. (2016). Ecotourism revisited: Last twenty-five years. *Czech Journal of Tourism*, *5*(2), 135–154. doi:10.1515/cjot-2016-0008

Dinç, A., & Kocan, N. (2012). *Ecotourism Planning of Murat Mountain (Usak, Kutahya/Turkey). International Conference on Applied Life Sciences (ICALS2012), Turkey, September 10- 12, 2012*. Retrieved from http://creativecommons.org/licenses/ by/3.0

Dyer, P., Aberdeen, L., & Schuler, S. (2007). Tourism impacts on an Australian indigenous community. *Tourism Management*, *24*(1), 83–95. doi:10.1016/S0261-5177(02)00049-3

Emmanuel, A., Nsor, C. A., & Boadi, S. (2017). The socio-cultural impact of ecotourism on park-adjacent communities in Ghana. *African Journal of Hospitality, Tourism and Leisure*, *6*(2), 1–14.

Fadahunsi, J. T. (2011). Application of Geographical Information System (GIS) Technology to Tourism Management in Ile-Ife, Osun State, Nigeria. *Pacific Journal of Science and Technology*, *12*(2), 274–283.

Fennell, D. A., & Dowling, R. K. (2003). Ecotourism Policy and Planning. CABI Publishing.

Gheorghe, C. (2013). Ecotourism: Definition and Concepts. *Journal of Tourism Studies and Research in Tourism*, *15*, 56–60.

Jamal, T., & Stronza, A. (2009). Collaboration theory and tourism practice in protected areas: Stakeholders, structuring and sustainability. *Journal of Sustainable Tourism*, *17*(2), 169–189. doi:10.1080/09669580802495741

Lai, P. H., & Nepal, S. K. (2005). Local perspectives of ecotourism development in Tawushan Nature Reserve, Taiwan. *Journal of Tourism Management*, *27*(6), 1117–1129. doi:10.1016/j.tourman.2005.11.010

Li, W. (2004). Environmental management indicators for ecotourism in China's nature reserve: A case study in Tianmushan Nature Reserve. *Tourism Management*, *25*(5), 559–564. doi:10.1016/j.tourman.2003.06.001

Loach, K., Rowley, J., & Griffiths, J. (2017). Cultural sustainability as strategy for the survival of museums and libraries. *International Journal of Cultural Policy*, *23*(2), 186–198. doi:10.1080/10286632.2016.1184657

Niñerola, A., Sánchez-Rebull, M. V., & Hernández-Lara, A. B. (2019). Tourism research on sustainability: A bibliometric analysis. *Sustainability*, *11*(1), 1377–1387. doi:10.3390u11051377

Nunnally, J. C., & Bernstein, I. H. (1994). The Assessment of Reliability. *Psychometric Theory*, *3*, 248–292.

Theingthae, S. (2017). Sustainability of Community Based Ecotourism Development after the Impact of Tsunami Disasters: Comparison between Buddhism Community and Muslim Community in Phuket Province, Thailand. *Journal of Tourism Research & Hospitality*, *6*(4), 1–10. doi:10.4172/2324-8807.1000175

Thomas, K. D., William, N. B., & Trotz, M. A. (2014). The Sustainability of Ecotourism Activities: Development of An Accessible, Applicable and Efficient Tool for Assessment in the Caribbean Region. *The West Indian Journal of Engineering*, *36*(2), 51–62.

Tosun, C. (2001). Challenges of sustainable tourism development in the developing world: The case of Turkey. *Tourism Management*, *22*(3), 289–303. doi:10.1016/S0261-5177(00)00060-1

Wall, G. (1997). Is Ecotourism Sustainable? *Environmental Management*, *21*(4), 483–491. doi:10.1007002679900044 PMID:9175538

Zhong, L. Z., & Liu, L. (2017). Ecotourism development in China: Achievements, problems and strategies. *Journal of Resources and Ecology*, *8*(5), 441–448. doi:10.5814/j.issn.1674-764x.2017.05.001

ADDITIONAL READING

Batra, G. S., & Dangwal, S. C. (1999). *Tourism Promotion and Development*. Deep and Deep Publications Pvt. Ltd.

Beeton, S. (2006). *Community development through tourism*. Landlinks Press. doi:10.1071/9780643093881

Brian, B., Robyn, C., & Chris, C. (2016). *Worlwide destinations: The Geography of Travel and Tourism* (7th ed.). Routledge.

Jyothirmaye, R. S., Venkateswara, R. B., & Durga, R. S. (2012). Rural & Eco Tourism in India: Problems & Prospects Environmental Impacts on Tourism. *International Journal of Business and Management*, *1*(1), 1–5.

Kamra, K. K., ... (2002). Basics of Tourism: Theory, Operation and Practice. New-Delhi: Kanishka Publishers.

Langroudi, M., & Hassan, S. (2003). *Rural planning focusing on Iran*. Jahad Daneshgahi Press.

Mathieson, A., & Wall, G. (1982). *Tourism: Economic, physical and social impacts*. Longman House.

Murphy, R. E. (1985). *Tourism: A Community Approach*. Methuen.

Sharma, S. (2008). Tourism, Technology and the Entrepreneur: A Case Study from Northeast India. *ICFAI University Journal of Business Strategy*, *5*(4), 71–80.

KEY TERMS AND DEFINITIONS

Attraction: A place, event, building or area which tourists want to visit.

Carrying Capacity: The amount of visitor activity that a site or destination can sustain.

Community: The group of people meets each other because they live nearby, not because they particularly want to be together. In a community, the members often have shared values and symbols.

Ecotourism: Responsible travel to natural areas which conserves the environment and sustains the livelihood of local people.

Infrastructure: Construction needed to support economic development.

Sustainable Development: Development that meets the needs and aspirations of the current generation without compromising the ability to meet those of future generations.

Sustainable Tourism: Sustainable Tourism leading to management of all resources in such a way those economic, social, and aesthetic needs can be fulfilled while maintaining cultural integrity, essential ecological processes, biological diversity, and life support systems.

Tourism Impacts: Effects, which may be felt because of tourism-associated activity.

Tourism Stakeholders: Individuals who have a vested interest in development, including community members; NGOs; planning, and government officials; hotel owners, tour operators, guides, transportation providers, and representatives from other related services in the private sector.

Chapter 7
Tourism as a Strategy for Geo-Education and Cultural Valorization of Territories:
Promoting Ecocultural Values and Geoconservation in Estrela Geopark, Portugal

Gonçalo Poeta Fernandes
https://orcid.org/0000-0003-1278-0369
Polytechnic Institute of Guarda, Portugal

Emanuel Castro
Estrela Geopark Association (AGE), Portugal

Rosa Branca Tracana
Polytechnic Institute of Guarda, Portugal

ABSTRACT

The mountains represent territories of exceptional environmental and heritage value, holding vital resources and functions for humanity. The conservation of its geo and biodiversity is decisive, given the growth in its demand, in particular for tourism and leisure activities. In addition to their immediate physical perception, they are spaces of science and knowledge and a resource for geoeducation to promote the natural and cultural values held. The creation of geoparks constituted a new paradigm in which geosciences, territories, their heritage (material and immaterial) and the ways of life of their communities are disseminated. In this context, geoparks are based on integrated territorial approaches to the enhancement, preservation and promotion of education, biodiversity, geological heritage, tourism and scientific research, among others. Tourism, as a strategy for geoeducation and territorial valorisation, is in the mission of Estrela Geopark, promoting educational programs with initiatives for the conservation of heritage and promotion of eco-cultural resources.

DOI: 10.4018/978-1-7998-5691-7.ch007

INTRODUCTION

Tourism assumes an important role in the management and conservation of natural areas, so the link between protected areas and tourism is vital for its sustainability and enhancement as a territorial asset. This mutual relationship is complex, sometimes the opposite, but tourism is always a critical component to consider when planning and managing protected areas, such as potential geoparks. Sustainable tourism practices in protected areas represent long-term commitments and ways to implement the geoconservation process, stimulating learning, formal and non-formal, for territorial enhancement.

The mountains are reservoirs of resources and services that are fundamental in the regulation of natural cycles, in biodiversity, in the preservation of cultural values associated with local ways of life, as natural museums of geomorphological history, as living labs and teaching resources for sustainability. At the same time, they are territories of increasing demand for tourism and leisure activities, which must be constituted with means of promoting geoeducation and geoconservation. In Portugal, mountain areas of high environmental/natural value have been classified as protected areas, based on their management by a conservationist logic that tends to preserve the environment and the implementation of a heritage design of the territory often in conflict with the communities. The frequent mismatch between global needs for nature protection and local needs for socioeconomic development has, in many situations, hampered the management of such spaces, as it may disqualify their perception and the activities established by their people. These positions require specific attention and coordinated the various stakeholders, especially management entities, able to create conditions for sustainability and qualification of these spaces, valuation of their products, greater ability to leverage the resources held as goods and services value to society and the economy.

The philosophy in the creation of geoparks focuses on the development of networks for an exchange of experience and joint promotion of concept and each member of the network. The three main areas of activity of a Geopark are (i) conservation of geological heritage; (ii) education for sustainability, and (iii) tourism and local development. In this sense, the Geoparks introduce a great responsibility in the creation/development of economic, tourist and social value. Their potentialities may be transversal in helping to create territorial value in a given geographical area. On the other hand, their potentialities may be appropriate to tourism based on the development strategies for the welfare of the community and enhancement of the territory, promoting a sense of belonging and collaborative work.

Thus, the Estrela Geopark (EG) introduces a great responsibility in the development of economic, tourist and social value. Its integration into a global network will allow the dissemination of its values and heritage, fostering its conservation and projecting the Estrela Mountain to a wider market, allowing the emergence of new products and services based on nature and local ways of life. In summary, EG seeks the development of an integrated and collective development strategy, allied to the valorisation of heritage, fostering geoconservation, education, and tourism.

In this context, this chapter seeks to analyse the importance of mountain territories, in respect of the resources and functions, it hosts, the relevance of its heritage, the educational value associated with its geological history, biodiversity, and ways of life. The role of Geoparks is acting as instruments of territorial valorisation, dissemination of geosciences, and promotion of sustainability strategies. Estrela Geopark's object of research is identifying its strategies of geoconservation and geoeducation, the initiatives of promotion of territorial and cultural value associated with their communities. It promotes the systematisation of good practices and actions built between the different actors of the territory, in the promotion of heritage, mountain lifestyle, and the geoconservation, in a sustainability strategy. In this context, tourism takes centre stage as a strategic process for the enhancement of the Estrela Mountain destination, promoting initiatives and developing geoeducation programs for sustainable tourism, committed to the territory and its communities.

MOUNTAIN TERRITORIAL PERCEPTION: FROM DISTANT AND UNKNOWN SPACE TO TOURIST DESTINATION

The mountains have always played a prominent role in the organisation and dynamics of space: for the attraction exercised in the conquest of the highest (which has plagued man throughout the ages), for its strategic importance as a place of wide views (primordial for the defence of the territory) and by the exploitation of its resources and reserves (originating specific ways of life) by the knowledge of its genesis and geodynamics (by the meaning for the geosciences), depending on the natural conditions. The topographic and climatic characteristics, the forms of land use and the exploitation of resources imply a complimentary articulation with the adjacent areas, for which the mountain has several aspects and meanings over time (Table 1). With the expansion of knowledge, use and exploitation of the Mediterranean mountains, new concepts were established, which went from the old mythological vision to the current perception of areas of great ecocultural value, the target of sustainable management, and ordering policies and classification as heritage territories. Thus, in its evolution and perception, it is possible to differentiate five stages linked to different forms of occupation and territorial valorisation: restrictions and mystery, achievement and challenge, refuge and exploration, decline and marginalisation, and valorisation and preservation.

The mountain for its degree of integrity and the geodiversity and biodiversity held takes on the connotation of purity and natural refuge. These characteristics, which value these territories in environmental terms, are the basis of their protection, given the growing demand, particularly for activities related to tourism and leisure in the last decades. Nowadays, the mountains, the best-preserved ecosystems on the European continent, represent territories of exceptional environmental value, assuming themselves as spaces for protection and laboratories for bio and geodiversity (Ariza et al., 2013; EEA, 2013). These heritage elements become references for the attraction of visitors and their valorisation as tourist assets, promoting geoconservation initiatives.

In the last decades, the concern for the rational use and occupation of the mountain has taken on new shapes, given the growing demand and the recognition of the value of its resources and functions (Gloersen et al., 2016; Kapos et al., 2010). Due to their natural characteristics, they have been subject to management based on protection assumptions, to safeguard their potential and perpetuate them within a natural setting, where the action of man, especially at the tourist level is reduced in terms of repercussions environmental issues. These policies promoted classification as protected areas, subject to management plans and regulations, which seek to preserve and enhance them. This situation has led to the preservation of certain territories and heritage elements, but it has also mortgaged its development, due to the constraints to economic activities and forms of land use, promoting demographic abandonment and landscape disqualification (Azevedo et al., 2016; Fernandes, 2018).

This evolution has profoundly changed the role of the mountain in the economic and territorial system, causing the loss of traditional uses and functions and the entry of others, with an important spatial impact, introducing organisational changes and different resource valuations (Dax, 2019; Fernandes, 2015). Natural, cultural and landscape resources, secondary to the traditional system, have become the

Table 1. Stages of evolution and perception of the mountain in Portugal (Adapted by Fernandes, 2015)

Phases	Characteristics	Period
Restrictions and mystery	Period of a mythical and supernatural vision, associated with ignorance, physical magnificence, isolation and expression of climatic and geological processes.	Until the 18th century, due to lack of knowledge and poor accessibility.
Conquest and challenge	Increasing accessibility and expanding the knowledge of these spaces, together with the exploitation of their natural resources and enjoyment by the most privileged social classes. The great scientific activity on these territories has made it possible to clarify their dimension and the interaction of the physical elements that structure them, with the man now using his resources.	From the end of the 18th century to the 19th century, thanks to the widening of access routes and knowledge of natural resources.
Refuge and exploration	Areas of refuge, which are home to a great diversity of species and integrity of natural resources, of particular interest and strategic significance for the low areas. The abundance and quality of its waters, flora, and subsoil boosted the appearance of places of exploration and settlement of the population.	Late 19th and early 20th century.
Decline and marginalisation	The population regression and traditional activities, linked to the agro-silo-pastoral system, originate a crisis in the use and perception of these spaces, as well as a rupture in the established balances. The economic and social disarticulation is accompanied by an increase in tourist practices and support equipment.	From the second half of the 20th century.
Valuation and preservation	The growing demand for an exogenous population, especially for recreational and leisure activities, together with the recognition of their vulnerability and ecocultural appreciation, justifies the adoption of protection instruments.	From the 1980s onwards, in the face of the growing ecocultural appreciation of these territories.

main factors in the organisation and perception of wide mountain territories. The multifunctionality of these spaces implies the synergistic combination of different functions, that is, between primary and immaterial productions, linked to landscapes, environment, leisure, leisure, and sports, among others (Figure 1).

In Portugal, the transformations that affect the uses of mountain and structure, in the present, are the result of an old process, but with a more significant evolution that started in the first half of the 20th century. In a retrospective approach, we can consider that, until the nineteenth century, these mountains were characterised by their barrier effect, isolation, seasonality, and mastery of agropastoral activities. To the pastoral activity is added forest exploitation, which harms the former, due to the pressures and impositions created, in particular, by the process of reforestation initiated by the Estado Novo, giving rise to migrations, abandonment of agricultural land and expansion of uncultivated land and forestry (Sande Silva, 2007; Azevedo, 2016; Aguiar et al., 2016). Since the 1950s, mountain areas have witnessed the exodus and ageing of their populations, the transformation of their economic structure, often associated with communitarian practices, and the retreat of industrial activity. This disarticulation between the relation and use of the land, along with its ecological value, led to the protection of these territories (Cunha, 2003; Cunha & Cordeiro, 2008; Fernandes, 2018). Over the last decades, the mountain has been perceived based on its environmental and cultural values, which imply the revaluation of its territory and resources and an attendant increase in demand for tourist activities. These changes were generated, mainly, by the appropriations of the economic system and the most privileged social classes, resulting in new activities and land uses, conflicting and competitive concerning agro-pastoralism, contributing, over time, to the restructuring of the productive system and the traditional social and spatial organisation on which the mountain was based.

Figure 1. Plurifunctionality of mountain territories and the role of tourism in its valorisation (Authors elaboration)

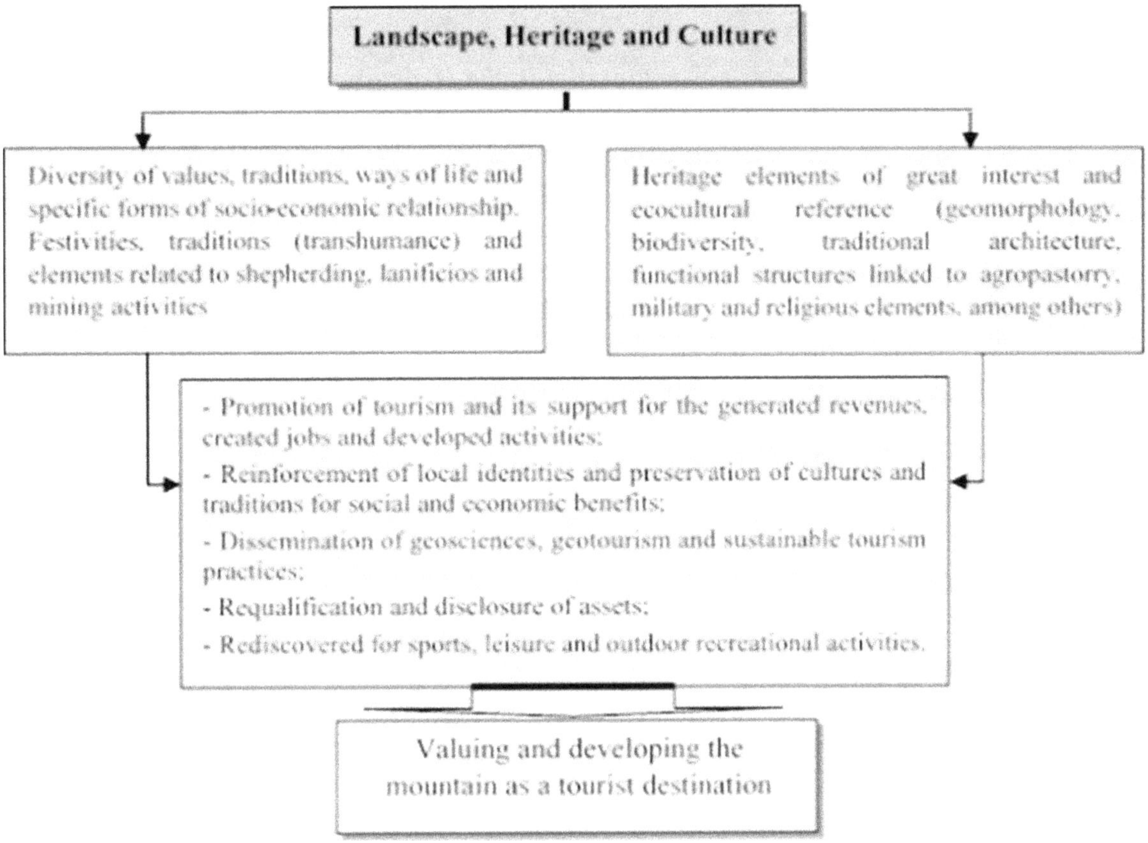

The environmental and economic valuation of goods and services, that mountain tourism provides to its communities, taking into account the characteristics of ecosystems and physical structure, constitutes a challenge for their ordering and management. The concern for the conservation of bio and geodiversity and local culture, as a heritage and resource for development, has been stimulated and promoted, namely through:

- Implementation of a conservation and protection statute, in particular, associated with its natural characteristics;
- Development of planning plans and creation of reserve statutes, seeking to preserve their ecology and traditional agricultural uses;
- Involvement of communities and key strategic actors in governance;
- Promotion of support for geoconservation and the development of programs for the conservation of forest resources;
- Encourage the recovery of buildings for new uses and functions, associated with tourism and the well-being of communities;
- Increase of information and signalling of areas of greater environmental sensitivity.

It becomes crucial to consider the potential and opportunities that these territories have in terms of perception, ecology and landscape, and socioeconomic interest (Table 2).

Table 2. Potentialities and opportunities for tourism in the Mediterranean mountain areas (Authors elaboration)

Perception	Ecological	Socioeconomic
• Favourable conditions for recreational and leisure activities; • Enhancement of the natural and cultural integrity of the mountains; • Strategic recognition of water, forest and mineral resources; • Increasing tourist demand associated with the natural dimension.	• Ecocultural value of landscapes; • Diversity of ecosystems; • Existence of species of fauna and flora that enhance cultural interest and the preservation of certain areas; • Large water reserves; • Enhancement of environmental quality.	• Diversity of resources capable of stimulating the offer of services related to sport, leisure and tourism; • The emergence of a system of social and environmental values that emphasise the areas of natural and cultural integrity; • Natural and cultural heritage to be valued and promoted; • New sources of income related to tourism and leisure; • The appearance of investors in non-traditional areas and functions; • Quality of agropastoral products • Industrial tradition and crafts.

Tourism has been playing a prominent role in the current dynamics of mountain areas, and is responsible for major changes in socioeconomic organisation and land use, promoting new opportunities, generating flows with different motivations and enhancing the ecocultural identities of these territories (Drexler, 2016; Ferrnandes, 2020; Meyer-Cech & Pröbstl, 2013).

In this context, tourist activities should foster a deeper knowledge of cultural heritage, providing its valuation and conservation, allowing the valorisation of local products and services, geoconservation, and the well-being of communities. Tourism will also be able to promote and value productions associated with the agro-silvopastoral system, cultural events, restoration of buildings, and expansion of trade and service activities, making it possible to secure the population, generate employment, and complementary activities (Debarbieux et al., 2018). The appropriation of the mountain for new uses and users, supported on a large scale by tourism and its activities, has allowed the recovery of real estate, expansion of hotel offers, improvement of accessibility, expansion of services and the promotion of new functions associated with natural resources (river beaches, nautical activities, climbing areas, animal species observation, mountain sports, geocaching, geotourism, etc.). This growth requires greater responsibility on the administrative entities and greater cooperation between the different actors, allowing the construction of strategies that make tourism compatible with the environment and the diverse functions of the mountain (EU, 2008; Strickland-Munro, Allison, & Moore, 2010). The tourism literature acknowledges that better collaboration and coordination of activities can enhance the performance of both individual actors and destinations (Pansiri, 2008).

GEOTOURISM IN TOURIST DIFFERENTIATION AND HERITAGE VALUE IN THE MOUNTAIN REGIONS

The mountains represent, at present, the best-preserved ecosystems of the European continent, holding a diversity of species and functions that give them a fundamental ecocultural value for humanity (CE, 2004). This enhances their preservation and, at the same time, strategies capable of promoting the valuing articulation between traditional activities related to agropastoral and forestry, tourism, and leisure activities.

Geotourism corresponds to the set of tourist activities related to the geological resources of territory and with all the scientific and environmental dimension that these resources represent, giving expression to a demand that seeks to obtain in loco knowledge of the geological elements, their formation, the landscapes and their relationship with local communities (Dowling, 2009, Sekhar et al., 2019). The valorisation of geological resources (geological element focuses on geology and landscape and includes both 'form', such as landforms, rock outcrops, rock types, sediments, soils and crystals, and 'process', such as volcanism, erosion, glaciation, …) and consequently the geographical area of its location, in particular those of international relevance, has promoted the classification of spaces of significant patrimonial value, the greatest example being the Global Geoparks of Unesco (Mignon, 2018). In this context geology and geotourism induce collective strategies for territorial valorisation, promoting scientific research, geoconservation, development of materials and didactic services, promotion of tourism services, and the involvement of local communities in geology and tourism valorisation (Newsome & Dowling, 2010). The relationship between the territory (elements of natural heritage), the scientific and didactic value of resources and the interest of communities in their preservation and generation of well-being, contribute to a complementary value-enhancement process. Geotourism, as a contact and learning activity with the territory, presents an active component, associated with journeys, routes, and rails, the reason why it fits healthy physical activity. This kind of tourism should be operated by various actors, including public and private entities, scientific institutions, local associations or non-governmental organisations, expanding its capacity to contribute to the development and sustainability (Ehsan, et al., 2012; Mignon, 2018).

The concept of geotourism thus fosters the idea that to fully understand and appreciate the environment, the abiotic elements of geology and climate should first be known, since these determine the biotic elements that exist there. By extension, these two components of the environment influence the cultural landscape of how people have lived in the area in the past, as well as how they live there today. This approach forms the basis of a holistic understanding of the environment and its components and, therefore, provides the resident and tourist population with a greater connection to the environment in which they live or which they are visiting (Newsome & Dowling 2018).

Thus geotourism enables the geoconservation and environmental education of the various social nuclei covered – both residents and visitors (Rodrigues & Neto de Carvalho, 2010). Geotourism is geologically based (based on the earth's geoheritage), sustainable (i.e., economically viable, community-enhancing and fosters geoconservation), educative (achieved through geointerpretation), locally beneficial and generates tourist satisfaction. It is a form of tourism that generates sustainability, both from the valorisation and conservation of the resources that support it and the services required, promoting tourism that allows the conservation and qualification of resources, the involvement of the local population and the promotion of economic activities, expanding services of collective benefit (Dowling, 2009). The involvement of local communities is relevant, both in stimulating the process of valorisation and expansion of the destination and increasing the quality of the tourist experience. Local communities can become involved in geotourism operations and the provision of knowledge, services, facilities, and products (Dowling, 2010; Shekar et al., 2019). Geotourism can also generate income for resource conservation management in addition to social and cultural benefits. The contribution may be through: financial means, scientific investigation, didactic manuals, and can also improve the quality of life to the host community by increasing equipment and support the construction of a territorial brand.

In this context, geotourism is a form of tourism based on natural heritage, including geological, botanical, or archaeological aspects (which structure the landscape), where the concept of sustainable development plays an essential role, due to the holistic effects generated on the territory. In other words, it can be described as the component of natural geodiversity of significant value to humans, including scientific research, education, and cultural development (Figure 2). Geodiversity and Geological Heritage are scientific and didactic resources, support of habitats, ecosystems and landscapes, socioeconomic engines for the sustainable development of rural areas and a fundamental part of the culture of man and his activities.

Figure 2. Structuring elements of Geotourism and territorial sustainability (Authors elaboration)

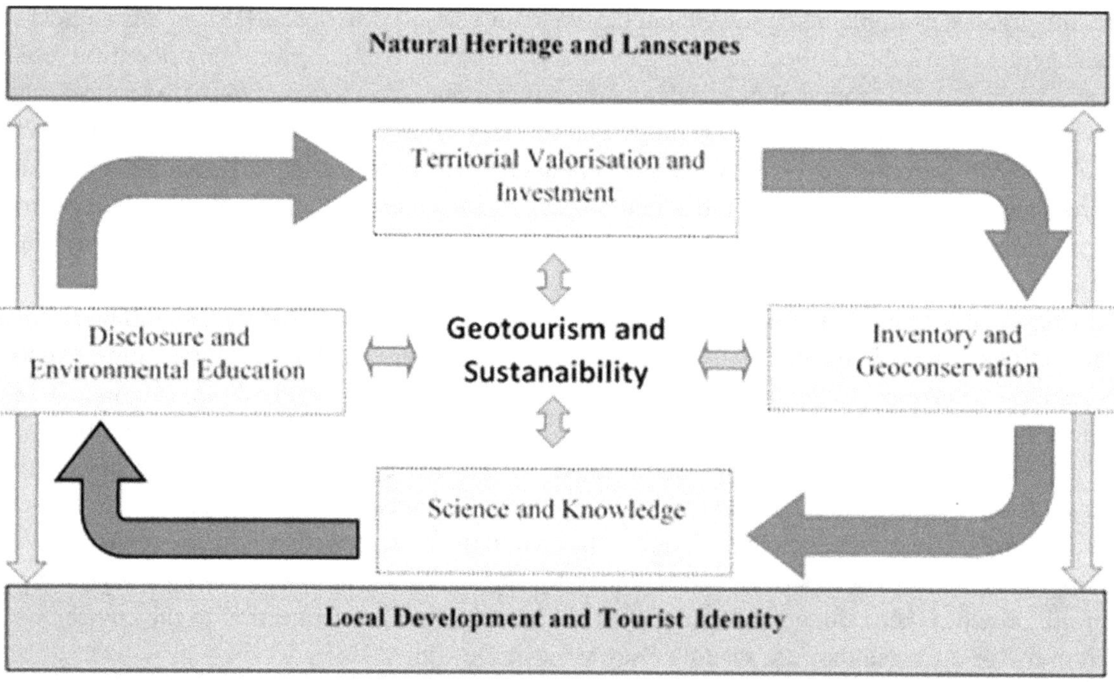

Geodiversity allows the construction of the identity of a place, arising both as a representation (an ideal that reveals meaning) and a material existence (the reality of living conditions) (Harner, 2001). It should be noted that geology appears integrated, with cultural and biological aspects, promoting a composite approach of destiny, where conservation aspects intensify and promote the preservation of geoheritage elements so that tourists can enjoy, learn and disseminate knowledge. These dimensions promote the creation of means and channels for the promotion of geological and geomorphological aspects, integrated into the landscape and local ways of life, promoting the development of itineraries, their dissemination as resources of the destination/region, promoting their conservation and assuming a didactic dimension of natural laboratory, supporting the visitation with new orientations and public (Ehsan et al., 2012; Hose et al., 2012). It is important to point out that upstream of a tourism strategy is a territorial strategy based on the valorisation of endogenous resources and downstream should be the tourism products created from the potentialities and resources that these areas have. The articulation between geodiversity, biodiversity, history, and local culture not only increases geotourism potential but also diversifies and complements the offer, leveraged development strategies, and social welfare.

The development of geotourism has multifaceted benefits including the following: it helps the local communities in creating new job opportunities, work as a useful platform for educational activities starting from school children to higher studies, research purposes and helps the remote places to get connected to other parts of the world through rail and road (Mignon, 2018; Shekhar et al., 2019). The overall development of the local population in terms of health, education, and prosperity can be secured, this is its sustainability.

GEOPARKS, GEOTURISM AND GEOEDUCATION. STRATEGIES OF THE ESTRELA GEOPARK IN THE PROMOTION OF SUSTAINABLE TOURISM

Estrela Mountain is located in Central Portugal and is part of the Iberian Central System, a mountain range that extends from Guadarrama, north of Madrid, to Montejunto, northeast of Lisbon. The Estrela Mountain is the highest in mainland Portugal, rising to 1993 meters at Alto da Torre, is a mid-altitude mountain range of small size when compared to other European mountains. However, it shows high values of geodiversity, as supported by the proposed geossites and by the remarkable development of Pleistocene glacial landforms and deposits (AGE, 2017).

The concept of geopark emerged at the end of the 20th century in Europe, with the creation of the European Network of Geoparks. A geopark is a geographically delimited territory with a sustainable development strategy based on the conservation of the geological and geomorphological heritage in association with the other elements of the natural and cultural heritage and local ways of life (Eder & Patzak, 1998). It seeks, through the protection and enhancement of the geological heritage, to improve the living conditions of the population, promoting, in an integrated way, the endogenous values and their tourist appropriation (Brilha, 2009; Stueve et al., 2002).

UNESCO Global Geoparks promote sustainable local development mainly through geotourism. To stimulate the geotourism in the area, a Global Geopark must have visibility. Visitors, as well as local people, need to be able to find relevant information. As such, UNESCO Global Geoparks need to provide information via a dedicated Website, leaflets, and detailed map of the area that connects the area's geologically significant sites (UNESCO 2004, Zouros & Mc Keever, 2008)). The key to capitalising on the potential benefits offered through geotourism development is to maximise the opportunities and minimise the adverse impacts through the protection of natural heritage. Geotourism has assumed a growing significance in tourism practices and geological heritage valorisation, as a result of a progressive concern of the scientific community and society in general about the natural heritage (including geological heritage and geodiversity). In this sense, and simultaneously with a concern of the public power and administrations (at different levels), landscapes and geology become a tool to promote economic development and territorial sustainability.

Geoeducation and geotourism have gained relevance in the projects of tourism development, above all in low-density territories. This new paradigm of "looking" at the territory has allowed the construction of a new narrative about the enhancement of endogenous resources, the landscape, and the policies of territorial tourism development (Frey et al., 2016). Education and science play a key role in the conservation and maintenance of geological and cultural diversity, stimulating research in the territory and actively contributing to the existence of a network of joint initiatives (publications, information exchange, conferences, joint projects, etc.) in a holistic view. In this scope the Estrela Geopark develops an action looking for the implementation of an International Scientific Advisory Committee, defining scientific hot topics and their links with societal needs and regional development. They also implement a grant system for research projects, promoting science education and outreach.

In this regard, Estrela Mountain is a geographically defined territory with resources of geological and geomorphological and ecocultural interest, capable of fostering the qualification of the territory and its capacity for new valuations and flows to the region, enhancing the sustainability of resources, fostering the economy and raising social welfare. The geoeducation strategy and territorial appreciation of the Estrela Geopark will boost recognition of the diversity of glacial marks, the richness of its ecocultural heritage, biodiversity and ways of agropastoral life, that make the Estrela Mountain a unique place for promoting a strong brand with tourist, heritage and cultural potential (Gomes et al., 2017).

The Estrela Geopark (composed territorially by the municipalities of Guarda, Seia, Gouveia, Celorico da Beira, Fornos de Algodres, Manteigas, Belmonte, Covilhã and Oliveira do Hospital, encompassing an area of 2,737.72 km2 and housing 171,668 inhabitants) have the mission to contribute to the protection, enhancement of the natural and cultural heritage (Figure 3). It will promote the perspective of further development and dissemination of scientific knowledge, promoting tourism and sustainable development of the territory. Parallel to the tourism valorisation, educational/pedagogical contexts are developed, as well as the scientific deepening, given the natural laboratory that these spaces contain, fostering new interests, new audiences, and demands. As Brilha (2009) defends *Geoparks are in a privileged position to play this role of promoters of education in Geosciences for sustainable development, directed to all types of public*.

The diversity of glacial vestiges, the wealth of its patrimony, and the characteristics of its landscape make unique geography out of the Estrela Mountain whose structured approach will allow the creation of a strong brand with touristic, patrimonial and cultural potential. The classification of the Estrela Geopark by the European and Global Geopark Networks would bring a new vision on the tourism of this territory and its sustainability, introducing a holistic view in the space, based on its geography, on its cultural references and its brand.

The Estrela Geopark presents a selection of 35 Glacial and fluvioglacial geosites (figure 4) showing the effects of glacial erosion and accumulation across spatial scales, from landscape to micromorphological features. The selected geosites also allow the identification of the importance of glacial geoheritage to the present-day landscape organisation and dynamics, showing that the Geopark is alive and dynamic, with the knowledge on its geodiversity being a key for the good management and territorial planning. The existence of a large variety of lithologies and petrographic types of granites inside the territory provides an added value to the periglacial heritage. In this category, some geosites showing the effects of slope processes, mainly debris flow activity (relict and contemporary) were also included, essentially due to their pedagogical significance relating to geohazards.

Figure 3. Geopark map - Location, altimetry and administrative division (Associação Geopark estrela-AGE)

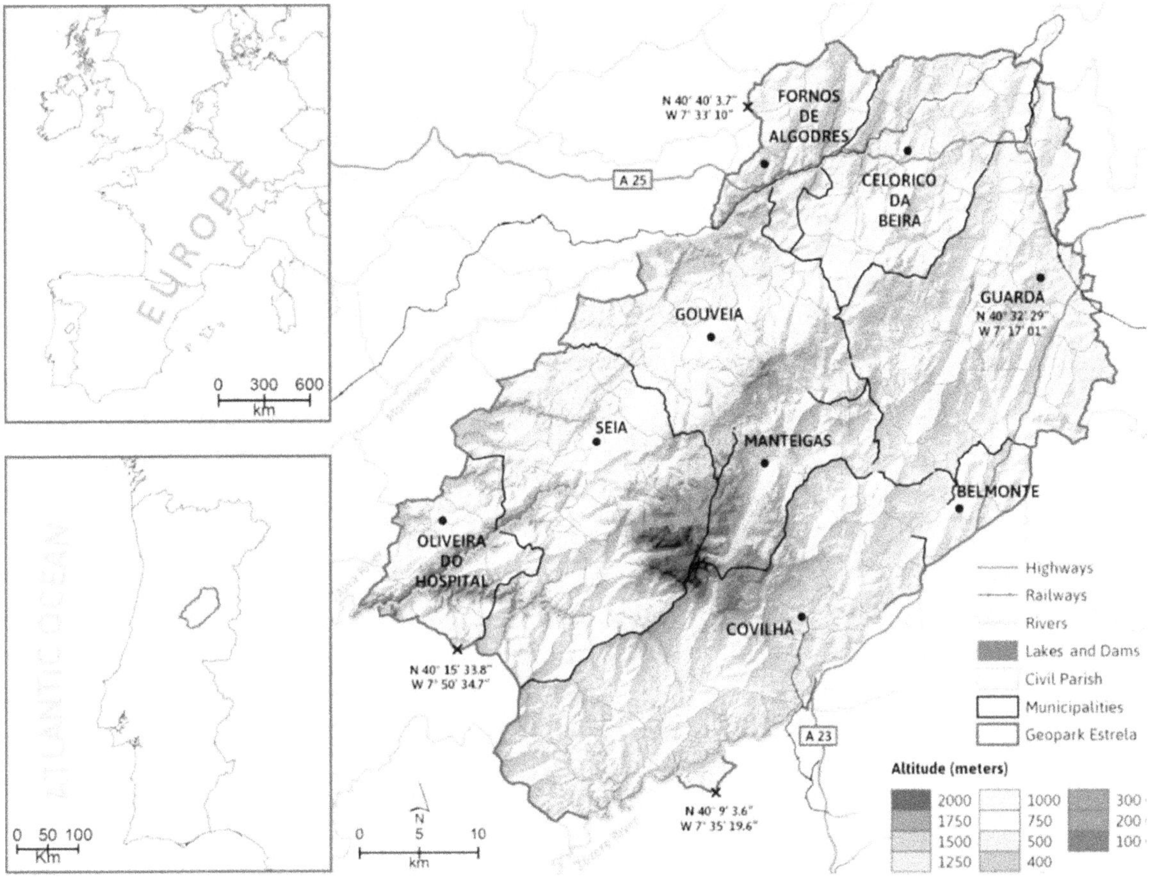

The heritage of the Estrela Geopark, especially the architectonic, has developed in close connection with geological traits of the territory. Granite marks the architectonic (historical and cultural) heritage, especially outstanding in castles, civil and religious buildings. Slates and schists are also used mainly in the metasedimentary areas, originally reflecting lower-income communities (Table 3).

Industrial heritage associated with textile (wool) and dairy activities are key for understanding the cultural background of the people from the Estrela Mountain and the evolution of the landscape, due to the links between economic activities and land-use dynamics. On the other hand, mining played a very large role in the economy of the region and is still of significance in some areas. The mining heritage, which has also been included in some geosites, shows the direct relation between society and geological resources, being especially significant to promote education and discussion on sustainability.

Geopark Estrela has the mission of contributing to the protection, valorisation and dynamisation of the natural and cultural heritage, with special emphasis on geological heritage, intending to deepen and disseminate scientific knowledge, promoting tourism and sustainable development of the Estrela Mountain. On the whole, Geopark seeks the development of an integrated development strategy, allied to enhance heritage, fostered geoconservation, education, and tourism, supported the construction of development strategies for the well-being of the community.

Figure 4. Representation of heritage elements – geosites – According to their typology/genesis (Associação Geopark Estrela- AGE)

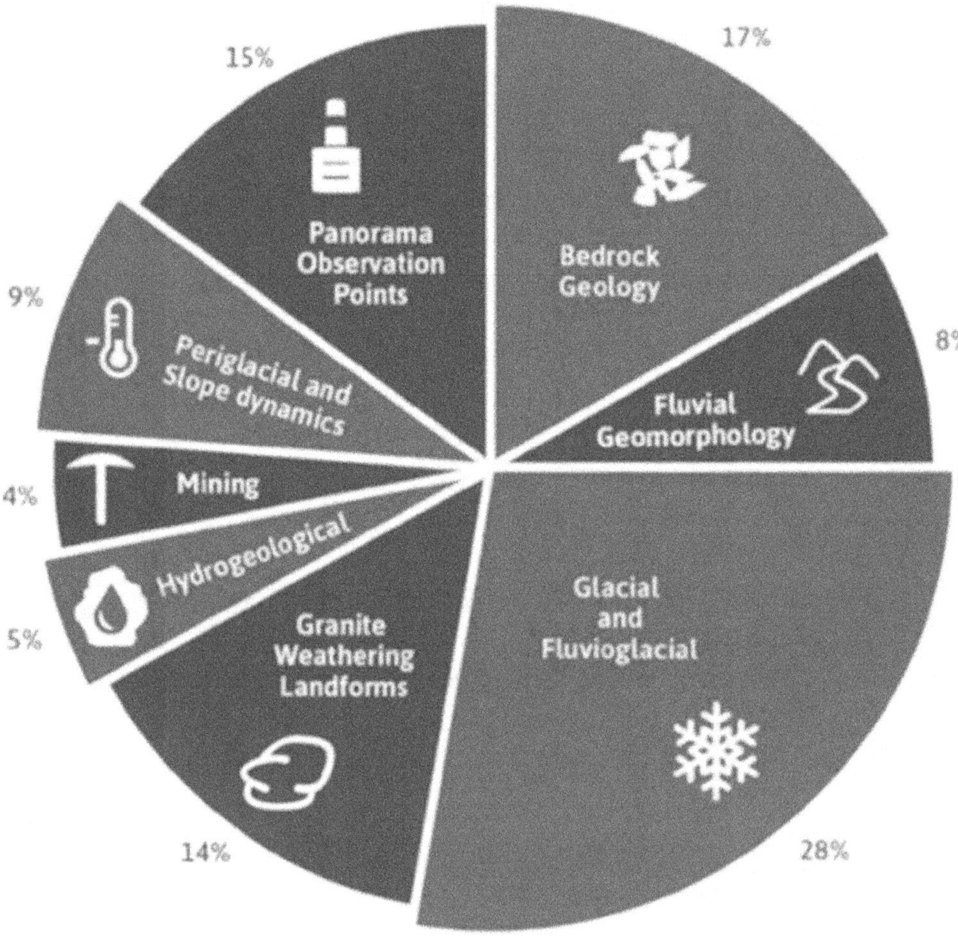

Most visitors are however limited to roadside stops for photos or shopping, concentrating more time in the villages, but not spending more than one or two nights in Estrela Mountain, with most of them still not searching for nature-related activities. Hence, despite the good existing trail network, it's rare to meet more than two to four people in a daylong walk in the plateau. This fact is essentially cultural, and relates to the lack of walking tradition by the Portuguese, and is also due to the relatively small number of foreign tourists. Hence, the margin for growth of geotourism, nature and adventure tourism in the region is large. While this shows the high potential for an increase in visitor numbers after obtaining the UNESCO brand, it will also pose major challenges for managing geosites and limiting their degradation. Therefore, the AEG is providing to the municipalities and the Natural Park, detailed management and monitoring plans for the geosites, especially for those with more potential for visitation.

Education is the basis of geotourism and a pillar for heritage sustainability and their local communities. Tourists are learning according to the didactic interpretation tools that are being offered. The more explicit the phenomena are and the more appealing the interpretation is, the more efficient the dissemination of geology becomes. On the other hand, a citizen who may have had the possibility of having studied geology is more conscious and interested in the practice of geotourism. Furthermore, geotourism allows for the promotion of geological patrimony through the sensitisation and the promotion of geology that becomes essential for a concerted geoconservation. For this to happen, awareness is important. In this way, geotourism assumes a fundamental role in sustainable local development, whereas leisure activities can involve environmental education (Hose, 2005; Silva, 2014).

Geoparks provide an opportunity for education through touristic availability of natural and cultural resources, in particular, that are geologically and geomorphologically significant. Geoeducation is focused on the knowledge of the history of our planet, as well as the influence of abiotic nature on human development (Farsani, et al., 2012; Hose, 2012). The educational promotion strategies using regional heritage resources, in an integrated approach designed in coordination with the programmatic frameworks of the different teaching cycles, are stimulating educational resources for the teaching process outside the classroom, generating experiences and learning *in situ*. The possibility of extending knowledge outside the school environment is of particular importance to the development of children and adolescents (Świercz and Smorzewska, 2015). In this collaborative context, the articulation of schools and higher education institutions is encouraged, in a synergistic connection with Geopark in the construction of active and experimental models, generators of cross-sectional learning, and territorial valorisation.

Table 3. Heritage elements of geopark territories and assets in the tourist strategy to promote the destination and its sustainability (Authors' elaboration)

Civil	Civil and architectural heritage, evidenced by the materials used, frontage architecture and functional structures. The organisation of the settlement, the water management structures, and the division of properties.
Religious	Diversity of elements, with different meanings and monumentalities associated with the Catholic religion and the Jewish presence, generating in some municipalities elements of international interest, materialised in the existing architecture and symbolic elements.
Military	Composed of forts, castles, walls, and bridges, dating from the Roman Empire, as a defensive structure and control of the territory of the mountains over the adjacent areas.
Archaeologic	A diverse set of elements that show the human presence in these territories and the forms of their appropriation since prehistory, Rome, and Medieval times. Some elements have great preservation and historical and scientific value in understanding the occupations and forms of human presence.
Mining	Mining complexes, which represent different extractive moments and historical contributions, with significant and specific importance for the local economy, technological evolution, and historical preservation.
Textile	Sets of factories and wool processing plants, revealing the strong link between agropastoral, the abundance of water, and the holding of specialised knowledge in the production of woollen yarns and fabrics. Some of these units are still active and others of historical relevance turned into Museums (Real Fabrica de Panos-Covilha).
Rural	Composed of traditional activities linked to agropastoral, structures for dividing the property and respective walls, ponds and dams, set of utensils and mechanism that last over time and represent the specific mountain ways of life
Festivities and Traditions	Religious and pagan festivities, many associated with vegetative cycles, or solar moments (solstices and equinoxes). Promotion of local products, crafts and mountain-specific traditions (sacred and profane)
Pastoralism	A striking element of Serra da Estrela's economic activity, giving meaning to two regional symbols, Cheese and Dog. The cheese corresponding to a valued gastronomic product and recognised for its organoleptic characteristics. The dog is a symbol for the protection of herds, their conduct, and an element of struggle and support for shepherds and the most isolated communities on the mountain.

The educational programs aim to empower communities of scientific knowledge, sensitivity, and attitudes for the preservation and enhancement of ecocultural resources and sustainable tourism practices, given the fragility that many of the resources present. In this context, pedagogical paths and programs were structured for the different levels of education, which aim to stimulate curiosity and interest in geosciences, contributing to the conservation and enhancement of the natural and cultural heritage of Estrela Geopark. Brilha (2009) and Mignosn (2018) defended educational interventions using geoparks that can contribute to promoting significant and relevant learning on geology and geoconservation, as well as to stimulate curiosity and interest for visiting other geoparks. These educational initiatives can serve as a vital link between scientific work occurring in the field and the people living in the region or community (Komoo et al., 2010, Azman et al., 2016). The practices developed allow the dissemination of geodiversity, biodiversity and mountain lifestyles, stimulating interest in the discovery and knowledge of the Geopark's territory. They promote tourism attitudes and practices with protection and preservation concerns, enhanced resource preservation, and sustainability (Hose, 2005; Bagri & Kala, 2016). Different interpretations are established, with transversal character and levels of conceptualisation, according to the different cycles, in a process of coherence with the curricular goals and looking for an evolutionary systematisation of the issues related to resources and tourist infrastructure, leisure and heritage. At basic education level management and sustainability of resources, impacts of tourism and sustainable development are addressed. For secondary education, analyses and reflections on tourism and spatial planning, environmental policies, and their relationship with tourism are promoted, as well as new opportunities for the rural world (Table 4).

Table 4. Educational Programmes and tourist Content in Pedagogical Routes according to the different levels of education. (Associação Geopark Estrela - AGE, 2018)

1st and 2nd cycle of Basic Education	3rd Cycle of Basic Education	Secondary education
Tourism and local environment	**Tourism and sustainability**	**Tourism and land use planning**
• Identify tourist attraction factors (beaches, natural parks, spas, monuments); • Recognise tourist infrastructure and equipment in the region (hotels, campsites, restaurants); • Discuss the advantages and disadvantages of tourism for development; • Reflect on leisure and tourism activities in the Serra da Estrela region; • Promote the preservation of heritage and the importance of protected areas.	• Sustainable resource management; • Natural resources and uses. - Classification and forms of exploitation; • Nature protection and conservation (Protected Areas); • Main tourist products and associated practices; • Main impacts of tourism; • Sustainable development in tourism and development.	• Environmental enhancement and protected areas; • New opportunities for rural areas; • Cultural heritage and ecocultural value of landscapes; • Tourism and territorial enhancement; • Environmental policies and tourism. • Tourism and territorial cohesion in mountain areas.

In terms of education and sustainability in tourism, the Estrela Geopark, with the development of these pedagogical programmes, promotes:

1. Approach to local communities and territorial enhancement in connection with municipalities;
2. Development of protocols with all the Schools in the dissemination of Estrela Geopark;
3. Creation of didactic materials to support visits and definition of routes;
4. Development of interpretive panels for visitors to Estrela Geopark;

5. Promotion of congresses and training courses aimed at teachers, educators and researchers;
6. Development of the Education and Science Network in conjunction with higher education institutions, research centres, and the school community;
7. Technical monitoring of tourist and study visits and awareness of sustainable practices.

These activities, adapted to groups and ages, include recreational actions dedicated to geodiversity (educational games, photography, contests) and more formal initiatives such as field classes, workshops, seminars, and laboratory, promoting multidisciplinarity (Table 4). In this way, aspects of biodiversity, culture, and ways of life in the region are integrated, promoting a broad and rigorous knowledge of Estrela Mountain and raising awareness of the sustainable tourist appropriation of its resources by disseminating geosciences. The most efficient way to achieve geoconservation is to increase public awareness about the value of geoheritage sites (scientific, aesthetic, educational, tourism, economic, intrinsic) through promotion and interpretation via geotourism (Burek & Prosser, 2008; Hose, 1995, 2008, 2012; Hose & Vasiljevic, 2012).

Scientific research, courses and the use of geosciences in activities for groups and finally geotourism are activities that require a decreasing amount of scientific transfer but, conversely, provide more economic benefits and possibilities of development and well being for the communities (Meléndez et al., 2011). The arrival of tourists can, in effect, promote the construction of access infrastructures, hotels, restaurants, tourist facilities, marketing products, and souvenir trade.

They are identified as logical dissemination of knowledge, strengthening geoconservation attitudes, development of resources, and guidance in analysing the functions and effects of their use on the environment and community. The logic of action, education practices, and geoconservation are systematised for sustainable tourism and for responsible behaviour and sense of belonging with the territory and its values. Geotourism seeks to attract the public and generate a socioeconomic benefit, based on the creation of a tourism infrastructure to support some elements of the geological heritage present in a region. To do this, it must be planned and must be locally beneficial in socio-environmental and economic terms (Dowling, 2009).

The richness of its heritage, the diversity of glacial traces, biodiversity, its ways of life make Estrela Mountain a unique geography whose structured approach will allow the creation of a strong perception and identity with tourists, heritage, and cultural potential. In the context of the sustainable development of the Estrela Mountain territory and based on the principle of conservation of the natural and cultural heritage, the Estrela Geopark intends to implement projects for the enhancement of natural spaces; to develop a network of Science and Education for Sustainability; to promote projects of environmental education; to strengthen relations with the community; to improve tourism; and to promote actions that increase the well-being of local communities. In general terms, the Estrela Geopark takes a collaborative position and supports the promotion of knowledge and education, encompassing different processes and actions, supported in the territories as didactic resources and in the development of materials and programs that allow a structured knowledge of the territories' histories, its geo-resources, geoconservation practices and the promotion of sustainable tourism strategies (Figure 5).

These activities, adapted to groups and ages, include recreational and recreational actions dedicated to geodiversity (educational games, photography, contests) and more formal initiatives such as field classes, workshops, seminars and/or laboratory, promoting multidisciplinarity. In this way, aspects of biodiversity, culture and ways of life in the region are integrated, promoting a broad and rigorous knowledge of Serra da Estrela and raising awareness of the sustainable tourist appropriation of its resources by disseminating

Figure 5. Systematisation of the role of Estrela Geopark in promoting geotourism and geoconservation (Authors elaboration)

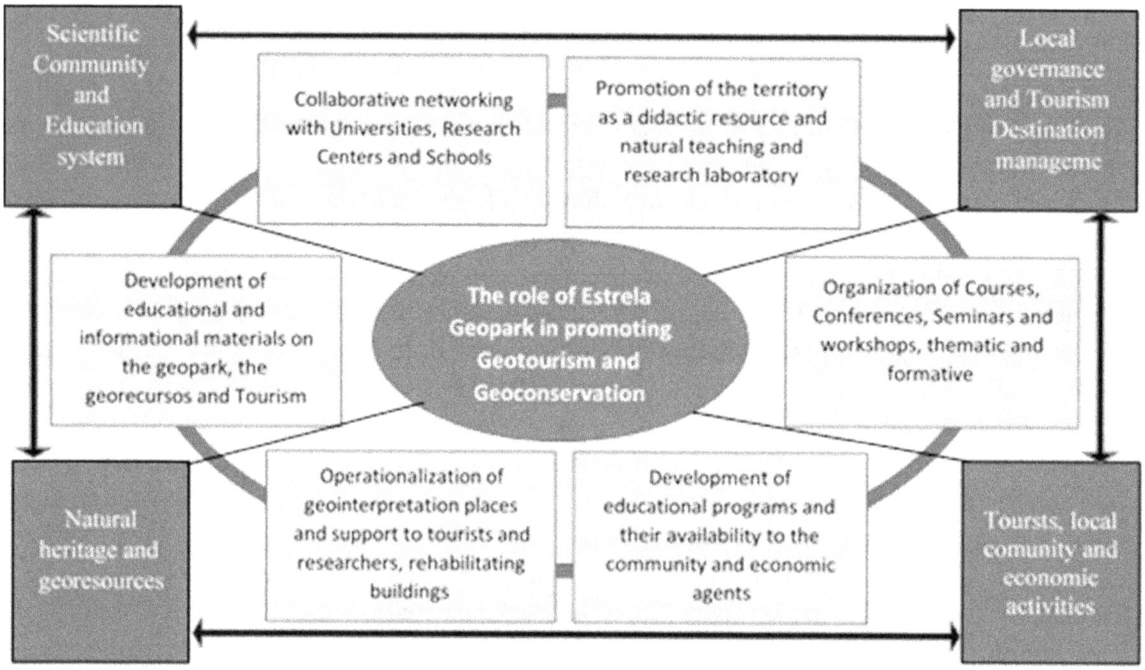

geosciences. The most efficient way to achieve geoconservation is to increase public awareness about the value of geoheritage sites (scientific, aesthetic, educational, tourism, economic, intrinsic) through promotion and interpretation via geotourism (Hose, 2012; Hose and Vasiljevic, 2012, Miśkiewicz, 2016).

Scientific research, teaching geology, courses and the use of geology in activities for groups and finally geotourism are activities that require a decreasing amount of scientific transfer but, conversely, provide more economic benefits and possibilities of development and well being for the communities (Meléndez et al., 2011). The arrival of tourists can, in effect, promote the construction of access infrastructures, hotels, restaurants, tourist facilities, marketing products, and souvenir trade.

CONCLUSION

The mountain territories have undergone significant transformations in their forms of use, perception and appropriation, for which tourism and leisure activities are the main agents of change. The Portuguese mountains, despite the trajectory of depopulation and disarticulation of traditional productive activities, have become known due to a growing appreciation of the natural heritage and the demand for tourism activities, reorientation in their functions and the management concerns of local administrative entities.

The classification of Estrela Mountain as a Geopark has in its mission to contribute to the protection, enhancement, and dynamisation of the natural and cultural heritage, with special emphasis on the geological heritage, to deepen and disseminate scientific knowledge, promoting tourism and the sustainable development of the Estrela Mountain. The relationship between geoconservation, science and tourism must be established in the different directions, enabling an integrated approach to the territory, in the natural and human dimensions and their interactions, promoting knowledge, raising awareness of heritage values and building attitudes and tourist practices that value the destination and enable their sustainable use. The educational programs, the several interpretative routes, seminars, workshops and training courses, the close relationship with the schools, the synergies with the tourist actors, among others, allow the development of new relationships of proximity and tourist valorisation, considering the patrimonial value and its sustainability.

Overall, the Geopark fosters the development of an integrated development strategy, coupled with geo-conservation, education and tourism, supporting the construction of developing strategies for community well-being and a value of participation in the tourist promotion of the region. The implementation of a geoconservation strategy based on scientific criteria and the creation of a science and education network for sustainability allows us to deepen the knowledge of the territories in a transversal way, enhancing the connection with Universities and Research Centers. The active articulation with the municipalities and the Schools allows us to monitor the interventions, adapt joint strategies, and promote sustainability initiatives in the appropriation of this mountain territory classified by UNESCO. The increasing tourist attraction supported in the geosites of the Estrela Geopark has allowed the dissemination of the destination (reducing the effects of seasonality), the promotion of geosciences, the development of collaborative strategies with educational institutions and local stakeholders. The geoconservation and the use of geoheritage for the promotion of sustainable development, especially in the most isolated areas of the Estrela Geopark, dissemination of knowledge on geoheritage to promote education, healthy lifestyles, and territorial cohesion is at the core of the AEG strategy. In this context, it had been built: collaboration networks that can promote a deeper knowledge of the territories and their resources; shared geoeducation and geoconservation strategies; incremented the involvement of communities; promoted a guided visit and tend to be more responsible with the existing heritage and ways of local life.

REFERENCES

AGE. (2018). *Programas Educativos Estrela Geopark (Ensino Básico e Secundário).* http://www.geoparkestrela.pt/application/views/assets/documents/doc_conteudos/Programas_educativos_Ensino_Basico_18_19.pdf

Aguiar, C., Rodrigues, O., Azevedo, J., & Domingos, T. (2016). Montanha. In Ecossistemas e Bem-Estar Humano. Avaliação para Portugal do Millenium Ecosystem Assessment. Lisboa, Escolar Editora.

Ariza, C., Maselli, D., & Kohler, T. (2013). Mountains: Our Life, Our Future. Progress and Perspectives on Sustainable Mountain Development from Rio 1992 to Rio 2012 and Beyond. Bern: Swiss Agency for Development and Cooperation (SDC), Centre for Development and Environment (CDE).

Azevedo, J. C., Cadavez, V., Arrobas, M., & Pires, J. M. (2016). Sustentabilidade da montanha portuguesa: realidades. In J. C. Azevedo, V. Cadavez, M. Arrobas, & J. M. Pires (Eds.), *Sustentabilidade da Montanha Portuguesa: Realidades e Desafios* (pp. 9–38). IPB.

Azman, N., Halim, S. A., Liu, O. P., Saidin, S., & Komoo, I. (2010). Public Education in Heritage Conservation for Geopark Community. *Procedia: Social and Behavioral Sciences, 7*(C), 504–511. doi:10.1016/j.sbspro.2010.10.068

Bagri, S. C., & Kala, D. (2016). Residents' attitudes toward tourism development and impacts in Koti-Kanasar, Indroli, Pattyur tourism circuit of Uttarakhand, India. *PASOS Journal of Tourism and Cultural Heritage, 14*(1), 23–39. doi:10.25145/j.pasos.2016.14.002

Brilha, J. B. R. (2005). *Património Geológico e Geoconservação: a conservação da Natureza na sua vertente geológica*. Palimage Editores http://www.dct.uminho.pt/docentes/pdfs/jb_livro.pdf

Brilha, J.B.R. (2009). A Importância dos Geoparques no Ensino e Divulgação das Geociências. *Geologia USP, 5*, 27-33.

Burek, C. V., & Prosser, C. D. (2008). The history of geoconservation: An introduction. Geological Society, London, Special Publications. *Geological Society of London, Special Publications, 300*, 1–5. doi:10.1144/SP300.1

CE. (2004). *Mountain Areas in Europe: Analysis of mountain areas in EU member states, acceding and other European countries*. https://ec.europa.eu/regional_policy/sources/docgener/studies/pdf/montagne/mount1.pdf

Cunha, L. (2003). A montanha do centro português: espaço de refúgio, território marginal e recurso para o desenvolvimento local. In *Território, Ambiente e Trajectórias de Desenvolvimento* (pp. 72–85). Centro de Estudos Geográficos, Universidade de Coimbra.

Cunha, L., & Cordeiro, A. M. (2000). A Montanha em Portugal – Uma perspectiva geográfica. Reunión de Expertos en Desarrollo Rural: Regiones Periféricas Europeas, las Áreas de Montaña.

Dax, T. (2008). *The role of mountain regions in territorial cohesion: a contribution to the discussion on the Green Paper on territorial cohesion*. Euromontana. https://www.euromontana.org/wpcontent/uploads/2014/08/final_dax_report_april2008_enlight.pdf

Dax, T. (2019). Development of mountainous regions: smart specialisation approaches as a means to overcoming peripheralisation. In I. Kristensen, A. Alexandre Dubois, & J. Teräs (Eds.), *Strategic Approaches to Regional Development. Smart Experimentationin Less-Favoured* (pp. 52–67). Routledge.

Debarbieux, B., Oiry Varacca, M., Rudaz, G., Maselli, D., Kohler, T., & Jurek, M. (Eds.). (2014). Tourism in Mountain Regions: Hopes, Fears and Realities. In Sustainable Mountain Development Series. Geneva, Switzerland: UNIGE, CDE, SDC.

Debarbieux, B., & Rudaz, G. (2015). *The Mountain: A Political History from the Enlightenment to the Present*. University of Chicago Press. doi:10.7208/chicago/9780226031255.001.0001

do Ambiente, A. (n.d.). *Sistema Nacional de Informação de Ambiente*. https://sniamb.apambiente.pt/

Dowling, R. (2009). Geotourism's contribution to Local and Regional Development. In C. Neto de Carvalho & J. C. Rodrigues (Eds.), *Geoturismo & Desenvolvimento Local* (pp. 15–37). Idanha-a-Nova.

Dowling, R. (2011). Geotourism's global growth. *Geoheritage, 3*(1), 1–13. doi:10.100712371-010-0024-7

Dowling, R. (2015). Geotourism's contribution to sustainable tourism. In M. Hughes, D. Weaver, & C. Pforr (Eds.), *The Practice of Sustainable Tourism: Resolving the paradox London* (pp. 207–227). Routledge.

Drexler, C., Braun, V., Christie, D., Claramunt, B., Dax, T., Jelen, I., Kanka, R., Katsoulakos, N., Le Roux, G., Price, M., Scheurer, T., & Weingartner, R. (2016). *Mountains for Europe's Future—A Strategic Research Agenda.* Bern, Switzerland, and Innsbruck, Austria: Mountain Research Inititative (MRI) and Institut für Interdisziplinäre Gebirgsforschung (IGF). http://mri.scnatweb.ch/en/resources/publications/mri-publications/2573-mountains-for-europe-s-future-a-strategic-research-agenda/file

EC. (2004). *Mountain areas in Europe: Analysis of mountain areas in the EU Member States, acceding and other European countries.* Stockholm: Nordregio. https://ec.europa.eu/regional_policy/sources/docgener/studies/pdf/montagne/mount1.pdf

Eder, W., & Patzak, M. (2004). Geoparks – geological attractions: A tool for public education, recreation, and sustainable economic development. *AUC Geographica*, *51*(2), 235–246. doi:10.18814/epiiugs/2017/v40i4/017036

EEA. (2010). *Europe's ecological backbone: recognising the true value of our mountains. Copenhagen.* European Environment Agency. https://www.eea.europa.eu/publications/europes-ecological-backbone

Ehsan, S., Leman, M., & Begum, R. (2012). Geotourism: A Tool for Sustainable Development of Geoheritage Resources. *Advanced Materials Research*, *622-623*, 1711–1715. doi:10.4028/www.scientific.net/AMR.622-623.1711

EU. (2008). *Green Paper on Territorial Cohesion*. European Commission.

Farsani, N. T., Coelho, C., & Costa, C. (2012). Geotourism and geoparks as gateways to socio-cultural sustainability in Qeshm rural areas, Iran. *Asia Pacific Journal of Tourism Research*, *17*(1), 30–48. doi:10.1080/10941665.2011.610145

Fernandes, G. (2012). Políticas públicas e equidade territorial em zonas de montanha: O exemplo da Cordilheira Central. In Respuestas de la Geografía Ibérica a la crisis actual, XIII Coloquio Ibérico de Geografía. AGE e APG, Santiago de Compostela.

Fernandes, G. (2015). *Geografia, cultura e desenvolvimento turístico em territórios de montanha: Das perspetivas às logicas de atuação e governação territorial.* Coleção Instituto Politécnico da Guarda.

Fernandes, G. (2017). Tourism, ecocultural heritage, development strategies, and territorial valorisation in the border mountains of the Iberian central mountain range. *Revista Egitania Scientia*, *17*, 67–81.

Fernandes, G. (2018). Resilience and territorial dynamics in the mountains in the extreme western areas of the Iberian Peninsula Range: Challenges of governance, territorial cohesion, heritage and tourism. In E. Gomez-Pellón (Ed.), *Rural worlds, social sustainability, and local landscapes in the globalisation era. Case studies in southern Europe* (pp. 265–288). Thomson Reuters Editorial ARANZADI.

Fernandes, G. (2020). Dynamics of tourism development and resilience in the medium mountains of Portugal: Challenges for the sustainable management in mountains of the Central Mountain Range. In D. Kala & S. Bagri (Eds.), *Global Opportunities and Challenges for Rural and Mountain Tourism* (pp. 263–282). IGI Global. doi:10.4018/978-1-7998-1302-6.ch015

Frey, M. L., Schafer, K., Buchel, G., & Patzak, M. (2016). *Geoparks: A Regional European and Global Policy*. Geotourism. doi:10.1016/B978-0-7506-6215-4.50014-2

Gløersen, E., Price, M., Borec, A., Dax, T., & Giordano, B. (2016). Cohesion in Mountainous Regions of the EU—Research for REGI Committee, European Parliament, Directorate-General for Internal Policies Policies Department B. In *Structural and Cohesion Policies, Regional Development, IP/B/REGI/IC/2015_175*. Brussels, Belgium: European Parliament. https://www.europarl.europa.eu/RegData/etudes/STUD/2016/573420/IPOL_STU%282016%29573420_EN.pdf

Gomes, H., Fernandes, M., Castro, E., & Vieira, G. &. (2017). The (Serra da) Estrela Aspiring Geopark (Portugal): Preserving geoheritage, while promoting science and its links to local communities. *Geophysical Research Abstracts* Vol. 19. EGU General Assembly.

Hose, T. (2005). Geoturismo europeo. Interpretacíon geológica y promoción de la conservación geológica para turistas. In D. Barretino, W. A. P. Winbledon, & E. Gallego (Eds.), *Patrimonio geológico: conservación y gestión*. Istituto Tecnológico Geomineiro de España.

Hose, T. (2011). The English origins of geotourism (as a vehicle for geo-conservation) and their relevance to current studies. *Acta Geographica Slovenica*, *51*(3), 343–359. doi:10.3986/AGS51302

Hose, T. (2012). 3gs for modern geotourism. *Geoheritage*, *4*(1-2), 7. doi:10.100712371-011-0052-y

Hose, T., & Vasiljevi, D. A. (2012). Defining the nature and purpose of modern geotourism with particular reference to the United Kingdom and South-East Europe. *Geoeheritage*, *4*(1-2), 25–43. doi:10.100712371-011-0050-0

Kala, D., & Bagri, S. C. (2018). Barriers to local community participation in tourism development: Evidence from mountainous state Uttarakhand, India. *Tourism (Zagreb)*, *66*(3), 318–333.

Kapos, V. (2000). Developing a map of the world's mountain forests. In Forests in sustainable mountain development: a state-of-knowledge report of 2000, (pp. 4-9). IUFRO Series N. 5 Wallingford, Reino Unido, Cab International.

Korner, C., & Ohsawa, M. (2005). Mountain systems. In R. Hassan, R. Scholes, & N. Ash (Eds.), *Ecosystems and Human Well-being: Current State and Trends* (Vol. 1, pp. 2–23). Island Press.

Meléndez, G., Fermeli, G., Moreira, J. C., Escorihuela, J., & Basso, A. (2001). What do we mean when we say Geotourism International Congress Geotourism in Action, 1, 2011, Arouca. Proceedings Arouca: Geopark Arouca, 97–100.

Messerli, B. (2012). Global change and the world's mountains. *Mountain Research and Development*, *32*(S1), 55–63. doi:10.1659/MRD-JOURNAL-D-11-00118.S1

Meyer-Cech, K., & Pröbstl, U. (2013). *Sustainable tourism in mountainous regions. In Tourism as a Challenge WIT Transactions on State of the Art in Science and Engineering* (Vol. 73). WIT Series on Tourism Today.

Migon, P. (2018). Geoheritage and geotourism. A European perspective. The Heritage Matters Series. Doi:10.2478/logos-2018-0017

Miśkiewicz, K. (2016). Promoting geoheritage in geoparks as an element of educational tourism. In A. Szponar & S. Toczek-Werner (Eds.), *Geotourism: organisation of the tourism and education in the geoparks in the Middle-Europe Mountains*. University of Business in Wrocław.

Mokhtari, D., Roostaei, S., Khodadadi, M., Ahmadi, M., Ebrahimi, O., & Shahabi, H. (2019). Evaluation of the Role of Environmental Education in Manesht and Ghelarang Geotourism Destination, Iran. *Journal of Quality Assurance in Hospitality & Tourism*, 203–218. doi:10.1080/1528008X.2019.1616039

Newsome, D., & Dowling, R. (2018). Geoheritage and Geotourism. In Geoheritage: Assessment, Protection, and Managemen Book. Elsevier Inc. doi:10.1016/B978-0-12-809531-7.00017-4

Pansiri, J. (2008). The effects of characteristics of partners on strategic alliance performance in the SME dominated travel sector. *Tourism Management*, *29*(1), 101–115. doi:10.1016/j.tourman.2007.03.023

Rieder, P., & Wyder, J. (1997). Economic and political framework for the sustainability of mountain áreas. In Mountains of the world: A global priority. Carnforth, Reino Unido, Parthenon.

Rodrigues, J. C., & Neto de Carvalho, C. (2010). Património geológico no Geopark Naturtejo: Base para uma estratégia de geoturismo. *E-Terra*, *18*(11), 1–4.

Shekhar, S., Kumar, P., Chauhan, G., & Thakkar, M. (2019). *Conservation and Sustainable Development of Geoheritage, Geopark, and Geotourism: a Case Study of Cenozoic Successions of Western Kutch*. Geoheritage. doi:10.100712371-019-00362-5

Silva, A. (2014). *A importância dos georrecursos para a divulgação e ensino das Geociências* (Dissertação de Mestrado). Universidade Nova de Lisboa.

Strickland-Munro, J. K., Allison, H. E., & Moore, S. A. (2010). Using resilience concepts to investigate the impacts of protected area tourism on communities. *Annals of Tourism Research*, *37*(2), 499–519. doi:10.1016/j.annals.2009.11.001

Stueve, A. M., Cooks, S., & Drew, D. (2002). The Geotourism Study: Phase I. Executive Summary. Washington, DC: Travel Industry Association of America.

Świercz, A., & Smorzewska, E. (2015). Selected examples of interactive teaching methods in the Centre of Geoeducation in the city of Kielce (Poland). *Procedia: Social and Behavioral Sciences*, *174*, 680–686. doi:10.1016/j.sbspro.2015.01.601

Zouros, N., & Mc Keever, P. (2008). The European Geoparks Network. *Proceedings. 3rd International UNESCO Conference on Geoparks*, 11-12.

Chapter 8
Ecotourism as a Trigger for Economic Development: A Resource-Based View of Calabria

Giuseppe Umberto Cantafio
University of Sunderland in London, UK

Vipin Nadda
University of Sunderland in London, UK

ABSTRACT

Peripheral Mediterranean regions, such as Calabria, have an urgent need to revitalise the economy through better use of natural resources. The increase in tourism can bring the sought after foreign exchange, resulting in the development of small and medium local enterprises. The chapter aims at evaluating the potential of the Calabria region. This was done by carrying out an analysis of a resource-based view of the aforementioned region. A list of potential nature-based tourism resource sites was identified and discussed. This study highlighted key findings and recommendations to policymakers.

INTRODUCTION

Many studies in the past have utilised different frameworks for assessing resources for nature-based tourism (Priskin, 2001; Güngör. & Bozyigit, 2011; Ataberk & Baykal, 2011; Widawski et al., 2018; Yu & Yin, 2018). Fewer studies have been framed around the case study of the Calabrian region in Southern Italy since it is not one of the most famous destinations for international tourists if compared to other Italian regions. The present study aims to evaluate the potential of the Calabria region in terms of touristic assets. The aim will be achieved by means of three objectives: producing tangible and measurable outcomes as a result of the Resource-Based View; analysing the current literature of the major topics linked to the themes of tourism and ecotourism; providing policy recommendations and critical thinking around the abovementioned themes.

DOI: 10.4018/978-1-7998-5691-7.ch008

The researchers hypothesise that there is a massive unrealised potential in terms of tourism and this study will attempt at proving this hypothesis by utilising frameworks available in literature in order to analyse the current situation of the geographical area taken as a scope of the study. More specifically to the existing gap in the literature, an attempt will be made to fill this gap, by providing a background on the major topics linked to ecotourism, and by offering a discussion on Economic Development, Sustainable Development, and Sustainable Tourism Development. In the present study, some theoretical background on Resource-Based View (RBV) will be offered as well. Subsequently, the chapter will provide a brief insight about an economic outlook of southern Italy and is particularly framed around the Calabrian region. Afterwards, the RBV approach will be applied to the particular case study of Calabria followed by some critical policy recommendations and conclusions.

LITERATURE REVIEW

Before deepening the topic of ecotourism, it is important to situate the present work in the setting of prior studies and academic works relating to the broader themes of economic development, sustainable development, sustainable tourism development, and then narrowing down the discourse to the theme of ecotourism. The following sections will offer a critical investigation of existing literature consistent with appropriate topics, validating how the present study will detect the gap in the literature, and outlining the academic and conceptual background of the study.

Economic Development

The description of economic development, on the word of Todaro (1995), is a growth in living standards, improvement of the people self-esteem needs and a free and just society. He suggested the Human Development Index as the most accurate assessment framework for economic development, in which he considered literacy rates, quality of life, life expectancy rates, and other indicators. These parameters, according to him, have a comprehensive impact on productivity and ultimately could lead to Economic Growth. The main aim of economic development is to increase the economic prosperity of a region through several initiatives that may involve job creation, and job retention, as well as tax base improvements and improved quality of life. As per there is not just a unique definition of economic development, therefore there will be wide-ranging approaches, plans, policies, and programs for realising successful economic development. Societies vary in their physical form, geographic settings, and political situations. Each society, for this motivation, will need a customised path for achieving its economic development (Economic Development Reference Guide).

Sustainable Development

The Brundtland Commission (1987) described sustainable development in terms of meeting the needs of future generations; in this day and age, there is more consciousness on the fact that the necessities of future societies may be dissimilar across countries and across time. An emergent and important interpretation of sustainable development focuses on conserving and increasing the opportunities open to people in all the countries around the world (Serageldin & Steer, 1994). The World Bank has been among the first institutions that incorporated comprehensive wealth inside sustainability analysis (Jarvis et al., 2011). Inside technical reports, some components of comprehensive wealth have been estimated, for instance, the values of natural, human and reproducible capital.

Sustainable Tourism Development

It can be found in literature as a linkage between sustainable development and tourism. The clustering of similar activities encompassing conservation of the territory and the creation of attractions in less developed areas can inspire towards major cooperation and partnerships between local firms and between communities in local and bordering regions. This new sustainable way of spurring the local tourism by living the natural attractions and responsibly utilising ecological resources can act as a vehicle for stimulating the local economic development (Briedenhann & Wickens, 2004).

Given the economic struggle that many peripheral European countries have faced in the last decade, there has been a delayed catch-up from the economic crisis of 2008 that swept all over the world. Under this perspective, sustainable tourism can be seen as an alternative developmental strategy for the economic and social regeneration of rural areas in peripheral Europe. This has been experienced in the past decades in other areas of Europe, such as Eastern Europe. These areas faced in the 1980s and in the first part of the 1990s a rapid rise in rural unemployment and tourism has been identified as a catalyst to fuel economic growth, increase the viability of underdeveloped regions and recover the standard of living of local communities (Simpson et al., 1998).

Sustainability has become a major concern of recent times concerning the appropriate use of resources at any tourism destination. One of the reasons for such a situation is the lack of analysis of impacts of development on the ecosystems that form the resource base for tourism products (Duncan & Dangerfield, 1999). Sustainable tourism development is a multidimensional concept as it can be an agent for socio-cultural enhancement, economic growth (Hunter, 1995) and environmental protection (Valentine, 1993), etc. to mention a few. It seeks long-term growth and development through the sustained availability of resources for future generations based on equitable distribution of harnessing opportunities. One of the integral aspects of sustainable tourism is that it is supply-led (Goodwin, 1996; Botterill & Pearce, 1995), the focus being on the management of various impacts unlike mass tourism which is more of supply led, being driven by tourist motivations and demand (Cooper et al., 1993). Thus, sustainable development is more focused on long term economic as well as ecological sustainability along with human life equity (Woodley, 1992). So it is an effort to find a balance between economy, environment and society (Butler, 1998).

It has always been a challenge to develop universal principles of sustainable tourism development as the sustainable systems are not generalised at the global scale, but are adapted to local conditions (Meyer & Helfman, 1993). Eco-tourism development provides benefits to various communities in terms of environment, culture and economics (Eshetu, 2014). Unlike other types, it provides a chance to the tourists to experience non-exploitative use of natural resources in such a way that it helps sustain the local economy as well. The modern aspect of sustainability is based on the understanding of management that preclude the exploitation of resources. Many approaches are available to explore sustainability, but here is an overview of the resource-based view (RBV).

Ecotourism

Ecotourism (sometimes referred to as Ecotourism or Sustainable Tourism) became a widely used term in the early 1980s when Hector Ceballos Luscurain coined the term ecotourism in 1983 as 'nature-based travel to a relatively undisturbed area' and Boo (1990), intended it as: "tourism that involves travelling to relatively undisturbed or uncontaminated natural areas with the specific objective of studying, admiring, and enjoying the scenery and its wild plants and animals, as well as any existing cultural manifestations (both past and present) found in these areas. It is also defined as a form of tourism that encompasses the discovery of the fragile, unspoiled, and relatively intact natural areas, all seen under a sustainable perspective, with the involvement of public authorities and national trust bodies, small and medium enterprises, and not-for-profit businesses, where activities have a low impact on the natural ecosystems.

Some other authors have provided more of a community-based definition for ecotourism, as per Cusack and Dixon (2006) it needs to be differentiated from nature-based tourism, which it helps instead in sustainable rural development and pushes for the protection of biodiversity and makes sure that conservation economically can be viable for local communities. All these processes and different definitions have a red thread; they can represent a small-scale alternative to standard commercial (mass) tourism.

The aim of ecotourism might be not only strictly related to the creation of profit for the firms. On the contrary, ecotourism can serve as a form of education to the traveller, providing funds for ecological conservation, directly helping the economic development and political empowerment of local communities, and promoting respect for diverse cultures and human rights. Since the 1980s environmentalists have thought about ecotourism as a critical practice, to allow future generations to live the same experience to destinations relatively undamaged by anthropic activity.

The process triggered by the implementation of ecotourism in a region could be viewed as a cycle, which is self-sustained, once that it starts working at full capacity. Eco-Tourism relies on good conservation and maintenance of the natural environment and natural resources, and it can work if there is the safety of the places and on the good status of infrastructures and services, which need to be maintained regularly to attract tourists. Also, the incoming revenues generated from tourists by utilising public transportation and other services, and shopping at local retailers, can serve as a financial stimulus to the local economy. The more tourists will arrive in the region, the more the returns for retailers will result; this will increase the tax revenues of the local government that in turn will invest in safety, infrastructures and keeping a good natural environment.

The previous section of the chapter has provided a solid theoretical foundation on which to lay on the study on ecotourism. Having analysed and synthesised a wide-ranging selection of appropriate related topics in literature has helped us to define what is ecotourism and identifying the need for it. The review of literature has supported us to construct a rational framework for the investigation, defending the thesis made by the researchers, focusing the gaps in the literature, and demonstrating how the study will contribute to current knowledge. The review has assisted to position the study within the current academic settings of present on-going discourse in the field of ecotourism.

Figure 1. Economic development cycle linked to eco-tourism (adapted from Cantafio, 2017)

TOURISM RESOURCE ANALYSIS

Resource-Based View

Resource-based theory focuses upon the various tangible as well as intangible resources like physical, financial, human, technological, skills, experience, reputation etc. that are utilised by the organisations to develop various products and services to be offered to satisfy customers/tourist needs (Barney, 1991). RBV defines the capability of an organisation to act through its resources (Olszak, 2014). The sustainable capability can be achieved through valuable, rare and inimitable resources. According to Teece et al. (1997), the main role of strategic management of any organisation is to adapt the resources and competencies to match the changing environment.

Here is the list of some of the tourism resources (Nadda, 2009), which if appropriately used, can help in sustainable tourism development.

- **Natural Resources**

Landscape, Mountain, forest, outstanding sunrise and sunset scenes, Vantage points other landscape aspects, special geological characteristics, Rivers, streams, waterfalls, lakes and reservoirs, hot and cold and mineral water springs, Temperature, amount of sunshine, precipitation, humidity, seasonality etc. Climate comfort/discomfort index. Forest, forest types, unique vegetation patterns, plants of special significance (aesthetic, medicinal, aromatic, edible), uniqueness and diversity of wildlife mammals, birds, reptiles, fishes and butterflies etc., National parks, wildlife sanctuaries and other wildlife reserves.

- **Human-made Resources**

Religious centres of local, regional and national significance. Religious performances, observances, and rituals, settlement patterns, Folk traditions, folklore and legends, Local cuisines, Art & Architecture, Sculpture and painting.

Folkdances/music and musical instruments, Local craftsmanship, religious, religious can cultural, specific local fairs, commercial/trade/craft fairs, various popular festivals and made of their celebration.

- **Historical/ Archaeological heritage**

Monumental heritage-forts, places, temples and mosques of historical and artistic value, Ancient ruins, Museums, Excavation sites/places of archaeological importance, Sites of important historical events, Places associated with legendary figures, Historical parks/gardens.

Commercial industrial scientific and educational institutions of special tourist interest.

Industrial centres, important centres of trade and Scientific and educational institutions, other centres of tourist interest.

- **Distinctive local occupational patterns as an attraction**

Are parts of this category the agricultural implements & practices, Horticulture, dairy, poultry etc.

- **Sports/Recreation/Entertainment facilities**

Facilities/opportunities for adventure and pleasure seekers, water sports: canoeing, kayaking rafting, rowing paddle boating, swimming, fishing, golf course, trekking, trailing, hiking, climbing, cycling and driving opportunities etc., opportunities for aero-sports via ballooning, hang gliding, paragliding etc.

- **Other recreation and entertainment facilities**

Health resorts, hill resorts, city; parks, theme parks, zoos, gardens, picnic sports etc., other pleasure pursuits like camel riding horse riding and elephant riding.

- **Other tourist Plant Facilities**

Highways, roads, airways etc. communication & public transport facilities.

- **Human Resources**

People, including local people hospitality, behaviour, communication, quantity of staff working in tourism service, quality of staffs working in tourism service

THE CASE STUDY OF CALABRIA

Calabria is an Italian southern region. Taking an outlook at Italy, it is possible to appreciate a unique economic situation, with a dramatic economic divide between Northern and Southern regions, also called the "Mezzogiorno", namely the "Midday". Southern regions present several socio-economic characteristics comparable to other Mediterranean regions (Cantafio, 2017).

If we analyse in depth the economic figures of Centre-Northern Italian regions, they present a Purchasing power adjusted GDP per capita of €33,700, almost comparable with Germany, which shows a GDP per capita of €35,800 (Eurostat, 2019). Conversely, according to the latest data from Eurostat (2019) the "Mezzogiorno" economic indicators are more similar to some poorer regions like peripheral Greece. For instance, Southern Italy regions indicate a GDP per capita of €18,900, while Greece showed a GDP per capita of €21,000. Calabria presented the lowest GDP per capita of Italy, at €17,200.

Other particular striking facts are referring to the post-economic crisis period 2009-2013, in which the "Mezzogiorno" lost 39,500 firms; almost 10,000 of them were belonging to the manufacturing sector. A similar situation has been observed in other Mediterranean areas, and still, most of the Mediterranean areas have not fully recovered from the period of recession experienced after the economic crisis.

Moreover, according to CENSIS (2018), Italy is the second last country in Europe regarding the standings of attracting foreign investments, with an over-all of 20.3% of GDP in 2017, ahead of just Greece.

Sustainable Ecotourism practices in Calabria

Ecotourism can represent a real opportunity for the peripheral European regions, like Calabria. This region has numerous natural assets, which can be considered as Ecotourism destinations. In the medium and long run, this could work as a trigger for local economic development and so having a spillover effect on the other industries of the region (e.g. services, accommodation, and foodservice industry). Something has been done in the last years in this sense, and different activities can be comprised in the ecotourism network.

In Calabria are present some examples of ecotourism destinations: in Zagarise, an area in the province of Catanzaro, where has been installed a natural park theme that consists of a trail on the trees. The park is called "Orme nel Parco", which means "footprints in the park", and it is the first eco-experiential park in Calabria, surrounded by a scenic forest of beech trees of three hectares, in one of the most charming and unspoiled sceneries of the Sila National Park, which it is renowned for its clean air. Moreover, the uniqueness of this location is highlighted by the fact that allegedly people in this park can breathe the purest air in Europe.

To access the park, in the days of regular opening, it is not necessary to book. Entrance to the park is free, you pay only the activities that you choose to perform and the services they are entitled, the cost and duration of which may vary.

In the last few years, numerous centres for rafting are sprouting in the region, especially in the northern area of the region, on the border between the regions of Calabria and Basilicata, on the Lao river, inside the Pollino, which is a massif in the southern Italian Apennines. Rafting and white-water rafting are recreational outdoor activities, which use an inflatable raft to navigate a river or other body of water. This is often done on white-water or different degrees of rough water, and generally represents a new and challenging environment for participants. Another activity that can be done in these kinds of centres is the canyoning, the sport of jumping into a fast-flowing mountain stream and allowing oneself to be carried downstream at high speed.

Figure 2. "Orme nel parco" natural theme park in Zagarise, Calabria. (Source: viviparchi.eu, 2015)

Canyoning is an extremely popular summer thrill in Scotland, in Spain on the Pyrenees, in Madeira (Portugal), in the Alps, but in recent periods the phenomenon is spreading all around Italy on the Apennines, and the Pollino massif is part of them.

Figure 3. People rafting on the Lao River (Paese24.it, 2015)

Another destination that can be mentioned inside the ecotourism network is the Hang loose beach, a 15,000 sq.mt. Sports Park on the Tyrrhenian coast of Gizzeria Lido, in the province of Catanzaro. It has a beach club, terraces on the sea, restaurant-pizzeria, beach bar, green areas and mini club for children, along with the fields of beach volleyball, street basketball and a spectacular skate park; a wide range of opportunities to customers to choose is offered, to spend their days relaxing or full of fun.

Different water sports can be practised on the beach (kite surfing, windsurfing, stand up paddle, sailing, canoeing).

These destinations can be the right way for releasing the potential that the Calabrian natural assets possess and still have not shown their maximum power. Only if embedded into a network they can spur local economic development and have a spillover effect on the other industries of the region.

Figure 4. Kitesurfs in Gizzeria Lido. (calabria.jblasa.com, 2015)

SOLUTIONS AND RECOMMENDATIONS

A revitalisation of the economy could be carried out in several ways. Initially, this could be done by means of implementing tax exemptions to business owners that can demonstrate sustainable and feasible management approaches of the firms or by creating tax exemption zones just for businesses with environmental sustainability purposes.

As suggested by the OECD (2019), there are many ways in which markets can better serve conservation aims, and Eco-tourism is one of them:

- Making better use of environment-related taxes, tradable permits and other market-based approaches to managing natural resources.
- Developing enhanced approaches to take ecosystem services into account during the policy-making process, which affects natural resources.
- Boosting the development of markets for products and services delivered to the customer in a sustainable manner (e.g. eco-tourism), by clearly defining property rights, disseminating information, and establishing institutions to enforce contracts.

Local entities should be really sensitive to this theme, because the importance of this resource can trigger the development of the whole region, and in turn leveraging into all the other industrial sectors.

Figure 5. The Calabrian network of main transportation hubs. A similar concept could be applied to other eco-tourism destinations in the region.(portalecalabria.com, 2015)

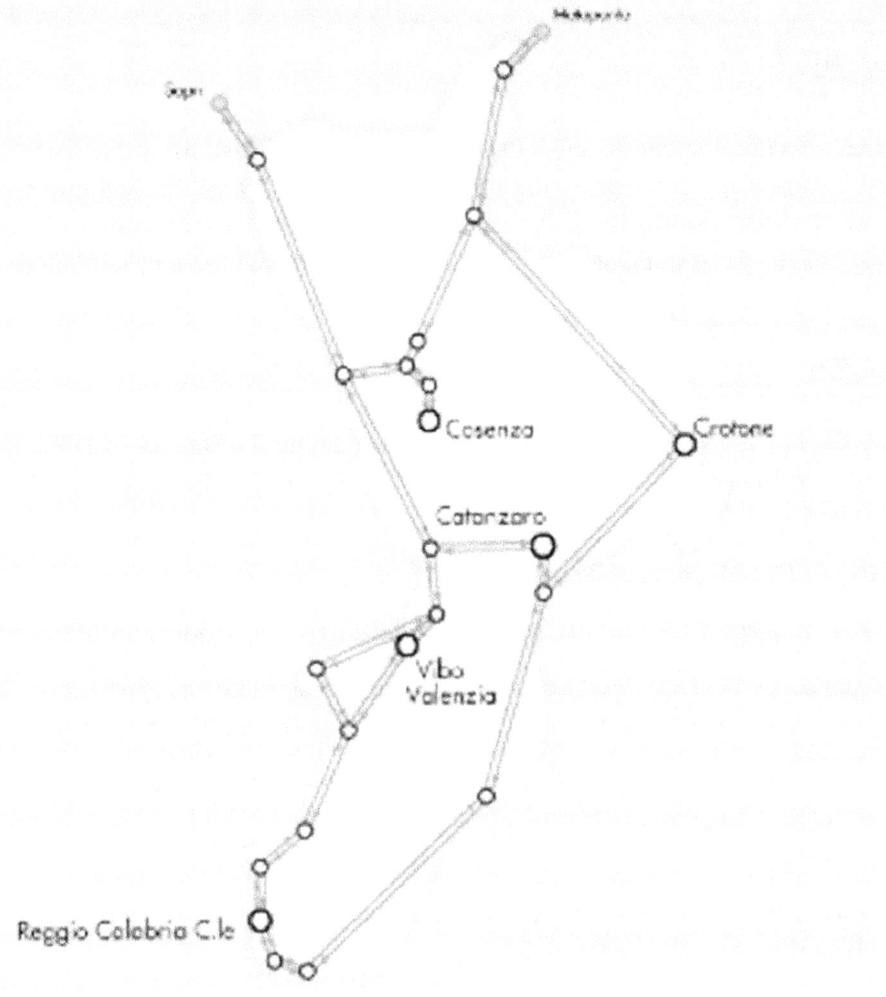

FUTURE RESEARCH DIRECTIONS

The authors recommend future researchers to do similar research in different rural areas, lagging-behind regions, or developing countries. Future research will focus on the most important stakeholders of sustainable tourism in Calabria, by collecting further data by means of interviews, aiming to identify further challenges and providing further suggestions and viable solutions for the development of best practices linked to sustainable tourism.

It will be useful for future researchers to find the differences and connections of the sustainable tourism strategies used in Calabria with other different areas of the globe. This will help future researchers to recognise that every region should have their ad-hoc strategies to be successful, and this might be a useful insight as well for the hospitality industry linked to sustainable tourism in other areas of the globe. Future researchers are also suggested to use different types of primary data collection, which could be more effective to achieve better results, other than a face-to-face interview if they can do so. By doing that, this will help readers get more information and have a more holistic view about the topics that might help the businesses linked to sustainable tourism to grow further.

CONCLUSION

Sustainable tourism development undoubtedly leaves lasting impressions on the social, economic and environmental health of the community through environmental protection, wildlife conservation and socio-economic development. However, the need of the hour is to have coordinated efforts on the part of various stakeholders for sustainable implementation of various tourism development projects.

REFERENCES

Ataberk, E., & Baykal, F. (2011). Utilisation of natural and cultural resources of Dikili (Izmir) for tourism. *Procedia: Social and Behavioral Sciences*, *19*, 173–180. doi:10.1016/j.sbspro.2011.05.121

Barney, J. (1991). Firm resources and sustained competitive advantage. *Journal of Management*, *17*(1), 99–120. doi:10.1177/014920639101700108

Boo, E. (1990). *Ecotourism: the potentials and the pitfalls* (Vol. 1). World Wildlife Fund.

Bottrill, C. G., & Pearce, D. G. (1995). Ecotourism: Towards a key elements approach to operationalising the concept. *Journal of Sustainable Tourism*, *3*(1), 45–54. doi:10.1080/09669589509510707

Briedenhann, J., & Wickens, E. (2004). Tourism routes as a tool for the economic development of rural areas—Vibrant hope or impossible dream? *Tourism Management*, *25*(1), 71–79. doi:10.1016/S0261-5177(03)00063-3

Brundtland, G. H. (1987). Our common future—Call for action. *Environmental Conservation*, *14*(4), 291–294. doi:10.1017/S0376892900016805

Butler, R. (1998). Sustainable tourism-looking backwards in order to progress? *Sustainable Tourism: A Geographical Perspective*, 25-34.

Calabria.jblasa.com. (2015). Available at: https://calabria.jblasa.com/wp-content/uploads/2015/10/Gizzeria-lido.jpg

Cantafio, G. U. (2017). *Merging Economic Development with Clusters Theory and Environmental Sustainability* (Unpublished PhD thesis). Reggio Calabria, Università degli Studi di Reggio Calabria Mediterranea.

Ceballos-Lascurain, H. (1993). Ecotourism as a worldwide phenomenon. *Ecotourism as a Worldwide Phenomenon,* 12-14.

Censis. (2018). *Rapporto sulla situazione sociale del paese*. Angeli.

Cusack, D., & Dixon, L. (2006). Community-based ecotourism and sustainability: Cases in Bocas del Toro Province, Panama and Talamanca, Costa Rica. *Journal of Sustainable Forestry*, *22*(1-2), 157–182. doi:10.1300/J091v22n01_09

Eshetu, A. (2014). Ecotourism as a viable strategy for livelihood diversification and sustainable natural resource management in Ethiopia (from eco-development paradigm point of view). *Journal of Environmental Science and Water Resources*, *3*(2), 40–52.

Eurostat. (2019). *Purchasing power adjusted GDP per capita*. http://https://ec.europa.eu/eurostat/databrowser/view/sdg_10_10/default/table?lang=en

Goodwin, H. (1996). In pursuit of ecotourism. *Biodiversity and Conservation*, *5*(3), 277–291. doi:10.1007/BF00051774

Güngör, S., & Bozyigit, R. (2011). Tourism potential of Gazipasa (Antalya-Turkey). *Procedia: Social and Behavioral Sciences*, *19*, 231–239. doi:10.1016/j.sbspro.2011.05.128

Hunter, C. J. (1995). On the need to re-conceptualise sustainable tourism development. *Journal of Sustainable Tourism*, *3*(3), 155–165. doi:10.1080/09669589509510720

Jarvis, M., Lange, G. M., Hamilton, K., Desai, D., Fraumeni, B., Edens, B., ... Ruta, G. (2011). *The changing wealth of nations: measuring sustainable development in the new millennium*. Academic Press.

Meyer, J. L., & Helfman, G. S. (1993). The ecological basis of sustainability. *Ecological Applications*, 569–571. PMID:27759309

Nadda, V. (2009). *Exploring adventure tourism potential in central Himalayas*. VDM publishers.

OECD. (2019). *Policies to enhance Sustainable Development*. Available at: https://www.oecd.org/greengrowth/1869800.pdf

Paese24.it. (2015). *Regata del Pollino. Rafting sul fiume Lao per una sfida all'ultimo... gommone | Paese24.it*. Available at: http://www.paese24.it/28384/pollino/regata-del-pollino-rafting-sul-fiume-lao-per-una-sfida-all ultimo-gommone.html

Portalecalabria.com. (2015). Available at: http://www.portalecalabria.com/site/servizi/trasporti_calabria.asp

Priskin, J. (2001). Assessment of natural resources for nature-based tourism: The case of the Central Coast Region of Western Australia. *Tourism Management*, *22*(6), 637–648. doi:10.1016/S0261-5177(01)00039-5

Serageldin, I., & Steer, A. (Eds.). (1994). *Making development sustainable: from concepts to action*. The World Bank. doi:10.1596/0-8213-3042-X

Simpson, F., Chapman, M., & Mahne, L. (1998). Partnership Approaches to Tourism and Rural Development in Post Socialist Europe, the Experience of Notranjski Kras, Slovenia. *Rural Tourism Management: Sustainable Options International Conference*.

Teece, D. J., Pisano, G., & Shuen, A. (1997). Dynamic capabilities and strategic management. *Strategic Management Journal*, *18*(7), 509–533. doi:10.1002/(SICI)1097-0266(199708)18:7<509::AID-SMJ882>3.0.CO;2-Z

Todaro, M. P. (1995). *Reflections on economic development.* Books.

Tyler, D., & Dangerfield, J. M. (1999). Ecosystem Tourism: A Resource-based Philosophy for Ecotourism. *Journal of Sustainable Tourism*, *7*(2), 146–158. doi:10.1080/09669589908667332

Valentine, P. S. (1993). Ecotourism and nature conservation: A definition with some recent developments in Micronesia. *Tourism Management*, *14*(2), 107–115. doi:10.1016/0261-5177(93)90043-K

Viviparchi.eu. (2015). *Parco avventura orme nel parco-divertimento-zagarise-catanzaro*. Available at: https://www.viviparchi.eu/strutture-dettaglio.asp?catalogueProdId=356&catalogueCatId=3&catalogueClassId=

Widawski, K., Jary, Z., Oleśniewicz, P., Owczarek, P., Markiewicz-Patkowska, J., & Zaręba, A. (2018). Attractiveness of protected areas for geotourism purposes from the perspective of visitors: The example of Babiogórski National Park (Poland). *Open Geosciences*, *10*(1), 358–366. doi:10.1515/geo-2018-0028

Woodley, A. (1992). Tourism and Sustainable Development: the Community Perspective. In J. G. Nelson, R. Butler, & G. Wall (Eds.), *Tourism and Sustainable Development: Monitoring* (pp. 135–147). Planning and Managing.

Yu, Q., & Yin, J. (2018). Evaluating urban eco-tourism resources and environment: A case study in Shanghai Chenshan Botanical Garden, China. *E&ES*, *121*(3), 032008. doi:10.1088/1755-1315/121/3/032008

Section 3
Gender Equality and Minorities

Chapter 9
Promoting Gender Equality and Women Empowerment in the Tourism Sector

Georgia Papadopoulou
https://orcid.org/0000-0002-9170-4278
Al Ghurair University, UAE

ABSTRACT

Gender equality and women empowerment constitute a longstanding topic since the issues have never been fully addressed, although some progress has been made. The tourism sector is one area that seems to have more opportunities for women and, thus, less gender inequalities. However, cases of discrimination are still experienced. Hence, this chapter explores the situation of gender equality and women position in the field of tourism and establish the causes of the problem in the sector. The opportunities for females and avenues of achieving fairness and empowerment are also discussed.

INTRODUCTION

Gender equality and women empowerment are among the most discussed topics dating back to the late 19th Century and up to now (Lorber, 2001). Although significant progress has been achieved, generally, females are yet to attain the same levels as males in most sectors across the world, meaning that the struggle is still present. On balance, tourism is one of the fields that can meet the needs for gender equality and empowerment of women since it has the proper opportunities through informal, formal, and indirect employment channels (Ferguson, 2011). However, just like in general workplace situations, females continue to face discrimination in the tourism sector (Sigüenza, Brotons, & Huete, 2013).

The number of academic articles addressing gender equality, tourism, and education has increased in recent years, as widely recognised in the literature, without gender equality there cannot be sustainability (Ferguson, 2011). Nonetheless, the effect of these results on tourism industry policy and practice at all costs has not been adequately dedicated (Nunkoo, Hall, Rughoobur-Seetah, & Teeroovengadum,

DOI: 10.4018/978-1-7998-5691-7.ch009

2019). There are several evaluations of gender and culture in tourism, but this is not the case with a lack of gender-responsive policies. In tourism and the Sustainable Development Goals (SDGs), it is vital to support gender equality (UNWTO, 2017). The global sustainable environment, reflected in the SDGs, requires that gender equity and tourist development be incorporated within SDGs (UNWTO, 2017). Academically, little attention has been paid either to clarifying the relevance for "gender responsible tourism" of the SDGs or to taking account of the broader context of gender-based sustainable tourism development. Few scientists considered the more comprehensive policy environment or established links with or intersecting with structural barriers. Eliminate all forms of violence against women and tackle emotional and financial support and domestic work are the key structural obstacles to gender equality.

The 2017 United Nation (UN) declaration as the Year of Sustainable Tourism for Development shows the importance of tourism as an industry that can support sustainable development, including human rights (UNWTO, 2018). Yet, while it is widely recognised that gender equity is an integral part of sustainability, very few tourist facilities have accepted diversity and representation for women as tourism projects, and strategies on the ground are developed that's also inconvenient. Sometimes initiatives aimed at improving the experiences of women who work in the tourism industry should not be based on a sound interpretation of gender, which is a requirement for significant women's roles.

Therefore, the main objective of this chapter is to explore the opportunities for promoting gender equality and empowerment in the tourism sector. To achieve this goal, the current situation surrounding gender equality and empowerment in the field was investigated. Besides, the reason behind gender inequalities, particularly in tourism, will be thoroughly reviewed. At the end of the chapter, opportunities for women in the tourism sector were discussed as well as the solutions to achieve gender equality and women empowerment in the sector.

BACKGROUND

The background of this chapter discusses the potential of the tourism industry in raising the employment opportunities, followed by the gender inequalities within the tourism industry. The background of this study also discusses the prior studies conducted in the tourism industry on women empowerment. At the end of this section, the authors discussed women empowerment and gender equalities.

Potential of the Tourism Industry

Looking at what entails the tourism sector, it appears to be wide and diverse. It is sometimes categorised as tourism and hospitality or tour and travel industry. As noted by Baum (2013), International Labour Organization (ILO) categorises hospitality and tourism as one sector. In defining and understanding tourism sector, the industry's value chain comes in. The tourism industry operates in vast inter-linked and related industries in addition to a range of auxiliary industries that serve people and tourists for various purposes (Camilleri, 2017). The tourism value chain includes tour and travel services such as transportation, booking and reservation, food and beverage (F&B), hospitality, accommodation, local travel, tourism destinations and products (Camilleri, 2017). In other words, the tourism field very wide comprises of different sub-sectors: leisure services, transport and travel agencies as well as hotels and accommodation. Hence, in this work, it is in intended that "travel and tourism", "Tourism", and "Tourism and hospitality" alongside "sector" and "industry" terminologies might be used interchangeably. Moreover, the tourism sector contributes significantly to the labour market as it contributes to about 10% of employment globally World Travel & Tourism Council. (2019). Employment in this sector is facilitated by low entry barriers, flexibility, and part-time work. With such conditions, the tourism sector provides job opportunities for women and has the potential of contributing to the overall economy significantly.

Gender Equality in the Tourism Industry

The general situation is that more men than women are in the labour force. For instance, a report by ILO had indicated that globally, 76.1% men compared with 49.4% women participate in the labour force (ILO, 2017). The situation varies across regions with Arab nations having the largest disparity. However, women spend more time in unpaid work compared to men. For example, about 21.7% of women globally provide full-time unpaid care compared to only 1.5% of men (ILO, 2017). At the same time, the likelihood of mothers being employed is generally low in comparison to that of the fathers. Some aspects of gender inequalities in the professional setting include unequal pay, skewed job promotion, sexual harassment, and lack of institutional support (The World Bank, 2017). Regarding salary, it has been shown that women earn less than men; for example, according to The World Bank (2017) report, on average they receive 60%-75% of what their male counterparts have, which concerns even such regions as the US, where females are considered to be more educated. This gap in the payment levels is even wider when part-time and informal sectors are considered. When it comes to job prospects, women have low promotion rate compared to men, which is further demonstrated by the fact that they are dismally represented in top management positions (Stamarski & Son Hing, 2015). In addition, it has been proven that less than 10% of the CEOs of leading organisations are females (The World Bank, 2017). Hence, lack of the representatives of this gender at high positions against the backdrop of those already in the lower levels is explained by the fact that they are forced to believe that top management is reserved only for men. The tourism sector has been described as one of the fields that are likely to have more opportunities to women, hence, contributing to fewer inequalities. However, discrimination is still present in this sphere. Besides, although females constitute the majority of employees according to International Labour Organization (ILO) (about 55% of workers in the global tourism industry and certain regions), their proportion is as high as 70%), they generally earn less, and hold lower positions compared to men (Baum, 2013). All these issues are also experienced in the tourism sector, despite it being considered a feminine field, which requires more women in key decision-making positions.

Studies on Gender Inequalities in the Tourism Industry

Gender equality is crucial to economic growth and sustainable development (ILO, 2020). There are various studies conducted on gender inequalities in different settings with different measures to assess gender inequalities within the tourism industry. However, the issues identified in this study cannot be overlooked because the focus in this chapter does not discriminate the issues according to the regions and did not include all factors that might affect gender equality and women empowerment across the tourism sector. For instance, Baum (2013) conducted a study to identify the issues and challenges that women encounter in the tourism and hospitality industry. The study employed a qualitative method and specifically focused on hotels and catering sub-sectors. It was established that occupational segregation, wage parity, and discrepancies in career growth and development opportunities are among the main issued facing women in the sector.

Do Le (2017) used descriptive qualitative method to investigate the reasons why women are underrepresented in the senior management in the hospitality industry, particularly the hotel sector. According to him, statistics are clear that there is a scarcity of women in senior decision-making positions in the hospitality industry. However, the study mostly focused on developing countries. Some of the reasons identified by this study as the causes of underrepresentation of women in the senior management include job segregation, outright discrimination, lack of family support, few women role models, work-life conflict, and being excluded from communication networks.

Similarly, Nishtar, Mustafa and Nufile (2017) analyse the socioeconomic issues associated with the women's role in the hospitality industry. The study was based on secondary sources and focused on general global perspectives. Some of the issues identified and analysed in this study include gender discrimination, women exploitation, an overrepresentation of women in low-status jobs, gender stereotyping in jobs, and informal job activities of women in this industry. The study concludes that gender equality and women empowerment can be achieved if the highlighted issues are addressed.

Another qualitative study by KC (2013) investigated the roles and challenges faced by women in the hospitality sector. Some of the challenges identified included lack of support from relatives, lack of a secure job, and balancing family roles and work. This study is, however, not generalisable because it was based in Nepal where women are considered to be of lower status than men, which can create an extreme picture.

Bazazo et al. (2017) explored the level at which different barriers affecting women in the tourism sector prevent them from achieving senior positions. In the study, they focused on societal barriers, internal business structural barriers, personal barriers, situational and governmental barriers. The study utilised primary research, including 200 women in the tourism and hospitality industries. Their findings indicated that all the highlighted barriers had moderate to low levels of impact. Nonetheless, it was important to note that governmental barriers and internal business structural barriers were identified as the leading barriers, respectively. Under the governmental barriers, "inadequate reporting and publicising of information related to "glass ceiling" issues" and "lack of strong and consistent equal opportunity monitoring and law enforcement" (pg. 57) were leading hindrances for women to achieve high positions in the tourism sector. Concerning internal business structural barriers scarcity of women in the top management making men dominate senior positions was the factor that had a high impact in hindering women in progressing to senior positions. Although in most countries especially in the Western countries, governmental barriers are not rampant; this study by Bazazo et al. (2017) shows an important point regarding the internal structural barriers, which are experienced by women in the tourism sectors in different regions.

An article by Sabina and Nicolae (2013) discuss the gender-related trends in a tourism destination. Moreover, they identify that women make up to 90% of wage employees in the tourism sector in some regions. Nonetheless, they note that women are the main players in various tourism sectors, but they are often discriminated against. In addition, they highlight aspects that affect women negatively in tourism; these include sexual objectification, human trafficking, poor wages, prostitution, sexual harassment, and dangerous working conditions. Also, they point out several barriers facing women in the tourism sector. The article advocates for gender equality and women empowerment in the tourism sector. In that line, Sabina and Nicolae, (2013) suggest that it through women empowerment in the tourism sector that economies can achieve their full competitiveness.

Unlike other studies that have shown gender inequalities in the tourism sector employment, a study by Maleka et al. (2018) in South Africa indicated otherwise. The study utilised a quantitative methodology and included adventure tourism employees. Focusing on different factors such as remuneration, health and safety, labour relations, basic conditions, and physical equipment, the researchers did not find discrepancies in employment conditions between male and female employees.

Gender Equality and Women Empowerment through the perspective of Sustainable Development Goals

The Sustainable Development Agenda 2030, adopted by the United Nations General Assembly in September 2015, included 17 SDGs and their corresponding 169 objectives. They are aimed at respecting everyone's human rights and achieving equality for women and girls, and as well as empowering them (UNWTO, 2015). SDG 4 and SDG 5, in particular, addressed gender equality inclusion. Inclusive and equal quality education is given by SDG4, for example, and meaningful education is encouraged for everyone. Yet SDG 5 is designed to achieve equality for women and girls to inspire both men and women (UNWTO, 2015).

It is a problem, and seldom does it occur to create a curriculum in tourism from a gender perspective, although there are many research related to women than men in tourism sector (Inui, Wheeler, & Lankford, 2006). The analysis of the positions, perspectives and achievements of women in tourism or recognition of barriers preventing women from reaching their potential are largely missing in tourism education (Pritchard, 2018). Capacity for community tourism education is often restricted by cultural factors, reproductive labour function and shortcomings in education and in basic education.

Concerning the seamless reduction of poverty, the incorporation of a gender perspective in tourism represents a clear and concrete action (Tucker & Boonabaana, 2012). If a gender-based approach and empowerment of women are not taken into account, the tourist model, including those promoting them as less damaging, will not be sufficiently sustainable and accountable. It means that all legislation, planning, management and analysis instruments relating to tourism are updated.

"Gender equality and women empowerment" is one of the sustainable development goals (SDG-5). Gender equality refers to the assurance of the same opportunities for both men and women. It entails equal and full rights to enjoy social, cultural, economic, and political benefits (United Nations, 2020). Women empowerment is a critical ingredient in gender equality promotion, focusing on the identification and remedying imbalances of power and allowing women to independently manage their lives (United Nations, 2020). It has been extensively demonstrated that when women are provided with equal access to decent work, education, economic decision-making, and political representation societies will accrue many benefits, economies will become more sustainable, and overall humanity will be positively impacted (ILO, 2017).

The stereotypical standpoint that is associated with gender inequalities has its roots from long-held traditional sexist believes where men are seen as leaders and women as subordinates. Furthermore, until recently, in most countries, men education was prioritised over that of women while women were excluded from public activities such as politics. Education was used to disqualify women from participating in important decision-making positions. Nowadays, women are educated even exceeding men in some developed countries. However, this has not stopped gender inequality from being experienced particularly in occupation and workplace.

Occupational sex-typing is a fundamental component of women's economic drawbacks and is the process by which specific occupations are defined as male and others are defined as female (Cohn, 2017). According to the International Labour Organization (2020), women around the world are still facing additional obstacles to access employment and to acquire decision-making positions in certain sectors. This horizontal and vertical gender segregation of employment, combined with the unequal distribution of unpaid work, like household and children activities, introduces differences in working conditions such as the gender pay gap and the over-representation of women in part-time jobs.

Women continue to be under-represented in traditionally male-occupied categories, and within similar categories, women are consistently paid below men, even if women's educational attainments are just as good or better than those of men in similar occupations (ILO, 2018). A research conducted by the International Labour Organization, which covered some 70 countries and about 80% of wage employees worldwide, it is shown that on average women currently continue to be paid approximately 20% less than men.

According to the Workplace Gender Equality Agency (2020), many countries have made significant progress towards gender equality recently, particularly in education, health and female workforce participation. Thus, the gender gap in the workforce is still dominant. Women, in Australia particularly, continue to earn less than men, are less likely to advance their careers as far as men and accumulate less retirement or superannuation savings while "motherhood myth" states that it is a woman's natural duty and privilege to bear children, and, that this role is the ultimate representation of femininity. This is accepted by women worldwide in a way that the associated costs of pregnancy and motherhood are often not considered by women themselves (Uusimäki and Johansson, 2020).

BARRIERS FOR WOMEN IN ENTERING THE LABOUR MARKET

Gender-biased Human Resources' Practices

As noted, in some cases gender inequality in the workplace can be due to institutionalised discrimination. Institutional gender discrimination happens where human resource policies are crafted in a way that they are biased against a particular gender. For example, the score of "face time" in measuring performance to reward worker based on their total time at the workplace (Stamarski & Son Hing, 2015). Taking that women tend to utilise flexible work arrangements due to their primary care roles, their "face time" scores are lower and subsequently are promoted or rewarded at a lower rate compared to men (Stamarski & Son Hing, 2015). This is an indication that some Human Resources (HR) practices directly contribute to gender inequality in the workplace.

In cases of internal recruitment, there seems to be a realistic chance of female employee to get up the career ladder. Nevertheless, as shown by Bickford and Glancy (2010), there are drawbacks in the tourism and hospitality industry using this method. This is because "word of mouth" is a common messaging channel for internal opportunities, which relies on personal contacts, unlike internal advertisements that are a bit formal. This is reiterated by Baum (2013), who notes that in the hospitality industry, referrals and connections are common in the promotion system rather than formal written criteria and advertisement. Given that women are horizontally segregated and do not have a strong informal network, this process of word of mouth acts as a factor in hindering their career progress in the sector.

Again, gender discrimination is exercised at the personal level by some HR decision-makers. It can happen during recruitment and assignment of roles, rewarding and job promotion, training opportunities and termination (Bazazo et al., 2017). Specifically, some jobs are considered as male-type, and thus in the recruitment process, there is an outright bias against women. Furthermore, based on the motherhood myths, women, who are competent in the arguably male field may be perceived as lacking warmth that is expected of a woman, hence being discriminated (Verniers & Vala, 2018). On the other hand, those who are seen as having warmth but low in masculine competence are considered as unqualified for demanding jobs but instead are considered for low position jobs. This indicates a cycle of bias and discrimination that hinders the achievement of gender equality and women empowerment in the workplace.

Occupational Segregation

Gender-based segregation is one factor that creates inequality in occupation and workplace (Correll, 2004). First, jobs that are considered female mostly have short job ladders, formal opportunities are fewer, and access to training is limited. Moreover, informal resources, such as social networks are negatively affected when there is gender segregation. Opportunities' information, visibility, and informal mentorship are often gained through occupational social networks, which are limited among women; this is because women are likely to network with fellow women in the same marginalised positions. It is common to have men occupying higher and managerial positions, while women are mostly taking lower levels; this forms a vertical gender pyramid (Kalev & Deutsch, 2018). The presence of obvious and subtle barriers that hinder women from occupying higher occupational levels is known as "glass ceiling", which is a form of vertical segregation. Vertical segregation is a situation when women are hindered from reaching higher managerial positions. It is also referred to as "stinky floor," meaning that females are stuck in lower seats, and it takes longer for them to be promoted to better prospects (Baum, 2013). The tourism sector of organisational management is blamed for gender segregation since the process of recruitment applies certain stereotypes to categorise jobs that fit for women and vice versa.

At the same time, there is horizontal segregation as females are overrepresented in the low-skill jobs, which is sometimes contributed by the typical recruitment steps. The horizontal segregation explains for relatively lower pay for women compared to men even when they have the same qualifications and in the same occupation. Nonetheless, it has also been shown that it is not uncommon for women at prestigious seats to earn less than men. For example, a Portuguese study by Costa, Breda, and Carvalho (2011) has indicated that despite males representing below 40% of the personnel in F&B section of the tourism sector, they represent the majority (60%) in top management positions. One reason for this situation may be explained by the occupational gender segregation. As has earlier been noted, women are overrepresented in lower job categories. For example, they mostly occupy such positions as customer service, cleaning, and certain clerical duties that have lower payment and rewards and, while men mainly have greater responsibilities or scientific and "object" occupations that have high payments and rewards (The World Bank, 2017). By the same token, it is common to find men dominating tour guide and driver's occupations which are among the lucrative jobs in the tour and travel industry due to tips and opportunities to travel.

It is important to note that segregation rates have remained high despite high levels and diverse education among women. Actually, men cannot take all the blame for this trend, but instead, the female-dominated occupations have become resistant to integration. For instance, due to relatively low pay and rewards, female-dominated occupations remain unattractive to men (Weeden, Newhart & Gelbgiser, 2018). In addition, men have been cultured not to venture in occupations that are considered feminine. Still concerning horizontal segregation, women dominate in the people-oriented occupations while men are in object-oriented occupation.

Interestingly, the tourism sector is considered a feminine field, but males still hold most of the powerful positions and enjoy higher average pay. Furthermore, as noted by Kogovsek and Kogovsek (2015), in the tourism sphere women are regarded as better managers who outperform their counterparts. It is explained that they have transformational leadership skills that are much needed in the sector, which indicates that more females must obtain top positions in the tourism field for it to be successful. However, the persistence of gender segregation and masculine leadership perceptions remain the main challenges to achieve the goal.

Occupational segregation is an important indicator of inequality. It has been shown that on average, women with the same education levels are likely to earn less than men because of the occupational paths that they take (Weeden, Newhart & Gelbgiser, 2018). Therefore, it can be said that a decrease in occupational segregation would result in a decline in pay gap and more women getting into senior positions. However, this is yet to be achieved despite more women entering into traditionally male-dominated occupation and managerial positions.

Lack of Women Role-Models

Role models play a significant part in inspiring those in the lower ranks and those who are not yet in the employment to focus on holding important positions in any industry. Therefore, lack of role models can discourage some group of people from aspiring to attain higher positions. This is the case among women as women, in overall, constitute only about 30% corporate board members indicating hostility towards women (Bazazo, Nasseef, Mukattesh, Kastero & Al-Hallaq, 2017). The hostility is probably due to the high number of men in the boards that has maintained its cohesiveness and which can only be neutralised by having more women in the boardrooms. However, the entry of women in those senior decision-making boards is blocked by the male-dominated management aloofness (Sabina and Nicolae, 2013). Therefore, this lack of visible female role models at senior positions serves as a disadvantage to women. By the token, there is a perception that women at senior positions are mostly single or who does not value the traditional family system. Such perceptions make women in the lower levels to shy away from positions as they find it difficult to embrace such attributes.

Women might sometimes have different leadership styles from men. However, it is not uncommon to find the "masculine" leadership styles, which are practised by men, being considered as the ideal ones. The assumption is that women are to put double the effort as men to progress. At the same time, women leaders who are seen to have masculine leadership qualities are labelled as mean and bossy. Furthermore, unlike men, where qualities such as confidence, vision, and entrepreneurship are considered valuable, for women, they are associated with arrogance and self-promotion. The indication is that there is a clash between a strong powerful woman and an "ideal woman". In the same line, due to their small number in top-level boards, women in senior levels are forced to conform with male opinions since they are the majority. Attempts to stand strong and tough on their positions are likely to be countered with rejection from peers or even being referred to as "iron maiden" (Bohnet, 2016).

Moreover, in some cases, women have to balance between competence and likeability. This is especially when women decide to abide by female stereotype; for example, they incorporate cooperation, sympathy, and care in their leadership strategy. According to McKinsey (2013), the tendency of women to utilise participative strategy in decision-making are seen to lack decision-making ability. Consequently, they are seen as likeable, they are not accorded rightful respect. Also, they are pressured to prove the effectiveness of the feminine style of management and, when performance is not observed, they are considered less competent compared to men.

According to the researchers, having female executive members produce fewer inequality in terms of gender in recruitment, promotion and retention (Hoobler, Wayne, & Lemmon, 2009). This gives a company a better opportunity to hire and maintain skilful employees. The workplace environment often profits from a strong female presence. A large US tech-based company, for an example, brought woman CEO to reduce its high turnover rate of men and women. The CEO decided to promote two senior women as top executives in the roles where they could manage the formerly exclusively managed boardrooms (Adams, Gupta, Haughton, & Leeth, 2007).

Lack of Family Support

Just like any other labour sector, family support is essential in encouraging women to venture in the tourism sector. However, in most regions, especially in developing countries, the traditional systems, where the man has power over women, are still in place. In such situations, women are less advantaged than men, and they cannot make career decisions independently. Furthermore, it has been noted that in some cases where women are better educated, earning more than men, or even holding senior positions compared to men, men are likely to develop tension and feelings of competition. In such cases, to compensate for their achievement and handle men's ego, some women "volunteer" to do more homework (Bohnet, 2016). This implies that the lack of genuine support from the husband is another barrier for women to pursue higher and successful positions in the labour market.

THEORIES AND MYTHS ON CAUSES OF GENDER INEQUALITY IN THE WORKPLACE AND OCCUPATION

Economic Capital Theory

One of these frameworks is an economic capital principle, and one of its tenets argues that males and females seek jobs that maximise household economic utility (Kalev & Deutsch, 2018). Hence, the latter choose low-paying positions that allow them to balance with their domestic work (Kalev & Deutsch, 2018). Another aspect of the theory explains that females tend to opt for less successful jobs to spend more time with family. Another point is that employers contribute to outright discrimination against women as their liability (Kalev & Deutsch, 2018). This individual-based theory may be true because there are limited occurrences since females yearn for equal job opportunities and salaries. Moreover, such a framework is noted as one way of trying to justify gender inequality in the workplace and occupations (Verniers & Vala, 2018).

Division of Labour Theory

Furthermore, it is important to mention the division of labour theory, which argues that organisations have traditionally continued to categorise jobs as masculine and feminine. The latter positions, which are allocated to women, are less advantageous than the ones that are given to men. The "female" jobs have different complexity levels, conditions, and payment formula from the "male" jobs that are discriminative on women (Kalev & Deutsch, 2018). Notwithstanding, due to long –time traditional division of labour, there exists self-stereotyping among women concerning employment and domestic duties. This is evidenced by a survey by McKinsey's (2016): women "automatically" overexert themselves with domestic work and men with career work. In the report, it was indicated that more women than men believe that women are responsible for all housework. Similarly, more women claim that woman's job is less demanding than that of man. Interestingly, more men (45%) claim that women perform more housework out of their choosing (McKinsey, 2016). However, men's assertion is a misinterpretation, since in actual sense may not be choosing to do a lot of housework it is that they do not have other choice and are ready to sacrifice for the success of men.

Ideal Worker Theory

By the same token, there exists the ideology of a perfect worker who operates all year long with no competing demands. It is almost obvious that females cannot constitute this type of personnel because they have domestic responsibilities, hence, being limited in job opportunities. Moreover, as explained by Kalev and Deutsch (2018), the stereotypical traits of an "ideal worker" are discriminative to women as they include aggressiveness, individualism, power, competitiveness, and authority, which are used to inappropriately evaluate this gender. It is common to think that women are seen to fit jobs with so-called feminine features, such as compassion and friendliness. It is argued masculinity is used by men to patronise other men who do not conform to masculinity and to harass women. In fact, it is noted that there is more likelihood of harassment towards women who are in supervisory positions working in male-dominated fields. Consequently, the wellbeing, career attainment, and performance of the women are negatively affected. Moreover, when women complain about the harassment they are likely to face isolation, retaliation, or even being redeployed from their position.

Motherhood Myth

Motherhood myth is used to justify inequality against women in the workplace. Justification is the social or psychological process that is utilised as an opportunity of expressing genuine prejudice while avoiding any form of sanction. This is achieved and sustained through legitimising myths that support inequality. Such myths are founded on the conservatism, social stereotypes, meritocracy and other related ideologies. At the core of the justifications are traditional gender roles that openly insist on power asymmetry between genders. Furthermore, in trying to justify discrimination against women, it is argued that the interdependence between men and women is achieved by having established gender roles (Verniers & Vala, 2018). By the same token, motherhood myths have significantly contributed to workplace and occupation discrimination against women. For example, in some settings, it is asserted that marriage is the most fulfilling and meaningful relationship for an adult; thus, women are forced to take marriage as the priority over employment. In that line, employed mothers are depicted as abandoning their natural motherly duties or putting their family relationship at risk. Such myths act as barriers for women to seek jobs that are permanent, well-paying, or powerful positions in the workplace.

CHALLENGES FACED BY WOMEN IN THE TOURISM SECTOR

Despite tourism being a sector that is more attractive to women compared to the overall economic sector, it has several challenges to women. The challenges include gender stereotyping, exploitation, informality in employment, employment constraints (concerning better and decent work), and low job status.

Low Work Status

Regarding low work status, apart from occupying lower positions, there is direct discrimination of women in the tourism sector. For instance, women are more likely to be laid off. The tourism and hospitality sector is characterised by demand fluctuations. As indicated by Baum (2013), women are more likely to be laid in the non-peak period compared to men, implying that they are not equally valued. On another perspective, women are taken as reserve labour that is utilised when there is a shortage of male labour. In that line, women are expected to exit when their services are no longer needed. Since women are seen as good at unpaid childcare and housework, they are viewed that they do not require much monetary compensation for their labour. Therefore, they are taken advantage of by the firms where try to make great profit margins possible by employing the cheapest labour.

According to ILO (2017), workers are entitled to better and decent work. This means work that considers the employees' safety, welfare, freedom of expressing concerns, personal development, participation in decision-making, and take into account equal opportunities and treatment for different genders.

Three distinct forms of discrimination linked to the gender-based work status are typically highlighted in the literature on the labour market inequality, including allocative discrimination, appraisal discrimination and discrimination in employment (Petersen & Saporta, 2004; Raajpoot, Sharma, & Chebat, 2008). Allocative discrimination is common to people who have limited access, whether on joining or in terms of career progression or desirable roles within their organisations or other organisations. Attributive discrimination, therefore, refers to unequal treatment of women in decision-making regarding recruitment and promotion, which, in turn, results in women in employment, facilities and occupations with comparatively low incomes. Typically, low profits, as well as a limited career growth in salaries, defines female employment. Distinctive employment is thus practically the same as unequal work status (Petersen & Saporta, 2004).

Low Wages

Past studies on gender disparity have demonstrated that gender-based wage inequalities in the labour market exist and persist (Kim, 2000). Several scholars have attempted to explain how gender diversity within a workforce, at the organisational level and the work levels impacted the inequalities (Blau & Kahn, 1994). In comparison, the wage level in the labour market is poor for women, who are mostly working with significantly lower wages. In the study by Pfeffer and Davis-Blake (1987), it was found that women received a lower wage proportion compared to men. The observational evidence presented by (Huffman & Velasco, 1997) suggested that relatively high numbers of women in managerial roles in an organisation are not exactly receiving the wage of managers. While both studies were conducted in qualitative mode, the findings are mostly compatible with the more general finding that high women's representation in a certain group may act as a negative signal about the relative value of the work performed by women in this category and the wage received by them.

Generally, the women pay is relatively lower than that of men. This is explained by horizontal and vertical segregation. By being overrepresented in the lower job categories, the cumulative average pay of women is low. Besides, it has been shown that women at the same positions with men generally get lower pay. Until the recent years, there was a different level of educational attainment between men and women, and that could explain this gap in payment as education plays an important role of earnings (Tyson, 2019). Some other factors that could explain the difference in the wages are that women choose different occupations than men and women are more likely to work part-time.

Exploitation

Regarding exploitation, women are likely to face physical, mental, and sexual exploitation in the tourism sector. Moreover, there are reports of women being exploited through prostitution and human trafficking in addition to sexual exploitation. Although in some perspective, it can be argued that women engage in tourism-related commercial sex as a way of earning a living, it cannot be justified as a benefit of tourism to women but rather as sexual exploitation. Furthermore, it put women at risk of sexually transmitted diseases, including HIV/AIDS and being addicted to hard drugs (Sabina & Nicolae, 2013). As reported by the European Institute for Gender Equality (2020), sexual exploitation in the tourism sector is evident in different countries. For example, in one survey in Finland, it was shown that about 38% of participating tourism sector employees had been sexually harassed by customers with the majority of the victims being women. Moreover, women employees in the tourism sector are more likely to be sexually harassed by colleagues, managers, as well as customers.

According to one study, it has been estimated that 38% of females have been abused in their workplaces globally (Chatterjee, 2018). The related factors have contributed to this situation, for example, it has been noticed that women working in hotels who depend on tips are more likely to be sexually harassed (Waudby & Poulston, 2017). Thus, it implies that they face many challenges in the sector, which is a problem that needs to be addressed. Also, it has been noted that some organisations, specifically in the tourism and hospitality sectors, have deliberate sexist marketing strategies. For example, some organisations, in their objective to attract heterosexual male clientele their hiring policy focuses on young, good looking women. This is outright discrimination against women, but it is "justified" by the marketing strategy. In such organisations, it is not uncommon to find women dress code being sexualised, for example, revealing uniform, high heels, and high-level grooming (e.g. make-up). However, it is rare to find such specifications for male employees.

One of the factors that make the tourism sector seem favourable for women is the informal opportunities; however, the informality of employment can be a hindrance for women empowerment. Tourism-related informal opportunities such as the making of souvenir, home hosting, and catering do not require formal qualifications and are highly flexible. Notwithstanding, the informal employment cannot empower women since it has not a clear path of growth.

Why Tourism is Still a Suitable Sector for Women

Tourism, despite the issues mentioned above, remains the most women-centric sector. It has been established that the representatives of this gender constitute 60% of managers in travelling (The World Bank, 2017). Moreover, women's participation in the tourism sector is significantly higher compared to the general economy in many countries. Particularly, in the G20 countries, women represent about 46.4%

employment share against 43.3% in the overall economy (World Travel & Tourism Council, 2019). In some countries such as Mexico, there is a big difference in the share of women in tourism sector in comparison with the overall economy (50.8% v 36.4%) (World Travel & Tourism Council, 2019). However, in Arab countries and India, the women employment in the tourism sector is way below that of the general economy.

Some of the reasons for these statistics have been pointed in the preceding sections, such as the available form of work-from-home and part-time jobs. Other factors include lower start-up financial requirements for some tourism businesses, which is enabled by digital platforms, for instance, vacation rentals by Owners (VRBO) and Airbnb. Besides, personal and hospitality skills are more emphasised and serve as an advantage to women (The World Bank, 2017).

RECOMMENDATIONS BY PRIOR RESEARCHERS

According to Remington and Kitterlin-Lynch (2017) in a country like the US despite women breaking the glass ceiling in the tourism sector management, women still face obstacles for their career advancement in the sector. Thus in their study, they investigated the successful women experiences in the sector to establish on how gender barriers can be overcome. Similar to other studies elsewhere, advancement challenges-related issues were the most common. These include systematic barriers, inadequate support systems, organisational commitment, work-life balance, and scarcity of female role models.

To achieve leadership advancement, Remington and Kitterlin-Lynch (2017), based on their findings, indicate appropriate and relevant strategies to break the identified barriers should be employed. The suggested strategies include participative leadership, flexibility in the workplace, promotion initiative, presenteeism, and working harder. Concerning participative leadership, it is stated that subordinates should be included in the decision-making process. Participative leadership has been shown to enhance trust and commitment in an organisation and increasing the employees' yearning to get to higher positions. This is beneficial particularly to women in the tourism sector because they constitute the majority of subordinate employees in the sector. In the same line, it has been shown that it is women in leaders who practice participative leadership more than men. Therefore, in a tourism organisation where men still dominate senior management, this strategy may still have challenges to implement. Regarding promotion initiative, the study by Remington and Kitterlin-Lynch (2017) indicated that the participants asserted that women need to take a proactive approach in managing their careers and consequently obtaining promotions. They suggest that women should not be passive in matters concerning their promotion where they only think their work and accomplishments would be recognised. They add that women in tourism should take steps such as active networking, development and maintenance of relationships that are essential in career advancement. This is backed by Ng et al. (2005), who added that women need to actively obtain more professional and educational experience to boost their promotion chances.

Concerning working harder and presenteeism, it is not a recommendable strategy to break the glass ceiling by women in tourism. This is against the findings from other studies that indicate women have to work extra harder than men since they are trying to balance work and home duties. It is asserted that by trying to work extra hard and to look more presentable than expectation only enhances the existing stereotypes that "women are weaker and thus they need to put extra effort to prove their ability". Instead, it is suggested that women should, at any appropriate means possible, refuse to apologise for their accomplishments. This implication is that women should be the core advocates for gender equality and empowerment by standing strong on what is appropriate and not giving in to gender stereotypes and systemic gender discrimination.

Similar to Remington and Kitterlin-Lynch (2017), Carvalho et al. (2018) conducted a study among women leaders to understand what can be done to achieve gender equality and women empowerment in the tourism sector. The solutions are categorised into public policies, organisational measures and individual strategies. Some of the suggested public policies include quotas, subsidised day-care, and awareness-raising actions. Organisational measures include workplace childcare facilities, hours banking, and flexible work arrangements. Individual strategies include assertiveness, education, and commitment.

SUGGESTIONS AND RECOMMENDATIONS

Based on the extensive review of the literature and the current scenarios of gender discriminations from the theoretical point of view, this study has reached on to suggest several approaches to increase gender equality in the tourism industry. As demonstrated in the UNWTO sustainable development goals of inclusive employment opportunities without gender discrimination, this study highlighted a few factors that can drive the objective of achieving sustainable development in the tourism sector. Therefore, the solutions to achieve gender equality and women empowerment in the tourism sector are given below.

Network and Mentorship Programs

It has been noted that women fail to get to senior positions in the tourism sector because of the lack of strong informal networks as men. This is particularly due to horizontal segregation that put women in lower status jobs that have a dead-end progression path. It is, therefore, recommended that women be provided with opportunities to network and find ways they can advance in the field. Such a program can include occasional conferences where women are taught about leadership and, at the same time, get a chance to share their experiences. However, the program should not purely include women, but men also need to be involved; this is because there are quite a several men who are for gender equality and women empowerment - there are women who are mentored by women. Furthermore, trying to exclude men is likely to exacerbate gender segregation.

Quota System

The quota system is where a certain number or percentage is deliberately reserved for a particularly disadvantaged group. In this case, women being the disadvantaged group in the top management they are supposed to give a certain minimum percentage (for example 25%, or 30%) of the top leaders such as in the board of directors in the tourism industry. The quota can be voluntary or mandatory from the policy perspective. The mandatory quotas are often legislated, and organisations are penalised if they fail to adhere. Also, government institutions, where the quota system has been legislated, are required to strictly follow the legislation. On the other hand, in a voluntary quota, there is a room of justifying in case of non-compliance. As indicated in a report by Smith (2018), the quota system has several probable positive impacts. First, the number of women in the top management boards is increased. With gender diversity, there is an improvement in the decision-making process- this is particularly important in the tourism sector where women make the largest proportion of employees. Then, by having a significant number of women in the top management, it is likely to motivate women in the lower levels to advance in their career.

Nevertheless, on the flip side, the quota system is controversial and has been taken with a caveat. The quota system may lead to less experienced women joining senior management boards when there is a thin supply of qualified women (Bohnet, 2016). This would have negative effects or no effect on the organisations, hence, making it meaningless to have extra women in the senior management. The quota may then make sense there is the availability of qualified women, but they have been side-lined due to stereotypes (Hopp, 2015). Notwithstanding, organisations may just apply the quotas to show that they are abiding with the law but fail to involve the women in decision-making. It is stated that quota does not always lead to gender equality in the workplace; it is a necessary thing in where other methods of increasing women in leadership seem not to materialise.

Reforming HR Practices

As earlier noted, biased HR practices contribute to gender inequality in the tourism sector. For example, it is common to find a gender-stereotypical description in job advertisements, for example, compassionate, supportive, interpersonal, which are stereotypical for "female" jobs, and assertive, ambitious, competitive, which are stereotypical for "male" jobs (Gaucher, Friesen, & Kay, 2011). At the same time, it is not uncommon to find gender-specific pictures representing specific jobs in the advertisements such as for receptionists, secretaries, and housekeeping women images are mostly used while for senior chef, food & beverage manager, finance, and tour guide men images are used. Such advertisement strategies make women lose confidence when applying for such "masculinised" jobs as they feel unwelcome. It is, therefore, important for tourism organisation to use gender-neutral ads or include women images in traditionally "male" jobs to attract more women across different jobs.

Organisations should have a clear and transparent payment formula for both men and women during recruitment so as prevent gender pay gaps (Cone et al., 2020). Also, organisations should be clear about wage negotiations to give women an equal chance as men (Gray & Shankar, 2018). Furthermore, tourism organisations should move away from using "face time" in their consideration for promotions. Given that due to their unique work-life imbalances, women are unfairly affected when the "face time" formula is applied. Therefore, by abandoning such a formula and considering the overall performance can work better for women.

Job Flexibility

It has been demonstrated that one challenge for women in the tourism sector and in the labour market, in general, is work-life balance (Beham, Drobnič, Präg, Baierl, & Eckner, 2019). Even in the best civil advancement, it is almost impossible to make women independent from home. With this reality, to achieve gender equality and promote women career advancement, the best thing is to do is to have a system that accommodates women and allows them to have work-life balance without being "penalised". Job flexibility is the solution, the common "flexibility" offers include working in shifts, job rotation, and job sharing, this is mostly due to the around the clock, seven days of the week nature of the tourism industry (Meekes & Hassink, 2020). Nonetheless, to cater for the special problems facing women other flexibility measures need to be included; such measures include time-banking, compressed hours, and at-home working. Regarding bank of time, it is common for the tourism sector, during peak season, to experience a shortage of workers forcing the available employees to work overtime and even lack day off. Usually, the extra time that they have is compensated through monetary payments. However, through an agreement between the organisations and workers, the overtime can be compensated by the extra off time during the non-peak season. This way, it will enable women to achieve work-life balance without feeling pressured.

When it comes to compressed hours, it is a situation where workers are allowed to complete their weekly working hours within a shorter time so that they are left with extra days in which they can carry out their other life activities. Such an option applies to women working in tourism and hospitality sector where the operation is 24/7, which can be strenuous, especially to married women and women with small children (Goldin, Kerr, Olivetti, & Barth, 2017).

The other option is working at home. This includes utilising technology for being posted at home. Technologies such as teleconferencing can enable women, especially those in management positions, to participate in decision-making even without availing themselves physically (Rayalwar, 2020). Furthermore, with technology, some jobs such as bookings can be conducted at home. At the same time, large tourism organisations can post female workers in their home country or region for them to have time for family life. This is because sometimes women are hindered to successfully work in the tourism sector by the cultural and societal systems that prevent them from travelling to distant places away from home.

Dealing with Sexual Harassment and Objectification

It has been noted that sexual harassment and objectification of women is rife in the tourism sector. This vice is common in different sections of the sector, but it is highest in the hotels' section where it has reached to an extent of being normalised as pointed out by Kensbock, Bailey, Jennings and Patiar, (2014). Also, as identified by Cheung, Baum and Hsueh, (2017), tourist sector workers such as those working as tour leaders are ignorant of the sexual harassment policies, and thus they do not know how to deal with sexual harassment. Another thing is that, in the tourism and hospitality business, it is viewed that "the customer is always right" and female workers, when confronted by the customers with sexual harassment behaviours, are confused on what to do and often fail to report such incidences. Besides, as earlier noted, sexist policies such as requiring young women working in the front line to look "sexy" by wearing revealing uniforms and body make-ups, are used by some tourism and hospitality organisations; this is an objectification of women.

One strategy to deal with sexual harassment is to have strong policies against the vice. The policies should have strong backing by the government and trade unions. It should be clear and visible to all employees and guests indicating that any such inappropriate behaviours amount to serious discrimination against women. Also, the policy should include the consequences that culprits would face. For tourism organisations that use sexist strategies to recruit and assign women duties, there should be robust policy from the government and trade unions against such practices and penalties set for any organisation that does not comply with the policy. It is also important to regularly teach women about their rights and make them aware of the channels that they can use to report cases of sexual harassment.

Supporting Women to run Tourism Businesses

Owning and running businesses is one of the greatest ways that women can be empowered. Notwithstanding, despite women being the majority in the tourism labour force, they are underrepresented in tourism business ownership. As cited by Serafimova and Petrevska (2018), women represent 29% of entrepreneurs in the European Union, which is similar with the rest of the developed countries where one-third of business organisations are owned and managed by women. In the developing world, the proportion of businesses ownership and management by women is much lower. Some of the constraints that contribute to the low participation of women include inadequate funding, lack of support, and cultural barriers. Of course, there are some significant differences in the challenges that women face in tourism entrepreneurship between developing and developing countries. For example, in the developing countries women have less access to capital, low rate of land ownership, and more bound by cultural beliefs that constraint them to run tourism businesses compared to their developed country counterparts.

Notwithstanding, tourism can be a source of women empowerment if they are given the opportunity and supported to start a tourism business. In developing regions, such as Africa, it has been shown that, with culturally compatible initiatives, women can run tourism businesses. For example, in Kenya women groups are formed that promote women in tourism enterprises such as cultural tours where women take tourists through their culture and traditional practices (Ampumuza et al., 2008). The women run empowerment programs through tourism have been shown to enhance other areas of women empowerment, for example, medical programs involving women, women of the marginalised communities joining colleges, and getting into areas that had been traditionally dominated by men such as driving. In countries like Nepal, rural women are being empowered through tourism, where there are being encouraged to work as independent tour guides practices (Ampumuza et al., 2008). In general, regardless of where women are (either in developing or developed regions), women require both governmental and non- governmental interventions to overcome barriers that hinder them from getting involved in the tourism entrepreneurship at the same rate as men.

Other Strategies

Considering childcare as one of the out-of-work activities that constraints women in employment, it would be helpful if organisations considered providing childcare services for working mothers (Legazpe & Davia, 2019). For instance, organisations should have in-company childcare, where mothers can drop and pick their children without worrying much about their wellbeing (Shet & Verma, 2019). In addition, for breastfeeding mothers, they can be having several minutes' breaks to breastfeed their young ones. Moreover, childcare services such as kindergarten can be availed closer to the places of work. All these would ensure that women do not utilise a lot of time out of work and that they are able to concentrate on the job just like men. Nonetheless, according to Baum (2013), the consideration for the childcare services should be well-thought-out for not to look like giving special favour to women, given that there are some men who have childcare issues. Therefore, in such cases, such men should be included.

Another strategy is to conduct regular mental training for women. The training should be done by women who already on the leadership positions (Shet & Verma, 2019). The female leaders would share experience and insight in the industry with women in lower employment levels. For example, they should tell them how to overcome gender stereotypes in the industry (Macias-Alonso, Alkhaled, Paunova, & Villeséche, 2018), how to balance work-life conflict, and how to deal with sexual harassment. The mental training would also allow women to know their rights and reject the victim cards.

At the individual level, women should refuse to be judged based on their gender but on their performance and accomplishments. They should not shy away from negotiating for their pay and venturing in male-dominated occupations. After all, as time goes by, men are becoming more respectful to principled women who have become successful in the "masculine" labelled occupations.

CONCLUSION

The tourism sector is one of the largest female employers, but it is not immune to gender inequalities, just like other spheres. They constitute the majority of personnel from lower positions, and their pay is generally worse than that of men. Furthermore, females are more likely to face workplace challenges, such as sexual harassment. Notwithstanding, gender equality and women empowerment can be achieved when there is up-to-date information about discrimination and systemic efforts to bring the balance. The first step in addressing the employment issues is examining wages, gender segregation in tourism jobs, constraints that do not allow women to acquire high positions. Moreover, there should be training opportunities to make them accessible to different seats in the tourism sector. In addition, as it has been shown, some females seek jobs positions that they can share with domestic responsibilities. In summary, this chapter calls for further investigation of the equality and empowerment issues in the tourism field with updated empirical data for the purpose of achieving the goal.

REFERENCES

Adams, S. M., Gupta, A., Haughton, D. M., & Leeth, J. D. (2007). Gender differences in CEO compensation: Evidence from the USA. *Women in Management Review*, *76*(6), 830–846. doi:10.1108/09649420710743662

Ampumuza, C., van der Heijden, F., & Hendriks, N. (2008). *Women Empowerment through tourism*. Retrieved from http://videa.ca/wp-content/uploads/2015/08/Women empowerment-through-Ecotourism.pdf

Baum, T. (2013). *International perspectives on women and work in hotels, catering and tourism*. Retrieved from https://www.ilo.org/wcmsp5/groups/public/---dgreports/---gender/documents/publication/wcms_209867.pdf

Bazazo, I., Nasseef, M., Mukattesh, B., Kastero, D., & Al-Hallaq, M. (2017). Assessing the Glass Ceiling Effect for Women in Tourism and Hospitality. *Journal of Management and Strategy*, *8*(3), 51. doi:10.5430/jms.v8n3p51

Beham, B., Drobnič, S., Präg, P., Baierl, A., & Eckner, J. (2019). Part-time work and gender inequality in Europe: A comparative analysis of satisfaction with work–life balance. *European Societies*, *21*(3), 378–402. doi:10.1080/14616696.2018.1473627

Bickford, N., & Glancy, S. (2010). *The case for change: Women working in hospitality, leisure, travel and tourism*. United Kingdom: People 1st. Available at: http://www.women1st.co.uk/system/assets/files/000/000/004/origi-nal/Women_1st_Case_For_Change_Executive_Summary_Novem-ber_2010.pdf?1342707875

Blau, F. D., & Kahn, L. M. (1994). Rising wage inequality and the US gender gap. *The American Economic Review*, *84*(2), 23–28.

Bohnet, I. (2016). *What works: Gender Equality by Design*. The Belknap Press of Harvard University Press. doi:10.4159/9780674545991

Camilleri, M. (2017). The Tourism Industry: An Overview. *Tourism, Hospitality & Event Management*, 3–27. doi:10.1007/978-3-319-49849-2_1

Carvalho, I., Costa, C., Lykke, N., & Torres, A. (2018). What Could be Done? Tourism,Gender (In) Equality and Women Managers' Views of What Should Change. *Proceedings of the International Conference on Gender Research.*

Chatterjee, R. (2018). *A new survey finds 81 percent of women have experienced sexual harassment.* Retrieved from https://www.npr.org/sections/thetwo-way/2018/02/21/587671849/a-new-survey-finds-eighty-percent-of-women-have-experienced-sexual-harassment

Cheung, C., Baum, T., & Hsueh, A. (2017). Workplace sexual harassment: Exploring the experience of tour leaders in an Asian context. *Current Issues in Tourism*, *21*(13), 1468–1485. doi:10.1080/13683500.2017.1281235

Cohn, S. (2017). The Determinants of the Division of Labor between Men and Women in Paid Employment in the Global North and South: How Occupational Sex-typing Informs the Study of Gender and Development. *Sociology of Development (Oakland, Calif.)*, *3*(1), 1–23. doi:10.1525od.2017.3.1.1

Cone, E. B., Westerman, M. E., Nguyen, D.-D., Stern, K. L., DesLoges, J., & Koo, K. (2020). Gender-Based Differences in Career Plans, Salary Expectations, and Business Preparedness Among Urology Residents. *Urology*. Advance online publication. doi:10.1016/j.urology.2020.04.123 PMID:32540301

Correll, S. J. (2004). Constraints into preferences: Gender, status, and emerging career aspirations. *American Sociological Review*, *69*(1), 93–113. doi:10.1177/000312240406900106

Costa, C., Breda, Z., & Carvalho, I. (2011). Gender inequalities in tourism employment: The Portuguese case. *Revista Turismo & Desenvolvimento*, *15*(1), 37–52.

Do Le, T. T. (2017). *The reasons and solutions for the under-representation of women in the hotels' senior management* (Thesis). Submitted to Helsinki Metropolia University of Applied Sciences. Available at: https://www.theseus.fi/bitstream-handle/10024/129302/Do

European Institute for Gender Equality. (2020). *Tourism: Relevance of gender in the policy area.* Retrieved from https://eige.europa.eu/gender-mainstreaming/policy-areas/tourism

Ferguson, L. (2011). Promoting gender equality and empowering women? Tourism and the third Millennium Development Goal. *Current Issues in Tourism*, *14*(3), 235–249. doi:10.1080/13683500.2011.555522

Gaucher, D., Friesen, J., & Kay, A. C. (2011). Evidence that gendered wording in job advertisements exists and sustains gender inequality. *Journal of Personality and Social Psychology*, *101*(1), 109–128. doi:10.1037/a0022530 PMID:21381851

Goldin, C., Kerr, S. P., Olivetti, C., & Barth, E. (2017). The expanding gender earnings gap: Evidence from the LEHD-2000 Census. *The American Economic Review*, *107*(5), 110–114. doi:10.1257/aer.p20171065

Gray, M. C., & Shankar, S. (2018). *The role of informal wage negotiations and formal promotion processes in explaining the gender wage gap: Centre for Social research & Methods*. Australian National University.

Hoobler, J. M., Wayne, S. J., & Lemmon, G. (2009). Bosses' perceptions of family-work conflict and women's promotability: Glass ceiling effects. *Academy of Management Journal*, *52*(5), 939–957. doi:10.5465/amj.2009.44633700

Hopp, K. J. (2015). How Effective Are Gender Quotas in Achieving Meaningful Change for Women? A Case Study of Argentina. *Interstate - Journal of International Affairs*, *2015/2016*(1). http://www.inquiriesjournal.com/a?id=1238

Huffman, M. L., & Velasco, S. C. (1997). When more is less: Sex composition, organisations, and earnings in US firms. *Work and Occupations*, *24*(2), 214–244. doi:10.1177/0730888497024002005

International Labour Organization (ILO). (2017). *Reducing gender gaps would significantly benefit women, society and the economy*. Retrieved 15 February 2020, from https://www.ilo.org/global/about-the-ilo/newsroom/news/WCMS_557266/lang--en/index.htm

International Labour Organization (ILO). (2018). *What lies behind gender pay gaps. Global Wage Report 2018/19*. Retrieved 7 June 2020, https://www.ilo.org/wcmsp5/groups/public/---dgreports/---dcomm/---publ/documents/publication/wcms_650553.pdf

International Labour Organization (ILO). (2020). *Labour statistics on women*. Retrieved 7 June 2020, from https://ilostat.ilo.org/topics/women/

Inui, Y., Wheeler, D., & Lankford, S. (2006). Rethinking tourism education: What should schools teach. *Journal of Hospitality, Leisure, Sport and Tourism Education*, *5*(2), 25–35. doi:10.3794/johlste.52.122

Kalev, A., & Deutsch, G. (2018). Gender inequality and workplace organisations: Understanding reproduction and change. In *Handbook of the Sociology of Gender* (pp. 257–269). Springer. doi:10.1007/978-3-319-76333-0_19

Kc, S. (2013). Roles and Challenges of Women in Tourism Sector of Western Nepal: A Micro- Ethnographic Study. *Nepal Tourism And Development Review*, *2*(1), 32–59. doi:10.3126/ntdr.v2i1.7379

Kensbock, S., Bailey, J., Jennings, G., & Patiar, A. (2014). Sexual Harassment of Women Working as Room Attendants within 5-Star Hotels. *Gender, Work and Organization*, *22*(1), 36–50. doi:10.1111/gwao.12064

Kim, M. (2000). Women paid low wages: Who they are and where they work. *Monthly Labor Review*, *123*, 26.

Kogovsek, M., & Kogovsek, M. (2015). *Hospitality and tourism gender issues remain unsolved: A call for research*. Retrieved from http://www.quaestus.ro/en/wp-content/uploads/2012/02/kogovsek.metka-1.pdf

Legazpe, N., & Davia, M. A. (2019). Women's Employment and Childcare Choices in Spain Through The Great Recession. *Feminist Economics*, *25*(2), 173–198. doi:10.1080/13545701.2019.1566754

Lorber, J. (2001). *Gender inequality*. Roxbury.

Macias-Alonso, I., Alkhaled, S., Paunova, M., & Villeséche, F. (2018). Women, Gender Stereotypes, & Leadership: Overcoming the Negatives or Accentuating the Positives? *Proceedings of the Seventy-eighth Annual Meeting of the Academy of Management*. 10.5465/AMBPP.2018.16184symposium

Maleka, M., Ntanjana, A., Tshipala, N., & du Plessi, L. (2018). Employment condition differences based on gender: A case of adventure tourism employees in Gauteng. Academic Press.

McKinsey & Company. (2013). *Women matter 2013: Gender diversity in top manage-ment: Moving corporate culture, moving boundaries*. Author.

McKinsey & Company. (2016). *Women matter 2016: Reinventing the workplace to un-lock the potential of gender diversity*. Author.

Meekes, J., & Hassink, W. H. (2020). *Fired and pregnant: Gender differences in job flexibility outcomes after job loss*. Academic Press.

Ng, T. W., Eby, L. T., Sorensen, K. L., & Feldman, D. C. (2005). Predictors of objective and subjective success: A meta_analysis. *Personnel Psychology*, *58*(2), 367–408. doi:10.1111/j.1744-6570.2005.00515.x

Nisthar, S., Mustafa, A. M. M., & Nufile, A. A. M. (2017). An Analysis of Socioeconomic Issues In Relation To the Role of Women in Travel and Tourism Sector: A Study on Global Perspectives. *The International Journal for Economics and Business Management*, *7*(1), 282–293.

Nunkoo, R., Hall, C. M., Rughoobur-Seetah, S., & Teeroovengadum, V. (2019). Citation practices in tourism research: Toward a gender conscientious engagement. *Annals of Tourism Research*, *79*, 102755. doi:10.1016/j.annals.2019.102755

Petersen, T., & Saporta, I. (2004). The opportunity structure for discrimination. *American Journal of Sociology*, *109*(4), 852–901. doi:10.1086/378536

Pfeffer, J., & Davis-Blake, A. (1987). The effect of the proportion of women on salaries: The case of college administrators. *Administrative Science Quarterly*, *32*(1), 1–24. doi:10.2307/2392740

Pritchard, A. (2018). Predicting the next decade of tourism gender research. *Tourism Management Perspectives*, *25*, 144–146. doi:10.1016/j.tmp.2017.11.014

Raajpoot, N. A., Sharma, A., & Chebat, J.-C. (2008). The role of gender and work status in shopping center patronage. *Journal of Business Research*, *61*(8), 825–833. doi:10.1016/j.jbusres.2007.09.009

Rayalwar, A. P. (2020). Role of ICT in Women's Entrepreneurship Development. *Studies in Indian Place Names*, *40*(55), 32–37.

Remington, J., & Kitterlin-Lynch, M. (2017). Still pounding on the glass ceiling: A study of female leaders in hospitality, travel, and tourism management. *Journal of Human Resources in Hospitality & Tourism*, *17*(1), 22–37. doi:10.1080/15332845.2017.1328259

Sabina, J., & Nicolae, J. (2013). Gender Trends in Tourism Destination. *Procedia: Social and Behavioral Sciences*, *92*, 437–444. doi:10.1016/j.sbspro.2013.08.698

Serafimova, M., & Petrevska, B. (2018). Female Entrepreneurship in Tourism: A Strategic Management Perspective in Macedonia. *Journal of Applied Economics and Business*, *6*(1), 21–32.

Shet, R. R., & Verma, P. (2019). Spiritual Intelligence and Mental Health: Study of Working Mother. *Global Health & Medical Tourism (GloHMT).*

Sigüenza, M. C., Brotons, M., & Huete, R. (2013). *The evolution of gender inequality in tourism employment in Spain.* Academic Press.

Smith, N. (2018). *Gender quotas on boards of directors.* Retrieved from http://womensboardaward.dk/wp-content/uploads/2019/02/2018-12-06-Nina-Smith-Gender-quotas-on-boards-of-directors.pdf

Stamarski, C., & Son Hing, L. (2015). Gender inequalities in the workplace: The effects of organisational structures, processes, practices, and decision makers' sexism. *Frontiers in Psychology*, *6*(5), 34–54. doi:10.3389/fpsyg.2015.01400 PMID:26441775

The World Bank. (2017). *Women and tourism: Designing for inclusion.* Retrieved from https://documents.worldbank.org/curated/en/401321508245393514/pdf/120477-WP-PUBLIC-Weds-oct-18-9am-ADD-SERIES-36p-IFCWomenandTourismfinal.pdf

Tucker, H., & Boonabaana, B. (2012). A critical analysis of tourism, gender and poverty reduction. *Journal of Sustainable Tourism*, *20*(3), 437–455. doi:10.1080/09669582.2011.622769

Tyson, L. (2019). *An economist explains why women are paid less.* Retrieved from https://www.weforum.org/agenda/2019/03/an-economist-explains-why-women-get-paid-less/

United Nations. (2020). *United Nations: Gender equality and women's empowerment.* Retrieved from https://www.un.org/sustainabledevelopment/gender-equality/

UNWTO. (2015, September 28). *UNWTO welcomes the adoption of the Sustainable Development Goals.* UNWTO. Retrieved from http://media.unwto.org/press-release/2015-09-28/unwto-welcomes-adoption-sustainable-development-goals

UNWTO. (2017). UNWTO tourism highlights 2015 edition. UNWTO.

UNWTO. (2018). *Tourism and the Sustainable Development Goals – Good Practices in the Americas.* UNWTO.

Uusimäki, L., & Johansson, K. (2020). The Vulnerability of Pregnancy and the Motherhood Myth. In L. Henderson, A. Black, & S. Garvis (Eds.), *Re)birthing the Feminine in Academe. Palgrave Studies in Gender and Education. Palgrave Macmillan.* doi:10.1007/978-3-030-38211-7_8

Verniers, C., & Vala, J. (2018). Justifying gender discrimination in the workplace: The mediating role of motherhood myths. *PLoS One*, *13*(1), e0190657. doi:10.1371/journal.pone.0190657 PMID:29315326

Waudby, B., & Poulston, J. (2017). Sexualisation and harassment in hospitality workplaces: Who is responsible? *International Journal of Culture, Tourism and Hospitality Research*, *11*(4), 483–499. doi:10.1108/IJCTHR-10-2016-0102

Weeden, K. A., Newhart, M., & Gelbgiser, D. (2018). *Occupational segregation. Stanford Center on Poverty and Inequality.* Retrieved from https://inequality.stanford.edu/sites/default/files/Pathways_SOTU_2018_occupational-segregation.pdf

Workplace Gender Equality Agency. (2020). *About workplace gender equality*. Australian Government. Retrieved from https://www.wgea.gov.au/topics/about-workplace-gender-equality

World Travel & Tourism Council. (2019). *Economic Impact 2019 World*. Retrieved from https://www.wttc.org/-/media/files/reports/economic-impact-research/regions-2019/world2019.pdf

World Travel & Tourism Council. (2019). *Travel and Tourism: Driving Women's Success*. Retrieved from https://www.wttc.org/-/media/files/reports/2019/driving-womens-success-2019.pdf

Chapter 10
In Pursuit of a Better Life?
High-Skilled Migration in Portugal

Nina Szczygiel
University of Aveiro, Portugal

Iwona Staniec
Lodz University of Technology, Poland

ABSTRACT

Nowadays, people enjoy the sense of freedom by moving from one place to another, seeking personal and professional growth, and satisfaction with life. Borders have become fluid. In the global world, education and employment seem to be within easy reach. People are pursuing lifelong learning and training to adjust to a changing job environment and maximise career opportunities. However, migrating qualified individuals can face specific difficulties in their way. The present research aims to study high-skilled migration in Portugal. For this purpose, a qualitative approach, with a focus group methodology, was applied. Two focus groups were conducted with foreign-born individuals residing in Portugal. The chapter aims to add to the understanding of how highly qualified migrants make decisions on migration, and how they perceive the job market, professional development, and career growth opportunities in Portugal.

INTRODUCTION

Over the past decades, there has been an interest in the study of migration and high-skilled migrants have been the focus of some of the investigations. Not surprisingly, global economies and societies open the world into a new dimension and bring new professional and personal development opportunities. The world is changing, and so are the people and their interpretations of life, happiness, and fulfilment.

Immigrants are the backbone of developed economies. As employees and employers, taxpayers, and active members of the community contribute to the receiving country political, social, economic, and cultural life. A frequently held view, supported by research, suggests that migrating individuals are highly entrepreneurial and more likely than natives to own a business (OECD/European Union, 2015; Lofstrom,

DOI: 10.4018/978-1-7998-5691-7.ch010

& Wang, 2019). In Portugal, numbers of self-employed foreign-born residents have been increasing over the years. There were 47.200 self-employed immigrants living in Portugal in 2007 what represented 5.4% of all self-employed in the country, whereas figures from 2016 reached 59.800 self-employed immigrants that constituted 9.8% of the self-employed. Majority of them were from non-EU (European Union) countries and worked in wholesale and retail trade, repair of motor vehicles and motorcycles sectors, and accommodation and food services. The entrepreneurship rate was the highest among the Chinese, but individuals coming from the USA, Brazil, England, and Germany were also found to be very active (OECD/European Union, 2018).

The motivation in this research stems from these increasing migrations flows at international and national levels in which high-skilled migration is emerging as a new wave in migration trends. Migration of skilled workforce is said to have diverse effects on technology transfer and innovation, economic development, and welfare. At the individual level, migration opens the mind to the increasingly global world and can have a positive impact on personal and professional development. It is, however, not free from challenges. Migrants can be confronted with a series of questions that may hinder their integration in the host country and the potential occupational advancement. The obstacles include recognition of academic degrees and professional qualifications, and social preconceptions.

This article examines high-skilled migration by the means of two focus groups conducted with non-native individuals living in Portugal. It focuses on the analysis of the decision-making elements and dynamics of migration: motives for migration, the choice of the host country and a potential remaining/return. It aims to analyse highly qualified migrants' perceptions about the labour market, employment and professional development opportunities and career progression.

BACKGROUND

Portugal is a country located on the Iberian Peninsula along the Atlantic coast with a range of different landscapes and unique cultural heritage. It is known from its gastronomy, sacred sites and *fado*. The climate varies substantially from one region to another and is influenced by the latitude and proximity to the sea.

The Portuguese economy has been steadily recovering from the crisis and expanding since 2014. The inflation rate was 1.0% in 2018 (Pordata, 2019). The unemployment rate has been continuously decreasing since the end of 2013 and was 7.0% in 2018, which was lower than a year before by 1.9% (AICEP, 2019). The reduction of unemployment has been related to the expansion of the Portuguese economy. The service sector is currently responsible for about 75% of the GDP (Pordata, 2019) and employs 68.3% of the total workforce (Eurostat, 2019). The shift towards a service economy has a growing tendency and corresponds to what is happening in other European countries. Among other tertiary activities, tourism plays an important role in the national economy. Portugal has successfully managed to combine its rich natural resources with competencies in innovating the tourism sector. In 2017, Portugal won the award for Europe's Leading Tourism Destination at the World Travel Awards positioning itself as a quality destination. The total contribution of travel and tourism to GDP in 2017 was 17.3% and is forecast to rise up to 20.5% in 2028 (World Travel & Tourism Council, 2018).

In 2018, annual GDP growth was 2.1% (World Bank, 2019). GDP per capita was 19,841 euro (Ministério de Economia, 2019). This corresponds to 23,397 PPS (purchasing power standard) as compared to 30.967 PPS EU-28 average (Pordata, 2019). The average annual family income in 2017 was 12,924

PPS, 19th in the rank of the EU-28 (income in Luxemburg ranked first was 33,811 PPS and in Romania that was the last on the list it was 5,915 PPS) (Pordata, 2019). The average salary in 2019 was 25,516 PPS while the EU-28 was 37,093 in 2018. The average minimum monthly (annual/12 months) salary in Portugal was 807 PPS, about a half of that paid in Luxembourg (Pordata, 2019). In 2017, the at-risk-of-poverty risk rate was from 16.9% to 18.3% (among demographic groups), and 36.9% of the Portuguese population was considered unable to afford unexpected expenses (Pordata, 2019).

In 2018, Portugal had 10,276,617 inhabitants concentrated in the mainland around two cities: Lisbon (506,000) and Porto (215,000). It is estimated that until 2080 the population will decrease to 7.9 million (INE, 2019). Data from 2011 and 2012 relate the tendencies in population change with natural change and migration balance, both negative in the case of Portugal (Moreira, 2014). Natural change balance in 2018 remained negative (Pordata, 2019). Over the last years, however, migration to Portugal has been increasing, steadily and considerably. Migratory balance in 2018 was 11,570 and positive (Pordata, 2019). The number of annual registered immigrants increased from 17,554 in 2013 to 43,170 in 2018. The previous countries of residence of the main immigrating populations were Brazil (24%), France (13%), United Kingdom (13%), Angola (8%) and Switzerland (5%). The number of persons coming from the European Union (EU) has doubled (INE, 2019; OECD, 2019c). Foreign population constitutes 4.1% of the total population of the country (Pordata, 2019).

Migration and Migration Flows

People have migrated since ever and for several reasons, but no commonly accepted agreement on the formal definition of migration has been reached (IOM, 2010). Human migration is the movement of people from one place to another, where 'the place' is regarded as a geographic area. Terminological ambiguity has further consequences in how different groups of people motivated by different factors are examined. Sometimes, the term 'migrant' is used as an umbrella to designate those moving voluntarily and involuntarily, but there is a substantial difference between them which is why UNHCR recommends referring to asylum-seekers or refugees as such (UNHCR, 2015). According to the United Nations (2018), migration implies changing a country of usual residence. Migration can be driven by economic, social, political, or environmental motives, among others. These motives can make a person want to leave their country ("push factors") or move to a country ("pull factors") (Allutis, Bishaw, & Frank, 2014). Migrants are individuals looking for a job or seeking new life experiences abroad. A self-initiated migration is a personal decision; however, it is made under social and economic circumstances (Stilwell et al., 2004).

There were 272 million international migrants worldwide in 2019. It is estimated 3.5% of the world's population lives outside of the country they come from (IOM, 2019). Patterns of migration vary from region to region; for example migrating individuals from Northern America tend to live outside their region of birth (IOM, 2019).

Portugal has been recently marked by its migration flows. Both the rates of emigration and immigration have been elevated. Traditionally, the rates of emigration in Portugal used to be high and are now the highest among the EU countries as a proportion of the population. It is estimated that 20% of the Portuguese population (about 2 million) lived abroad in 2016 (Carrilho, & Perista, 2016). This tendency has been accompanied by fast-growing immigration, making Portugal a residency destination for foreigners. Immigration, as opposed to emigration, is a relatively recent phenomenon and has reached substantial numbers since the mid-1970s (Peixoto, & Sabino, 2009). According to the OECD, Portugal received about 40.000 new long-term or permanent immigrants in 2017, an increase of over 20% from the year before. More recent data suggest further rise: between January and September 2018 58.562 new residence permits were issued, and between January and September 2019, the number was 82.929 (Sampson, 2019).

Economic and Social Development Aspects in Migration

Among some underlying motivations of migration, voluntarily migration is, therefore, someone's attempt to improve some aspects of life. Whether the decision is made with the idea of personal fulfilment or professional growth, and immediate or long-term career plan investment is a matter of individual choice. In conventional terms, one might expect that the destination the individual chooses shall provide better living conditions and improve their quality of life (Bartram, 2014).

Standard of living is a measure of material welfare of a country. Its real meaning are goods and services people of some geographic area, usually a country, can buy over some period of time. For cross-country comparisons, GDP data need to be converted in order to produce data expressed at purchasing power standard. However, as an aggregate measure, GDP cannot provide a practical indication on wealth distribution among the population. Standard of living is confined to economic conditions and does not comprise other factors fundamental to human life, such as happiness and dignity.

Economic measures can and should be complemented by others that comprise different aspects of life and therefore provide a more comprehensive approach to what life actually is. One of such measures is the quality of life that reflects the general well-being of individuals and societies. WHO defines quality of life as "individuals' perception of their position in life in the context of the culture and value systems in which they live and in relation to their goals, expectations, standards, and concerns" (The WHOQOL Group, 1995, p. 1403). It reflects "an individual's overall perception of and satisfaction with how things are in their life" (Wood-Dauphinée et al., 2002, p. 137). Perception of quality of life is subjective, changes over the lifespan and is multidimensional (Cella, 1994). Indicators of the general quality of life consider wealth and economic conditions indicators, as well as health, education, social environment, and social networks.

Traditional and Contemporary Views on Migrations

Patterns of contemporary migrations and the profile of migrants may differ from how they were conventionally seen. International migratory movements were traditionally associated to the idea of low- and semi-skilled workers who would escape political instability, poverty and unemployment in low-income countries and move to more wealthy countries in pursuit of better living conditions. Immigrants would complement the labour market occupying jobs found unappealing to the natives and that would require a limited skillset. Specific professions and occupational groups resolved structural labour shortages in sectors such as construction and maintenance, agriculture, fishing, and domestic service. Will it be because of little education and low qualifications or limited language proficiency and struggles with competencies recognition, immigrant workers would more often hold so-called 3D-jobs: dirty, dangerous, and demanding (difficult) (Orrenius, & Zavodny, 2013). The phenomenon is now diversifying. While high-income countries are still drawing an influx of immigrants attracted by better economic opportunities, other tendencies in developed nations have arisen. Air travel advancements and development of information and communication technologies have made the international movement cheaper and less burdensome (Luthra, Platt, & Salamonska, 2016). Social contacts and meaningful relationships with family and friends abroad mitigate the psychological costs of migration. Here, international migration is seen as part of the life cycle and source of personal and professional experience. Individuals who move do it in large part voluntarily, whether for economic, family or study reasons. Those moving for job-seeking purpose are qualified and often driven by a combination of economic and lifestyle motives. Migration is a choice rather than a necessity.

For some people, a place of residence is expected to provide a more fulfilling life experience overall and is accounted for its contribution to one's well-being. A concept of lifestyle migration is focused on motivations that make some people choose certain places and what migration offers to some people in those places (Benson, & O'Reilly, 2016). Lifestyle migrants are motivated by the idea of a "good life", and while living a good life can mean different things to different people, there has been some consistency in things contributing to the quality of life in lifestyle migrants such as moderate climate, safe environment, slower pace of life, and healthy lifestyle (Torkington, 2010; Osbaldiston, Picken, & Denny, 2020). Hoey (2005) argued that for migrants "the choice made of *where* to live is consciously, intentionally also one about *how* to live' (p. 615). Lifestyle migration routes resemble model summer travel itineraries: it is oriented toward already developed warm weather and mild climate tourism destinations (Torkington, 2010). Migration destinations are hoped to be places that allow an individual (re)create their identity (Benson, & O'Reilly, 2016) and are continuously evaluated and compared to the known (Torkington, 2010). It this sense, migration is perceived as an ongoing process and not a goal in itself.

Another modern times tendency of recent origins but already well established in Portugal is (international) retirement migration. With many forms it might take, one of the most common refers to retired individuals moving temporarily or permanently to other places in search of improved quality of life (Gustafson, & Cardozo, 2017). Retirement is now seen as a long-term goal and a hard-earned reward. Higher life expectancy means a longer retirement period to enjoy, and increasing numbers of people start preparing in advance retirement strategies and resources. While the socio-demographic profile of retired migrants varies, the group is regarded as wealthy, affluent and still privileged (Gustafson, & Cardozo, 2017). Weather and climate play an important role in their decision, and the choice of the destination is in large part directed to favourable weather condition locations (Kahveci, Karacan, & Kosnick, 2020). A well-developed infrastructure, tourism attractions and activities all over the year have attracted retirees to regions along the Mediterranean coastlines (Hudson, & Hudson, 2020).

High-Skilled Migration

There is no internationally recognised system of classification of skills. The term "skilled" is usually translated as formal education (that is associated with qualifications). The first is easy to measure (for example, in years of study) as opposed to acquired skills (Csedő, 2008). This taxonomy considers, therefore, two factors, the level of education and professional qualifications. Following this rationale, a high-skilled migrant is a person with a tertiary education level or its equivalent, but the group is evidently not homogeneous (Salt, 1997).

Academics, researchers, other higher education professionals and students belong to the group suggested by Salt (1997) in his classification of high-skilled migrants, among corporate specialists, consultants, sportspeople and artists. This migration can be temporary and last from short to longer periods and go to a permanent stay. Mahroum (2001) suggested another classification, based on five groups, each following its own motivation and logic for mobility: (1) senior managers and executives; (2) engineers and technicians; (3) scientists; (4) entrepreneurs; and (5) students.

International study experience can significantly change the direction of life, and recent data clearly show a high interest in studying abroad (86%) (Knight-Grofe & Rauh, 2016, p. 43). People declare that the experience abroad has influenced their academic path (65%) or career path (71%) (Knight-Grofe & Rauh, 2016). In the case of university students, the choice of the host institution for international experience seems to be strongly influenced by the distance between countries, opinion of peers, economic factors, language proficiency, and academic performance (Szczygiel, 2018). Batterbury and Toscano (2018) have distinguished three types of students over the last fifteen years. The first group has had a broad professional experience in the field and is now looking for a change in the career path. The second group is "seeking validation" at the postgraduate level. This group often comprises international students taking a degree in the English language abroad. The third group transits into master directly from an undergraduate level. Not only did these students have different experiential progression, but they also come from different backgrounds expressed in terms of socio-cultural background, experience, and funding. Doctorate students, post-doc researchers and academics do mobility as part of their professional activities or change city/country for work-related purposes.

It is estimated that around 65% of those who have moved to another country, belong to the managerial and specialist group and, consequently, can be classified as highly qualified (Stilwell et al., 2003). Salt (1997) referred to two basic concepts in the migration of high-skilled staff, brain exchange and brain waste. Brain exchange illustrates the dynamics and consequences of human movement for sending and hosting countries and societies. Migration implies a flow of expertise between the country of origin and the country of destination. Evident unidirectional flow is referred to as "brain gain" or "brain drain". The departure of well trained and qualified individuals is translated as a loss for the sending society and its potential of development ("brain drain"). Brain gain refers to the way these individuals will contribute human capital to the host society. Brain waste exemplifies a detrimental consequence of deskilling. In other words, the job does not allow the individual applying the skills and experience they have.

Well educated and highly qualified migrants moving to another country bring with themselves their education, skills, talent, and potential. If returning to the country of origin at some point, they take additional skills and knowledge that can help the economy grow, create new ventures and new jobs.

METHODS

This was a qualitative study based on the focus group methodology. Two focus groups (FG) were conducted in October 2019 with high-skilled foreigners as defined by Salt (1997) living in Portugal. The motivation for the study was to get a comprehensive idea of what drove highly qualified individuals to come and to stay in Portugal and how they perceived the work environment in the country. The objective was to examine the rationale behind the decision to migrate and the perceptions about the professional ecosystem in Portugal. The choice of the country, city or institution to study or work, options and preferences driving the decision and opinions about the job market were the focus of the analysis.

Focus group 1 (FG1) was conducted in Portuguese and focus group 2 (FG2) was conducted in English. Participation in the study was voluntary. All participants provided written consent. The sessions were recorded with the permission of all participants and subsequently transcribed for the purpose of the analysis.

RESULTS

Characteristics of Participants

Focus group 1 had 8 participants, of Brazilian and Columbian nationality, with ages ranged between 27 and 41, and educational attainment between bachelor and PhD degree. The session was conducted in Portuguese. Table 1 provides a description of the participants of the first focus group.

Table 1. Focus group 1 description

ID	Age	Gender	Country of origin	Education attainment	Field of study/work	Field of study/work at present
FG1_M1	36	M	Brazil	master	Communication and design	Cultural studies – PhD student
FG1_F1	34	F	Brazil	master	Psychology	Cultural studies – PhD student
FG1_F2	34	F	Brazil	master	Marketing and advertising	Administration
FG1_M2	37	M	Brazil	master	Education	Education – PhD student
FG1_F3	27	F	Columbia	bachelor	Design	Design
FG1_F4	37	F	Brazil	PhD	Journalism	Information science
FG1_F5	31	F	Brazil	PhD	Tourism	Tourism
FG1_M3	41	M	Brazil	master	Social communication. Tourism	Information technician. Factory worker

Source: Own elaboration

There were five participants in the focus group, originally from Serbia, Russia, Poland, and Spain, aged from 28 to 33, educational attainment between master and PhD degree. That session was conducted in English. Description of the participants of the second focus group is provided in Table 2.

Table 2. Focus group 2 description

ID	Age	Gender	Country of origin	Education attainment	Field of study/work	Field of study/work at present
FG2_F1	33	F	Serbia	master	Languages and literatures. Philology	Language education
FG2_F2	32	F	Russia	PhD	Marine sciences. Chemistry, ecology	Customer support
FG2_F3	28	F	Spain	master	Scenic arts. Geology	Scenic arts
FG2_F4	28	F	Poland	master	Croatian linguistic	German teacher. Folk musician
FG2_F5	32	F	Serbia	master	Tourism management	Tourism management

Source: Own elaboration

Reasons for Migrating, for Choosing the Destination Country and for Remaining in Portugal

International experience, at any stage of education or career pathway, is relevant for broadening skills and qualifications necessary for understanding complex processes and structures underlying contemporary business relations. When questioned about the reason to come to Portugal, most participants of both focus groups did not have doubts, it was related to education. During the discussion, they revealed more details of their decision. For instance, FG1_F1 acknowledged she had never left the place she was born, and it was something she really wanted to try. She used the opportunity to pursue a doctorate degree in defence of her decision. Both FG1_M1 and FG1_M2 came to accompany their wives, but they did want to have the experience to live in another country. FG1_M1 had considered doing a PhD, although not necessarily abroad. In the case of some participants, the first contact with the country was within a master or doctoral program they enrolled in. Others studied over a semester or year under an international mobility program. In one case, it was a tourism experience that later converted into a decision to reside, supported by ongoing social integration and professional opportunities.

When making the decision about the destination, participants were certain Portugal was the country of choice for the international experience. In the case of international student mobility programs, in which interinstitutional agreements determine the extent to which mobility is possible, countries in the application are placed in order of preference. For participants who came the first time under a student exchange program, Portugal was the first choice on that list. FG1_F3, whose objective was doing a master's degree, chose the university specifically because of the reputation of the program, something she could not find either in her home or neighbour countries. Some similarity between Spanish and Portuguese language was an additional reassuring factor. On the other hand, FG2_F2 confessed she did not select the country itself, but the doctoral program and a Portuguese university was part of the international consortium in charge. She found out she was coming to a Portuguese higher education institution when she earned a PhD fellowship.

Participants who remained in Portugal did it because of personal/family reasons and self-realisation. As time passed by, they found a fit in society and lifestyle. Independently on whether their plans involved staying or returning to the country of origin, participants shared some expectations for the future in which they could make use of acquired knowledge and skills, for a good quality of life and a better life overall.

Perceptions About the Labour Market and Perspectives for Career Progression

In the eyes of participants, experience abroad was an investment in their future, at the beginning always with a term. Plans could change over time: "I finished the scholarship, but I returned, opportunities came up, and I ended up staying" (FG1_F5). Interestingly, the same participant complemented in regard to the personality and cultural aspects of her integration: "I actually managed to adapt very well, because I was always like this and I was seen as rude there… here, no one notices".

In the opinion of participants, the situation of the job market in Portugal was concerning, "even for the Portuguese" (FG1_F3). Participants referred to a difficulty of qualifications recognition that migrant communities could be facing. Even with a validated and recognised diploma, recognition of skills and competencies seemed problematic. There had been issues of contractual instability and low salaries when compared to several other EU countries, therefore, simply "having a job does not mean that the market is good" (FG1_F3). As an example, FG1_F4, PhD, currently postdoctoral researcher, never managed to work in the field. She worked in the catering sector, always thanks to endorsement. Although from a different field, the experience of FG2_F2 was similar: "the area of marine sciences is not easy, but it's been three years since my PhD, and I still haven't worked where I want. I work in customer service in English. It's not a difficult job, but that's not what I dedicated all these years for". On the other hand, perceptions depended on expectations and working conditions. FG1_F2, who did not work in the field either, drew attention to the fact that although financial conditions could be improved, she was generally happy with what she was doing and the working environment. In that spirit, FG1_F5 noted that observing the world around, and people's lives and trajectories made her realise how fortunate she was, and that had been making quite a good progress in her professional life.

Perceptions about career progression and professional advancement depend deeply on the place and moment in life. From an immediate perspective, participants' opinions focused on limited job opportunities offering application of skills and competencies in the field, constrains in skills recognition and limited market size. Participants seeking settlement in the host country praised the stability, at all levels. Despite challenges, "it is somewhat better to be unemployed in Portugal than in Brazil" (FG1_F4). "I used to earn very little there and so do I in here. It is not the same. I have a chance to get European citizenship" (FG2_F1). Support networks, with family and friends included, and safety were of high value. Participants who at the time of the interview had their return scheduled, hoped for career advancement in the long term as a future reward of the investment. This cluster shared the belief that the future would bring positive developments. The opportunity to work and share experiences with others from the field in the international context was bringing another dimension to their professional growth. Several of the members of FG1 perceived their education as an ongoing process and living abroad was seen as a "tremendously rich" life experience (FG1_M3) contributing to personal and professional development in a variety of aspects.

In this logic, a job offer that entailed a relocation to another country would need to meet some requirements, of which opinions varied among participants. Factors raised in the conversation included good weather, tolerance, the stability of employment, and attractive working conditions. To some participants salary and compensation, in general, were found more relevant than to others what had to do with their current situation and priorities in life: "I already have my life set up here so it would have to be a very good proposal (FG1_F5). On the other hand, FG1_M2 hold that having somewhere to come back put him in a very comfortable position. With that safeguard in mind, he was able to evaluate job opportunities through the climate they provided and fulfilment and growth perspectives, and look for what mattered the most to him, which was having a life experience in another place.

DISCUSSION

Educated and skilled workforce, from and outside of Europe can contribute to the growth and competitiveness of European economies by employing knowledge and skills acquired in European universities and companies in Europe. Higher education institutions play an important role in delivering international study programs and advocating the exchange of students and research staff.

There has been a permanent increase in international mobility flows over the years. The OECD projects that by 2025 the number of international students will reach eight million (OECD, 2017). Europe is an attractive destination in education and training. Across OECD countries, the level of incoming mobility has been, on average, at 6% (OECD, 2019b). European students prefer to stay and study in the

EU countries. They account for 24% of mobile students in all OECD countries but 42% in the OECD countries within the EU (OECD, 2019b). Many European countries actively promote their higher education offer. Students are more likely to enrol in foreign higher education institutions for more advanced education levels. In OECD countries, international students account for 6% of tertiary programs but 22% of doctoral programs (OECD, 2019b).

Portugal positions itself as an attractive and affordable destination for international education, research, and mobility. The higher education system offers high quality, comprehensive education in multiple programs and fields of study. Research institutes provide expertise, knowledge, and innovation. International mobility programs, including the Erasmus program, have been a success, with increasing numbers of incoming students. In the academic year of 2009/2010 Portugal received 7,385 students while in the academic year 2016/2017 the number reached 14,306 (European Commission, 2018). Foreign or international students accounted for 4% in the first and 8% in the second cycle of studies, and 27% in the doctoral programs, the latter higher than the OECD and EU-23 average (OECD, 2019b). Results of this research support these data. The reason of the first contact has been in the case of most of the participants related to education and to pursue the study whether at the master's or doctoral level or as an international or mobility program student. Participants whose educational experience ended decided to extend their stay.

The number of immigrants has been increasing in Portugal for three consecutive years and has reached its peak in 2019. "There have never been so many foreigners living in Portugal": over 480,000 non-nationals live now in Portugal with a residence permit and the highest number of applications for Portuguese citizenship in the last five years was registered in 2018 (Público, 2019). These tendencies seem in line with the results of this study. Participants did express their need for integration and participation in the community, in the sense of civic engagement and citizenship.

Attitude toward a country depends on where the person is coming from and going to: participants with the return objective tended to perceive the experience with clearly delineated beginning and end, and regarded it an investment. Those who extended the stay developed coping strategies to face difficulties and obstacles of daily life.

Data suggest that in G-20 countries, foreign-born are more likely to be employed than their native-born peers. On the other hand, educated immigrants are more likely to be unemployed than their native-born peers (OECD, 2019a). This is because, in migrant populations, skills tend to be depreciated and undervalued. This goes in line with the perceptions of participants of this study and difficulties in the job market they have experienced.

When considering a job opportunity abroad at the time of the interview, factors that sprang from the discussion were openness and tolerance, organisational climate, and stability, and financial conditions of the employment. The top five countries in the global HDI ranking are Norway (0.953), Switzerland (0.944), Australia (0.939), Ireland (0.938) and Germany (0.936). Portugal was placed 41st in 2017 (0.847), with 81.4 years of life expectancy at birth, 16.3 years of schooling, 9.2 means years of schooling, 27,315 gross national income per capita (PPP), having increased one position from the year before (UNDP, 2018). As compared to some other EU countries, Portugal does not rank high in economic indicators, but it is recognised in its quality of life components. Living in Portugal has been associated with family relations and networks, interpersonal relations, social support, and lifestyle (Sardinha, 2014; Szczygiel, & Santana, 2018).

Evidence suggests that, in high-skilled migration, the motivations are complex, and the economic dimension does not stand alone. A study from Finland done in two waves, in 2008 and 2010, found that money was indeed an important motive for moving to another country because the salary in 70% of the cases was higher than the one offered at home. While the standard of living was a factor in consideration, there were others. It is argued that qualified migrants value factors such as lifestyle and seeking adventure. Madison (2006) calls it "existential migration" because they never really created the bonding with the country in which they were born. He argues that "rather than migrating in search of employment, career advancement, or overall improved economic conditions, these voluntary migrants are seeking greater possibilities for self-actualising, exploring foreign cultures in order to assess their own identity" (cited by Koikkalainen, 2011, p. 55). In the present study, migrating to Portugal was associated to self-realisation, since the most common reason for migration indicated by interviewees was education and family/personal reasons. The decision itself was not driven by strictly economic or employment seeking motives. Instead, participants believed that gained knowledge, skills and abilities would contribute to their development and future career advancement.

Modern times migrants could be referred to as 'modern nomads'; they represent a generation that has adopted a relaxed, urban lifestyle and does not easily settle down. This generation is affine to the digital world and navigates the job market in search of flexibility, freedom and healthy environments (Sánchez-Hernández, González-López, Buenadicha-Mateos, & Tato-Jiménez, 2019). They look for professional growth and development, seek satisfaction and accomplishment but not at the cost of long-term work-life balance (Oishi, & Hamada, 2020). They are constant experience-seekers in different dimensions of life. They love to travel and widen their perspectives. Getting to know new tourism destinations can actually be an inspiration to migrate, as Williams and Hall (2002) recall in their chapter discussing tourism, migration and "tourism-informed mobility". Inspired by their tourism experiences, people may create affinity with the destination to the point they envision relocating. As residents abroad, immigrants actively work on the promotion of tourism of both, the sending (Balli, Balli, & Jean Louis, 2016) and the receiving country among their extended social networks as they share life moments and experiences on social media and host family and friends (Dwyer, Seetaram, Forsyth, & King, 2014; Griffin, & Dimanche, 2017).

CONCLUSION

Participants of the study have been, in most cases, either current or former international and mobility students and the principal motives to migrate to Portugal were education, self-realisation, and family issues. In the case of the previous contact with the country, that appeared to be a factor richly contributing to the decision to reside abroad and in Portugal in particular. Interviewees' perceptions about the job market were pessimistic because of past experience and current professional situation. Own experiences in the job market and other high-skilled migrants' experiences they were aware of were tough, ranging from validation of academic degrees and diplomas, through recognition of qualifications to job opportunities far beyond education level and competencies. However, two distinct groups arose when analysing career opportunities. The participants planning to return to their home country felt more positive and confident about the future. They also tended to perceive the experience of migration, which they acknowledged as uneasy times, as temporary and investment in the future. Those who had lived in Portugal longer and who had settled down were unsure and worried about that matter. In general, a job opportunity abroad was seen as potentially worth consideration if providing a good working environment, employment stability and complying with the standards of tolerance, receptivity, and balance. Financial rewards were found relevant, while they had not been on the top of priorities when moving to Portugal.

The aim of this research was to examine high-skilled migration in Portugal. The research offers insights on how qualified immigrants perceive the national job market, employment situation, growth and development opportunities and career progression. With migration on the rise around the globe and in Portugal, it is hoped this study can provide its contribution into understanding on how highly qualified individuals make their decisions on migration and perceive the world around when searching for education, job and a better life in the future.

ACKNOWLEDGMENT

This work was financially supported by the Research Unit on Governance, Competitiveness and Public Policy (UIDB/04058/2020), funded by national funds through FCT - *Fundação para a Ciência e a Tecnologia*.

REFERENCES

Aluttis, C., Bishaw, T., & Frank, M. W. (2014). The workforce for health in a globalised context – global shortages and international migration. *Global Health Action*, *7*(1), 23611. Advance online publication. doi:10.3402/gha.v7.23611 PMID:24560265

Balli, F., Balli, H. O., & Jean Louis, R. (2016). The impacts of immigrants and institutions on bilateral tourism flows. *Tourism Management*, *52*, 221–229. doi:10.1016/j.tourman.2015.06.021

Bartram, D. (2014). Migration and quality of life in the global context. In W. Glatzer, L. Camfield, V. Møller, & M. Rojas (Eds.), *Global Handbook of Quality of Life - Exploration of Well-Being of Nations and Continents* (pp. 491–503). Springer.

Batterbury, S., & Toscano, M. (2018). Seeking justice through interdisciplinary environmental education at postgraduate level: Lessons from Melbourne, Australia. *Revista Internacional de Educación para la Justicia Social*, *7*(1), 141–156. doi:10.15366/riejs2018.7.1.007

Benson, M., & O'Reilly, K. (2016). From lifestyle migration to lifestyle in migration: Categories, concepts and ways of thinking. *Migration Studies*, *4*(1), 20–37. doi:10.1093/migration/mnv015

Carrilho, P., & Perista, H. (2016). *Portugal: High and rising emigration in a context of high, but decreasing, unemployment*. Eurofound. Accessed 31 May 2020, from https://www.eurofound.europa.eu/publications/article/2016/portugal-high-and-rising-emigration-in-a-context-of-high-but-decreasing-unemployment

Cella, D. F. (1994). Quality of life: Concepts and definition. *Journal of Pain and Symptom Management*, *9*(3), 186–192. doi:10.1016/0885-3924(94)90129-5 PMID:8014530

Csedő, K. (2008). Negotiating skills in the global city: Hungarian and Romanian professionals and graduates in London. *Journal of Ethnic and Migration Studies*, *34*(5), 803–823. doi:10.1080/13691830802106093

Dwyer, L., Seetaram, N., Forsyth, P., & King, B. (2014). Is the migration-tourism relationship only about VFR? *Annals of Tourism Research*, *46*, 130–143. doi:10.1016/j.annals.2014.03.009

European Commission. (2018). *ERASMUS+ 2017 in numbers Portugal*. Accessed 30 November 2019, from https://ec.europa.eu/programmes/erasmus-plus/about/statistics_en

Eurostat. (2019). *Employment in services*. Accessed 15 October 2019, from https://ec.europa.eu/

Griffin, T., & Dimanche, F. (2017). Urban tourism: The growing role of VFR and immigration. *Journal of Tourism Futures*, *3*(2), 103–113. doi:10.1108/JTF-10-2016-0036

Gustafson, P., & Cardozo, E. L. (2017). Language use and social inclusion in international retirement migration. *Social Inclusion (Lisboa)*, *5*(4), 69–77. doi:10.17645i.v5i4.1133

Hoey, B. A. (2005). From pi to pie: Moral narratives of noneconomic migration and starting over in the postindustrial Midwest. *Journal of Contemporary Ethnography*, *34*(5), 586–624. doi:10.1177/0891241605279016

Hudson, S., & Hudson, L. (2020). *A Worldwide Guide to Retirement Destinations*. Cambridge Scholars Publishing.

INE. (2019). *Estatísticas Demográficas - 2018*. Lisboa: Instituto Nacional de Estatística, IP.

IOM. (2010). *Mainstreaming Migration into Development Planning. A Handbook for Policy-Makers and Practitioners*. Geneva: Global Migration Group. Accessed 1 June 2020, from https://publications.iom.int/es/books/mainstreaming-migration-development-planning-handbook-policy-makers-and-practitioners

IOM. (2019). *World Migration Report 2020*. International Organization for Migration.

Kahveci, C., Karacan, E., & Kosnick, K. (2020). Tactical mobility: navigating mobile ageing and transnational retirement between Turkey and Germany. A comparison between Turkish-German and German retirees. *Journal of Ethnic and Migration Studies*. Doi:10.1080/1369183X.2020.1724532

Knight-Grofe, J., & Rauh, K. (2016). *A World of Learning: Canada's Performance and Potential in International Education 2016*. Canadian Bureau for International Education.

Koikkalainen, S. (2011). Highly skilled Finns in the European labor market: Why do they move abroad? In E. Heikkilä & S. Koikkalainen (Eds.), *Finns Abroad: New Forms of Mobility and Migration* (pp. 46–61). Institute of Migration.

Lofstorm, M., & Wang, Ch. (2019). Immigrants and entrepreneurship. *IZA World of Labor*, *85*(2). Advance online publication. doi:10.15185/izawol.85.v2

Luthra, R., Platt, L., & Salamonska, J. (2016). Types of migration: The motivations, composition, and early integration patterns of "new migrants" in Europe. *The International Migration Review*, *52*(2), 368–403. doi:10.1111/imre.12293

Ministério de Economia. (2019). *Principais Indicadores Económicos de Portugal. Gabinete de Estratégia e Estudos*. Accessed 17 November 2019, from https://www.gee.gov.pt/pt/publicacoes/indicadores-e-estatisticas/principais-indicadores-economicos-de-portugal

Moreira, H. (2014). Estrangeiros e nascidos no estrangeiro residentes em Portugal, CPLP em destaque. *Revista de Estudos Demográficos*, *53*, 91–116.

OECD. (2017). *Education at a Glance 2017: OECD Indicators*. Paris: OECD Publishing. Accessed 29 November 2019, from https://www.oecd-ilibrary.org/education/education-at-a-glance-2017_eag-2017-en

OECD. (2019a). *2019 International Migration and Displacement Trends and Policies Report to the G20*. OECD, IOM, UNHCR. Accessed 1 December 2019, from https://www.oecd.org/migration/mig/G20-migration-and-displacement-trends-and-policies-report-2019.pdf

OECD. (2019b). *Education at a Glance 2019: OECD Indicators*. Paris: OECD Publishing. Accessed 30 November 2019, from https://www.oecd-ilibrary.org/education/education-at-a-glance-2019_f8d7880d-en

OECD. (2019c). *International Migration Outlook 2019*. Paris: OECD Publishing. doi:10.1787/c3e35eec-en

OECD/European Union. (2015). *Indicators of Immigrant Integration 2015: Settling*. Paris: OECD Publishing; doi:10.1787/9789264234024-

OECD/European Union. (2018). *Inclusive Entrepreneurship Policies: Country Assessment Notes*. Portugal 2018.

Oishi, N., & Hamada, I. (2020). Quality of life in Japan and emigration: The perspectives of Japanese skilled immigrants in Australia. In M.-Ch. Tsai & N. Iwai (Eds.), *Quality of Life in Japan: Contemporary Perspectives on Happiness* (pp. 193–214). Springer. doi:10.1007/978-981-13-8910-8_9

Orrenius, P. M., & Zavodny, M. (2013). Immigrants in risky occupations. In A. F. Constant & K. F. Zimmermann (Eds.), International Handbook on the Economics of Migration (pp. 214-226). Cheltenham: Edward Elgar Publishing Limited. doi:10.4337/9781782546078.00019

Osbaldiston, N., Picken, F., & Denny, L. (2020). Exploring emotional reflexivity in British lifestyle migration to Australia. *Population Space and Place*, *2328*(5). Advance online publication. doi:10.1002/psp.2328

Peixoto, J., & Sabino, C. (2009). *Portugal: Immigration, the labour market and policy in Portugal: Trends and prospects*. IDEA Working Papers, nº 6 April 2009. Citizens and Governance in a Knowledge-based Society.

Pordata. (2019). *Retrato de Portugal na Europa*. Lisboa: Fundação Francisco Manuel dos Santos. Accessed 30 November 2019, from https://www.pordata.pt/ebooks/PT_EU2019v20191020/mobile/index.html

Público. (2019). *Nunca houve tantos estrangeiros a viver em Portugal: são mais de 480 mil*. 28 de junho de 2019. Accessed 25 November 2019, from https://www.publico.pt/2019/06/28/sociedade/noticia/ha-480-mil-imigrantes-viver-portugal-maior-numero-ha-registo-1878007

Sampson, D. (2019). *Portugal the country of choice for immigrants*. The Portugal News. https://www.theportugalnews.com/news/portugal-the-country-of-choice-for-immigrants/51452

Sánchez-Hernández, M. I., González-López, Ó. R., Buenadicha-Mateos, M., & Tato-Jiménez, J. L. (2019). Work-life balance in great companies and pending issues for engaging new generations at work. *International Journal of Environmental Research and Public Health, 16*(24), 5122. doi:10.3390/ijerph16245122 PMID:31847460

Sardinha, J. (2014). Lifestyle migrants in Central Portugal. Strategies of settlement and socialization. In M. Janoschka & H. Haas (Eds.), *Contemporary Geographies of Leisure, Tourism and Mobility: Contested Spatialities Lifestyle Migration and Residential Tourism* (pp. 174–189). Routledge.

Stilwell, B., Diallo, K., Zurn, P., Dal Poz, M. R., Adams, O., & Buchan, J. (2003). Developing evidence-based ethical policies on the migration of health workers: Conceptual and practical challenges. *Human Resources for Health, 1*(8), 8. Advance online publication. doi:10.1186/1478-4491-1-8 PMID:14613524

Stilwell, B., Diallo, K., Zurn, P., Vujicic, M., Adams, O., & Dal Poz, M. (2004). Migration of health-care workers from developing countries: Strategic approaches to its management. *Bulletin of the World Health Organization, 82*, 595–600. PMID:15375449

Szczygiel, N. (2018). Integrating international experience in business education. In M. Zakrzewska, & R. Gruenwald (Eds.), *Proceedings of 11th ICEBE & 7th ICIE & PEESA III International Conference on Engineering and Business Education, Innovation and Entrepreneurship, and Capacity Building in Higher Education* (pp. 347-352). University of Szczecin.

Szczygiel, N., & Santana, S. (2018). 'Call me if you need': Social support experiences under economic and social change. *Social Work in Health Care, 57*(9), 794–810. doi:10.1080/00981389.2018.1508112 PMID:30118651

The WHOQOL Group. (1995). The World Health Organization quality of life assessment (WHOQOL): Position paper from the World Health Organization. *Social Science & Medicine, 41*(10), 1403–1409. doi:10.1016/0277-9536(95)00112-K PMID:8560308

The World Bank. (2019). *Country profile*. Accessed 25 November 2019, from https://data.worldbank.org/country/portugal

Torkington, K. (2010). Defining lifestyle migration. *Dos Algarves, 19*, 99–111.

UNHCR. (2015). *Handbook for Emergencies* (4th ed.). Office of the United Nations High Commissioner for Refugees.

United Nations. (2018). *Measuring International Labour Mobility*. UNECE.

United Nations Development Programme. (2018). *Human Development Indices and Indicators: 2018 Statistical Update*. New York: UNDP. Accessed 26 November 2019, from http://www.hdr.undp.org/sites/default/files/2018_human_development_statistical_update.pdf

Williams, A. M., & Hall, M. C. (2002). Tourism, migration, circulation and mobility: The contingencies of time and place. In A. M. Williams & M. C. Hall (Eds.), *Tourism and Migration: New Relationships Between Production and Consumption* (pp. 1–52). Kluwer Academic Publishers. doi:10.1007/978-94-017-3554-4_1

Wood-Dauphinée, S., Exner, G., Bostanci, B., Exner, G., Glass, C., & Jochheim, K. A. (2002). Quality of life in patients with spinal cord injury-basic issues, assessment, and recommendations. *Restorative Neurology and Neuroscience*, *20*(3-4), 135–149. PMID:12454362

World Travel & Tourism Council. (2018). *Travel & Tourism Economic Impact. 2018 Portugal.* Author.

ADDITIONAL READING

Bailey, A., & Mulder, C. H. (2017). Highly skilled migration between the Global North and South: Gender, life courses and institutions. *Journal of Ethnic and Migration Studies*, *43*(16), 6, 2689–2703. doi:10.1080/1369183X.2017.1314594

Czaika, M. (2016). *High-Skilled Migration in Times of Global Economic Crisis.* IMI Working Papers Series No. 126, Discussion Paper 16.16. Oxford: International Migration Institute.

Föbker, S. (2019). "This is not a career move" - accompanying partners' labour market participation after migration. *Comparative Migration Studies, 7*(6) [online]. doi:10.118640878-018-0104-4

Kerr, S. P., Kerr, W., Özden, C., & Parsons, Ch. (2016). Global talent flows. *The Journal of Economic Perspectives*, *30*(4), 83–106. doi:10.1257/jep.30.4.83

McDonald, J. T., & Worswick, Ch. (2015). High-skilled immigration in a globalised labor market. In B. R. Chiswick & P. W. Miller (Eds.), *Handbook of the Economics of International Migration* (Vol. 1, pp. 537–583). North-Holland.

Nathan, M. (2014). The wider economic impacts of high-skilled migrants: a survey of the literature for receiving countries. *IZA Journal of Migration, 3*(4) [online]. doi:10.1186/2193-9039-3-4

Shirmohammadi, M., Beigi, M., & Stewart, J. (2018). Understanding skilled migrants' employment in the host country: A multidisciplinary review and a conceptual model. *International Journal of Human Resource Management*, *30*(1), 96–121. doi:10.1080/09585192.2018.1511615

Skeldon, R. (2018). High-skilled migration and the limits of migration policies. In M. Czaika (Ed.), *High-Skilled Migration: Drivers and Policies* (pp. 48–64). Oxford University Press. doi:10.1093/oso/9780198815273.003.0003

KEY TERMS AND DEFINITIONS

High-Skilled Migration: The movement of qualified individuals between home and host geographic area.

International Student Mobility: The spatial movement of students to receive education outside of their home country. Can be done under an interinstitutional agreement or independent.

Lifestyle: A way of life of an individual, group or culture. Includes attitudes, behaviours, habits, moral standards, and economic factors, among others.

Migration: The movement of people from their home geographic area to a host geographic area.

Migratory Balance: The difference between the number of persons who have entered and left the territory during a given time period (usually one year).

Motivation: A driving force creating an individual willingness to perform an action.

Quality of Life: A general perception of the well-being of an individual resulting from his/her evaluation of positive and negative features of life.

Skill: One's ability to use acquired knowledge to perform a specific action and produce concrete results.

Social Integration: A process during which a new member of a social structure assimilates its culture, organisation and values.

Section 4

Local Communities' Participation in Tourism Development

Chapter 11
Community-Based Tourism and the Sustainable Development of Rural Territories

Nadine Heleno
University of Aveiro, Portugal

Filipa Brandão
https://orcid.org/0000-0002-9323-4572
GOVCOPP, University of Aveiro, Portugal

Zélia Breda
https://orcid.org/0000-0002-5882-063X
GOVCOPP, University of Aveiro, Portugal

ABSTRACT

Nowadays, trends in the tourism sector are changing, and tourists are increasingly looking for an authentic experience. They wish to escape from mass tourism and urban spaces, as well as their awareness for sustainability is increasing. In response to this trend, community-based tourism in rural settings has the potential to meet the demand requirements, while ensuring benefits to the territory and community of rural destinations. The present chapter aims to present the dynamics of community-based tourism and how it can promote regional development in rural, low-density regions, in accordance to the Sustainable Development Goals, and with high participation of residents in the several areas of the tourism industry.

INTRODUCTION

Nowadays, new trends in tourism demand are emerging. Tourists are increasingly looking for authentic experiences and wishing to escape from mass tourism, as well as their awareness for sustainability is increasing. In response to this trend, community-based tourism (CBT) has the potential to meet demand requirements while ensuring benefits to destinations and their communities. CBT contributes to both

DOI: 10.4018/978-1-7998-5691-7.ch011

tourism and global sustainability, being an alternative form of tourism to develop a destination in a sustainable way. However, for this, it is crucial that all stakeholders are aware of the opportunities and responsibilities that arise from it, and take actions that maximise the positive impacts and minimise the negative ones.

This chapter aims to discuss the development of tourism destinations under the Sustainable Development Goals (SDGs), especially rural areas, involving the community in the process and allowing it to be favoured, at various levels, through tourism. The importance of the application of community-based tourism in these areas is justified by the fact they are facing new challenges that have caused a rural decline. The crisis in the agricultural sector, triggered by industrialisation and urbanisation, led to negative impacts, such as unemployment and depopulation, which have caused the abandonment of land and changes in the landscape, as well as have increased the risk of fires and other forms of devastation (Moric, 2013). As a result, and in most rural areas, there is weak development and decline at the economic, social and environmental level (Lane, 1994). Iorio and Corsale (2010) argue that tourism is an effective means for socio-economic development, but also for the regeneration of rural areas, especially those affected by the decline in agriculture, because there is a growing demand for non-resort based tourism and rural destinations (Lane, 1994).

Since rural areas are characterised by economic, socio-cultural and ecological fragility, and have great tourism potential, the development of tourism in these areas must necessarily follow the principles of sustainability. If there is sustainable supply management in response to growing demand, there are many benefits that rural areas can achieve. The economy will be diversified and will allow the development of local businesses, which will reduce poverty levels and enable economic stability; there will be an increase in the appreciation of the natural and cultural heritage on which efforts will be concentrated to preserve and conserve; as well as positive social impacts, such as population settlement and its participation in the sector, which will allow fighting against inequality and empowering rural communities.

Community-based tourism can contribute to the sustainable development of rural areas since, intrinsically linked to the SDGs, local development is achieved where CBT is applied under the best practices. This growing contribution on the part of this form of tourism is achieved due to the growing demand for authentic and unique experiences that allow escaping from routine and mass tourism destinations that do not offer experiences that visitors increasingly value.

The community must be involved in the planning, development and supply of tourism products, and must be the main beneficiary of the development of a CBT project. This is because profits are obtained by welcoming tourists to the communities in which part of the revenues must be applied in order to develop the community and offer maintenance and protection of the heritage. The main focus are the community, which is made up of members who share the same culture and are interconnected at the economic and social level, being geographically circumscribed, and the endogenous resources that must be protected as they are the attractions for visitors.

Thus, the tourism product offered must be focused on competitive management, resource conservation, and the creation of benefits for the community. In order for the product to be of quality and to meet consumer expectations, it is also necessary to know the motivations of tourists so that the various stakeholders promote strategies and policies that guarantee the competitiveness of rural destinations. In order to respond positively to the great competitiveness in the tourism market, and to obtain such benefits, in addition to a sustainable approach, it is necessary to apply management, planning and marketing strategies that make it possible to obtain competitive advantages. The creation of partnerships and networks play an important role in the development of competitive advantages at the regional level, since they have the role of facilitating access to knowledge, resources, markets and technology, and encourage innovation. Thus, it is crucial to interconnect the public sector, the private sector and communities when planning and developing tourism in order to gain a competitive advantage in a sustainable way and that allows an integrated development of the region.

As aforementioned, this work, which is conceptual in nature, aims to identify the premises of community-based tourism in rural areas and to understand how community tourism in rural areas contributes to sustainable development. To meet this goal, the chapter is organised as follows: first, the literature on community-based tourism is reviewed, namely in what concerns its concept and trends, stressing its development model, impacts and demand, following which its implementation in destinations is addressed. Subsequently, the role of CBT in achieving the SDGs is described in more detail. Finally, conclusions are reported, and suggestions for further research and applications are highlighted.

COMMUNITY-BASED TOURISM

Concept and Trends

Community-based tourism is developed on a fundamental idea: tourism planning and development involving host community participation (Hall, 1991; Russell, 2000). Despite this acknowledgement, the concept of community-based tourism differs from author to author regarding the scope of community participation. Although there is no universal definition, there are aspects common to the various concepts, as well as differentiating aspects. Some authors claim that the community should be the owner and manager of CBT projects, while other authors include the participation of external parties, thus accepting different levels of involvement and partnership agreements (Giampiccoli, & Kalis, 2012; Pawson et al., 2017). Due to the flexibility of the concept, Goodwin and Santilli (2009) highlight the lack of rigour in limiting the scope of community-based tourism. This may lead to the emergence of obstacles in implementing and developing related projects.

Giampiccoli and Kalis (2012) describe CBT as a strategy for community development that must be achieved by increasing the self-confidence and empowerment of community members, through measures of sustainability and promotion and conservation of local culture. However, as mentioned, it foresees the participation of external actors so that they can provide facilitating platforms for the community to be able to make decisions and implement actions independently, and to be able to control its development. For Asker et al. (2010), CBT is a form of small scale tourism that has as a premise the integral management of tourism businesses by the community and for the benefit of it, being essential that the community feels that it owns the projects and that it is involved in the CBT initiatives, both in the planning and decision-making phases and in the development of operations. As a result, locals should guide the way tourism is managed in their place of residence (CBI, 2020).

Bearing that in mind, authors such as Goodwin and Santilli (2009), Kayat (2014), and the World Wildlife Foundation International (WWF International, 2001) also approach this form of alternative tourism as a community-managed and/or community-owned tourism that aims to distribute benefits to the community involved. Planning at the community level is also advocated by Simmons (1994), who emphasises this as an essential aspect to guarantee the satisfaction of visitors and obtain benefits for residents. Additionally, Coriolano et al. (2009) point out that CBT must be characterised not only by community participation but also by integrated management of local productive combinations that will strengthen pre-existing economic activities, such as agriculture, fishing, and crafts, for example.

Brohman (1996, p. 60) appears to present the most complete definition of CBT by arguing that:

Community-based tourism development would seek to strengthen institutions designed to enhance local participation and promote the economic, social, and cultural well-being of the popular majority. It would also seek to strike a balanced and harmonious approach to development that would stress considerations such as the compatibility of various forms of tourism with other components of the local economy; the quality of development, both culturally and environmentally; and the divergent needs, interests, and potentials of the community and its inhabitants.

Thus, this form of alternative tourism is based on the premise of the involvement of the community and its participation in the planning and management of tourist activities. This premise is essential for the tourism industry since residents, when involved, allow the development of projects and the generation of benefits for the community (Blackstock, 2005; Goodwin & Santilli, 2009; Kayat, 2014). Murphy (1985) stresses the importance of community involvement in arguing that the values and visions of community members should be incorporated into tourism planning processes and in the implementation phase.

Despite the lack of agreement in defining CBT, Goodwin and Santilli (2009) portray related projects as having specific characteristics, such as bringing benefits to individuals or households of the community, as well as collective benefits. These include the creation of assets which are used by the community as a whole, namely roads, schools, clinics, etc.; conservation initiatives with community and collective benefits; joint ventures with the community and/or collective benefits, including an anticipated transfer of management; community-owned and managed enterprises; private sector enterprises with community benefits; product networks developed for marketing tourism in a local area, community enterprise within a broader co-operative; and private sector development within a community-owned reserve. Mtapuri et al. (2015) add that CBT products allow for longer stays, thus demanding for suitable infrastructures, facilities and other supporting services; are authentic; facilitate the "full immersion" experience and the understanding of local context, which requires proper activities; tourism expenditure should go to CBT entities and not to operators external to the community; and should include specific natural, human-made and cultural attractions related to the local context. Ardianto (2012) stresses the need for CBT to enable tourists to discover local habitats and wildlife, celebrating and respecting traditional cultures, rituals, and wisdom.

Blackstock (2005) points out that literature, from the perspective of community-based tourism, fails when:

1. it adopts a functional approach to community involvement because community-based tourism seeks to ensure the longevity of the tourism industry and does not focus on social justice, it aims to obtain economic benefits while discarding the importance of local empowerment;
2. when it treats the local community as a homogeneous group, assuming that interests are shared and that there is a consensus among all members of the community;
3. when it neglects structural restrictions on local control of the tourist industry. Local control does not automatically translate into participation in decision-making processes since that participation is dependent on those in power at the local level. It is necessary to understand the structural divisions within the communities, as well as to understand the external barriers to local participation and control.

There are several obstacles to the success of CBT. According to Giampiccoli and Kalis (2012), these obstacles can go through the performance of power structures with traditional characteristics that promote unbalanced power, and this imbalance of power can occur at the level of gender and/or ethnicity. Also, internally, unconsciousness and lack of knowledge about the tourism sector undermines the positive development of projects and unequal access to local opportunities. Deficiency in marketing capacity and the peripheral nature of communities also constitute obstacles to success and to obtaining benefits through the application of CBT. Developing companies that offer benefits to the local community, while protecting the indigenous environment and culture and being economically viable, is a key challenge for this form of tourism that aims to promote sustainability (Simpson, 2009).

Development Model

One of the obstacles to CBT is traditional power structures that promote unbalanced power (Giampiccoli & Kalis, 2012). Blackstock (2005) argues that it is necessary to dismantle the barriers imposed on community participation, as well as it is necessary to allow collective responses to problems that arise at the local level. To this end, governments, private sector, and non-governmental organisations must support the local community to become an active participant (Kayat, 2014). Bearing this in mind, and emphasising the importance of the community for the tourism industry, Okazaki (2008) presents a model of community-based tourism that is based on Arnstein's (1969) ladder of citizenship participation model and the evolutionary model of tourism partnerships proposed by Selin and Chavez (1995). In the model presented by Okazaki (2008), aspects such as participation, redistribution of power, collaboration process, and social capital are included. This model aims to support the evaluation of a community's participation in the tourism planning process and to help understand the necessary actions to be taken to increase community involvement in tourism.

Arnstein's ladder of citizen participation (1969) includes eight steps, which are considered the levels of participation, distributed by three categories (Figure 1). Through this model, it is possible to identify the level of participation of a community in the tourism industry, as well as to define which actions should be implemented for community participation to increase.

The three categories are designated as nonparticipation, tokenism, and citizen power, ranging from the absence of citizen involvement in planning activities to a high level of integration of the community in planning and decision-making processes. The evolutionary model of tourism partnerships (Selin & Chavez, 1995) allows assessing the partnerships and processes that occur within them. It consists of five processes, as shown in figure 2: background, problem definition, the definition of direction, structuring, and results.

There are several catalysts for partnerships such as crises, the intervention of intermediaries, a mandate, the fact that there is a common vision, strong leadership that allows the various interests to meet, and the existing incentives. The evolution of the cyclical nature of partnerships consists of the following three phases:

1. Configuration of the problem: the stakeholders appreciate the interdependence and realise that the solution to the problems needs a joint effort. However, this is not enough for a partnership to start since the benefits resulting from the partnership have a very relevant weight.
2. Configuration of the direction: the stakeholders establish the objectives and rules that allow them to examine the problems.

3. Structuring: the interactions between stakeholders are managed systematically and the formalisation degree diverges between partnerships.

Figure 1. The eight steps of the ladder of citizen participation (Arnstein, 1969)

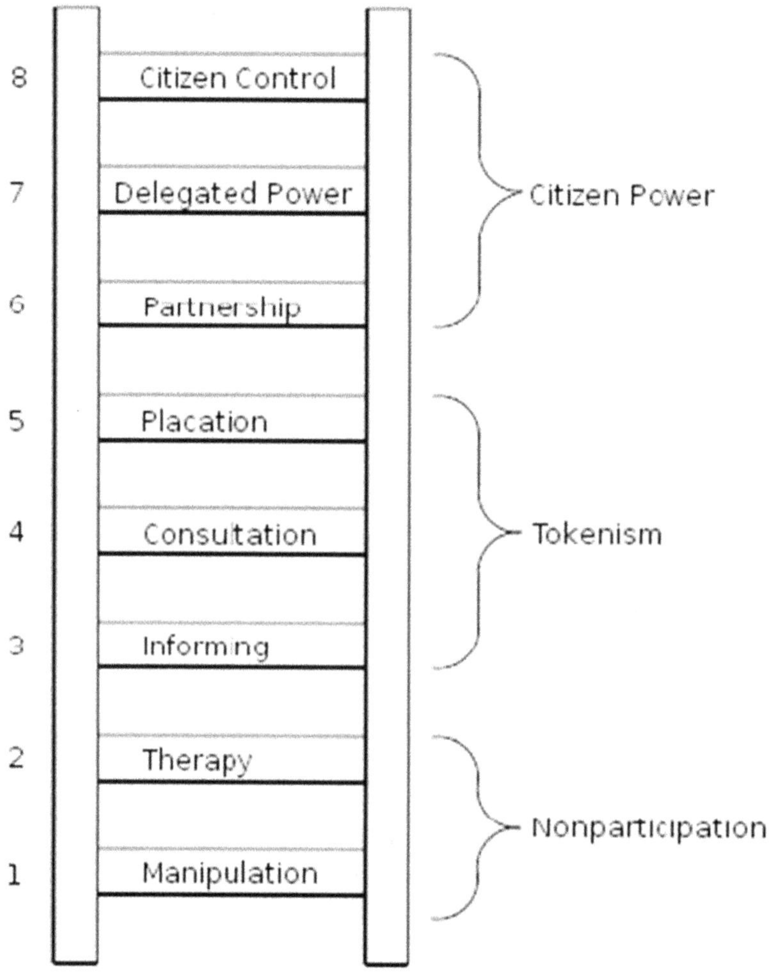

Once a partnership is established, results appear. The most common results are the programmatic ones, which, according to Selin and Chavez (1995), can be visiting centres, brochure, and interpretive trails, for example. Less tangible results can be improved relationships in community groups and the recognition of good work. Subsequently, the partnership undergoes a cyclical reassessment which, if it is prosperous, may result in an enlargement of the focus, however, if there is a loss of interest on the part of the partners, it may result in the extinction of the partnership.

Figure 2. Evolutionary model of tourism partnerships (Selin & Chavez, 1995, p. 48)

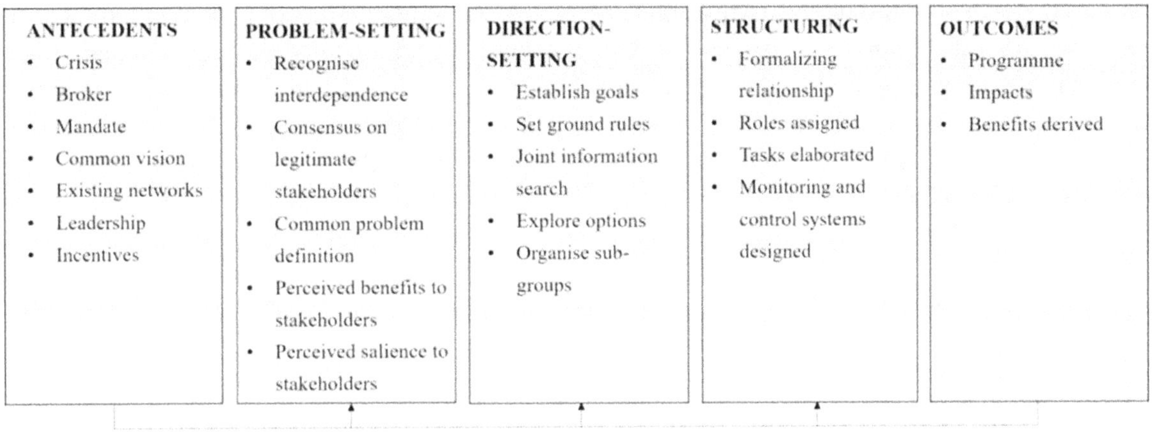

The bommunity-based tourism model developed by Okazaki (2008) is inspired in the two above-mentioned and is structured in a graphic composed of two dimensions (Figure 3). The horizontal axis includes the partnership processes of Selin and Chavez (1995) and, in the vertical axis, the community participation levels and the redistribution of power, according to Arnstein (1969).

Figure 3. Community-based tourism model (Okazaki, 2008, p. 513)

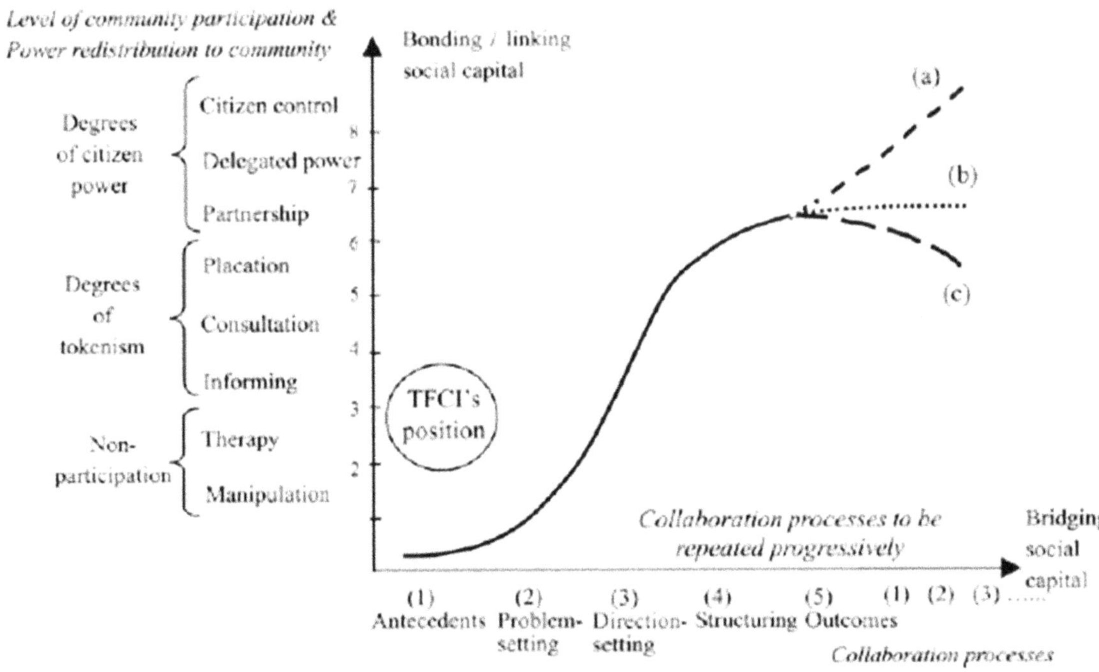

The graph's upward sloping curve is influenced by five propositions listed by Okazaki (2008). The first says that, when community participation is promoted, the redistribution of power is facilitated. The second says that if the collaboration process does not continue, community participation and progress in the redistribution of power also have no future. The third says that if there is no progress in terms of participation and the redistribution of power, there will be no promotion of collaboration. The fourth says that differences in power will weaken collaborations. Finally, the fifth says that social capital is gradually established in the process, and this contributes to increasing sustainability by creating synergies not only within the community but also between the community and other stakeholders in tourism.

The shape of the graph is dependent not only on internal factors but also on external factors. The tourism development phase, the economic, political, socio-cultural and environmental conjunctures, the availability of resources and access to them, the level of support of residents in the development of tourism, the existence of conflicts and the existence of facilitators of dialogue in the community are factors that influence the curve's path.

The curve determines the intersection between the vertical and horizontal axes and, once it indicates that the community has reached the partnership step on the vertical axis, three paths will occur and which are illustrated by the model in Figure 3. When there is an agreement among stakeholders to continue with community participation and the redistribution of power by the community, the graph curve will rise (a). The curve will be stagnant (b) if it is found that the stakeholders are satisfied with the level of participation achieved and with the redistribution of power carried out. If other stakeholders reject the power attributed to the community, or if the community is internally divided, the curve will move downwards (c).

Thus, this model allows knowing the level of involvement of a community in tourism and to determine the initiatives that could benefit community-based tourism. Okazaki (2008) also states that the applicability and usefulness of this model in more advanced stages of tourism development are undetermined.

Impacts

Community-based tourism, compared to mass tourism, has a lower need for infrastructure and services, which translates into reduced negative impacts at the local level since the modification of the space is less and because there is a valorisation of natural and cultural resources (Bursztyn et al., 2009).

When CBT is applied in a region and it succeeds, there are several and different positive impacts, whether direct or indirect, and individual or applied to the community. First, to be considered a case of success, according to Asker et al. (2010) in its Manual of Best Practices, the support for the development of the local economy should be verified through the diversification of the job offer, that is, there should be a diversification of the economic sectors in the region.

This form of tourism must be financially viable and demonstrate good practices concerning its management. Despite the challenge of ensuring economic viability, this aspect is essential for the community to achieve positive results; however, it depends on several factors such as the market, the product and relations with the private sector (Goodwin & Santilli, 2009; Manyara & Jones, 2007; Pawson et al., 2017; Salazar, 2012; Simpson, 2009; Supriharjoa et al., 2016). Therefore, to achieve economic viability and the extraction of benefits for the community, the tourism product must be competitive (Kayat, 2014). Additionally, it should not only demonstrate economic success, as it should also focus on social capital and community empowerment, where the respect and encouragement of equal participation of community members must be emphasised (Asker et al., 2010; Giampiccoli, & Kalis, 2012; Goodwin, & Santilli, 2009).

In addition, it is essential that the quality and safety of the tourist experience is ensured, as well as educating visitors about the culture and natural aspects with which they are coming into contact, which enables, among other measures, ecological sustainability, minimising negative environmental impacts, and heritage conservation and promotion (Asker et al., 2010; Goodwin & Santilli, 2009). Therefore, it is essential that tourist activities do not cause negative impacts not only at an environmental level but also at a cultural level, that is, tourist activities should not harm the traditions and way of life of the local population (CBI, 2020).

On the other hand, if the development of CBT is not evaluated, planned, and managed effectively, it can have negative impacts on society and the environment. Negative impacts will also be observed when it is shown that CBT's main objective is to obtain quick economic and financial benefits without considering the community (Asker et al., 2010).

After the successful implementation of CBT, positive impacts will emerge not only in the community by contributing to the overall well-being and the redistribution of economic benefits to vulnerable groups, but also in the conservation and promotion of environmental and cultural space (Asker et al., 2010; Brohman, 1996; CBI, 2020; Goodwin & Santilli, 2009; Supriharjoa et al., 2016).

Improving the well-being of the community is achieved through the potential of CBT for the development of the local economy through the creation of diversified employment for local people, the maximisation of the use of local goods and services, and the development of infrastructure. The environment benefits from increased awareness in favour of ecologically sustainable local development that pursues to minimise negative impacts through supportive policies and environmental strategies (Asker et al., 2010; Simpson, 2009). Local culture also benefits due to increased valorisation, and due to the promotion, conservation and enhancement of cultural heritage (Asker et al., 2010) being essential to assure respect for local culture and lifestyle (Giampiccoli & Kalis, 2012).

In addition to the benefits observed in the three pillars of sustainability, Simpson (2009) classifies the impacts in three categories: financial, non-financial and intangible. The first relate to the creation of jobs and to the emergence of diverse economic opportunities, which will benefit not only individuals, but families and the community as a whole. This is because, through these benefits, contributions can be applied to the development of community assets, such as schools, health clinics or even grinding mills, for example (Goodwin & Santilli, 2009; Simpson, 2009). Non-financial benefits are associated with decreasing the vulnerability of communities, developing skills, improving and increasing access to information, food security, and strengthening community organisations (Simpson, 2009). The less tangible benefits may range from renewed pride, empowerment, physical security, cultural benefits, optimism, to increased community participation in decision-making processes (Simpson, 2009). Community empowerment must be achieved at four levels: economic, psychological, social, and political (Scheyvens, 2002). Empowerment is crucial for the various members of the community to be involved in participation, and the results of that participation should lead to greater community empowerment (Kayat, 2014; Sutawa, 2012). In addition, Kayat (2014) also suggests that empowerment will fall mainly on women and young people who are directly involved in the development of tourism. However, Goodwin and Santilli (2009) state that the success of several applied projects is not properly monitored, which makes it impossible to quantify all the benefits of this form of alternative tourism for local communities. This happens when there is little effort in adopting measures to register or measure the benefits generated. Despite this, knowing the benefits is essential when integrating the community into the tourism planning processes since it has associated costs, and the community members themselves not only like it, but also need to know what benefits they will derive from the investment of their time and work in the development of the tourism industry at the local level.

It is also important to acknowledge that there are direct and indirect beneficiaries of CBT impacts. Direct beneficiaries are those who obtain employment, who are artisanal producers, tour guides or others who are directly involved in the provision of the tourist product, obtaining, to that extent, income through the activity they provide. Indirect beneficiaries are members of the community who benefit from the application of profits obtained through the success of CBT in the development of projects at the educational, health, safety, infrastructure, and other levels (Giampiccoli & Kalis, 2012).

Profiling the Demand

There are no data on international travellers practising CBT. However, and despite being a niche market in the hospitality industry (Aminudin et al., 2015), a study developed by CBI (2015) concludes that almost 50% of mainstream tourists are interested in some aspects of CBT. Globally, practitioners of CBT search for the authenticity of community life, as this type of product relies on close contact between visitors and hosts (Dolezal, 2011).

People participating in this alternative form of tourism are usually well-educated, with high-income, and are experienced travellers. Despite these common features, if one considers their interest, two types of tourists can be distinguished – hard travellers and soft travellers –, who are part of three groups – older generation, younger generations, and families with children – which, according to CBI (2020), have the following characteristics.

- Hard travellers include a small group interested in an authentic cultural experience. They value sustainability and do not care about discomfort since they want to experience the life of local people to the fullest regardless of what this may imply. To attract this segment, it is important to highlight the authentic experience in product marketing.
- On the other hand, soft travellers are a wider group that values culture and interaction with locals, also looking for a unique experience. However, the issue of comfort is considered. To attract this target segment, it is important to offer comfort through the quality of service provision, especially concerning accommodation.
- The older generation is a vast group, aged between 50 and 70 years, who have time and money to invest in tourist experiences. They are likely to travel in low season, and the adventure activities they seek to have a soft character combined with a certain luxury. Attracting this segment involves offering a unique and authentic experience that includes culture and soft adventure activities, as well as comfort.
- Younger generations' members are well educated and search for distant destinations. It is divided into two categories: those aged 18 to 30 who, despite having time to dedicate to tourism, have limited means to invest in it. They are called 'smaller budget, extensive time'. The book their activities directly with local service providers and community-based tourism product exert a high attraction to this group, because they are cheaper than the standard offer. This allows them to contribute significantly to local development. The second category is called 'high income, little time', and is composed of people aged between 31 and 50 years old who normally travel as a couple and for whom comfort is an important variable. To attract this target segment, it is necessary to

promote a unique experience and offer several options of accommodation and activities that can be combined (that is, so that a tourist package can be created that meets the visitors' budget). Volunteer projects or extended stay projects that involve activities with the local community can also be presented.

- Finally, *families with children* are a growing group that travel to make the trip educational for children, so interaction with locals is valued, as well as safety. To atract this market segment, it is necessary to highlight the safety of the destination, offer activities that can be interesting for both adults and children, promote contact between children and between the culture and lifestyle of the places through organised visits to schools, sports and other activities among children.

Role of CBT in Achieving Sustainable Development Goals (SDGs)

Tourism is an industry with a strong potential to contribute to the achievement of each of the seventeen SDGs, as defined by the United Nations (UN). In particular, one can highlight Goals 8 (inclusive and sustainable growth), 12 (sustainable consumption and production) and 14 (sustainable use of oceans and marine resources). Within these, there are three targets specifically mentioning tourism:

- Target 8.9: "By 2030, devise and implement policies to promote sustainable tourism that creates jobs and that promotes local culture and products";
- Target 12.b: "Develop and implement tools to monitor sustainable development impacts for sustainable tourism that creates jobs and promotes local culture and products."
- Target 14.7: "By 2030, increase the economic benefits to Small Island developing States and least developed countries from the sustainable use of marine resources, including through sustainable management of fisheries, aquaculture and tourism" (UNWTO, 2015).

Considering the CBT concept, it is clear that this form of alternative tourism has an important role for sustainable development and for the achievement of the SDGs since it allows to meet the goals established by the UN. That is, CBT is based on the implementation of measures towards sustainability, promotion and conservation of culture (Asker et al., 2010; Giampiccoli & Kalis, 2012; Goodwin & Santilli, 2009), as well as fostering the various local economic activities throughout valuing their interconnection with tourist activity (Brohman, 1996; Coriolano et al., 2009), promoting local products in the most different areas (handicrafts, fishing, gastronomy, others). The premise of community participation in tourism allows jobs to be created for residents, which also brings greater local economic sustainability. Thus, target 8.9 identified above, is achieved considering the premises of this form of tourism, when measures that generate jobs and that promote local culture and products are established.

Target 12.b is met if when there is the responsibility on the part of those involved in the local tourism sector to develop and implement tools that allow monitoring the various impacts of CBT at the local level. Such monitoring is essential not only to achieve the objectives sought by the UN in general, but in particular to understand the state of local tourism development, if it follows sustainable standards, and whether the impacts are being positive or negative. In other words, implementing monitoring tools assesses the sector's sustainability and the viability of CBT (Asker et al., 2010).

Regarding target 14.7, this is achieved when there is sustainable management of the tourism sector. This sustainable management is achieved by CBT since this form of tourism is based on community participation in tourism planning and development, as well as it values equal participation, empowerment, and increased self-confidence in the local community (Giampiccoli & Kalis, 2012). It also seeks to promote economic, social, and cultural well-being, as well as the development of other economic activities (Brohman, 1996), and environmental protection (Asker et al., 2010).

With regard to the contribution of CBT to the achievement of the 17 SDGs (Table 1), there should exist, in general: the creation of decent jobs for all; encourage the development of various local activities, mainly agricultural; implement safety and health measures; implement educational programs; ensure equal community participation in tourism planning and development; making tourists aware of environmental issues; manage natural resources in a sustainable way; promote the use of renewable energy; support the improvement of infrastructures and technologies; preserve and conserve heritage; guarantee and protect the rights of locals, tourists and employees.

Thus, Asker et al. (2010) conclude that CBT contributes to the sustainable development of the sector and the overall destination area and contributes to achieving SDGs. In short, this contribution is demonstrated when tourism:

- Supports local economic development through job diversification;
- Respects and encourages equal participation by the local community;
- Minimises negative impacts on the environment;
- Preserves and promotes cultural heritage and well-being;
- Educates visitors about culture and nature;
- Demonstrates good management practices;
- Ensures the quality and safety of the tourist experience for all individuals involved.

Implementation of CBT in Destinations

Implementing community-based tourism projects should be carefully thought, as its success involves the inclusion and participation of locals. This significant involvement of communities demands for a special attention to the development process. Bearing this in mind, Asker et al. (2010) advance a Best Practice Manual including the fundamental elements to consider when planning a CBT project, assuring that it follows CBT's principles and overall sustainability.

At an early stage of development, and according to the authors, it is crucial to implement community participation in planning the project. Thus, it will be up to the community itself to decide what type and level of involvement it intends to have with the project. After this phase of defining community involvement, it is also essential to define a project that goes according to the development projects already existing in the territory. The project should complement and support previously planned projects.

Identifying an appropriate organisational and management structure and ensuring adequate management systems are also important steps to be taken. This is because having an effective management team, maintaining standards and quality control, operating in accordance with regulations, increasing resilience, and ensuring constant monitoring and evaluation, are an asset for the success of the project. After these considerations in the initial stages of the process, it is necessary to define the financing and asset management mechanisms. In other words, it is required to define the sources of financing and the appropriate structures, share the benefits in an equitable way, manage the resources, manage the supply and demand in order to guarantee sustainability, prepare the community to manage the tourist offer, and develop capacity in order to motivate them for the effective management and delivery of the project.

Table 1. Role of CBT in achieving the Sustainable Development Goals (own elaboration based on UN-WTO & United Nations Global Compact Network Spain, 2016)

SDG	Role of community-based tourism
1. No poverty	• Create decent jobs for everyone, especially locals, women, young people and discriminated groups.
2. Zero hunger	• Stimulate local agricultural production
3. Good health and well-being	• Implement health and safety plans and measures in order to prevent accidents and health problems.
4. Quality education	• Implement educational programs for tourism employees and for tourists. • Implement community development projects. • Make society aware of the importance of practicing tourism responsibly.
5. Gender equality	• Offer decent jobs for women. • Ensure equal participation in tourism planning and development.
6. Clean water and sanitation	• Sustainably manage aquifer resources. • Ensuring effective sanitation management.
7. Affordable and clean energy	• Apply efforts to develop energy efficiency measures. • Increase the use of renewable energy.
8. Decent work and economic growth	• Offer decent work for local community members. • Promote the development of various local activities auxiliary to tourism.
9. Industry, innovation and infrastructure	• Invest in sustainable facilities and infrastructure, also available for locals. • Improve the technological capacity of destinations, mainly rural destinations, in terms of information and communication technologies.
10. Reduced inequalities	• Allow the involvement of everyone in tourism planning and development processes.
11. Sustainable cities and communities	• Build and adapt infrastructures accessible to everyone. • Preserve and conserve the local heritage. • Support the regeneration of marginal/peripheral areas.
12. Responsible consumption and production	• Implement measures for the efficient use of natural resources, through the sustainable management of food and waste, the proper selection of suppliers and products, and the promotion of local culture and products.
13. Climate action	• Reduce energy consumption and promote the use of renewable energy.
14. Life below water	• Promote integrated management of coastal areas, supporting the reduction of marine pollution and the sustainable use of marine resources in tourist activities.
15. Life on land	• Implement measures to combat deforestation and desertification through the creation of educational programs for tourists on the preservation of the environment.
16. Peace, justice and strong institutions	• Involve local communities in the various activities. • Offer decent jobs for everyone. • Take measures regarding the use of local resources. • Be transparent. • Guarantee and protect the rights of locals, tourists and employees.
17. Partnerships for the goals	• Promote cooperation between the most diverse actors, such as firms and businesses, civil society, governments, and educational institutions.

Another key and necessary aspect is product development and marketing. In the first stage, it is fundamental to define the product and know the target market in order to adapt the marketing strategies and add value to other tourist activities. Building relationships with various stakeholders in the tourism sector is another relevant aspect of supporting the CBT project, since strengthening strong relationships with key stakeholders is an important step towards the development of a strong basis for project development.

Of the relationships established, it is essential to value relations between communities and the private sector, as well as collaboration between communities and tourism authorities. Another very important dimension concerns the monitoring of CBT initiatives implementation and management. Twining-Ward (2007) advances a set of steps that allow this monitoring, namely: (i) planning and monitoring, (ii) scoping key issues, (iii) developing indicators, (iv) collecting data, (v) evaluating results, (vi) planning the response, (vii) communicating results; and (viii) reviewing and adapting.

COMMUNITY-BASED TOURISM IN RURAL DESTINATIONS

Tourism is relevant to the development of rural areas, especially when these areas are characterised by poverty and lack of economic opportunities, because when there is a combination between the agricultural, civic and/or cultural nature of these spaces, an attractive tourist asset appears that is enriched when the identity characteristics are well preserved, since they are able to attract tourists from urban spaces (Lane, 1994; Okech et al., 2012).

Okech et al. (2012) mention that rural territories have a high lack of viable options for economic activities that can be developed since they have few resources. For this reason, tourism is seen as an important alternative, because it is based on basic services and infrastructure that are easily developed and adjusted.

Martínez-Roget et al. (2015) also suggest that these areas have tourism potential as they enable a quality tourist experience, this being a positive factor for the attractiveness of tourism in rural areas.

As there is a capacity to attract visitors to these areas, rural tourism is considered an important agent for rural development by enabling the generation of net benefits for local residents, as well as increasing opportunities for the same residents to participate in the development of the region's tourist products (Iorio & Corsale, 2010; Okech et al., 2012). The increase of the participation of locals in tourism development is achieved through the creation of local businesses and firms, because in the specific case of tourism in rural areas, the nature of the tourism product involves services or goods from locally-owned companies, but also other economic activities such as handicrafts, the production and sale of traditional products, transportation services, recreation, guides, among others (DGADR, n.d.; Okech et al., 2012). The growth of rural tourism results from the increasing interest in heritage, traditional gastronomy, authenticity, concern with health, and contact with nature (DGADR, n.d.). These are extremely important resources that can be used by local communities to develop their own tourism businesses and CBT projects.

From the 1970s onwards, the negative effects of tourism in rural areas began to be acknowledged. As a result, this type of tourism is no longer seen as an economic activity free of harm to the environment (Lane, 1994). Thus, rural areas, and consequently rural tourism, are at risk due to environmental, economic, social, and cultural factors. Thus, it is essential that there is good planning and development so that there are no negative impacts that cause conflicts and/or destruction of the environment in which tourism is developed (Kayat, 2014).

Tourism has great potential to stimulate and revitalise rural areas, allowing the development of the territory and the rural economy that is expressed through (DGADR, n.d.):

- diversification of activities related to agricultural exploration;
- sustaining farmers' income;
- maintaining, creating, and diversifying jobs;

- development of new services;
- nature and landscape conservation;
- support for rural art and crafts;
- boosting cultural initiatives;
- recovery of historical heritage;
- increasing the role of women and the elderly;
- revitalisation of communities due to new dynamics, ideas, and initiatives

However, for this to occur, it is necessary to develop forms of tourism that operate under the principles of sustainability. Eusébio and Figueiredo (2014), list some aspects that must be fulfilled to achieve sustainability in a rural tourist destination which are:

- Sustaining the cultural characteristics of the host community;
- Contributing to improving the quality of life of residents;
- Stimulating the employment of residents in the economic activities that integrate the tourism industry;
- Sustaining, enhancing, and conserving the landscape and habitats, considering the defined carrying capacity limits;
- Stimulating a balance between the potential growth of tourism and the needs for conservation of the environment;
- Minimising negative environmental and socio-cultural impacts;
- Sustaining the satisfaction of visitors and residents;
- Sustaining the tourism industry.

When defining a sustainable tourism development strategy in rural areas, it is necessary that certain aspects are considered. One of the aspects is the assessment of local participation in the definition of strategies and decision making. This emphasises the need for community involvement in order to achieve sustainable tourism development, which is the basis of CBT. In this regard, Kayat's model (2014) portraying the structure for the sustainability of community-based rural tourism (CBRT) brings an important contribution by identifying the criteria that allow achieving that sustainable development (Figure 4).

The centre of this model is the local community and its involvement and commitment. The community must be supported through collaboration and partnerships, it must be financially empowered and trained, as well as it should be led. In this way, the community is able to manage, conserve and generate benefits for itself. Other aspects that Kayat (2014) considers as key to achieving sustainability are the need for the community rural tourism product to present competitive management, the conservation of resources, and the creation of benefits for the local community (economic, cultural, environmental and social benefits). Bearing this in mind, tourism in rural areas should adopt sustainable management, promoting respect for the distinctive characteristics of the region, so that the negative effects are minimised, that economic and social tensions are alleviated or even resolved, that the rural economy is sustainable to ensure a balance in economic growth through tourism, and that local communities and businesses are involved in decision-making processes and that they collect benefits from tourism development (DGADR, n.d.; Garrod et al., 2006; Lane, 1994).

Figure 4. A framework to understand the sustainability of community-based rural tourism (Kayat, 2014, p. 5)

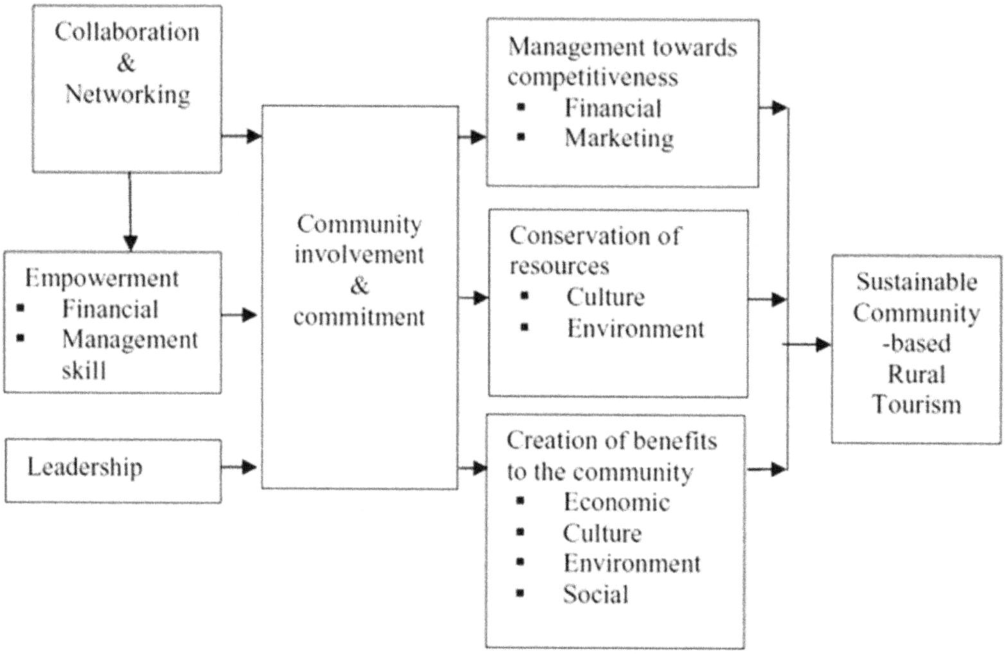

When a sustainable approach to tourism in rural areas is adopted, positive impacts are maximised, and negative impacts are significantly minimised. However, although rural tourism is considered an agent of rural development, there are some uncertainties regarding the economic impacts on local rural communities and regarding the improvement of the quality of life provided by tourist development in the region (Iorio & Corsale, 2010), not least because tourism can prove to be detrimental in certain areas while in others it can be an asset for positive local development (Lane, 1994; Silva & Leal, 2015). In order to guarantee added value, it is necessary to invest in the conservation and maintenance of rural capital, to invest in tourism businesses in rural areas, and a government intervention must be verified (Garrod et al., 2006).

Several studies have been carried out with the intention of emphasising how rural areas and communities can benefit from being involved in tourism (Iorio & Corsale, 2010). The impacts felt in the communities and perceived by them will be all the more positive, the more tourism is developed around endogenous resources, integrating the natural heritage and culture, as well as the local community (Eusébio & Figueiredo, 2014). Communities will benefit from the recreational opportunities and facilities that will arise, from the internal investment that can benefit the economy and the environment (Garrod et al., 2006).

In rural tourism, when there is a tourism promotion aimed at increasing the participation of locals in the management of the tourism product, in decision-making and in the tourist business in the region, there will be an increase in the economic and financial benefits distributed by the which will have positive effects on the alleviation of poverty in rural areas where tourism operates (Okech et al., 2012). Lane (1994), in his literature review, confirms tourism as an incredibly powerful tool for economic growth as it allows the transfer of capital, income and employment from industrial, urban and developed areas to non-industrial areas.

When the rural tourism offer is of quality and based on sustainability, it allows enriching the experience of tourists, generating increasing revenues that benefit local residents, while verifying, at the same time, low negative impacts on the environmental and socio-cultural level (Velázquez et al., 2010). With the adoption of measures that promote sustainability, there is a fight against the negative aspects of the rural environment. In general, the natural and cultural heritage is preserved and conserved, there is greater economic stability, it allows economic diversification, the reduction of poverty, the improvement of infrastructure, it allows the settlement of the population, and the fight against inequality through valuing equal participation.

CONCLUSION

The term 'community-based tourism' emerged in the 1990s, following the adoption of good practices for the development of sustainable tourism that were based on the concept of the triple bottom line, with a focus on three dimensions: environmental, social and economic. Thus, benefits should arise in terms of the economy, society and the environment in which the community is involved (Asker et al., 2010; Blackstock, 2005; Pawson et al., 2017). Since the CBT seeks to transform tourism into a more sustainable industry, in the context of the development of sustainable tourism, this form of alternative tourism has a relevant role whose value has been recognised for decades (Goodwin & Santilli, 2009; Kayat, 2014; Salazar, 2012). The need that small communities feel to overcome adversity makes community-based tourism the form of sustainable tourism that brings more permanent and longer-lasting benefits (Parent et al., 2009). This form of alternative tourism has as its main focus the community, which is made up of members who share the same culture and who are interconnected at an economic and social level, and are also geographically circumscribed (Giampiccoli & Kalis, 2012). It also has as main objective the involvement of the community in tourism (CBI, 2020; Goodwin & Santilli, 2009).

The aim of CBT initiatives is based on sustainability objectives that, in general, imply the conservation of environmental biodiversity and the reduction of poverty in the spaces and communities in which it is applied. Because it is so suitable for tourism in rural areas, the CBT and rural tourism, together with sustainable tourism and ecotourism, usually go hand in hand due to the similar objectives that are shared by them. In order to achieve them, it is essential to develop and plan tourism according to the guidelines for the protection and enhancement of the region's nature and culture, as well as for the economic benefit and well-being of the local residents (Asker et al., 2010).

The development of CBT aims at strengthening local economies, alleviating poverty, conserving nature, the environment and resources, and cultural development (Kayat, 2014; Sutawa, 2012). Although CBT brings benefits at the country level, Giampiccoli and Kalis (2012) argue that it should be understood as a major contributor to economic development at the local scale, that is, in the community in which it is exercised. Thus, it depends on the local community and the endogenous resources of the places where it is developed (Ismail et al., 2016), as these are the aspects that attract tourists to the destinations.

The growing interest of tourists in CBT allows, therefore, the increasing benefits for the communities in which tourism is developed. It has a positive impact on all the SDGs, contributing to achieving them. In addition, tourism also contributes as a tool to raise tourists' environmental and social awareness. However, for this more responsible and sustainability-oriented form of tourism to present benefits, it needs that the product offered is focused on competitive management, resource conservation and the creation of benefits for the community. It should be noted that these three aspects must be supported through the involvement and participation of the local community, because it and its empowerment are key to the development of sustainable tourism (Kayat, 2014; Sutawa, 2012).

REFERENCES

Aminudin, N., Yahya, W. S. Z., & Sumarjan, N. (2015). Supply analysis in the continuity of community-based tourism. In S. M. Radzi, N. Sumarjan, C. T. Chick, M. S. M. Zahari, Z. Mohi, M. F. S. Bakhtiar, & F. I. Anuar (Eds.), *Theory and practice in hospitality and tourism research* (pp. 247–251). CRC Press.

Ardianto, E. (2012). Dimensions of community-based tourism. A case study of surf community in Bali, Indonesia. *Journal of Tourism Research, 5*(B), 38-46.

Arnstein, S. R. (1969). A ladder of citizen participation. *Journal of the American Planning Association, 35*(4), 216–224.

Asker, S., Boronyak, L., Carrard, N., & Paddon, M. (2010). *Effective community-based tourism: A best practice manual.* Sustainable Tourism Cooperative Research Centre.

Blackstock, K. (2005). A critical look at community-based tourism. *Community Development Journal: An International Forum, 40*(1), 39–49. doi:10.1093/cdj/bsi005

Brohman, J. (1996). New directions in tourism for the Third World. *Annals of Tourism Research, 23*(1), 48–70. doi:10.1016/0160-7383(95)00043-7

Bursztyn, I., Bartholo, R., & Delamero, M. (2009). Turismo para quem? Sobre caminhos de desenvolvimento e alternativas para o turismo no Brasil. In R. Bartholo, D. G. Sansolo, & I. Bursztyn (Eds.), *Turismo de base comunitária: Diversidade de olhares e experiências brasileiras* (pp. 76–91). Letra e Imagem.

CBI. (2015). *Tailored intelligence study: Community-Based Tourism (CBT) in Bolivia.* Ministry of Foreign Affairs of the Netherlands CBI - Centre for the Promotion of Imports from developing countries.

CBI. (2020). *What are the opportunities for community-based tourism from Europe?* Retrieved March 20, 2020, from https://www.cbi.eu/market-information/tourism/community-based-tourism/community-based-tourism-europe

Coriolano, L. N., Araújo, A. M., & Vasconcelos, F. P. (2009). *Arranjos produtivos locais do turismo comunitário: Atores e cenários em mudanças.* EDUECE.

DGADR. (n.d.). *Fator de desenvolvimento rural.* Retrieved March 14, 2020, from https://www.dgadr.gov.pt/diversificacao/turismo-rural/fator-de-desenvolvimento-rural

Dolezal, C. (2011). Community-based tourism in Thailand: (Dis-)illusions of authenticity and the necessity for dynamic concepts of culture and power. *ASEAS – Austrian Journal of Southeast Asian Studies, 4*(1), 129–138.

Eusébio, C., & Figueiredo, E. (2014). Turismo e desenvolvimento sustentável de destinos rurais. In E. Kastenholz, C. Eusébio, M. J. Carneiro, & J. Lima (Eds.), *Reinventar o turismo rural em Portugal: Cocriação de experiências turísticas sustentáveis* (pp. 51–58). UA Editora.

Garrod, B., Wornell, R., & Youell, R. (2006). Re-conceptualising rural resources as countryside capital: The case of rural tourism. *Journal of Rural Studies*, *22*(1), 117–128. doi:10.1016/j.jrurstud.2005.08.001

Giampiccoli, A., & Kalis, J. H. (2012). Community-based tourism and local culture: The case of the Mamapondo. *Pasos Revista de Turismo y Patrimonio Cultural*, *10*(1), 173–188. doi:10.25145/j.pasos.2012.10.017

Goodwin, H., & Santilli, R. (2009). *Community-based tourism: A success?* ICRT Occasional paper 11, 1-37.

Hall, C. M. (1991). *Introduction to tourism in Australia: Impacts, planning and development*. Wesley and Longman.

Iorio, M., & Corsale, A. (2010). Rural tourism and livelihood strategies in Romania. *Journal of Rural Studies*, *26*(2), 152–162. doi:10.1016/j.jrurstud.2009.10.006

Ismail, M. N. I., Hanafiah, M. H., Aminuddin, N., & Mustafa, N. (2016). Community-based homestay service quality, visitor satisfaction, and behavioral intention. *Procedia: Social and Behavioral Sciences*, *222*, 398–405. doi:10.1016/j.sbspro.2016.05.192

Kayat, K. (2014). Community-based rural tourism: A proposed sustainability framework. *SHS Web of Conferences, 12*.

Lane, B. (1994). What is rural tourism? *Journal of Sustainable Tourism*, *2*(1–2), 7–21. doi:10.1080/09669589409510680

Manyara, G., & Jones, E. (2007). Best practice model for community capacity-building: A case study of community-based tourism enterprises in Kenya. *Preliminary Communication*, *55*(4), 403–415.

Martínez-Roget, F., Moutela, J. A. T., & Núñez, J. C. E. (2015). Chaves do êxito do turismo em espaço rural (TER): Evidências a partir da lealdade turística na Rede das Aldeias do Xisto. *Revista Portuguesa de Estudos Regionais*, *40*, 65–81.

Moric, I. (2013). The role and challenges of rural tourism development in transition countries: Montenegro experiences. *Turizam*, *17*(2), 84–95. doi:10.5937/Turizam1302084M

Mtapuri, O., Giampiccoli, A., & Jugmohan, S. (2015). Community-based tourism affinity index: A visitor"s approach. *African Journal of Hospitality, Tourism and Leisure*, *4*(2), 1–13.

Murphy, P. E. (1985). *Tourism: A community approach*. Methuen.

Okazaki, E. (2008). A community-based tourism model: Its conception and use. *Journal of Sustainable Tourism*, *16*(5), 511–529. doi:10.1080/09669580802159594

Okech, R., Haghiri, M., & George, B. P. (2012). Rural tourism as a sustainable development alternative: An analysis with special reference to Luanda, Kenya. *Revista de Cultura e Turismo*, *6*(3), 36–54.

Parent, S., Klein, J. L., & Jolin, L. (2009). Le développement communautaire local et le tourisme communautaire: Une analyse conceptuelle comparative. *Journal for Communication Studies*, *4*(1), 73–89.

Pawson, S., D'Arcy, P., & Richardson, S. (2017). The value of community-based tourism in Banteay Chhmar, Cambodia. *Tourism Geographies: An International Journal of Tourism Space. Place and Environment*, *19*(3), 378–397.

Pawson, S., D'Arcy, P., & Richardson, S. (2017). The value of community-based tourism in Banteay Chhmar, Cambodia. *Tourism Geographies: An International Journal of Tourism Space, Place and Environment*, *19*(3), 378–397.

Russell, P. (2000). Community-based tourism. *Travel & Tourism Analyst*, (5), 89–116.

Salazar, N. B. (2012). Community-based cultural tourism: Issues, threats and opportunities. *Journal of Sustainable Tourism*, *20*(1), 9–22. doi:10.1080/09669582.2011.596279

Scheyvens, R. (2002). *Tourism for development: Empowering communities*. Prentice Hall.

Selin, S., & Chavez, D. (1995). Developing an evolutionary tourism partnership model. *Annals of Tourism Research*, *22*(4), 844–856. doi:10.1016/0160-7383(95)00017-X

Silva, L., & Leal, J. (2015). Rural tourism and national identity building in contemporary Europe: Evidence from Portugal. *Journal of Rural Studies*, *38*, 109–119. doi:10.1016/j.jrurstud.2015.02.005

Simmons, D. G. (1994). Community participation in tourism planning. *Tourism Management*, *15*(2), 98–108. doi:10.1016/0261-5177(94)90003-5

Simpson, M. C. (2009). An integrated approach to assess the impacts of tourism on community development and sustainable livelihood. *Community Development Journal: An International Forum*, *44*(2), 186–208. doi:10.1093/cdj/bsm048

Supriharjoa, R. D., Rahmawatia, D., Santoso, E. B., Setiawana, R. P., & Pradinie, K. (2016). Factors influencing community-based heritage sustainability in Kampung Kemasan, Gresik. *Procedia: Social and Behavioral Sciences*, *227*, 498–502. doi:10.1016/j.sbspro.2016.06.106

Sutawa, G. K. (2012). Issues on Bali tourism development and community empowerment to support sustainable tourism development. *Procedia Economics and Finance*, *4*, 413–422. doi:10.1016/S2212-5671(12)00356-5

Twining-Ward, L. (2007). *A toolkit for monitoring and managing community-based tourism*. School of Travel Industry Management, University of Hawai.

UNWTO. (2015). *Tourism and the Sustainable Development Goals*. UNWTO.

UNWTO & UN Global Compact Network Spain. (2016). *The tourism sector and the Sustainable Development Goals: Responsible tourism – A global commitment*. UNWTO & UN Global Compact Network Spain.

Velázquez, G., Valderrama, J., Ruiz, J., Martínez, O., Morales, L., Verján, R., & Flores, J. (2010). Community-based tourism and the development of local small businesses in rural Baja California, Mexico. *WIT Transactions on Ecology and the Environment*, *142*, 653–663. doi:10.2495/SW100591

WWF Internacional. (2001). *Guidelines for community-based ecotourism development*. Retrieved March 14, 2020, from https://wwf.panda.org/wwf_news/?12002/Guidelines-for-Community-based-Ecotourism-Development

Chapter 12
Events as Sustainable Experience for Local Communities of a Tourist Destination

Cecília Lopes Lobo
University of Aveiro, Portugal

Rui Costa
GOVCOPP, Universidade de Aveiro, Portugal

ABSTRACT

There is an increasing concern for public stakeholders to develop sustainable tourism strategies, in order to satisfy the needs of today's society, without compromising future generations. Considering the key dimensions of sustainable tourism, this chapter aims to analyse the well-being of a destination's local community and its relationship with tourist events. This exploratory research used a qualitative approach by employing a semi-structured interview with the most relevant stakeholders of the destination. It is possible to conclude that local events can have different positive impacts on the quality of life of the residents. Greater support for local political strategies, better feedback, and word-of-mouth, translating into a positive and sustainable promotion of the destination by the local community are results that can be expected from organising events also designed for the residents of the destination, as a sustainable and inclusive experience.

INTRODUCTION

Tourism is one of the largest sectors of the world economy, having increasingly significant weight in several dimensions, such as job creation, export earnings, and a country's gross domestic product, as well as cultural and social dimensions. Equally growing, the event industry, is constantly developing nowadays, translating into an increasingly present trend: event tourism. This contributes significantly to

DOI: 10.4018/978-1-7998-5691-7.ch012

leisure and business tourism in a destination, being often associated with the tourist development of the territory and, consequently, with the analysis of its relationship with tourist demand, visitors.

However, events held at a destination also have a major impact on residents and an influence on their quality of life. This article presents a study that analyses the importance that events can have on the quality of life of residents of the city of Viseu (Portugal), showing that events can positively influence the destination's local communities. These conclusions make it possible to bridge the current topic of sustainability and, consequently, sustainable tourism. Sustainable tourism is guided by three main dimensions: the socially just, the economically sustainable, and the ecologically correct. In social terms, visitors are considered, but above all the receiving community, that is, residents of the tourist destination.

Thus, this chapter analyses the relationship between the events of a destination and how their organisation can be taken into account as a sustainable strategy, since these will have implications for the lives of residents, as well as for the way residents perceive them, whether local events can help to promote the destination and, consequently, have economically competitive advantages.

THEORETICAL FRAMEWORK

Events and Tourism

Event tourism has recently shown notable growth in society, leading to an increase in research on this topic. In fact, events, currently organised on an ever-increasing scale, become a more significant investment in the tourism development strategies for destinations at various levels, being fundamental a greater focus on event management (Getz & Page, 2007; Getz & Page, 2014; Morgan, Pritchard, & Roger, 2004; Nicholson & Pearce, 2001; Panfiluk, 2015; Xiang & Petrick, 2006).

Characterised as leisure or entertainment activity programs (Quinn, 2009); presentations, celebrations or specific performances of a cultural or business nature (Bowdin, Johnny, William, Rob, & Ian, 2012) or unique and spatio-temporal phenomena that involve interactions between different stakeholders (Getz, 2008), the truth is that there is not only one definition of what events are. There are different perspectives on their definition, which are added and completed in different aspects.

In addition to the promotional strategy of the tourist destination, events are also sometimes used by local governments as a form of community and cultural development strategy, which are called "local events" in the literature. This type of event, usually aimed at family leisure and entertainment, is organised for the local community, allowing an appreciation of the locality as well as its residents (Bowdin et al., 2012; Janiskee & Carolina, 1996).

Through the existing literature on the theme of event tourism, it is possible to see clearly that the events end up merging with tourism, bringing various impacts, insofar as: "festivals and events are effective instruments in attracting new visitors, as well as making repeat visits due to the advantage of differentiation" (ETC/WTO, 2005, p. 44). Economic impacts are often the most analysed, however, these are not the only ones identified by the authors. In fact, events can also have an impact on the social, cultural, and environmental perspective (Panfiluk, 2015), and in certain situations, these can have both negative and positive results. The economic impacts are the most mentioned by the case studies. These are generally studied by analysing the increase in the number of jobs, the increase in local taxes, the increase in the number of overnight stays due to events, among others (Getz, 2014; Getz & Page, 2007; OECD, 2018; Panfiluk, 2015). They end up being the easiest impact measurement, which often leads to other types of impacts not being considered and studied. In fact, events can be an important tool in the economy of the destination, becoming a core business, as well as a strategy of local governments to benefit the local economy.

The main economic impacts, referred to in the literature, as being associated with the holding of events in tourist destinations are the direct and indirect expenses made by visitors who travel to the destination because of a certain event on the spot, namely spending on local businesses, in food & beverage or in the hotel industry for possible stays in the destination itself, for example. Another positive impact that can happen in a destination thanks to the events can be related to the development of tourist and non-tourist infrastructures in the city, as a more effective response to the tourist demand of a destination.

Regarding the social perspective, this is approached mainly from the perspective of the social impacts that events can cause to residents, with several authors highlighting the importance that events can have at the level of the local community receiving the event. Authors consider that event tourism can have an impact on community development, through several aspects, such as the improvement of local infrastructure or the participation of residents through volunteering (Getz & Frisby, 1988) thus creating a feeling of involvement and enhancement of its heritage. Some prominent social impacts are presented in the literature, such as community development, long-term promotional benefits, an increased number of jobs, which can be considered both a social and an economic impact,, and finally, the feeling of self-pride and belonging that events can trigger in residents.

Destination Events and Residents/Local Communities

As mentioned, the social impacts of the events are mostly associated with the local community of the destination. Several authors consider that, at present, the cultural attractions of a destination, as well as the events held there, play a very significant role in stimulating the tourist development of a region. The events, as previously mentioned, make it possible to offer different experiences to visitors to a destination, thus becoming a strategic tool as a cultural product, increasing the competitive advantages of a destination and combating problems in the tourism sector, such as seasonality.

Compared to this dimension, it is also possible to look at this theme not from the tourists' point of view, but the residents themselves. In fact, in addition to the events having the capacity to reinforce the positioning of tourist destinations, the events can also make the destination appealing places not only to visit, but also to work and live there (Valle, Mendes, & Guerreiro, 2012).

Thus, several communities currently develop an agenda of events and festivals with the main objective of providing leisure and cultural experiences to the residents of the community, as well as stimulating their development at different levels (Gursoy, Kim, & Uysal, 2004). In fact, in addition to the benefits on the demand side, there are several impacts that events can bring to residents of a destination, mentioned by several authors.

Gursoy & al. (2004) summarise the main advantages that events can bring to a local community, referred to in the literature, which are presented in figure 1.

The previous figure summarises the relationship that exists between local events, residents, and the destination. In fact, there are countless dimensions mentioned in the literature that demonstrate that there are indeed impacts of events on residents and the community, in addition to the most studied economic factors. There are social impacts of great importance, which deserve to be analysed and which, however, are often excluded.

Figure 1. Relation between the residents and the local events (Own elaboration)

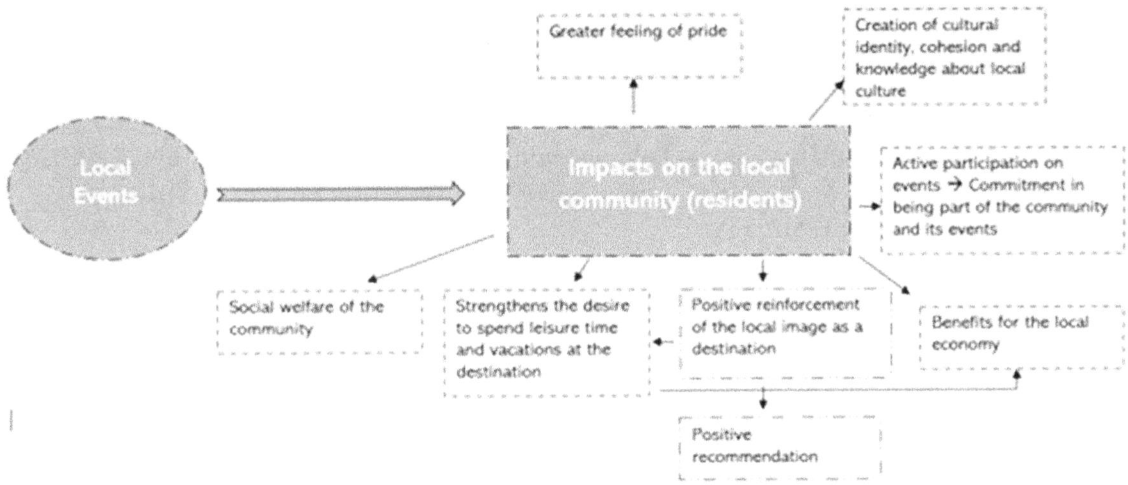

With regard to residents of a destination, events may have the ability to stimulate their sense of pride and belonging, thus also contributing to the creation of cultural identity, as well as the creation of knowledge about the local culture itself (Valle et al., 2012). In addition to social well-being, it is mentioned in the literature that events can stimulate the desire to spend more leisure time and vacations at their own destination, which can consequently translate into economic benefits, stimulating the local economy (Capriello & Rotherham, 2011; Gursoy et al., 2004). In addition, Gursoy et al. (2004) refer to the fact that events have the capacity to create a feeling of pride and belonging among the residents, thus contributing to social well-being as already mentioned.

According to a study by Valle et al. (2012), the results indicate that the experience of residents of the Algarve region in the events of the Allgarve program is important for the formation of the image that they have of the region and of the events themselves. Thus, it is important to conclude that event organisers must also take into account the residents of the destination when planning the event agenda, since their participation, involvement and support are essential to ensure the success of events, especially more local events (Valle et al., 2012),), as well as the success of an improvement in the quality of life of residents of a tourist destination, associated with leisure and well-being.

METHODOLOGY

Since events are a fundamental element of the tourism supply in the city of Viseu, it was considered pertinent to analyse the possible influence of local events on residents of this destination, since these are the main participants and consequent target audience of this type of tourist attraction. Thus, to collect and analyse data related to this study dimension, the present investigation was based on a qualitative methodology, having been conducted interviews with five agents from the territory of Viseu in order to understand their perceptions, what would be the importance of the events for the quality of life for Viseu residents. We opted to use a qualitative analysis to the local stakeholders of the territory since the interviews allow to obtain more complete and personal answers about the questions posed and also allow to obtain diversity and richness of content.

Among the selected universe, five stakeholders of different types were considered, to obtain a greater representation of the various local stakeholders existing in the territory. They were also selected to work with as much diversity as possible, interviewing from public entities to the private sector. For a better organisation of responses and to maintain their anonymity, the agents interviewed were identified by:

1. Public Entities: EPu1; EPu2; EPu3;
2. Private Entities: EPr1; EPr2.

Regarding the questions asked in the interviews, it is possible to classify them in two main categories of analysis:

1. The relationship between the events and the residents of the destination: participation, word of mouth and quality of life;
2. Residents, the appreciation of the destination and the consequent recommendation of the destination.

ANALYSIS AND DISCUSSION OF RESULTS

As previously mentioned, questions were asked to local stakeholders of the territory regarding the importance of residents in the participation of events and in the dissemination/word-of-mouth of the city and its events, as well as the importance that events can have in the residents' quality of life. Thus, the respondents' answers allowed to understand if the residents have any relevant role in this theme analysis, according to the opinion of the interviewed stakeholders.

From the point of view of EPu1, residents do indeed have an important role in the participation of city events, approaching a different view from the other respondents. In fact, this entity addresses this importance from the point of view of residents as employees and entrepreneurs. In other words, it considers that, for the organisation of events at a destination, the collaboration of local companies and associations, such as restaurants and accommodation, for example, which end up being hosts in this dimension, is essential.

Still, in this perspective, he says that if visitors are poorly received by residents and stakeholders, the image of the city ends up being compromised, as well as the image of the related events themselves. It is thus mentioned that it is important for residents, as an integral part of a destination's tourist offer, in this case of events, to feel satisfied with the dynamic surrounding of the destination. This fact will probably be related to the residents' well-being, their satisfaction and consequent support in the tourist management strategies of the destination, thus contributing to the reinforcement of a positive image of the destination, as well as a perspective of sustainability in city tourism, since the well-being of the local community is one of the essential dimensions for the practice of this type of tourism.

From the point of view of residents as the target audience of the events, EPu3, EPr1 and EPr2 consider residents to be a fundamental part, since "residents are the most important market segment of any marketing strategy defined by a municipality, because it is necessary that events have the capacity to make residents an integral and interested part of them" (EPu3). Sharing this vision, EPr2 reinforces, as already mentioned before, that residents are the first consumers when events are held at a destination, considering that they "should be organised outside, but inside too", taking into account paying attention to the needs not only of potential visitors, but also of the local community of Viseu, since they always end up being involved in the organisation of tourist events.

On the other hand, and from a private agent perspective, EPr1 says that it realises residents themselves often do not know the history of the city where they live, thus considering it important that residents actively participate in the events that take place in the city while visitor and connoisseur, as they will be able to bring in people from outside, whether family or friends, as well as enrich cultural and historical knowledge of their place of residence.

Regarding the dimension of the relationship between the events and the residents' quality of life, the responses also generally followed in the same direction. For EPu2, events can have a positive impact on the lives of residents, as they help to "reaffirm the own identity of the community, just as they help to sophisticate people, creating knowledge and also creating the feeling of "I belong to", this is also "a matter of pride for them". This vision is shared by EPr2, which believes that events can "make the city more cosmopolitan, more informed and, ultimately, a city of the future".

In the same line of thought, EPu3 and EPr1 consider that the events offer the possibility of a varied number of cultural activities, entertainment, and experiences to residents, thus translating into social well-being of the resident consumer. Another interesting aspect mentioned by EPr1 is the fact that this social well-being caused by the events may be related to greater support for the municipality's strategies, in relation to the city's tourist sector, as mentioned above.

On the other hand, EPu1 mentions not only positive impacts, but also possible negative impacts on the resident community: "there are always some negative impacts, since events can also affect the daily lives of residents, such as traffic cuts, noises at night. There are two weights on the scale, which always deserve the attention of those who concern them." Although it is the only entity referring to this other dimension, it is important to take this opinion into account for later conclusions of these results.

The culmination of these latest analyses leads to the dimension of the importance of residents in publicising the city and its events, and here, the opinion of respondents is entirely consensual: residents have a significant role in publicising the city and its events. For EPr2 and EPu1, it is relevant to say that positive tourist experiences both in the city and at events end up generating a positive means of speaking in their contact networks, as well as "a satisfied person discloses to another person, a dissatisfied person publicise ten" (EPu1), giving the idea that word-of-mouth has an important power that can be positive but also dangerous. In the same way, EP3 considers that residents have an important role in this dimension "since it is this audience that, when stimulated, can help in the projection and dissemination of word-of-mouth, attracting more attention to them with their family members and friends". For EPu3, "if residents feel an integral part and identified with the DNA of the destination where they live, they will naturally promote it positively and get involved in their activities".

For EPr2, this relationship is obvious. It also makes reference to events in this dimension, saying that "events must be created so that residents can adhere to them, so that their perception is shaped positively and so that a word from residents and surroundings is generated positively". This last analysis ends up demonstrating how this perspective can be related to the Sustainable Development Goal 11, which mentions the sustainability of communities. It is possible to verify a sustainable cycle with regard to the relationship between the positive impact of events on the residents' quality of life and the positive word-of-mouth it can generate, bringing benefits to the destination, but also to the local community itself, through positive economic impacts.

Figure 2 allows us to understand the obtained results from the stakeholders' interviews:

Figure 2. Framework of the interviews results (Own elaboration)

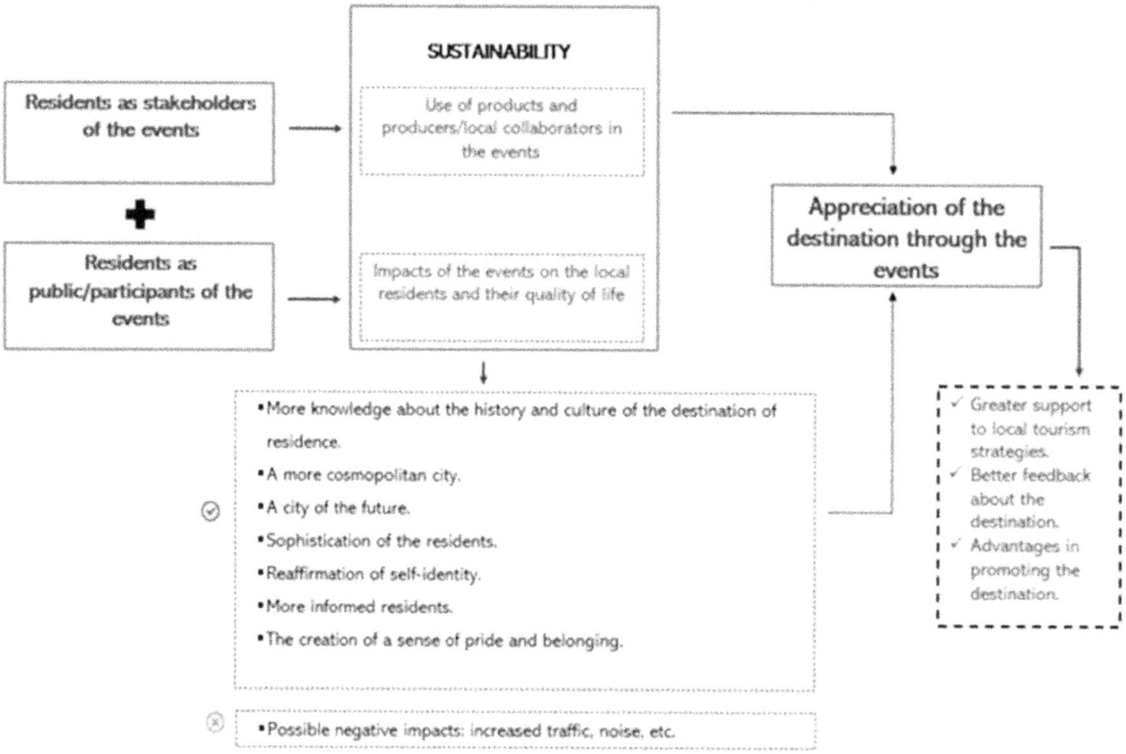

CONCLUSION

According to the analysis of the respondents' responses, it is possible to conclude that, as expected through the theoretical context on this theme, all respondents consider that events can have positive impacts on the quality of life of residents, and this in several ways mentioned. According to some responses, the events allow to reaffirm the identity of the community itself, as well as to create a feeling of belonging in the residents, which can also be a factor of pride. In addition, it is said that the events could offer a wide range of cultural and entertainment experiences, contributing to the social well-being of local consumers. Finally, it was also mentioned by the entities that the events contribute to the creation of knowledge in the residents, thus creating a more informed and cosmopolitan community. Only one interviewed entity mentioned the possibility of negative impacts on residents, such as increased traffic and annoying noise. Although it is important to take this opinion into account, it is important to realise that the positive impacts are mentioned more often by the interviewees.

This analysis refers, in fact, to one of the essential dimensions to consider in the theme of sustainability in tourism: the local community. In addition to the environmental and economic points of view, the social dimension has an extremely considerable weight in this area that is now under consideration. In addition to visitors, the local community must always be considered when developing sustainable local tourism policies. Their impact on their daily lives and their needs are sometimes a forgotten theme in the cycle that the tourism sector develops in destinations and in their communities, thus becoming a bigger problem.

Bridging the Sustainable Development Goals, the investigation of this article may provide guidelines on topics related to objective number 11 – making cities and communities inclusive, safe, resilient, and sustainable (United Nations, 2019). In fact, it is possible to verify, through the results of this chapter, that tourist events may have the capacity to create and strengthen not only economic relations in the destination, as well as positive social relations in the urban areas of the destination and also in the peri-urban areas. The fact that the events are organised not only for potential visitors but also for the destination residents themselves allows the creation of a feeling of belonging and satisfaction, promoting the participation of the urban and rural community in the city's events. In addition, the organisation of events with the collaboration of local entrepreneurs and producers is also a way of making tourist events a sustainable experience for residents of the destination, while also boosting local economic growth and the creation of possible jobs. This relationship could go against SDG number 8 – promoting inclusive and sustainable economic growth, full and productive employment, and decent work for all. The way events can enhance the value of local products by their resident producers is an extremely important issue that deserves attention for future research on the topic.

The positive relationship existing between tourist events and the residents of the destination is doubly advantageous, both for the local community itself and for the local tourism stakeholders. In fact, it is possible to see that, on the one hand, events are enhancers of social activities beneficial to the community, resulting in social well-being. On the other hand, and consequently, if the local community is satisfied and feels that the events are positive points for their daily lives and for the city, they will more easily be willing to develop a positive word of mouth, promoting the destination in a beneficial way, attracting new potential visitors.

REFERENCES

Bowdin, G., Johnny, A., William, O., Rob, H., & Ian, M. (2012). *Events Management*. Academic Press.

Capriello, A., & Rotherham, I. D. (2011). Building a preliminary model of event management for rural communities. *Journal of Hospitality Marketing & Management*, *20*(3–4), 246–264. doi:10.1080/19368623.2011.562411

do Valle, P. O., Mendes, J., & Guerreiro, M. (2012). Residents' Participation in Events, Events Image, and Destination Image: A Correspondence Analysis. *Journal of Travel & Tourism Marketing*, *29*(7), 647–664. doi:10.1080/10548408.2012.720151

ETC/WTO. (2005). *City tourism and culture - The European Experience*. European Travel Commision. Retrieved from www.etc-corporate.org

Getz, D. (2008). Event tourism: Definition, evolution, and research. *Tourism Management*, *29*(3), 403–428. doi:10.1016/j.tourman.2007.07.017

Getz, D. (2014). *The forms and functions of planned events: Past and future. The Future of Events and Festivals*. Academic Press.

Getz, D., & Frisby, W. (1988). Evaluating Management Effectiveness In Community-Run Festivals. *Journal of Travel Research*, 21–27. doi:10.1177/004728758802700105

Getz, D., & Page, S. J. (2007). Event Studies Theory, research and policy for planned events. In Development (3rd ed., vol. 134). Academic Press.

Getz, & Page, S. J. (2014). Progress and prospects for event tourism research. *Tourism Management, 52*, 593–631. doi:10.1016/j.tourman.2015.03.007

Gursoy, D., Kim, K., & Uysal, M. (2004). Perceived impacts of festivals and special events by organisers: An extension and validation. *Tourism Management, 25*(2), 171–181. doi:10.1016/S0261-5177(03)00092-X

Morgan, N., Pritchard, A., & Roger, P. (2004). Destination Branding : Creating the unique destination propostition. *Book*, 335.

Nicholson, R. E., & Pearce, D. G. (2001). *Why Do People Attend Events : A Comparative Analysis of Visitor Motivations at Four South Island Events*. Academic Press.

OECD. (2018). *OECD Tourism Trends and Policies, 2018*. Advance online publication. doi:10.1787/tour-2018-en

Panfiluk, E. (2015). Impact of a Tourist Event of a Regional Range on the Development of Tourism. *Procedia: Social and Behavioral Sciences, 213*, 1020–1027. doi:10.1016/j.sbspro.2015.11.520

Quinn, B. (2009). *Festivals, events, and tourism*. The SAGE Handbook of Tourism Studies. doi:10.4135/9780857021076.n27

United Nations. (2019). *About the Sustainable Development Goals*. Retrieved from https://www.un.org/sustainabledevelopment/

Xiang, L., & Petrick, J. F. (2006). A review of festival and event motivation studies. *Event Management, 9*(4), 239–245. doi:10.3727/152599506776771526

Section 5

Sustainability in Tourism Service Providers

Chapter 13
Communicating Sustainable Initiatives in the Hotel Industry:
The Case of the Hotel Jakarta Amsterdam

Vanessa de Oliveira Menezes
Unicentro, Brazil

Elena Cavagnaro
Stenden Hotel Management School, NHL Stenden University of Applied Sciences, The Netherlands

ABSTRACT

This study aims to gauge how to communicate sustainable initiatives in hotels, based on guests' perceptions, using the Hotel Jakarta Amsterdam as a case study. It was employed empirical research conducting semi-structured personal interviews with hotel guests by a non-probabilistic sample. Guests were randomly approached in the lobby of the hotel during August 2019. According to the data, 'A description on the digital distribution channels, such as Expedia, Priceline, Booking.com, etc.'; 'Signs inside the guestroom'; and an 'Informal chat by an employee when you check-in' were, by guests' perception, the favourite ways to communicating the sustainable initiatives developed by the hotel.

INTRODUCTION

Nowadays, sustainable practices, principles, and designs are not only of interest to hotel guests but rather something they have come to expect. This new scenario developed partly as the result of growing demand for environmentally responsible products and practices (Cho, 2014; Sohail, 2017; Pakdil, & Kurtulmuşoğlu, 2017), which, with the growth of environmental awareness over the past decades, has led to the creation of a group of sustainability-focused consumers (Shaw et al., 2005; Freestone & McGoldrick, 2008). To satisfy this group, while meeting the companies' organizational goals, such as cost savings and increased competitive advantage, hotels have been developing, investing in and implementing several sustainable initiatives in their facilities (Bohdanowicz & Martinac, 2003; Legrand, Sloan, & Chen, 2016, Menezes, 2015; Pakdil, & Kurtulmuşoğlu, 2017). However, investing in sustainable initiatives, when framed

DOI: 10.4018/978-1-7998-5691-7.ch013

within organizational objectives, is in vain if actions are not effectively communicated to guests and stakeholders, and if they do not meet the expectations of this audience. For this reason, understanding what guests consider important while analyzing how a business communicates its sustainable actions can provide a solid foundation for defining ways to engage and build customer satisfaction and loyalty.

From this scenario, this study aims to gauge how to communicate sustainable initiatives in hotels based on guests' perceptions, using the Hotel Jakarta Amsterdam as a case study. The research question was: what is the best channel to communicate the sustainable initiatives developed by the Hotel Jakarta Amsterdam to its guests to attract their attention and meet their expectations?

Hotel Jakarta Amsterdam is a sustainable 4-star hotel in Amsterdam, the Netherlands. It is part of the portfolio of WestCord Hotels, a Frisian family company that has 18 hotels in the Netherlands. Hotel Jakarta Amsterdam is one of the first energy-neutral hotels in the Netherlands, and it employs many other sustainable initiatives.

This study is relevant and justified by the fact that the content of sustainable communication has already been debated in academic studies (Font & McCabe, 2017, Hardeman, Font & Nawijn, 2017, Blose, Mack & Pitts, 2015, Goldstein, Cialdini, & Griskevicius, 2008), but the best ways to present this information have not been discussed. Therefore, this study bridges a research gap. Furthermore, not all hotels adequately communicate their actions to guests, as every person has their expectations regarding environmental sustainability. Therefore, research of this nature could rationalize hotel resources and thus make a practical contribution to the hospitality industry, reaffirming the importance of this study.

Theoretical Framework

Responsible practices have not only become a required standard, but also a resource for cutting costs, improving organizational performance, and meeting customer expectations. Giebelhausen et al. (2016) found that consumers participating in sustainability programs experience a positive emotional response, which increases customer satisfaction. The study of Ailawadi et al. (2014, cited in Brockhaus et al., 2017) had similar results, identifying that sustainability initiatives also increase loyalty behaviour.

It is interesting to note that willingness to choose more sustainable hotels can also be measured by a monetary variable such as the hotel rate. Studies indicate that 45% of guests agree to pay a higher rate to support sustainability programs (Susskind, 2014), a finding corroborated by Nielsen (2014) and Pakdil and Kurtulmuşoğlu (2017). On the other hand, Millar and Baloglu (2011), based on an empirical study, suggest that the vast majority of guests believe that sustainable hotels should not have a different rate from those that are not. Furthermore, according to Millar and Baloglu (2011), no matter how many guests demand a more sustainable approach from the hotel company, not everyone is willing to pay more for these initiatives, which means that opinions on the matter are not unanimous.

Many researchers strive to understand why guests choose a hotel with an environmental approach. Yadava et al. (2018) conducted a study of Indian travellers and found out the presence of green trust is a significant and necessary condition for choosing a green hotel. Additionally, other subjective reasons, such as biospheric values, attitude, and subjective norms were found in the study. However, Cho (2014) highlighted that while appropriate knowledge is a prerequisite for pro-environmental behaviour, most consumers currently lack this knowledge. Therefore, although guests are increasingly aware of environmental issues, many people remain ignorant of them.

Regarding the sustainable initiatives that are important to guests and their perception of them, Han and Chan (2013) conducted an exploratory study of tourist perceptions in sustainable hotels and showed that saving energy, using environmentally friendly materials and not using single-use consumables was perceived as what a hotel should do, and these data were partly corroborated by the study of Millar and Baloglu (2011). Recent research has shown that along the attributes related to accommodation; guests also consider proper food and employee relations as essential components of a sustainable offer (Cavagnaro, Melissen & Düweke, 2018), factors also identified in the research conducted by Tseng and Hung (2013). Therefore, each guest focuses on a different initiative and a different facet of sustainability, and these shows, above all, the relevance of understanding the guest's perception of environmental sustainability to communicate the hotels' initiatives in the best way possible. Thus, this information is important when it comes to understanding which initiatives are the most important to guests.

According to Legrand, Sloan, and Chen (2016), good communication of sustainable initiatives in hospitality has become a highly differential factor and creates a positive image of the company. Studies examining the influence of the communication of environmentally sustainable initiatives on hospitality have already been the subject of research within the academic sector (Kim & Kim, 2013, Gil-Soto et al., 2019). According to Font and McCabe (2017), the type of intervention or message conveyed to the guest can dramatically increase (or decrease) the effectiveness of environmentally sustainable initiatives. However, not all sustainability-related measures are communicated or visible to guests at hotel facilities, in-depth knowledge of how messages are perceived by guests and reflect their awareness and understanding of sustainability is lacking (Cavagnaro, Melissen, & Düweke, 2018).

Wehrli at al. (2013) attempted to identify the most effective messages to communicate sustainable initiatives to potential guests by comparing the effectiveness of the most emotionally targeted messages and the most rational messages and found that the rational messages were most effective. This shows that most consumers have a conscious view of sustainability and understand it as a necessity. According to Lee and Oh (2014), an effective and popular persuasion strategy, useful for designing the best message to influence people's behaviour, is the framing of the message, the presentation of information with different values, whether positive (gain), or negative (losses). This type of framing is known to interact with content factors, and this study shows interaction with three message content factors: (i) the regulatory focus of the message (i.e., prevention vs. promotion focus); (ii) the level of message interpretation (high and low level of interpretation); and (iii) message relevance (selfish or altruistic focus). However, according to Font and McCabe (2017), fear-based messages, including campaigns highlighting the possible consequences of climate change, have proved to be less effective than other types of messages, and the communication of these initiatives should focus on empowering consumers' capacities for change in order to be more effective.

Legrand, Sloan, and Chen (2016) highlight that the focus of the message must always be visible, easy to understand, relevant, and honest. Cavagnaro, Melissen and Düweke (2018) explain that guests are receptive to sustainability-related messages. Nevertheless, this communication should have an informative, encouraging, and positive focus, paying more attention to energy (economy and production), recycling, waste and waste management, and, particularly, the waste of food. These topics are the most obvious and worrying to guests. Goldstein et al. (2008), when conducting a study on towel reuse in a US hotel chain, found that messages highlighting that a high percentage of guests are in the habit of reusing towels produced a significantly higher towel reuse rate than messages that only reinforced the importance of preserving the planet. Thus, the normative signal, which we do not observe as standard in hotels, produced a significantly higher towel reuse rate than the standard for the industry, which reinforces the need for more research on the subject.

Scientific research has been conducted on why guests behave in a pro-environmental manner and the content of messages available at hotels for guests to make environmentally informed decisions. However, there is a research gap concerning which communication channels are the most effective in attracting guests' attention. A powerful message is useless if the communication channel is inefficient, which is why this study is dedicated to identifying the best ways to communicate the sustainable initiatives developed by hotels so that guests can be aware of the initiatives and thus make the decision to adopt more sustainable behaviours.

Methodological Procedures

It was employed empirical research conducting a semi-structured personal interview with hotel guests by a non-probabilistic sample. Guests were randomly approached by one of the researchers in the lobby of the hotel or in the restaurant where the breakfast was served. These locations were chosen based on their informal approach, accessibility, and level of comfort provided to the potential candidates (Longhurst, 2010).

The interview script was written in English. They had 16 questions and it had three sets of questions: the first one was dedicated to the guests' profile and marketing issues related to the reasons for the hotel selection; the second set asked about sustainability; and the third set was dedicated to understanding what might be the best way to communicate sustainable initiatives. The questions were based on an interviewed script conducted at Stenden Hotel in Leeuwarden, the Netherlands, in April 2019, where the constructs were tested by a pilot conducted in the same month.

On August 19 and 20 2019, 38 guests were randomly approached at the Hotel Jakarta Amsterdam. Twenty-two guests voluntarily agreed to participate in the survey. It was pre-defined that only hotel guests who had spent at least one night at the property would participate in the survey. The reason for this is that customers have more knowledge and can answer the defined script. The interview was held in English and Spanish. First, there was an informative talk about the purpose of the research, and guests were asked if they knew about the concept of sustainability. If they did not have any knowledge about it, the researcher explained sustainability according to the Triple Bottom Line definition (Elkington, 1994) and the Sustainable Development concept (OECD, 2011). The participants were informed about the anonymity of the survey and their personal information being safely protected. In addition, permission was requested to record the conversation on an Apple iPhone 5S mobile phone; five guests refused to have their interviews recorded and their requests were honoured. On average, the application of the questionnaire took from nine to 20 minutes.

All interviews were recorded, transcribed verbatim, and analyzed following a general tabulation where the numbers were presented according to its representation in the arithmetic mean and content analysis approach (Harding, 2013) which some answers were coded and interpreted. Each guest questioned also received a code (JG1, JG2, JG3, so on) and their words were used as quotes and added to the findings to reaffirm some results. All of the analysis and most of the data collection was done by the researchers.

For this document, only part of the results will be presented, directly related to sustainable initiatives and how to communicate it. This decision was taken to offer a more straightforward paper.

Findings

The guests were asked if they think the hotel should communicate what measures they take regarding sustainability; twenty-two guests said 'yes', they should do so, and two guests said 'no'. Even though the vast majority stated it would be interesting if the hotel communicated their initiatives, there is a kind of scepticism about guests' interest in this kind of information as the quote from JG5 when he explained that the hotel should communicate, but then he stated, "*Not everybody cares*". Among those who do not think the hotel should communicate what measures they take JG4 said "*not more than now*", suggesting the hotel has done enough.

The next question asked the guest to mark each of the initiatives listed on a scale from 1 to 10. A 1 meant that it was of no importance for them, and 10 meant that it was very important. The eleven alternatives were selected according to the technologies highlighted in other publications (Legrand, Sloan, & Chen, 2016; Pakdil & Kurtulmuşoğlu, 2017; Blose, Mack, & Pitts, 2015; Lee & Oh, 2014) and the initiatives developed by the hotel. The most important initiative for the hotel´s guest was 'Rainwater collected to spray the subtropical indoor garden' with an average grade of 9.0. This alternative was followed by 'PV panels that collect solar energy to heat the shower water' (8.9), 'Warmth and cold storage system that takes care of heating and cooling the hotel' (8.77), 'LED lighting' (8.72); 'High insulation in the whole building' (8.59), 'Use of the most sustainable materials in the building such as natural wood and concrete' (8.45), 'Charging points for electric cars' (7.77); and 'Environmental sustainable furniture' (7.18). The three least important alternatives from the guests' point of view were, 'Using organic ingredients in the restaurant' (6.95); 'Reuse of towels' (6.8), and 'Environmentally sustainable linen' (6.72).

Regarding the high scores of technologies such as 'PV panels that collect solar energy to heat the shower water', JG6 justified his choice by stating, "*Produce energy is beneficial for the planet in the long term*". This alternative also is supported by the article of Cavagnaro, Düweke, and Melissen (2018) when it is mentioned that renewable energy, are the ones that guests consider unique and differentiating. About 'LED Lightning', JG5 said, "*It's important because it's easy to do*", an opinion also shared by JG8. About the low scores for 'Using organic ingredients in the restaurant' many guests showed a lack of interest in organic food and JG12 said, "*Organic is not always good for the environment*", showing some kind of scepticism related to the importance of organic products for the planet. About 'Reuse of towels', five guests stated that they prefer to have the towels changed every day and one said that the daily change is related to hygiene reasons. Another guest said that in a luxury hotel, it is expected to have towels changed every day and JG13 said that the low score is due to priorities "*saving energy is more important than saving costs*". In this case, he assumed that the reuse of towels is done just to cut cost for the hotel, without thinking about the energy savings that it would be made not using washing machines or the environmental impact due to the decrease of chemicals products used in the washing process such as laundry detergent and softener released into the water.

The next question asked the guest to mark each of the social initiatives listed on a scale from 1 to 10. A 1 meant it was little or no importance for them and 10 meant that it was very important. The five alternatives in the question were selected according to the initiatives highlighted in other publications (Legrand, Sloan, & Chen, 2016, Cavagnaro & Curiel, 2012) and the initiatives developed by the hotel. The most important initiative for the guest was: 'Valorize and choose regional suppliers' (8.6); 'Fair Trade linen' (8.45); and 'Fair Trade furniture' (8.36). About 'Valorize and choose regional suppliers' JG7 justified her answer saying that this initiative shows "*a sense of community*", a similar answer was given by JG6 who said that the initiative shows the hotel cares. The two least important alternatives from the guests' point of view were 'Staff uniform made of sustainable fabric' (7.36); and 'Collaboration with the Tropenmuseum, a local who is providing Indonesian art pieces displayed at the Hotel' (7.1). About the last alternative, JG5 said "*this initiative is good just for the hotel branding*", showing that in his point of view, establishing a partnership with a local museum and giving space at the hotel to show their collection is not a selfless initiative on the hotel´s part. JG8 and JG13 justified their low score stating they don´t care about the initiative or museums. This shows that guests see only the museum itself and not the initiative that privileging a local museum that has a significant collection related to the hotel's theme can be more sustainable than bring pieces from other countries.

The next question asked if the guests would like to be informed about the initiatives developed by the hotel. Twenty-two guests said 'yes' and two guests said 'no'. Even though the vast majority stated they would like to be informed, not all guests paid attention to the communication features of the hotel or didn´t ask about them. This data could be related to the fact sustainability is just the fourth reason why the guests chose Hotel Jakarta Amsterdam.

The next question asked if the guest read or paid attention to the signs and the brochures regarding sustainability available inside the guestroom during their stay. Thirteen guests said 'yes' and nine guests said 'no'. When this question was verbalized, many guests primarily said 'no'. It was then that they realized they saw signs inside the guestroom, so they changed their answer to 'yes'. Therefore, sometimes this form of communication does not capture the guests' attention too often. Despite that, it is worth mentioning the high number of people who said 'no'. The data shows that even if there is communication about sustainability, guests often do not pay too much attention. This can happen because the current communication does not catch them or they are not interested. In this case, a brochure is available at the guestroom desk. This brochure discusses the sustainability of the hotel as well as the history of the hotel and the amenities they offer, and signals in the bedroom and bathroom. This last hypothesis conflicts with the previous question's answer when twenty-two guests said they would like to be informed about the sustainable initiatives developed by the hotel, which also arouses discussion if they said they would like to know or they just reinforce a behaviour expected by society, not a real interest on the subject. (Carrigan & Attalla, 2001).

Regarding how to communicate the sustainable initiates developed by the hotel, the last question asked the guest to mark each of the communication forms listed on a scale from 1 to 10. A 1 meant that they do not like it at all and 10 meant they like it very much. The six alternatives in the question were selected according to the initiatives highlighted in previous researchers (Lee & Oh, 2014; Gerdt, Wagner, & Schewe, 2019). The favourite ways to communicate were, 'A description in the digital distribution channels such as Expedia, Priceline, Booking.com, etc.' (8,77); 'Signs inside the guestroom' (8,01); an 'Informal chat by an employee when you check-in' (7,68); and an 'Informative folder inside the guestrooms' (7,41). The least two favourites ways of communicating were, 'Funny advertisements in the corridor of the guestrooms' (7,00), and 'Informative videos on the TV screen' (6,58). Regarding the most favourite ways of communicating, many guests stated they read the descriptions provided by the hotels on the distribution channels because, according to JG16, "*It is the best way to find the most interesting place to stay*". About the least favourite ways of communicating, guests justified their choice saying they wouldn´t pay attention to advertisements in the corridors (2) and the advertisements don´t fit with the concept of the hotel (1). About the informative videos, guests claimed they do not watch TV or they do not watch TV when they are travelling (8). However, two guests said this is the easiest way to catch their attention. Therefore, it can be an appealing way to communicate, but may not be suitable for all kinds of guests. JG18 suggested that the hotel communicate its sustainable initiatives by an email

Table 1. Score of different forms of communication (The Authors, 2019)

Communication forms	Score
A description in the digital distribution channels such as Expedia, Priceline, Booking.com, etc.	8,77
Signs inside the guestroom	8,01
Informal chat by an employee when you check-in	7,68
Informative folder inside the guest room	7,41
Funny advertisements in the corridor of the guestrooms	7,00
Informative videos on the TV screen	6,58

when Jakarta Hotel Amsterdam sends a welcome message to the guests after the booking confirmation. However, JG14 said he does not want to receive more emails and more information. Again, this way of communication can be an appealing form, but it may not be suitable for all kinds of guests.

The answers bring light to the definition of what might be the best way to communicate sustainable initiatives developed by the hotel, but it does not give a precise conclusion of the subject. It would be important to test by an experiment whether each of the ways listed as the most important ones in the guests' perception works in everyday life.

CONCLUSION

The study sought to verify how to communicate sustainable initiatives developed at Hotel Jakarta Amsterdam based on the guests' perceptions. According to the data, twenty-two guests think a sustainable hotel is important; however, only thirteen guests believe Hotel Jakarta Amsterdam takes care of environmentally sustainable initiatives; consequently, there is a gap between the message communicated by the hotel and the guests' awareness.

'Rainwater collected to spray the subtropical indoor garden', 'PV panels that collect solar energy to heat the shower water', and 'Warmth and cold storage system that takes care of heating and cooling the hotel' were the most important environmental initiatives from the guests' points of view. 'Valorize and choose regional suppliers, 'Fair Trade linen' and 'Fair Trade furniture' were the most important social initiatives. Therefore, it is suggested that the communication regarding sustainability at Jakarta Hotel Amsterdam gives priority to these initiatives.

Twenty-two guests would like to be informed about the initiatives developed by the hotel; however, only thirteen people read or paid attention to the signs or the brochures regarding sustainability available inside the guestroom during their stay. Thus, the communication forms need to be more attractive, different without being intrusive, and it needs to fit the concept of the hotel.

'A description on the digital distribution channels such as Expedia, Priceline, Booking.com, etc.'; 'Signs inside the guestroom'; and an 'Informal chat by an employee when you check-in' were the favourite ways to communicating, stated by the guests. However, it would be recommended to take an experiment with all these different types of communication to check which one would work at the hotel regularly.

There are some limitations to the study. The paper addresses a few samples of guests interviewed over a few days during the week in the summer season. Each day of the week and season of the year bring different kinds of guests to the hotel property; therefore, the findings cannot be generalized. To do so, it would be necessary to apply major research during different periods of the year. It would be necessary to be defined a different type of script, or more qualitative to get more data that could understand deeply the perception of the guests regarding sustainability or a quantitative one that will require a statistic sample to be representative. The interview script applied in this research represented a mix of both (quantitative and qualitative).

This research is part of a larger investigation focusing on the communication of sustainable initiatives in hotels. Suggestions for further investigations include more research at Hotel Jakarta Amsterdam in different periods of the year and try experiments with different ways of communication to check if each once could yield better results for the hotel. Also, this study or similar research could be replicated in other properties of WestCord Hotels.

REFERENCES

Bechtel, R. B., Corral Verdugo, V., & de Queiroz Pinheiro, J. (1999). Environmental belief systems: United States, Brazil, and Mexico. *Journal of Cross-Cultural Psychology*, *30*(1), 122–128. doi:10.1177/0022022199030001008

Blose, J., Mack, R. W., & Pitts, R. E. (2015). The Influence of Message Framing on Hotel Guests' Linen-Reuse Intentions. *Cornell Hospitality Quarterly*, *56*(2), 145–154. doi:10.1177/1938965514556480

Bohdanowicz, P., & Martinac, I. (2003). Attitudes towards sustainability in chain Hotels – results of a European survey. *Proceedings of the CIB International Conference on Smart and Sustainable Built Environment (SASBE2003).*

Brockhaus, S., Amos, C., Fawcett, A. M., Knemeyer, M., & Fawcett, S. E. (2017). Please clap! How customer service quality perception affects the authenticity of sustainability initiatives. *Journal of Marketing Theory and Practice*, *25*(4), 396–420. doi:10.1080/10696679.2017.1345594

Carrigan, M., & Attalla, A. (2001). The myth of the ethical consumer – do ethics matter in purchase behaviour? *Journal of Consumer Marketing*, *18*(7), 560–578. doi:10.1108/07363760110410263

Carrillo-Hermosilla, J., & Könnola, T. (2009). *Eco-innovation. When sustainability and competitiveness shake hands*. Palgrave.

Cavagnaro, E., & Curiel, G. (2012). *The three levels of sustainability*. Greenleaf.

Cavagnaro, E., Düweke, A., & Melissen, F. (2018). The host-guest relationship is the key to sustainable hospitality: Lessons learned from a Dutch case study. *Hospitality & Society*, *8*(1), 23–44. doi:10.1386/hosp.8.1.23_1

Cho, Y. (2014). *Different shades of green consciousness: The interplay of sustainability labeling and environmental impact on product evaluations*. Springer Science and Business Media Dordrecht.

Elkington, J. (1994). Towards the Sustainable Corporation: Win-Win-Win Business Strategies for Sustainable Development. *California Management Review*, *36*(2), 90–100. doi:10.2307/41165746

Font, X., & McCabe, S. (2017). Sustainability and marketing in tourism: Its contexts, paradoxes, approaches, challenges and potential. *Journal of Sustainable Tourism, 25*(7), 869–883. doi:10.1080/09669582.2017.1301721

Freestone, O. M., & McGoldrick, P. J. (2008). Motivations of the ethical consumer. *Journal of Business Ethics, 79*(4), 445–467. doi:10.100710551-007-9409-1

Gerdt, S. O., Wagner, E., & Schewe, G. (2019). The relationship between sustainability and customer satisfaction in hospitality: An explorative investigation using eWOM as a data source. *Tourism Management, 74*, 155–172. doi:10.1016/j.tourman.2019.02.010

Giebelhausen, M., Chun, H. E. H., Cronin, J. J. Jr, & Hult, T. M. (2016). Adjusting the warm-glow thermostat: How incentivizing participation in voluntary green programs moderates their impact on service satisfaction. *Journal of Marketing, 80*(4), 56–71. doi:10.1509/jm.14.0497

Gil-Soto, E., Armas-Cruz, I., Morini-Marrero, S., & Ramos-Henriquez, J. M. (2019). Hotel guests' perceptions of environmental friendly practices in social media. *International Journal of Hospitality Management, 78*, 59–67. doi:10.1016/j.ijhm.2018.11.016

Goldstein, N. J., Cialdini, R. B., & Griskevicius, V. (2008). A room with a viewpoint: Using social norms to motivate environmental conservation in hotel. *The Journal of Consumer Research, 35*(3), 35. doi:10.1086/586910

Han, H., Hsu, L. T., & Lee, J.-S. (2009). Empirical investigation of the roles of attitudes toward green behaviors, overall image, gender, and age in hotel customers' eco-friendly decision-making process. *International Journal of Hospitality Management, 28*(4), 519–528. doi:10.1016/j.ijhm.2009.02.004

Hardeman, G., Font, X., & Nawijn, J. (2017). The power of persuasive communication to influence sustainable holiday choices: Appealing to self-benefits and norms. *Tourism Management, 59*, 484–493. doi:10.1016/j.tourman.2016.09.011

Harding, J. (2013). *Qualitative data analysis from start to finish*. Sage Publication.

Kim, S. B., & Kim, D. Y. (2014). The effects of message framing and source credibility on green messages in hotels. *Cornell Hospitality Quarterly, 55*(1), 64–75. doi:10.1177/1938965513503400

Lee, S., & Oh, H. (2014). Effective Communication Strategies for Hotel Guests' Green Behavior. *Cornell Hospitality Quarterly, 55*(1), 52–63. doi:10.1177/1938965513504029

Legrand, W., Sloan, P., & Chen, J. S. (2016). *Sustainability in the hospitality industry: principles of sustainable operations* (3rd ed.). Routledge. doi:10.9774/gleaf.9781315690261

Longhurst, R. (2010). Semi-structured interviews and focus groups. In N. R. Clifford, S. French, & G. Valentine (Eds.), *Key methods in geography* (2nd ed.). Sage Publications.

Lopes, A. S., & Almeida, P. S. (2018). Socioenvironmental sustainability actions in hotel management. *Revista Brasileira de Ecoturismo, 11*(3), 378–398. doi:10.34024/rbecotur.2018.v11.6689

Menezes, V. O., & Cunha, S. K. (2017). Ecoinovação em redes hoteleiras globais: designs, barreiras, estímulos e motivações. In V. O. Menezes, M. B. Maçaneiro, & S. K. Cunha (Eds.), *Observatório de ecoinovação: aspectos teóricos e casos ilustrativos* (pp. 167–184). CRV.

Millar, M., & Baloglu, S. (2011). Guests' preferences for green guest room attributes. *Cornell Hospitality Quarterly*, *52*(3), 302–311. doi:10.1177/1938965511409031

Nielsen. (2014). *Global consumers are willing to put their money where their heart is when it comes to goods and services from companies committed to social responsibility*. Retrieved at: https://www.nielsen.com/us/en/press-room/2014/global-consumers-are-willing-toput-their-money-where-their-heart-is.html

OECD. (2011). *Education at a Glance 2011: OECD Indicators*. OECD Publishing.

Pakdil, F., & Kurtulmuşoğlu, F. B. (2017). Using quality function deployment for environmentally sustainable hotels: A combined analysis of customer and manager point of view. *European Journal of European Research*, *16*, 252–275.

Porter, M., & Van der Linde, C. (1995). Toward a new conception of the environmental-competitiveness relationship. *The Journal of Economic Perspectives*, *9*(4), 1995. doi:10.1257/jep.9.4.97

Rauwald, K. S., & Moore, C. F. (2002). Environmental attitudes as predictors of policy support across three countries. *Environment and Behavior*, *34*(6), 709–739. doi:10.1177/001391602237243

Schultz, P. W., & Zelezny, L. (1999). Values as predictors of environmental attitudes: Evidence for consistency across 14 countries. *Journal of Environmental Psychology*, *19*(3), 255–265. doi:10.1006/jevp.1999.0129

Shaw, D., Grehan, E., Shiu, E., Hassan, L., & Thomson, J. (2005). An exploration of values in ethical consumer decision making. *Journal of Consumer Behaviour*, *4*(3), 185–200. doi:10.1002/cb.3

Sohail, M. S. (2017). Green marketing strategies: How do they influence consumer-based brand equity? J. *Global Business Advancement*, *10*(2), 229–243. doi:10.1504/JGBA.2017.084607

Susskind, A. (2014). Guests' reactions to in-room sustainability initiatives: An experimental look at product performance and guest satisfaction. *Cornell Hospitality Quarterly*, *3*(55), 228–238. doi:10.1177/1938965514533744

Tseng, S. C., & Hung, S. W. (2013). A framework identifying the gaps between customers' expectations and their perceptions in green products. *Journal of Cleaner Production*, *59*, 174–184. doi:10.1016/j.jclepro.2013.06.050

Van Teijlingen, E., & Hundley, V. (2002). The importance of pilot studies. *Nursing Standard*, *16*(40), 33–36. doi:10.7748/ns2002.06.16.40.33.c3214 PMID:12216297

Vikan, A., Camino, C., Biaggio, A., & Nordvik, H. (2007). Endorsement of the new ecological paradigm: A comparison of two Brazilian samples and one Norwegian sample. *Environment and Behavior*, *39*(2), 217–228. doi:10.1177/0013916506286946

Wehril, R., Priskin, J., Demarmels, S., Kolberg, S., Schaff, D., Schwarz, J., Truniger, F., & Stettlerand, J. (2013). *How to communicate sustainable tourism products effectively to customers: An explorative study presented at the World Tourism Forum Lucerne*. Lucerne University of Applied Sciences and Arts.

Wiernik, M., Ones, S. D., & Dilchert, S. (2013). Age and environmental sustainability: A meta-analysis. *Journal of Managerial Psychology, 28*(7/8), 826–856. doi:10.1108/JMP-07-2013-0221

Yadav, R., Balaji, M. S., & Jebarajakirthy, C. (2019). How psychological and contextual factors contribute to travelers' propensity to choose green hotels? *International Journal of Hospitality Management, 77*, 385–395. doi:10.1016/j.ijhm.2018.08.002

Chapter 14
Accessible Tourism and Digital Platforms:
A Study Applied to the City of Viana do Castelo

Catarina Meira
Polytechnic Institute of Cávado and Ave, Portugal

Isabel Silva Martins
Polytechnic Institute of Cávado and Ave, Portugal

Bruno Barbosa Sousa
https://orcid.org/0000-0002-8588-2422
Polytechnic Institute of Cávado and Ave, Portugal

ABSTRACT

The disclosure of accessibility conditions in tourism supply encourages the demand for destinations. The fact that a destination seeks solutions in order to satisfy the needs and preferences of tourists, regardless of the nature and degree of their disability, fosters a spirit of social and corporate responsibility, contributing to a differentiated, competitive, and, at the same time, sustainable destination. Therefore, the present study aims to present the accessible resources in the city of Viana do Castelo, based on the Tur4ll platform. The objective is to analyse the tourism offer available and suitable for tourists with special needs (SN). By using a qualitative methodology, a diagnosis of the study context is presented, as well as the main constraints and limitations that are sometimes inherent to tourism with specific motivation.

INTRODUCTION

Tourism is assumed as one of the major engines with a more significant impact on the global development and growth of many economies. Year after year, there has been an increase in competition in this sector, although this competition is established between tourism destinations (Ferreira & Sousa, 2020).

DOI: 10.4018/978-1-7998-5691-7.ch014

To promote a tourist destination in the face of new tourist offers and new market segments is, nowadays, fundamental, at a time when tourism is a segmented activity with specific individual interests and that value experience (Ferreira & Sousa, 2020; Sousa & Silva, 2019). It is becoming more and more emergent to adapt territories to people with SN. For this to happen, it is imperative an awareness on the part of the agents of tourism supply (Alves et al., 2020). Soares and Breda (2014) state that the tourism sector, as a necessary element in a system of economic, social and cultural relations, must be prepared for people with needs at different levels, having as acquired the right to participate in the tourist activity in any community. The notion of accessible tourism is presented by Darcy, Cameronb and Pegg (2010) as a form of tourism that allows people with access needs, including mobility, sight, hearing and cognitive limitations, to practice tourism independently with equity and dignity through the provision of universally designed tourism products, services and environments (Malek & Costa, 2015; Brandão, Costa & Buhalis, 2018).

Increasingly, tourism is asserting itself as a multifaceted and geographically complex activity, in a logic of segmentation and market niches (Sousa & Rodrigues, 2019). The group of people with SN is quite diverse, even though today the almost indistinct use of the terminology "people with special needs" and "people with disabilities" is still common. In fact, the term "people with special needs" refers both to people with disabilities and not (Tomé, 2014). Regarding the breadth of this segment, Devile (2007) states that the group of people with SN is quite diverse and includes not only people with some kind of disability, but also elderly people, parents who carry prams, children, obese people, people who carry heavy luggage, as well as people with particular health problems, among others. However, a new paradigm may be emerging regarding people with SN in the field of accessible tourism. People with Autism Spectrum Disorder (ASD). The ASD is a neurological condition, often associated with other pathologies, and consequently compromising the individual's autonomy and functionality. The decision to choose this research arises with the interest of contributing with a project that allows presenting the city of Viana do Castelo and its resources accessible to people with SN, making this same information available in a digital platform "TUR4all", which aims to support and inform the tourist in the planning and organisation of a trip.

Thus, and in view of the above, the study intends to present inputs in the area of tourism (special needs) and in the area of marketing and consumer behaviour. In the next sections of the document, the main constraints and behaviours associated with the tourist with SN will be addressed, as well as the planning and management directed towards the sustainability of the accessible destination. In specific, TUR4all is the App for Accessible Tourism in the Iberian Peninsula. It provides information about accessible accommodation, restaurants, bars, monuments, museums, cultural centres, wine tourism, beaches, leisure activities, tourism offices, city tours in short, accessible destinations and experiences. The application enables people with special needs to plan their trips, with access to objective and accurate information about the level of accessibility of tourism resources (Alves, Sousa & Machado, 2020). TUR4all has been designed so that users can customise their search criteria according to their specific needs. This chapter focuses on accessible tourism and tourism for all. A case study will be presented and discussed. The authors also consider some implications for management, as well as give suggestions for future lines of research.

BACKGROUND

The Tourist with Special Needs: Profile, Motivations, Constraints and Behaviours

According to a study carried out in 2003 by the European Commission, there were about 50 million people with some form of disability at European level (Devile, 2009). In Portugal, according to data from the National Statistics Institute (2011), approximately 40.5% of people between the ages of 15 and 64 had at least one health problem or long-term illness, and 17.4% had difficulties in performing basic tasks. Vasconcelos (2015) states that according to the Technical Report conducted by Turismo de Portugal and the European Network Accessible Tourism (ENAT, 2014), the senior population is the one who contributes most to the tourist activity. This segment has considerable income and tends to spend more than other tourists, on an average of 80 euro per night, totalling between 700 and 1100 euro in their trips. Allen et al. (2014) refer to several studies (e.g., Alegre & Pou, 2003; Blazey, 1992; Nicolau & Mas, 2006) according to which, senior tourists show a greater tendency to longer stays, since they have no work or family obligations, which translates into a greater availability /opportunity to travel. For Devile (2009) tourists with disabilities tend not to travel alone, being accompanied by family, friends or specialised people, which translates into an economic benefit for tourism. The data revealed by Smith et al. (2013) show that only 29% of people with disabilities travel alone, while 52% do so accompanied. The same study shows that the tourist with SN has the same profile as a conventional tourist. However, the tourist with SN has a preference to travel by car, from the place of origin to the place of arrival. Another important characteristic is his willingness to travel during low seasons (Rabbit, 2014), which may contribute, in a positive way, to the tourist seasonality. The loyalty of these clients is a crucial point, as this market has fewer options to choose from, in finding an accessible destination that fulfils their requirements and needs (Machado, 2010; Machado, 2014; Alves et al., 2020).

As previously mentioned, and as a possible case study, people with ASD tend to have characteristics similar to other tourists with SN. However, the difference in the needs of these individuals is mainly due to intrinsic barriers and quite specific ones. As an example, an experience at the airport, where apparently simple actions such as waiting in a queue, answering questions, complying with procedures, among others, can translate into an enormous challenge for this population, and can generate socially less accepted or even aggressive behaviours. To avoid such constraints, it is important to anticipate certain situations and prepare people with ASD, showing videos, images, explaining what needs to be done and what you will find in that environment (Alves et al., 2020). Like any other niche market, people with ASD may become a potential market. Regardless of the degree of disability of the individual, tourists with SN not only have the same motivations, but also share the same feelings of most tourists (Macedo e Sousa, 2019). It is important to provide conditions that allow them to meet their motivations (Fontes and Monteiro, 2009). Neumann and Reuber (2004) carried out a comparative analysis of the main motivations of tourists with and without reduced mobility. The study concluded that those with some mobility conditioning had, as their main motivations, to experience nature, to recharge batteries, to improve their health, to enjoy a healthy environment and a feeling of relaxation, and anti-stress and pressure relief (Soares and Breda, 2014). Another study that analyses the motivations to participate in the tourist activity (Kastenholz, Eusébio, Figueiredo & Lima, 2011) reveals that those interviewed with SN are strongly motivated in order to overcome all kinds of barriers they may encounter. As for their motivations, in a trip, the most relevant aspects are related to pleasure and satisfaction obtained from experience, namely the feeling of pleasure, meeting new people and personal development at the social, physical and intellectual level (Kastenholz et al., 2011; Tomé, 2018).

The European Commission commissioned, in 2012-2013, a study to understand the demand for accessible tourism from travel behaviour in twelve European Union (EU) countries. The data collected from 2011 registered about 138.6 million people with access needs in the EU, 35.9% of these were disabled people aged 15-64 and 64.1% were older adults aged 65 and over. Among EU countries, the UK, France, Germany, Italy and Spain have the highest number of people with access needs, all over 10 million (Silva, 2017). In 2012, there were about 783 million trips within the EU - a gross amount of 786bn euro. A growth of 862 million trips is expected in 2020, equivalent to an average annual growth rate of 1.2%. The senior market was the one that contributed the most for that value, with around 204 billion euro, followed by other markets for people with disabilities, worth 148 billion euro. On average, people with SN in the EU travel with 1.9 companions (Silva, 2017). Other aspects of this market behaviour include information research. According to Ray and Ryder (2003), word-of-mouth information, the web and travel guides are the most used sources by people with reduced mobility.

Another study carried out in 2002 by Devile (2009), consisting of a sample of seventy individuals, showed that for people with reduced mobility, the importance of obstacles goes beyond the accessibility in accommodation. It was found that the three main factors considered in tourism activities are accessibility in accommodation, public spaces and transportation, which are responsible for 72.3% of responses. No less relevant, the lack of information corresponds to 12.1% of the total sample. This means that the lack and quality of information are aspects that should be understood and treated by agents in the tourism sector as an important factor. In this sense, accessibility should be transversal to all components of the tourism product. This practice refers to social responsibility in the sustainable development of tourism, valuing the promotion of inclusive tourism products, without neglecting aspects of equity and justice (Pinto da Silva et al., 2019). Such principles should be seen as a component of sustainable tourism (Devile, 2009).

Accessible Tourism Destination: Planning and Sustainable Benefit

Planning strengthens the compatibility between meeting economic and social needs and the need to preserve the environment, natural and cultural resources, in order to ensure the sustainability of the destination as a tourism product (Silva, 2016). Accessible tourism necessarily refers to social responsibility in the sustainable development of tourism, so for ethical and social reasons, it is essential to work in order to offer people with SN the same conditions of tourist offer. Darcy and Dickson (2009), refer that there is still some resistance from the tourism industry, regarding the adoption of the principles of universal design and staff training on people with SN. Accessible tourist destinations should take a systemic approach, involving all elements of the tourism product. The accessible tourism experience not only involves transportation, accommodation and accessible activities, it also involves the availability of adequate information and human resources prepared to serve this population (Devile, 2009).

Accessible tourism has been recognised worldwide (Tomé, 2014), not only for its social importance, but also for the associated economic potential. It is also an important strategy to ensure the sustainability of tourism (Coelho, 2014) and the qualification of destinations. When a destination allies itself with accessible tourism, identity is built around social responsibility and concern for the general well-being of the population. In fact, although there can be accessible tourism without the dimensions of sustainability, there is no sustainability without accessible tourism (Macedo & Sousa, 2019).

Michopoulou and Buhalis (2011, p. 291; 2013, p. 231), in their studies, mention four "enhanced", "intensified" and "generic challenges in the tourism sector" when creating accessible websites: interoperability, i.e. the ability of one system to interact and communicate with another (networked communication); content integration, a determining factor in consumers' purchasing decisions, through the detail, depth and type of information available; personalisation, i.e, generating individualised content, allowing each user to specify his or her requirements and interests in accessing the necessary information and in the desired format; and accessible design, considering the layout of websites and eliminating elements such as low colour contrast, graphics and animations. The results of Michopoulou and Buhalis (2011) usability revealed 20 essential requirements for eTourism accessibility, based on printing options, interface flexibility and content/information presentation. In their study these authors advocate changing font size and icons on web pages; constant automation of data entry; adding short explanatory notes; simplifying mechanisms for selecting and attaching maps and images; integrating accessibility filters (alphabetically classified) with other tourism filters throughout the process (Macedo & Sousa, 2019).

For long-term sustainability, it is necessary to include economic, environmental and social objectives, seeking profit opportunities through increased competitiveness and digital solutions (Vilas-Boas & Sousa, 2019). To this end, destiny can increase its degree of competitiveness through innovation, specialisation, investment, risk taking and productivity improvements, adoption of ethical practices and creation of strategic coalitions (Barbacena, 2012). ENAT, the European coordinator of Accessible Tourism policy, concludes that the objectives for accessible tourism should be considered through the rights of people with disabilities as regards access to the tourism sector. Furthermore, sustainability should also allow the contribution of both consumers and service providers, since tourism should support the diversity and richness of local and regional cultures (Nunes, 2011). Despite the progress already made, Portugal still has weaknesses and it is notorious the inability of the tourism sector to implement solutions that allow all tourists access, without barriers, to a range of tourism infrastructures such as accommodation, transport, monuments, among others. This incapacity results above all from a legal system that does not provide for sanctions to agents who, in non-compliance with the law, are obliged to adapt tourist facilities to people with SN, namely with reduced mobility (Machado, 2009). Considering that it is essential to develop an action plan for the implementation of accessibility that involves all stakeholders, Tourism of Portugal has developed the Management Manual for Accessible Tourist Destinations (2017), which recommends the adoption of management measures to achieve the objectives of the destination, namely, "accessibility should become a key element in the "mark of the destination" (p. 56).

While local communities and businesses are being encouraged to take more responsibility in management and planning processes (Coelho, 2014), the public sector plays a central role in the development of accessible tourist destinations. People with disabilities need special actions in a number of areas (e.g. accessibility to public spaces, adapted infrastructure and tourist information), which are usually managed by the public sector. However, in many of the destinations, the stakeholders in the tourism sector are not aware of the difficulties of this target audience.

Proposed Methodology and TUR4Aall Platform

The empirical study of this work was based on a case study, implemented through an exploratory and descriptive research, of a qualitative nature, with the objective of identifying the accessible resources existing in the city of Viana do Castelo, based on the Tur4all digital platform. The data collection took place between February and October, 2018, in the city of Viana do Castelo. Starting by understanding to what extent the existence of accessible resources, as a factor of attraction of the tourist destination, contributes to the tourist progress of the city of Viana do Castelo, the research followed the methodological path, based on four phases. The first phase was observation. Surveying and inventorying the city's potential accessible resources was the second phase. The third phase was the evaluation and synthesis of the existing resources. And finally, the fourth phase corresponded to the presentation of the research results, translated into the elaboration of a brochure presenting the accessible resources existing in Viana do Castelo.

1st phase: Scanning the tourist environment of the city of Viana do Castelo and the way in which the city managers are committed to issues related to accessibility in order to understand its potential for accessible tourism, based on the work already done by the municipality. A visit to the city, namely to the historical centre and its surroundings, made it possible to observe that in terms of accessibility, the municipality has been developing an extensive work in pedestrian areas.

2nd phase: Carrying out a primary survey on the accessibility conditions of the city's tourist resources, with the support of questionnaires provided by TUR4all. After surveying the resources, its managers were contacted, initially by telephone and later by email. In this one, the objective of the work was included, a brief characterisation of the TUR4all platform, as well as the request for authorisation to visit the site. Once the authorisation was obtained, the visits were made. The choice of resources was based on three criteria: The first was the identification of the main points of interest of Viana do Castelo. The second was the existence of accessible conditions. Finally, the choice of resources that could be used by all kinds of people with SN, aligned with the different types of tourism that the city promotes, specifically, Leisure/Culture tourism, Nature/Adventure tourism, Gastronomy and Sun and Sea tourism.

3rd phase: Evaluation and synthesis of existing resources, using the TUR4all platform. It corresponds to the synthesis of the results of the questionnaires carried out on each resource.

4th phase: Presentation of Viana do Castelo's accessible resources in a brochure, with the aim of filling the information gap for these audiences.

The TUR4all platform is being developed and operated by Accessible Portugal, a non-profit association at the service of Tourism Supply and Demand Qualification, with the support of Turismo de Portugal and the Vodafone Portugal Foundation, being based in the homologous TUR4All Spain. TUR4all is a resource for tourists, private companies and the public sector, with the purpose of promoting accessible tourism in Portugal. It consists of an informative tool available on a webpage and mobile application. Its main objective is to provide information and disseminate the accessible tourism offer in Portugal and Spain. This tool, pioneer in Portugal, allows the interaction with the end-user, dynamised in a model of social networks. Specifically, it seeks to provide objective information about the destinations' accommodation, restaurants and tourist activities and their accessibility. The platform is available in several languages: Portuguese, Spanish, Basque, Valencian, Galician, Catalan, French, English, German, Italian and Mandarin.

The objective of TUR4all is not to value or classify the equipment, but only to inform about its conditions, which allows tourists to make better choices, especially tourists with SN. In a way, it is an active community of users, allowing them to evaluate, score and comment on the accessibility of tourist resources, being at the same time a promotion channel for all accessible tourist establishments and destinations. This dynamic tool involves several actions that allow access to information about accessible resources in Portugal and Spain. In the provision of information on the resource, which includes the accessibility criteria, are involved technicians, users, companies and public entities, as well as tourist destinations. All this involvement allows users, on the demand side, to have knowledge of a varied supply and information on the accessibility conditions of tourist resources.

On the TUR4all platform, there are three ways to participate:

1. Users – users who register on the platform can access more extensive information regarding the user who is not registered on the platform, such as adapted transports, restaurants with Braille menus, among other options. They can also comment, score and recommend these resources or save them as favourites. It is also possible to communicate with other users who share the same needs or interests. In addition to TUR4all's technicians, users can also register information about the resources on the platform, identifying the accessibility characteristics they have, both in their area of residence and in carrying out new tourist experiences in a given destination. For that, it is necessary that the user registers himself in the platform or through the application.
2. Tourism companies – Tur4all provides support to tourism companies that can promote their unit in an accessible way, receiving guidelines to achieve a more adequate response to the market of people with SN. As a result, the companies in the sector improve their positioning and create a marketing channel that distinguishes them as socially responsible. As a form of recognition, these units are given the TUR4all label.
3. Public Entities and Tourist Destinations – Tourist resources under the management of public entities can also be part of the Tur4all network, as long as all its infrastructures are accessible and contribute to a better quality of life for tourists and the local population.

As a basic tool, the TUR4all platform provides questionnaires that are used to analyse the accessibility conditions of the resources, which contain a sequence of information on the access conditions, from outdoor and indoor spaces, adapted bathrooms, furniture, car parks for disabled people, among others. In addition to the conditions, specifics on the characteristics of the spaces are analysed, such as the height of the staircase, signage, door width and size of the manoeuvring spaces. To help with the answers, TUR4all also offers a User's Manual which provides useful clarifications about the measures and what is considered accessible, guiding the user, who visits the resource, when filling in the questionnaires. These are available to the user, registered in the platform, in PDF format or provided via email, in Word (offline).

CONTEXT AND DISCUSSION

Tourism for people with special needs, considered by many to be a niche market, today presents data that deserve reflection. In Portugal, 18% of the population has difficulties in carrying out basic activities, in Europe, the figure is around 17.6%, and, on a global scale, the figures do not differ much. If we add to these numbers the elderly population, children and accompanying persons, who also need adjusted contexts, this market could represent a very significant stain for tourism (Macedo & Sousa, 2019). Tourism for people with special needs, which is growing rapidly, represents a market segment with extreme potential, in which companies must invest. Especially because a company that is concerned with satisfying this market, will not only have a short-term return, but also reveal social and corporate responsibility, contributing to the creation of a differentiated, competitive and sustainable destination.

In this sense, the city of Viana do Castelo is beginning to affirm itself as a symbol of the North of Portugal for its traditions, culture, nature, as well as for the nautical activities that are increasingly present in the city. The Business Confederation of Alto Minho (CEVAL) states that the tourist flow has been transforming the dynamics of the city of Viana do Castelo, and much is due to the cultural and architectural heritage that the city presents, the proximity to the city of Porto, considered the best European destination, as well as the importance of the Caminhos de Santiago (Santiago Paths). In 2016 and 2017, the number of overnight stays registered in the hotel units increased from around 159 thousand to 175 thousand people. The markets that showed the greatest growth were Spain (27%), France (13%) and Germany (3.6%). The national market showed growth of around 16% (INE, 2016). In a generic survey (Pordata, 2016) on the number of tourists who visited or stayed in the Northern region, more than half of the tourists (54.9%) fall into the 25-64 age group. People over 65 represent 16% of tourists. 16.6% are tourists under 15 years old, which may mean that Viana do Castelo is a tourist destination for families. Along with the boost in the number of visitors, the tourist business in Viana do Castelo also showed, by the end of 2017, a growth in the order of 27% in relation to previous years, with a noticeable distribution throughout the year, which has contributed to the low seasonality of tourism in the city. As a result of this new dynamic, the municipality has been concerned with organising more events throughout the year in order to keep up with new market trends, and this process is fundamental for economic development.

The adaptation of the tourist destination offer, namely to receive people with SN, including elderly, sick, disabled people, children, among others, is beginning to affirm itself as an opportunity for the destinations (Fontes & Monteiro, 2009). Destinations, as well as tour operators, should position themselves in the field of accessible tourism in a sustainable and competitive way, facing the new challenges that the tourism industry faces, as well as implementing new tools and methods to support the agents of this sector in key strategies for responsible and ethical planning of a destination (Montes & Aragall, 2009). The implementation of a tourism destination based on accessibility is based on an organised and complete process. According to Montes and Aragall (2009) this implementation process goes through four phases: 1 – awareness phase; 2 –implementation phase; 3 – development and strengthening phase; 4 – differentiation and consolidation phase. Associated to these phases, there is a calendar for the transformation of a destination into a tourist destination for all (Montes & Aragall, 2009).

In view of the above, from the resident's point of view, the transformation phase to a tourist destination for all, states that the municipality of Viana do Castelo is in the implementation phase. As such, the city has been establishing an adequate environment for its products and services by creating the conditions of accessibility, not only for tourists, but also for the residents themselves. The interest of the political forces in creating the conditions of accessibility is clearly visible; however, the agents of the offer have developed and structured little accessible environments for all. This development only comes in the sense of responding to people with reduced mobility. In this sense, the city has currently been working on the requalification of the pedestrian areas of the historic centre of Viana do Castelo, promoting the quality and well-being of the community and visitors. The work already accomplished comes from the project "Viana do Castelo Cidade Saudável", strengthening a set of actions and interventions aiming at the elimination of the barriers that are placed daily to people with reduced mobility, passing through the physical barriers to social and cultural barriers. For that, the Viana para Todos project, associated to the Viana do Castelo Cidade Saudável project, allowed the definition of an accessible roadmap, created in the scope of the municipal policy and the Portuguese and European network of healthy cities, which aims at the promotion of health, namely the issues related to accessibility. Although the historic centre has good access conditions, the resources that include the tourist activity do not have accessibility conditions. In other words, access to restaurants, accommodation, shopping areas and transport is conditioned, which does not allow independent access.

Given the concept of accessibility, and as a consequence of the fieldwork, the criteria of accessibility are mostly related to the needs of access. It is not visible criteria that meet the needs felt by people with visual, hearing and intellectual disabilities, in the circle of tourist activity. All projects are important because they provide technical innovation for the development of Society, but some stand out for their scope and significance and for the way in which they help to build a more sustainable and fair future. Fundación Vodafone Portugal signed a Protocol agreement with Accessible Portugal to finance the customisation of the TUR4all project in Portugal, through which both a Web Platform and Mobile App (available for iOS and Android) were developed to provide information about accessible tourism resources in Portugal (Tur4all, 2019). Tourism is a strategic undertaking for Portugal, being its greatest economic activity in regional development, in the extent to which a multiplicity of departments and agents all work together as part of a network. According to Tur4all (2019), the "2027 Tourism Strategy – Think in the Future, Act in the Present" envisages the setting out of a strategic benchmark for the next 10 years, taking into consideration various changes and perspectives as well as opportunities and challenges arising from them.

CONCLUSION

The practice of accessible tourism encompasses any person, regardless of their physical condition or mobility, so everyone should have the same opportunity and right to enjoy it, without any kind of constraints. The satisfaction of the needs and demands of this market makes the destination more attractive and differentiated from other destinations. The development of tourist destinations that are more accessible to all citizens is increasingly becoming a strategic concern to ensure their competitiveness. Throughout this research, it has been possible to see that accessible tourism must be based on all the components of the tourism product, promoting joint effort and collaboration between the various agents involved in tourism, in order to develop a destination that guarantees accessibility for all. Despite the legislation in force and beyond the growing concern of all agents, the city of Viana do Castelo still has a long way to go in order to become an accessible and inclusive tourist destination. The effort that the Viana do Castelo City Hall has been developing in order to guarantee the sustainable development of the destination, around the accessible tourism, is mainly visible and directed to people with reduced mobility. Thus, the focus has been on the accessibility of pedestrian areas, monuments and services. As far as tourism companies and services are concerned, there is a clear lack of conditions and concern about accessibility criteria, and therefore urgent work needs to be done to raise awareness and implement management measures in order to highlight to these agents (resources) the social importance and economic benefit that accessible tourism can bring them, as well as to enhance the city of Viana do Castelo, as an inclusive and competitive tourist destination. The decrease in seasonality is also an important argument for the various agents to understand the importance of this tourism qualification.

According to Henriques (2009), the inclusion of people with disabilities in the tourist activity, as in any other activity, is a mission that should involve all managers. Viana do Castelo is already considered a strong tourist destination, with a significant weight in the region's economy. Its festivities, heritage and activities, namely those related to nautical activities in which the municipality has been investing, have resulted in a growing number of tourists. Despite this effort, it seems clear to us that the accessible tourist offer is still little known. Therefore, this work has been carried out with the purpose of promoting these resources. If a tourist with SN wants to visit the city, little information is available about the resources that present accessibility conditions. This tourist will only be able to organise his trip if he previously contacts the resources, asking about the requirements he/she needs, until he finds the one that meets his/her needs. In other words, the planning of the trip becomes particularly difficult and the destination becomes less attractive.

According to Souza et al. (2006), before making a trip, this segment should seek information about the place to visit, assessing the characteristics of the accommodation, such as location, the existence of lifts, accessible toilets, among other particularities, as well as the existing activities, transportation and medical services. The TUR4all platform arose from the need of the destinations, namely Viana do Castelo, to disclose their accessible resources. This dynamic platform, aims to facilitate and support the tourist with SN in the organisation and preparation of his trip. However, it is still in the development stage, limiting the accessible resources that are available in the platform, although it is daily updated, with the introduction of new resources, either inserted by the technicians of Accessible Portugal, responsible for its operationalisation, or by the Users. Having said that, in the short/medium term, TUR4all will make available more information about the accessible tourist resources in Viana do Castelo. Being a free promotion channel for companies, it is natural that as the disclosure is having positive impacts on the profitability of the companies, in the future there will be more interest from other resources, in establishing accessible conditions for all. It will be a benefit for all agents, for the destination and for people with SN.

FUTURE RESEARCH DIRECTIONS

This study is preliminary and exploratory research based on a quantitative approach. The conclusions are based on the researchers' perception derived from the Portuguese case study. With the insights from this study, it is expected that future work can contribute to the development of empirical studies to address the developed propositions. In specific, it is expected that future research explores the relationship between online relationship quality, e-trust, e-commitment, e-satisfaction and e-WOM (e.g. Tur4all in Portugal). For example, studies may look at a deeper understanding of the phenomenon in several social tourism contexts. It is important to reveal the individual decision that causes satisfaction and/or dissatisfaction, and shed light on the most determinant and critical attributes in explaining the overall social tourist experience, in the specific the case of digital marketing contexts. Studies could further develop a questionnaire to be applied to internet consumers in Portugal to capture a cross-sectional view of the relationships among the constructs gauging their impact.

REFERENCES

Alegre, J., & Pou, L. (2003). El turismo de sol y playa desde la perspectiva de los consumidores Europeos. Un nuevo desafío en el desarrollo turístico de Baleares. *Serie Estudios Regionales Islas Baleares*, 149-176.

Alén, E., Luis Nicolau, J., Losada, N., & Domínguez, T. (2014). Determinant factors of senior tourist length of stay. *Annals of Tourism Research*, *49*, 19–32. doi:10.1016/j.annals.2014.08.002

Alles, M., T., F. (2009). Turismo Accesible: Importancia de la Accesibilidad para el sector turístico. *Entelequia*, (9), 211-224.

Alves, G. M., Sousa, B. M., & Machado, A. (2020). The Role of Digital Marketing and Online Relationship Quality in Social Tourism. In J. Santos & Ó. Silva (Eds.), *Digital Marketing Strategies for Tourism, Hospitality, and Airline Industries* (pp. 49–70). IGI Global. doi:10.4018/978-1-5225-9783-4.ch003

Barbacena, H. (2012). *A globalização do turismo e a competitividade dos destinos turísticos* (Doctoral dissertation).

Blazey, M. A. (1992). Travel and retirement status. *Annals of Tourism Research, 19*(4), 771–783. doi:10.1016/0160-7383(92)90066-X

Brandão, F., Costa, C., & Buhalis, D. (2018). Tourism innovation networks: A regional approach. *European Journal of Tourism Research, 18*, 33–56.

Caldeira, M., Silva, N. & Nunes, F. (2016). Turismo acessível em Guimarães. *Oportunidade e Desafio para uma cidade inclusiva Brasil & Norte Holos, 4*, 341-357.

Coelho, C. (2014). *Turismo acessível para todos, um paradigma emergente e um desafio para a oferta turística – O caso dos espaços museológicos e empreendimentos turísticos de Cascais* (Master's thesis). Escola Superior de Hotelaria e Turismo do Estoril.

Darcy, S. (2006). *Setting a research agenda for accessible tourism.* Sustainable Tourism CRC.

Darcy, S., Cameron, B., & Pegg, S. (2010). Accessible tourism and sustainability: Adiscussion and case study. *Journal of Sustainable Tourism, 18*(4), 515–537. doi:10.1080/09669581003690668

Darcy, S., & Dickson, T. (2009). A Whole-of-Life Approach to Tourism: The Case for Accessible Tourism Experiences. *Journal of Hospitality and Tourism Management, 16*(1), 32–44. doi:10.1375/jhtm.16.1.32

Devile, E. (2009). O desenvolvimento do Turismo Acessível: Dos argumentos sociais aos argumentos de mercado. *Revista Turismo & Desenvolvimento, 11*, 39–46.

Devile, E. L., Jesus, F., & Cruz, A. I. (2011). O desenvolvimento de turismo acessível na lousã: Perspectivas dos agentes locais do sector do turismo. *International Conference on Tourism & Management Studies, 1*, 529-539.

Falls, V. (2013). *Turismo Acessível para Todos - Recomendações OMT.* Academic Press.

Ferreira, J., & Sousa, B. (2020) Experiential Marketing as Leverage for Growth of Creative Tourism: A Co-creative Process. In Advances in Tourism, Technology and Smart Systems. Smart Innovation, Systems and Technologies, (vol. 171. pp 567-577). Springer, Singapore Systems, Smart Innovation, Systems and Technologies 171. doi:10.1007/978-981-15-2024-2_49

Fontes, A., & Monteiro, I. (2009). O Projecto "Lousã, Destino de Turismo Acessível": um estudo de caso da aplicação de uma abordagem sistémica ao Turismo Acessível. *Revista Turismo & Desenvolvimento*, (11), 61-72.

Hassan, A., & Rahimi, R. (2016). Consuming "Innovation" in Tourism: Augmented Reality as an Innovation Tool in Digital Tourism Marketing. In Global dynamics in travel, tourism, and hospitality (pp. 130-147). IGI Global.

Kastenholz, E., Eusébio, C., Figueiredo, E., & Lima, J. (2011). Accessibility as a competitive advantage of a tourist destination - the case of Lousã. In Field Guide for Case Study Research in Tourism, Hospitality, and Leisure, Advances in Culture, Tourism & Hospitality Research book series. Emerald Publishers.

Macedo, C., & Sousa, B. (2019, July-September). A acessibilidade no etourism: Um estudo na ótica das pessoas portadoras de necessidades especiais. *PASOS. Revista de Turismo y Patrimonio Cultural, 17*(4), 709–723. doi:10.25145/j.pasos.2019.17.050

Machado, M. (2010). Evolução Urbana e Turistificação dos lugares: Um olhar sobre o Rio de Janeiro. *Revista Turismo & Desenvolvimento, 2*(13), 817–825.

Machado, M. (2014). Turismo e mobilidade reduzida. *Revista Turismo & Desenvolvimento, 4*(21), 365–376.

Machado, P. (2009). Turismo Acessível para Todos. *Revista Turismo & Desenvolvimento, 11*, 24.

Malek, A., & Costa, C. (2015). Integrating communities into tourism planning through social innovation. *Tourism Planning & Development, 12*(3), 281–299. doi:10.1080/21568316.2014.951125

Marujo, M. N., & Carvalho, P. (2010). Turismo, planeamento e desenvolvimento sustentável. *Turismo e Sociedade, 3*(2), 147–161. doi:10.5380/tes.v3i2.19635

Michopoulou, E., & Buhalis, D. (2011). *Technology Platforms and Challenges. Accessible tourism: Concepts and issues*. Channel View Publications.

Moniz, A. (2006). *Sustentabilidade do Turismo em Ilhas de Pequena Dimensão: O Caso dos Açores* (Master's thesis). Universidade dos Açores.

Montanari, M. G., & Giraldi, J. M. E. (2013). Competitividade no turismo: Uma comparação entre Brasil e Suíça. *Revista Brasileira de Pesquisa em Turismo, 7*(1), 92–113. doi:10.7784/rbtur.v7i1.564

Monteiro, A. (2017). *Evolução do Desenvolvimento da Oferta Turística do Turismo em Espaço Rural em Monsaraz*. Universidade Europeia.

Montes, R. & Aragall, F. (2009). Turismo Acessível ou Turismo para Todos? Sustentabilidade do Negócio. *Revista Turismo & Desenvolvimento*, (11), 141-145.

Moraes, A. (2006). Competitividade e sustentabilidade: É possível as cidades turísticas serem sustentáveis e competitivas ao mesmo tempo? *PASOS Revista de Turismo y Patrimonio Cultural, 4*(3), 443–446. doi:10.25145/j.pasos.2006.04.031

Nicolau, J. L., & Mas, F. J. (2006). The influence of distance and prices on the choice of tourist destinations: The moderating role of motivations. *Tourism Management, 27*(5), 982–996. doi:10.1016/j.tourman.2005.09.009

Oliveira, A. (2008). Planejamento participativo como instrumento de desenvolvimento turístico responsável. *Red de Revistas Científicas de América Latina, 8*(3), 22-28.

Pinto da Silva, F., Brandão, F., & Sousa, B. (2019). Towards socially sustainable tourism in cities: Local community perceptions and development guidelines. *Enlightening Tourism. A Pathmaking Journal, 9*(2), 168-198.

Schmetzke, A. (2012). Guest editorial: Accessibility of Web-based information resources for people with disabilities. *Library Hi Tech, 20*(2), 135–136. doi:10.1108/07378830210432534

Silva, C. (2017). *Turismo Acessível – Estudo de caso da cidade de Coimbra* (Master's thesis). Instituto Politécnico de Coimbra.

Soares, A., & Breda, Z. (2014). Destinos turísticos Acessíveis e Pessoas com Mobilidade Reduzida – Uma realidade? O caso da cidade de Aveiro. *Revista Turismo & Desenvolvimento, 3*(21/22), 383–396.

Sousa, B., & Rodrigues, S. (2019). The role of personal brand on consumer behaviour in tourism contexts: the case of Madeira. *Enlightening Tourism. A Pathmaking Journal, 9*(1), 38-62. doi:10.33776/et.v9i1.3597

Sousa, B., & Silva, M. (2019). Creative tourism and destination marketing as a safeguard of the cultural heritage of regions: The case of Sabugueiro Village. *Revista Brasileira de Gestão e Desenvolvimento Regional, 15*(5), 78-92.

Tomé, M. (2014). Turismo e Mobilidade. Reduzida. *Revista Turismo & Desenvolvimento, 4*(21/22), 365–376.

Tomé, M. (2018). Restrictive factors for the tourism. The perception of real and potential demand about the public security in Rio de Janeiro (Brazil). *Estudios y Perspectivas en Turismo, 27*(4), 968–984.

Vilas-Boas, V., & Sousa, B. (2019). The Role of Service Quality in Predisposition for Portuguese Online Commerce. *Quality - Access to Success, 20*(173), 12-17.

KEY TERMS AND DEFINITIONS

Digital Marketing: Is the marketing of products or services using digital technologies, mainly on the Internet, but also including mobile phones, display advertising, and any other digital medium.

E-Satisfaction: The contentment of a consumer with respect to his or her prior purchasing experiences with a given retail-oriented website.

Social Media: Is the collective of online communications channels dedicated to community-based input, interaction, content-sharing, and collaboration (e.g., websites and applications, forums, microblogging, social networking, social bookmarking, social curation, and wikis are among the different types of social media).

Social Tourism: Is tourism that provides economic opportunities via travel and tourism for persons who are economically weak or otherwise disadvantaged.

Chapter 15
Four Innovative Ways to Reduce Food Wasting in Hotel Restaurants

Manuel Au-Yong-Oliveira
https://orcid.org/0000-0002-2154-6171
GOVCOPP, University of Aveiro, Portugal

Monika Chodilíková
Tomas Bata University in Zlín, Czech Republic

Iveta Košárková
Tomas Bata University in Zlín, Czech Republic

Kristýna Pšeničková
Technical University of Liberec, Czech Republic

Patryk Lewandowski
University of Lodz, Poland

ABSTRACT

Tourism plays a very important role in society. However, it has many negative impacts, such as the increasing problem of food wasting in hotel restaurants, especially when providing all-inclusive/all-you-can-eat breakfasts and dinners. This study aims to propose some activities which could lead to food wasting reduction. Through a questionnaire (with 132 respondents, from various countries), the authors wanted to find out respondents' attitudes towards suggested solutions and this problem in general. As part of the questionnaire, four possible activities were proposed in order to help to reduce food wasting and contribute to better food security in today's world. The questionnaire results proved that respondents would appreciate such opportunities, therefore answering the research question. Fridges (in which people could put their untouched leftovers), samples (before people take the whole portion of food), special cards (containing the ingredients used for the preparation of the meal), and daily announcements (of food wasting in kilograms) are four possible ways to reduce food wasting in hotel restaurants.

DOI: 10.4018/978-1-7998-5691-7.ch015

INTRODUCTION

Society is currently becoming more and more aware of food wasting. Food is not only wasted during consumption, but also in the process of its preparation. It is a paradox that sometimes food wasting is visible in countries where at the same time people suffer from hunger. According to a study (ReFED, 2016, p.1), currently, up to 40% of food worldwide is being wasted. Still, the amount of food thrown away in hotels is usually underestimated or simply ignored. People generally assume that if there is a good restaurant in a hotel, all of the food is eaten and therefore no attention is being paid to this problem. That today's world is not aware of such a big problem is visible from the following statistics:

In a world where 850 million people are undernourished, global food supply per person approximates to 570 kg: roughly, 380 kg is consumed, 140 kg is lost in the production and 50 kg is wasted by consumers. Households generate 53% of the total food waste in Europe, the processing industry 19%, food services 12%, the primary production sector 11%, and the retail/wholesale sector 5% (Racz, Marchesi & Crnković, 2018, p.25).

Based on the evidence above, it is clear that hotel restaurants contribute significantly to this issue, to be more precise – by 12%. The aim of this paper is to highlight and come up with some activities which could possibly reduce food wasting in the hotel restaurant environment. In addition to this, it is necessary to determine whether people would be willing to participate in such programs. Considering that 12% is not a small number, dealing with food wasting in food services could possibly have a positive impact on solving the current situation.

This paper is divided into both theoretical and analytical parts. The theoretical part contains a literature review focused on three main areas – food-wasting as a tremendous problem worldwide, hotels as a source of food wasting and, last but not least, food-wasting reduction programs. The theoretical part of this paper was necessary in order to understand how enormous this problem really is. As for the analytical part, a questionnaire was conducted to receive quantitative data from people of many nationalities to find out their attitudes towards suggested activities and this global problem. The analysis of the collected data was done to answer the research question – would people be willing to participate in activities suggested in this paper, in order to reduce food wasting in hotel restaurants?

LITERATURE REVIEW

This chapter consists of three sub-chapters dealing with the topic of food wasting. Specifically, the problem of food wasting, in general, is mentioned, moving on to the concrete issues of countries which are wasting the most in connection with tourism (including some examples of hotels). Finally, the last sub-chapter is dedicated to food-wasting reduction programs.

Food Wasting as a Tremendous Problem Worldwide

Wasting food is considered to be one of the most enormous problems worldwide. 1.3 billion tons of wasted food per year all around the world are adequate proof of the statement. The food-wasting results from the carelessness or even intentional throwing of food into a dustbin (either because it is past its sell-by date or even fresh foodstuff) (Khader et. al., 2019). Food wasting does not enable people to accomplish food security worldwide which is a part of the Sustainable Development Goals. Mainly the issue is related to Goal 2, which is aimed at ending starvation, and securing and achieving food security. In addition to that, it also includes improving nutrition and supporting the sustainability of agriculture (Gil et al., 2019). However, this carelessness is considered not to be ethically correct because there is about one billion people who are in need and who starve. Besides this huge issue in society, wasting food has an impact on environmental pollution (including soil, water, as well as air) too, because it contributes to taking waste - packages and food itself - to a landfill (Ciulli, Kolk & Boe-Lillegraven, 2019).

Food wasting is connected to issues related to the environment, to the economy and to societal affairs, because a small awareness of this enormous problem leads not only to the pollution mentioned above but also to inequality in consumption and more (Racz, Marchesi & Crnković, 2018). The awareness of people worldwide is mostly insufficient. They usually do not see that there are diversified processes of harvesting, growing and making food ready, in general, behind their consumption. It is not just about throwing one bread away but many inputs (water, soil, fuel consumption, energy, various means of transport, people's efforts and much more) are used to get a final product. It could take just one second to get rid of a piece of cake, but there is a time-and-money-consuming process behind it (Racz, Marchesi & Crnković, 2018). Moreover, there are other negative reasons which impede the reduction of food wasting. See the following list of negative issues that make the process of food wasting reduction harder (Ciulli, Kolk & Boe-Lillegraven, 2019). It includes, for example:

1. Incomprehension of the meaning of expiration date.
2. Refusing to buy and sell a product with small defects made during its production but not hampering the use of a particular product.
3. Incurring expenses or companies in the food industry may be afraid of health problems after their food donations.
4. Throwing food away is less expensive than undergoing the donation process.
5. Companies are doing their business only in their own financial interest (Ciulli, Kolk & Boe-Lillegraven, 2019).
6. Wasting food during cooking.
7. Buying more food than a household is able to consume (Racz, Marchesi & Crnković, 2018).

To deliver food to those in need is considered to be one of the best options regarding the reduction of food wasting policy, when there is a failure of food waste prevention. Figure 1 shows the most and less desirable ways about what to do with the food surplus. If the food still meets the requirements of edibility, the surplus of food can be redistributed. In case the food is not suitable for humans, due to various conditions, but it still meets certain requirements, the surplus of food can be utilised for feeding farm animals. On the other hand, when food is not edible and suitable for either humans and animals at all, there are different options of how to deal with it, but the worst that can be done is landfilling (Garcia-Garcia, et al., 2017).

Hotels as a Source of Food Wasting

This part of the paper is focused on specific sources of food waste and certain areas where this problem has a very negative impact. People travel every day, whether internationally or within a country. One

Figure 1. The most appropriate/inappropriate ways of how to deal with the issue of food waste (based on Garcia-Garcia et al., 2017, p.2215)

of the most sought-after types of travelling is staying by the sea. People choose this type of relaxation to take a rest from the daily bustle. That is the reason why all-inclusive programs are offered by almost every hotel resort. Hotel guests, due to unlimited food consumption in hotel restaurants, are unable to estimate how much they actually eat. Portions of food that they take often remain untouched on the tables and then end up thrown into the trash by the staff.

Hotels want to offer their customers the best service, even though it leads to excessive wastage. According to eco-business experts, almost 25% of the food from the hotel kitchens' is thrown into the trash. For every dinner it is about 350 grams. If 350 grams of discarded food were multiplied by the hundreds of thousands of hotels and of course all the rooms involved, a terrible figure would be reached. Luxurious hotels, such as Sofitel and Pullman, discard 47 tons of food waste per year. Mid-range hotels such as Mercure and Novotel throw away 35 tons a year, and economy hotels like Ibis throw away 17 tons a year. All these hotels and many more belong to the multinational hospitality company AccorHotels, which has its hotels almost all over the world. However, unfortunately, together these hotels waste 35% of their purchased food, on average (Koh, 2017).

Most of these hotels mentioned above are located in countries such as Egypt and Tunisia. Tourists visiting these countries have the opportunity to see the beauty, luxury and enjoy the almost unlimited service of hotels but, on the other hand, they may also see the poverty surrounding these hotels. Egypt and Tunisia are among the countries where poor living conditions persist. The population of Egypt is growing rapidly and currently has reached more than 92 million inhabitants. Due to the declining domestic food production, the country is increasingly dependent on food imports from other countries. Excessive waste and food losses already occur during the harvest, storage, processing, transportation, distribution and consumption of food. Even the little domestic production that gets into hotels is often thrown away, either in the hotel kitchen during cooking or after, when it remains untouched on the tables (Food and Agriculture Organization of the United Nations, 2019).

Another example of this issue is the hotel complex located in the Bahamas. This hotel produces an incredible 36% of the total waste on the island (Curtis & Slocum, 2016). Moving on to other destinations, Turkey is worth mentioning as well. Hotels usually offer 5kg of food per guest, no matter whether they are able to consume this amount or not. However, guests are not able to consume this amount of food, and everything that is not eaten is wasted. According to Ali Akgun, the director of the Food Waste

Prevention Platform, food waste in Turkey costs an incredible TL 214 billion a year. Due to this high number, the Food Waste Prevention Platform is being actively used in order to reduce food waste in hotels – for instance, by informing hotel guests about the problem, using smaller sizes of plates or by training hotel staff (Daily Sabah Turkey, 2019).

It should be in the interest of every hotel to reduce food waste, as it could, at the same time, possibly cut expenses for buying ingredients. Food waste in hotels is the most perceived in the Asia-Pacific region, home to 800 million hungry people in the world. On the other hand, China and the Philippines are export-based countries in the food sector, which is quite a paradox. Winnows Pourrat, an expert in food-wasting prevention, says it is very important to get every employee to see the value of each meal that ends up thrown in the trash (Koh, 2017).

On the other hand, society has come to be more aware of the growing problem of food wasting. Many programs are being set up to support the fight against food wasting. Behind these programs stands, for example, the World Wildlife Fund (WWF) with the cooperation and support from The Rockefeller Foundation and the American Hotel & Lodging Association (AHLA). Famous hotel chains such as the Hilton, Hyatt, InterContinental Hotels Group, Marriott International, Hershey Entertainment & Resorts, Sage Hospitality and Terranea Resort are involved in this program (PR Newswire, 2017).

The pie charts below (Figure 2) provide information about the amount of food surplus that is donated where it is needed. The first chart shows that, in America, only 22.6% of the food surplus is donated to shelters or to other organisations. The second chart presents the amount of food donated by hotels in Europe, the Middle East and Africa (EMEA) – at only 36.8%. Asia and the Pacific have the largest percentage of donating food surplus – 47.4%. Globally, the amount of donated food is very small and stands at only 26.1% worldwide (Green Lodging News and Greenview, 2017).

Figure 2. The surplus that is subsequently donated (based on Green Lodging News and Greenview 2017, p. 18)

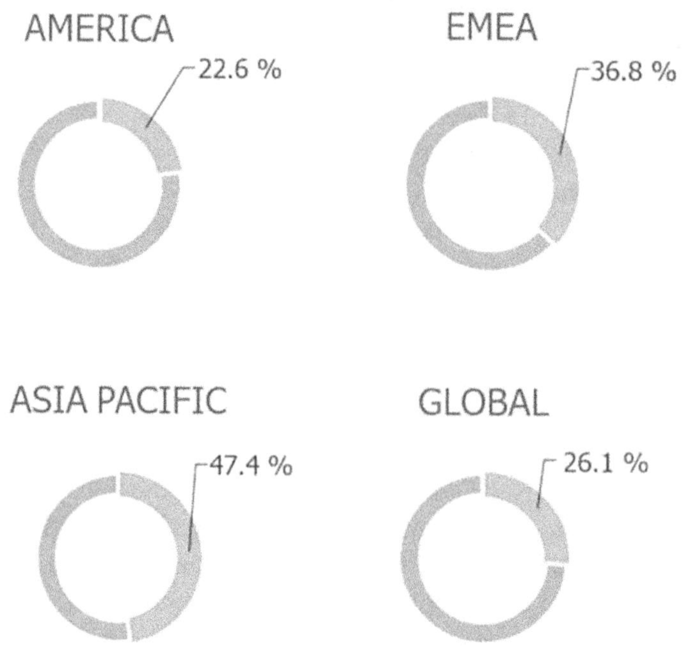

Food Wasting Reduction Programs

As has been already stated in this paper, food-wasting represents one of the most crucial problems of today's world. However, the number of existing programs concerning food waste reduction all around the world indicates that not just governmental institutions do not turn their back on this problem but also individuals – who are constantly coming up with new creative ideas on how to reduce food wasting and thus simultaneously help those in need.

This sub-chapter deals with food banks and other programs that are involved in solving this issue. One of the great examples of such a committed organisation is the European Food Banks Federation (FEBA) which is a network operating in 24 countries in Europe. Its objective is to gain food security through food waste reduction. This organisation tries to achieve this objective with the help of three areas of activity – firstly, employees are involved in raising the awareness of issues connected to hunger, malnutrition, food-wasting and to the existence of food banks. The second area is devoted to the spreading of innovative practices which contribute to ending hunger and to the expansion of the partnership. Furthermore, last but not least, the organisation is focused on coaching in order to develop food bank services, fundraising and the establishment of new food banks (European Food Banks Federation, 2019).

According to FEBA's 2018 annual report, 66% of food donated to their food banks comes from drink and food manufacturers, wholesalers, retailers or other foodservice providers who donate the food which would be otherwise wasted (European Food Banks Federation, 2019). In 2018, FEBA redistributed 781 thousand tons of food (it equals to approximately 4.3 million daily meals) to 45 700 different charities and thereby helped 9.3 million people in need. The fact that FEBA's network has 31,700 co-workers, of which 85% are volunteers demonstrates that society is concerned about these problems. The numbers stated above clearly prove that the existence of food banks immensely contributes to fighting against food wasting and hunger (European Food Banks Federation, 2019).

Another program which is definitely worth mentioning was founded by four university students of Maryland. It is called the Food Recovery Network, and currently it represents the largest American student movement fighting against food wasting and hunger. In their 2018 report, they point out that food wasting is a complex problem which directly contributes to various global issues, such as a climate change. It is necessary to bear in mind all the resources which are used in the production of meals as for instance water, soil or even human labour. In most cases, average meals travel many kilometres to reach their consumers, which annually produces hundreds of tons of carbon emissions (Food Recovery Network, 2019).

Expressed in numbers, in this network, there are more than 230 universities involved with the number of volunteers around 11,418 who so far have helped to recover more than three million pounds of food, and, as a consequence, they have reduced CO2 emissions by 2.6 million kilograms. The operation of this network is based on collecting food from different facilities such as school canteens, restaurants, farm markets, stores, etc. and afterwards donating it to food banks or to other organisations that help people in need (Food Recovery Network, 2019).

Donating food can sometimes be challenging if a business is based far away from more densely populated areas where there is no possibility to donate food. However, there are other options concerning how to deal with food surplus. Glenuig Inn represents one of these businesses which face this challenge, but, still, it has a zero-food waste. Employees gather food leftovers and organic waste such as paper napkins and after these are co-mingled with wood pellet. This mixture is then dried in a special machine with the output of a crumb which is used afterwards as a source of energy (Glenuig Inn, 2019).

METHODOLOGY

In order to determine the attitudes towards suggested activities and problems connected to food wasting in general, a questionnaire was used. People from various countries were asked to fill in the online questionnaire that was available on Google Forms. The purpose was to reach as many nationalities as possible; therefore, the questionnaire was shared via social networks such as Facebook, WhatsApp, Instagram and Email platforms. Due to the various nationalities that participated in the questionnaire, the English language was used. This method, as a source of quantitative data, was employed because of the necessity of having a large number of responses. The questionnaire had the title "Should hotel restaurants be involved in reducing food waste?" and consisted of sixteen questions divided into five different sections. There were three types of questions – open, closed and open-closed. The questionnaire was released on October 18, 2019, and the collecting of replies ended on October 23, 2019.

Concerning the literature review, various databases (e.g. Scopus, ProQuest, Springer), websites and annual reports were used in order to find out articles related to this topic. Documentation was chosen according to keywords, the reading through of abstracts, their attractiveness and dates of publishing – material older than 2016 was not considered. For demographic information see table 1, to find out how many women and men, what age groups and nationalities participated in the questionnaire.

Table 1. Demographics of the respondents (Proposed by the authors)

Gender		Age		Nationality	
Female	63.6%	17 and less	0.8%	American	3.03%
Male	36.4%	18-22	21.2%	Croatian	4.54%
		23-27	60.6%	Czech	43.18%
		28-34	6.8%	Polish	6.06%
		35-41	2.3%	French	3.03%
		42-49	4.5%	Italian	7.57%
		50-59	3%	Portuguese	3.03%
		60+	0.8%	Slovakian	3.03%
				Slovene	2.27%
				Other	24.26%

RESULTS

One hundred and thirty-two (132) respondents answered the questionnaire regarding the food-wasting topic. This sub-chapter deals with the results of the questionnaire. For detailed information, see the analysis below.

In the first question respondents were asked to express their opinions about if, according to them, tourists are aware of wasting food during their all-inclusive holidays (including all you can eat breakfast/half-board). 77.3% replied that, based on their opinions, tourists do not realise food is wasted in hotel restaurants. On the other hand, 22.7% of them responded that tourists are actually aware of food waste in such restaurants. Connected to this question, respondents were asked if they consider food-wasting as one of the most emerging problems in today's world. 84.8% agreed on that, whereas the rest of them (15.2%) do not perceive it as being serious.

In the next question, respondents should respond concerning if they have ever experienced tourists taking more food than they are able to eat. It is visible from the chart (figure 3) that everyone has experienced it before. Even 30.3% of respondents claim that they always see this issue during their vacations.

Figure 3. Have you ever seen someone in a hotel restaurant taking more food than he/she is able to eat? (Proposed by the authors)

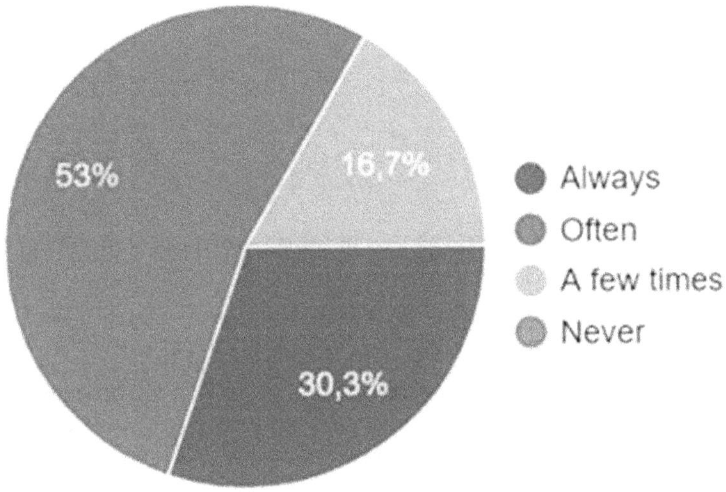

The purpose of the next question was to find out if respondents have ever done it themselves. 54.5% stated that they had done it a few times. At the same time, 33.3% have never done it. Figure 4 is a bar chart of countries where this problem occurs most frequently. The question was asked in the form of multiple-choice selection with the possibility to add their own answers. The bar chart shows that the majority of food wasting situations appear in countries with very well-developed tourism, including Egypt, Turkey, Tunisia, Italy, Spain and Greece. Each number in the bar chart represents how many times the particular country was selected to be highly likely to waste food in hotel restaurants.

The respondents were asked if they know of some programs or organisations that are dealing with the improvement of the situation of food wasting. Most of them, 71.2% selected the "No" option. Therefore, the topic of these programs was also included in the theoretical background of this paper. However, 28.8% claim that they are aware of these programs and they know about them. In the next question, they suggested some of these programs/organisations/options that help to reduce food wasting – such as Too good to go, Nesnězeno, Zachraň jídlo, food banks, public fridges on streets, Food sharing, Slow food, ZwickShare and others.

The next part of the questionnaire was focused on suggested activities which could possibly help to reduce food wasting in hotel restaurants. The purpose of this section was to find out whether people would be interested in such activities and if they would appreciate joining the process of reducing food wasting and improving this global issue.

Figure 4. In which country/countries? (Proposed by the authors)

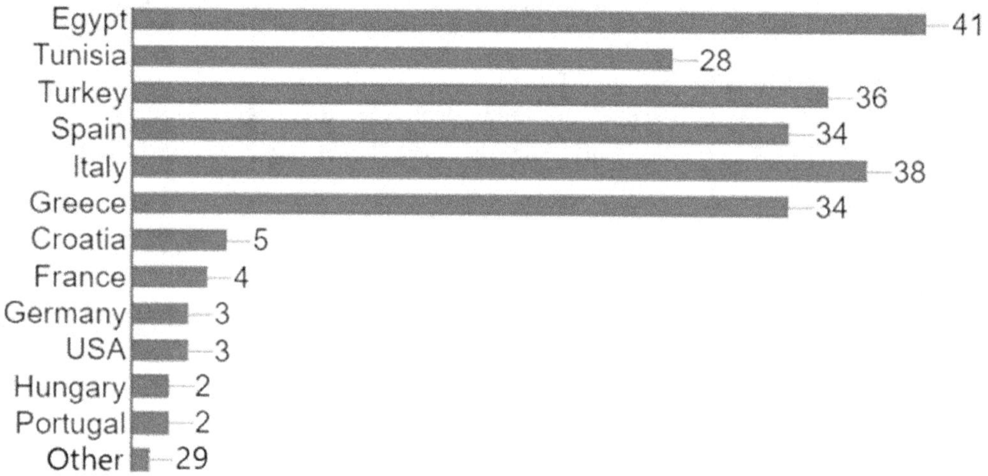

The first idea is that there would be a fridge in a hotel restaurant in which people could put their untouched leftovers that would be then delivered to those in need. 97.7% of the respondents expressed their interest in having such a program in a hotel, because they would be a part of the food-wasting reducing process while at the same time, they could help people suffering from hunger.

The second suggested solution consists of providing people just with a sample of food before they will take a whole portion of it. 90.2% would support this idea. Therefore, they would appreciate trying samples of food before eating a whole portion.

Another option is having a special card containing not only the name of a meal but also the ingredients – so that consumers would avoid eating something they do not like. 89.4% of the respondents believe that introducing such a description card would reduce food wasting. Only 14 people (10.6%) out of the 132 respondents stated that they do not see any connections between having such a card and reducing food wasting.

Last but not least, a suggested activity is about daily announcements of food wasting in a particular hotel restaurant in kilograms. The aim of this question was to discover whether it would have any impact on tourists' behaviour or not (Figure 5). 60.6% of respondents believe that such an idea could affect people's behaviour. In comparison to 12.1% who believe that this idea would not influence human behaviour and would not contribute to reducing food wasting in tourism, for 27.3% of the respondents it is hard to say.

CONCLUSION AND SUGGESTED SOLUTIONS

The aim of this paper was to propose some activities which could lead to food-wasting reduction and through the questionnaire, the authors wanted to find out respondents' attitudes towards suggested solutions and this problem in general. As part of the questionnaire, four possible activities were proposed in order to help to reduce food wasting and contribute to better food security in today's world. The results of the questionnaire proved that respondents would appreciate such opportunities; therefore, the research question has been answered. Please see below descriptions of the suggested activities.

Figure 5. Do you think that there would be any impact on your behaviour if hotels would announce daily food wasting in kilograms? (Proposed by the authors)

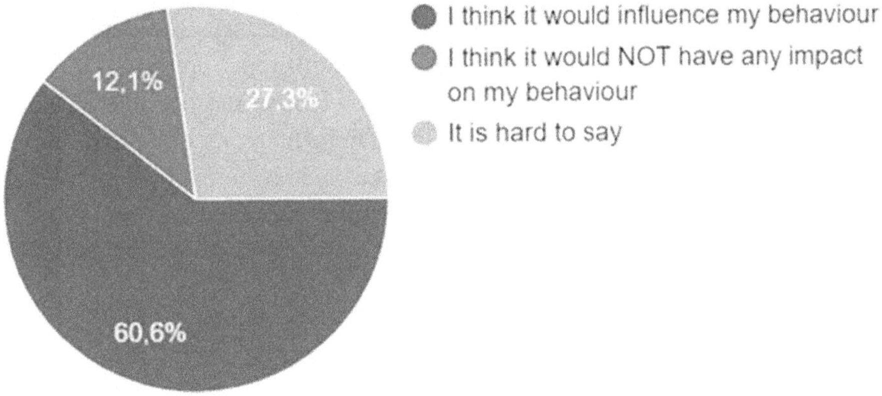

The first idea is related to the opportunity of having fridges in hotel restaurants in which people could put their untouched leftovers that would be then delivered to those in need. Thanks to this step, people would support solving two global problems that are food wasting and hunger. There would be staff responsible for controlling the condition of leftovers and preparing packages of food that would be delivered directly to either local shelters or food banks. Food banks would redistribute food among those in need or other organisations.

The next solution lies in providing consumers with samples of food before they take the whole portion of it. It means that there would be plates of samples for every meal so people could try just a little piece of food to decide whether they like the taste or not. These samples represent only a small amount of money for hotels, but it could be a big step forward to food-wasting reduction.

Having a special card (either a paper one or an electronic one that would even be customised, based on the languages of guests) containing not only the name of the meal but also the ingredients used for its preparation – in order to avoid eating something people do not like – is another option on how to reduce food wasting. As for the electronic version, consumers would click on their flags, and they would immediately see the ingredients that meals contain. Paper cards would better suit small hotels with guests of just a few nationalities.

Due to the huge amount of wasted food, hotel restaurants could adopt activity related to daily announcements of food wasting in kilograms. The authors believe that it could have a positive impact on human behaviour, which is also proven in the questionnaire. We argue that if consumers could see the number of kilograms that end up in the trash, they could start to think about this issue and consequently change their eating habits. In addition to that, there could also be a comparison of wasted food over a period of time (a day-to-day comparison).

To sum up the theoretical part, important terms were outlined to support the subsequent analytical part. The significant data were introduced to highlight how it is important to come up with innovative solutions that would help to find the proper way of eliminating food waste. It is visible from the questionnaire results that people are not really aware of programs dealing with food waste reduction; therefore, the sub-chapter named food-wasting reduction programs was added to the literature review.

The authors believe that implementing these activities in hotels would lead to influencing consumer behaviour in a positive way, with the consequence of reducing food waste. As was stated at the beginning of this paper, food services contribute to this problem by 12% (of the total food waste in Europe). The authors conclude that if hotels would adopt these ideas, they would influence people. Besides a positive impact on human behaviour during vacations, people could think about options of how to eliminate food waste when they return home. Therefore, it may be "only" 12% at the beginning, but it can reach far more. People could change their habits at home, resulting in influencing of up to 53% of households, as has been stated in the introduction. Each process must start somewhere and, step by step, a significant impact may be generated.

REFERENCES

Ciulli, F., Kolk, A., & Boe-Lillegraven, S. (2019). Circularity Brokers: Digital Platform Organisations and Waste Recovery in Food Supply Chains. *Journal of Business Ethics*, 1–33. Retrieved October 16, 2019, from. doi:10.100710551-019-04160-5

Curtis, K. R., & Slocum, S. L. (2016). Role of Sustainability Certification Programs in Reducing Food Waste in Tourism. *Journal of Developments in Sustainable Agriculture*, *11*, 1–7. doi:10.11178/jdsa.11.1

Daily Sabah Turkey. (2019). *Hotels fighting food waste to be recognised*. Retrieved October 7, 2019, from https://www.dailysabah.com/turkey/2019/04/15/hotels-fighting-food-waste-to-be-recognized

European Food Banks Federation. (2019). *Towards the Next Decade Together: FEBA Annual Report 2018*. Retrieved October 16, 2019, from https://www.eurofoodbank.org/images/cont/feba-annual-report-2018-final_file.pdf

Food and Agriculture Organization of the United Nations. (2019). *FAO in Egypt*. Retrieved October 5, 2019, from http://www.fao.org/egypt/programmes-and-projects/food-loss-waste-reduction/en/?fbclid=IwAR0i8NJyHSTnJKy0vdfeMq7PhbmkFvWc7rAX4-mC1yyEkKLV_oMvzNGldU8

Food Recovery Network. (2019). *2018 Annual Report*. Retrieved October 16, 2019, from https://static1.squarespace.com/static/555b5cf1e4b0864ccf1a0156/t/5c7041d141920266f5dce0ed/1550860780285/FY18+Annual+Report+for+Website.pdf

Garcia-Garcia, G., Woolley, E., Rahimifard, S., Colwill, J., White, R., & Needham, L. (2017). A Methodology for Sustainable Management of Food Waste. *Waste and Biomass Valorization*, *8*(6), 2209–2227. doi:10.100712649-016-9720-0 PMID:32010409

Gil, J. D. B., Reidsma, P., Giller, K., Todman, L., Whitmore, A., & van Ittersum, M. (2019). Sustainable development goal 2: Improved targets and indicators for agriculture and food security. *Ambio*, *48*(7), 685–698. doi:10.100713280-018-1101-4 PMID:30267284

Glenuig Inn. (2019). *Being Green*. Retrieved October 17, 2019, from https://glenuig.com/being-green/

Green Lodging News and Greenview. (2017). *Green Lodging Trends Report 2017*. Retrieved October 2, 2019, from https://greenview.sg/wp-content/uploads/2017/09/green-lodging-trends-report-2017.pdf

Khader, B. F. Y., Yigezu, Y. A., Duwayri, M. A., Niane, A. A., & Shideed, K. (2019). Where in the value chain are we losing the most food? The case of wheat in Jordan. *Food Security*, *11*(5), 1009–1027. doi:10.100712571-019-00962-7 PMID:31976025

Koh, H. (2017). *The unseen scandal of hotel food waste. Eco-business*. Retrieved on October 5, 2019, from https://www.eco-business.com/news/the-unseen-scandal-of-hotel-food-waste/

PR Newswire. (2017). *World wildlife fund, American Hotel & Lodging Association and the Rockefeller Foundation bring hotel brands together to prevent hotel food waste: Hotel properties kick-off 12-week pilot programs to reduce food waste in hotels*. Retrieved October 7, 2019, from https://search.proquest.com/docview/1879099480?accountid=17116

Racz, A., Marchesi, V. V., & Ivana Crnković, I. (2018). Economical, Environmental and Ethical Impact of Food Wastage in Hospitality and Other Global Industries. *European Journal of Bioethics*, *9*(1), 25–42. doi:10.21860/j.9.1.1

ReFED. (2016). *A Roadmap to Reduce U.S. Food Waste by 20 Percent: Executive Summary*. Retrieved October 10, 2019, from https://www.refed.com/downloads/Executive-Summary.pdf

Section 6

Education, Leadership, and Partnerships

Chapter 16
Leadership in Tourism:
Authentic Leaders as Facilitators of Sustainable Development in Tasmanian Tourism

Josh Rowlands
https://orcid.org/0000-0002-5935-1628
University of Tasmania, Australia

Matthew Knox
https://orcid.org/0000-0002-6092-071X
University of Tasmania, Australia

Tessa Campbell
University of Tasmania, Australia

Anna Cui
https://orcid.org/0000-0001-7199-3669
University of Tasmania, Australia

Luke DeJesus
University of Tasmania, Australia

ABSTRACT

This chapter proposes to discuss the application of authentic leadership as the 'vehicle' forward for tourism. Specifically, how authentic leaders in the tourism industry can help achieve the United Nations Sustainable Development Goals (SDGs), and why they are important to the Tasmanian tourism industry. As such, the authors propose the research question: How can authentic leadership enable the sustainable development of tourism in Tasmania? This chapter commences by exploring tourism in Tasmania and the related leadership gap found in the industry, followed by a brief explanation of our critical review method. The literature review then examines how tourism, a diverse industry, has the potential to contribute to the United Nation's SDGs. The authors aim to demonstrate how sound authentic leadership behaviours among tourist vendors facilitate ethical employment practices and economic growth in Tasmania. Finally, the chapter explores the possible implications of a synthesis of authentic leadership and sustainable development in the context of Tasmania.

DOI: 10.4018/978-1-7998-5691-7.ch016

INTRODUCTION

Sustainable development within tourism is a topic of debate throughout Australia. Empty seats have been a common site for Australian tourism operators this summer (Chalmers, 2020). Australian news across the country echo similar sentiment "Tourism in Australia is taking a hit with the 'double whammy' of bushfires and coronavirus travel bans" (Whiley, 2020). In a time where tourism across the nation is experiencing extreme difficulty, from a combination of the Australian bushfires and COVID-19, the topic of economic, social, and environmental sustainability is at the forefront of tourism discourse. How can tourism, particularly ecotourism, respond to such hard times? While small business owners and tourism operators battle with the impact of natural disaster and disease outbreak, marketing campaigns urging Australians to holiday in affected towns and regional areas has been one strategy to support small business as they re-build. However, it is increasingly evident that a vehicle is needed to guide the industry towards a sustainable future.

Tasmania, the southernmost island of Australia, is one of the world leaders in sustainability and environmental preservation. Emerging research identifies that ecotourism emergence can be enabled through social entrepreneurs and authentic leaders (Knox et al., 2020). The relative isolation of the Tasmanian island from the recent events impacting the mainland of Australia creates an opportunity for the state to develop a unique tourism strategy to lead a sustainable future. Alongside the Great Barrier Reef and Uluru, Tasmanian wilderness was titled in the top three 'natural tourist icons' in Australia (Dowling, 2001). A multitude of sustainable developments across the island state gives Tasmania unique value and the credibility to explore how tourism can create inclusive, full and productive employment that is decent work for all, as well as generating sustainable economic growth.

This chapter looks to discuss the application of authentic leadership as the 'vehicle' forward for tourism. Specifically, how authentic leaders in the tourism industry can help achieve the United Nations Sustainable Development Goals (SDG's), and why they are important to the Tasmanian tourism industry. As such, the authors propose the research question: *How can authentic leadership enable the sustainable development of tourism in Tasmania?* This chapter commences by exploring Tourism in Tasmania and the related leadership gap found in the industry, followed by a brief explanation of the chosen critical review method. The literature review then examines how tourism, a diverse industry, has the potential to contribute to the United Nation's SDGs. This chapter aims to demonstrate how ethical employment practices, and economic growth in Tasmania is facilitated by sound authentic leadership behaviours among tourist vendors (Stratford, 2006; Adams, 2009). Finally, this chapter explores the possible implications of a synthesis of authentic leadership and sustainable development within the context of Tasmania.

Tourism in Tasmania

Many nations' government try to diversify their channels of revenue and growth within industries in anticipation of either a recession or minor economic collapse within a single or multiple sectors due to internal and external influences (Delios & Beamish, 1999). Tourism is an industry that offers an economic stabiliser in many countries due to its ability to stand alone in growth independently of the performance of other economic indicators (Cho, 2001). This independence can be drawn from the value that consumers and business receive from either natural or constructed sites, cultural experiences and uniqueness of the environment. Regions such as Portugal, Italy, and Vietnam share these significant characteristics, which explains in part their nations ability to have experienced sustainable growth in the tourism sector. Full employment can be defined as the condition in which virtually all who are able and willing to work are employed ceteris paribus (Mosler, 1997). Unemployment is widely used by economists as a measure of stability for a nation's economy as it allows an in-depth analysis of a populations current labour and wage situation. Recent papers, such as Jolliffe and Farnsworth (2003) and Ashworth and Thomas (1999), acknowledge that the condition a tourism sector may find itself in can have a direct or indirect effect on employment levels within a nation or state. This can be due to employment security and stable wage growth occurring in remote geographical locations, or unusual fields of work that would commonly not be available. Therefore, tourism as a sector can contribute to successfully achieving Sustainable Development Goal 8, as it has a positive effect on the economic indicators identified by the United Nations in 2015. This goal aims to promote sustained, inclusive, full, and productive economic growth.

It is important to note that Tasmania has not only these characteristics but has an evolving population and business scheme that is beneficial for the sector of tourism. The recent increase in the tourism industry has created, developed, and expanded jobs across the state of Tasmania. This can be demonstrated by the examples of Bridestowe Lavender Estate and Saffire Freycinet, located in Tasmania's North and East Coast, respectively. Areas and populations that previously thought could not produce sustained, inclusive employment are now thriving. The luxury 'eco-resort', Saffire exemplifies the appeal and profitability of the Tasmanian wilderness. Located on the isolated Freycinet Peninsula, Saffire provides various ecotourism attractions (e.g. kayaking and beekeeping) receiving worldwide recognition in both Forbes Magazine, and the Evening Standard's top international travel destinations (Connor, 2016; Fathom, 2019).

However, this sharp increase in tourism growth must be managed to ensure its ability to be sustained over long periods of time. The effect of Saffire's international popularity is a factor Tasmania's leaders must consider when avoiding the pitfalls of ecotourism, such as the effect of overcrowded attractions (Bramwell & Lane, 2011). From popular surfing beaches to hiking Mount Everest, natural attractions around the world are becoming increasingly impacted by today's growing population. Weaving through crowds of people is not the only issue with popular tourist destinations, as demonstrated in Indonesia. The appeal of Indonesia's climate and scenery resulted in large portions of untouched land being converted into resorts and malls, removing significant ecotourist attractions (Walpole & Goodwin, 2000). Managing the sustainable development of Tasmania's ecotourism attractions, requires the consideration of the economic, social, and environmental impacts. Saffire's consideration exists in the form of small and intimate experiences, with their newest attraction allowing small groups to participate in beekeeping activities in a controlled environment (Bestic, 2019).

Bridestowe Lavender Estate in Tasmania's north likewise shares considerations of sustainability following its influx in visitors after its scenery became a viral sensation. The Estate has continually aimed to improve efficiency and sustainability with wholistic practices. Bridestowe transitioned from using two million litres of water per year from a local river, to zero litres per year. Their farms complete self-sufficiency for water usage along with the installation of eighty solar panels in 2014, is a testament to their sustainability approach (Bridestowe Lavender Estate, 2020).

Boasting worldwide recognition for its natural attractions and a booming tourism industry, the small Australian state of Tasmania exemplifies the appeal of authentic experiences in tourism (Knox et al., 2020). Recent data identifies $44.3 billion per annum is spent on tourism in Australia (Tourism Research Australia, 2019a), with Tasmania's industry boasting the highest growth and consumption rates (Tourism Research Australia, 2019b). Ecotourism stands out as Tasmania's most profitable tourism sector, attracting twice the number of ecotourists than the rest of Australia (Matysek & Kriwoken, 2003). Ecotourism's appeal stems from the preservation of its relatively untouched natural environment, an ethos reflected in

Tasmanian society (Ford et al., 2009). The purpose of this chapter is to provide theoretical underpinnings as to how Tasmanian leaders can foster such sustainable development in the booming state.

Sustainable Tourism

Tasmania is at the forefront of sustainable tourism with negative and positive impacts of the industry being debated to determine the future of Tasmania (Ooi & Hardy, 2020). Twenty per cent of Tasmania is World Heritage listing (World Heritage Committee, 2007) and as such, attracts sustainable tourism that are leaders in their field.

Blue Derby Pods Ride has the vision to create a sustainable world with a driven purpose to provide guests with a moment of great realisation (Blue Derby Pods Ride, 2020). Their five core values: fun, inquisitiveness, adventure, trust and balance, act as the driving force for the vision and purpose. Like many Tasmanian businesses who engage in sustainable tourism, Blue Derby Pods Ride started with a sustainability-focused vision. This sustainable focus comes alive when its sustainable purpose shapes its business operations. By outlining what businesses' values are, scholars can discuss how these companies are impacting the community and their inhabitants (Miller & Twining-Ward, 2005).

Blue Derby Pods Ride is in Tasmania's North East. The area is well known for its vast areas of temperate rainforest with stunning Myrtle trees. The town of Derby was once centred on a thriving tin mine; however, more recently, it has become a mountain biking mecca with world-class trails (Ogilvie, 2017). Compared to forestry and mining, tourism has emerged as an environmentally sustainable alternative where natural beauty is the economic resource (Ooi & Hardy, 2020). Blue Derby Pods Ride, among many other tourism businesses, have been the saviour for the town of Derby. Yet, Derby is only one example of many small Tasmanian towns that have been saved through sustainable ecotourism.

The ability to develop and incorporate a sustainable tourism sector is key for economies that either depend on or aspire to improve their tourism industry. Using methods that assist low destructive impact on the environment, society, and local culture can help in generating future employment in a viable manner. This perspective on economic growth within an industry, such as tourism, can negate the possible negative effects from mismanaged regulatory oversight, seen throughout nations such as Indonesia and Jamaica (Wall 1993). The Tasmanian Walking Company (TWC) is an enterprise that is authorised by the Tasmanian Parks & Wildlife Service (Tasmanian Walking Company, 2020a). They offer complete guides through some of Tasmania's most beautiful areas, such as the Cradle Mountain, Bay of Fires and Wineglass Bay. TWC's vision is to enable people to connect and conserve the natural environment (Tasmanian Walking Company, 2020b). The company embodies numerous sustainable tourism characteristics by training guide leaders in sustainable and ethical practice as well as having built its famous architecturally designed eco-lodges on the Three Capes Track (Tasmanian Walking Company, 2020a). TWC partners with a range of sustainability organisations to instigate awareness, participation, and funding for: Carbonneutral, Wildcare Tasmania, Tasmanian Aboriginal Guide Training Program, Rapter Refuge, Save the Tasmanian Devil, Bonorong Wildlife Santuary and Conservation Volunteers Australia (Tasmanian Walking Company, 2020b). Therefore, TWC is a clear example of a firm's ability to encompass and incorporate techniques that sustain Tasmania's tourism sector.

Another prime example of sustainable tourism in Tasmania, is that of Josef Chromy Wines (JCW). Since their establishment in 2007 (Josef Chromy Wines, 2020), the winery has continued to demonstrate its ability to operate in a sustainable manner, addressing economic, social, and environmental issues. In 2019, JCW became involved in Wine Tasmania's 'Vin0' initiative, focusing on implementing environmentally sustainable practices at wineries and vineyards across Tasmania (Wine Tasmania, 2020). JCW utilises a state-of-the-art water recycling system allowing them to treat the water used on site, to then be reused for the irrigation of their crops (The Wine Syndicate, 2020). The winery uses environmentally conscious wine bottles, which use fewer resources during production and allowing for a greater transport yield, reducing the number of resources used during transportation. JCW also sells their grape marc onto third parties, to produce cattle feed (East Coast Stockfeeds, 2020), which ensures the production process of their wine has very little impact on waste production (The Wine Syndicate, 2020). In 2013, JCW installed a 100kW solar panel system allowing the winery to reduce their consumption of mains electricity by more than 50% (Bradford Energy Solutions, 2013); this decision has proven beneficial for the Tasmanian environment, as well as economically beneficial for JCW. Josef Chromy Wines is a clear demonstration of how tourism in Tasmania already has a strong foothold when it comes to sustainable practices, and many other establishments will look to JCW as a benchmark for achieving sustainable tourism in Tasmania.

The Leadership Gap

Within organisational settings, positive leadership behaviours result in motivated, comfortable and self-regulating followers (Luthans, 2002). Such followership encourages creativity and innovation, facilitating knowledge sharing to improve the goods and services being provided (Crawford et al., 2020). Likewise, the Tasmanian tourism industry provides benefits from relative economic success, impacting related sectors, by generating employment and investment in infrastructure (Tourism Tasmania, 2019b). Recent statistics support the notion that the tourism industry is a key component of Tasmania's economy (Tourism Australia, 2019b), as such this paper posits the fostering of positive leadership behaviours, similar to that exhibited by authentic leadership, to sustain the tourism industry's current success.

Over recent years there has been noticeable growth within the tourism sector in Tasmania, but the poor leadership and organisational culture demonstrated within its governing body (Baker, 2019; Denholm, 2019), Tourism Tasmania, may have contributed to the stagnation currently being experienced across the state (see Table 1). The stagnation is problematic, as the benefits of a strong visitor economy flow to other related sectors beyond tourism, which generate employment and investment in infrastructure that benefits the wider community (Tourism Tasmania, 2019b).

Table 1. Total visitors and visitor expenditure within Tasmania from 2014-2019 inclusive (Tourism Tasmania, 2014, 2015, 2015, 2016, 2017, 2018, 2019a)

Year (ending September)	2014	2015	2016	2017	2018	2019
No. of Visitors (million)	1.06	1.14	1.19	1.28	1.30	1.32
Variance (YoY)	*+6%*	*+7%*	*+4%*	*+8%*	*+1%*	*+2%*
Total Visitor Expenditure (billion)	$1.74	$1.92	$2.07	$2.30	$2.41	$2.53
Variance (YoY)	*+15%*	*+10%*	*+8%*	*+11%*	*+4%*	*+5%*

Leaders within the tourism industry of Tasmania acknowledge that nature-based tourism is a large and growing industry, which is predominantly dependent on government monitored areas of the environment (Buckley & Sommer, 2001; Tovar & Lockwood, 2008). Given the rise in demand of ecotourism in Tasmania, and the pitfalls of exploiting the natural environment, the authors propose that an ethos of sustainability is required to maintain the natural environment that fuels tourism in Tasmania. Without effective leadership, the conservation of Australia's environment may be at risk, which in turn would have a negative effect on Tasmania's tourism industry and the larger economy. The Tasmanian tourism industry contributes approximately $2.79 billion or 10.7 per cent to gross state product: the highest in the country. Additionally, it supports approximately 37,400 jobs or 15.6 per cent of total Tasmanian employment (Tourism Tasmania, 2019b). Given these statistics, effective leadership is a very important factor within the industry and state.

Environmental deterioration and unsustainable overdevelopment are perceived to be significant negative outcomes of tourism growth, which impacted communities located near tourist hot spots within Tasmania (Tovar & Lockwood, 2008). It also identified that environmental conservation is important to Tasmanian residents. There is a need for sustainable practices within the Tasmanian tourism industry and the importance placed by those who live within the state (Tasmanian Government, 2004; Tovar & Lockwood, 2008). The issues presented in this scenario may be addressed if the businesses in question are able to do the following. Firstly, be enthusiastic about becoming environmentally sustainable, and secondly seeking to work alongside the local community to resolve the issues caused by poor leadership within organisations.

With corporate social responsibility and sustainability becoming more important among the population (Crabb, 2019; Morgan, 2019), Australians and Tasmanians seek effective leaders within their government and local businesses. The leadership gap within Tourism Tasmania has brought scrutiny to the organisation, following the publication of articles in The Australian and the Australian Broadcasting Corporation, which identified poor organisational culture and poor leadership behaviours. For example, managers "had to be encouraged to say 'hi' to their staff at least once a week" (Baker, 2019) which demonstrates a poor leader-follower relationship given the leaders needed encouragement to be actively involved with their followers. The lack of emotional intelligence demonstrated by leaders within Tourism Tasmania (Baker, 2019; Denholm, 2019), would have a detrimental effect on the motivation of their employees. This, in turn, has a negative effect on the completion of job tasks as well as the respect and pride each employee has regarding their responsibilities within the workplace. This lack of motivation may lead to poor management of the natural environment, which is an integral part of tourism within Tasmania. If the environment is not adequately cared for, this may decrease the number of tourists visiting Tasmania, which will then decrease expenditure and negatively impact the local economy. It is evident from table 1 that Tourism Tasmania, who experienced growth between 2014-2017 and then stagnation for the following two years, and the discovery of the poor leadership occurred in 2016.

METHOD

This chapter applies the method of a critical literature review to explore the research question: *How can authentic leadership enable the sustainable development of tourism in Tasmania?* The objective of this chapter is to develop a critical perspective on the role of authentic leadership in enabling the sustainable development of tourism in Tasmania. To do so, this paper highlights the contemporary literature on authentic leadership, and its relationship with sustainable development in Tasmania's tourism sector. The rationale for engaging in a critical review over a rapid review is due to the importance of complete identification of the most significant items in Tasmanian tourism (Grant & Booth, 2009). A rapid review's completeness of searching is determined by time constraints and thus may lack key areas in the field of Tourism. This critical review method evaluates existing research for the purpose of adding to the current literature and expanding theoretical understandings on the topic of authentic leadership behaviours. Likewise, this method and scope reflect that of Knox et al.'s (2020) piece on ecotourism in Tasmania. This is due to the contribution it provides to the literature on tourism in Tasmania, as well as its application of authentic leadership behaviour.

LITERATURE REVIEW

Through good practices of authentic leadership, sustainable economic growth across Tasmania can enable inclusive, decent, and productive employment (Adams, 2009; Stratford, 2006). Inclusive, decent and productive employment are the combination of factors necessary to provide employees with a sense of belonging in a place. By pursuing realistic sustainable development goals, Tasmania will come one step closer to achieving a more effective sustainable growth (Whiteley, 1999). The following shall explicate the literature behind these conclusions, to provide a foundation for this proposed chapters contribution.

It has been demonstrated with the tourism industry that there have been exponential increases throughout the years, and this indicates that there is going to be upcoming growths for businesses and openings for leadership roles. This is where sustainable development will assist in the growing tourism industry. Sustainable development focusses on linking the relation between companies to the community, allowing them to grow together to accommodate for the growing tourism industry. With the communities and companies working hand in hand, the target market will be encouraged to spend more time in Tasmania with a higher economic multiplier occurring. To integrate and lead this movement of change into the Tasmania community, Tasmania will be the leaders of tomorrow. Leaders should possess various behavioural traits that encourage new growth and momentum for businesses and to promote this growth throughout the community too. With the increase in tourism activity in Tasmania, the authors believe that implementing successful authentic leaders into this booming industry in Tasmania will benefit companies and communities for a better future.

Sustainable Development

Sustainable development is a notion that lends itself to economics, ecology, and society. The term grew in popularity at the turn of the twenty-first century, with several scholars attributed to its development (Cash et al., 2003; Daly, 1990; Lélé, 1991). The most commonly used definition of Sustainable Development: (cited 1,609 times on Google Scholar, March 13, 2020) "Economic development that meets the needs of the present generation without compromising the ability of future generations to meet their own needs." Bossel, 1999, p.2; WCED, 1987).

Although sustainable development seems ideal, there are constraints to which limits the effectiveness of this method (Bossel, 1999). Firstly, constraints of physical conditions and laws of nature. Not everything is possible in laws of nature. Secondly, rules of logic, the laws of nature and rules of logic cannot be broken. This would imply that exemplified restrictions are unavoidable and that it may hinder the sustainable development system. Examples of these restrictions are minimum soil requirements for plant growth, or a maximum of energy gained from dams. This presents the constraint of accessibility; where the physical environment is the constraint, namely so: available space, the fertility of soil, waste absorption, rivers, atmosphere and the availability of renewable and non-renewable sources. Therefore, some of these are set limitations where you can count the depletion rate, and others are rate limitations where you can measure it by the output of energy or absorption rate. There is only one permanent source of energy support on the earth's surface, which is solar energy. Other energy sources such as fossil fuels are depleting and considered to be harmful to the environment. They can cause changes in the atmosphere. The idea of sustainable development would be to utilise solar energy for every use, this is considered as recycling, and recycling is, therefore, a quintessential requirement of sustainability.

Economic growth can have undesirable effects, such as the disruption of traditional cultures and damage to the environment. Considerations of ethics, ecology, and society are evident when determining whether economic growth is sustainable or not, as economic growth bears social and political consequences (Friedman, 2017). Despite these considerations, unsustainable economic growth can lead to imbalance, resulting in a short-term gain for a long-term loss. For example, the improvement of human resources, capital, natural resources, and technology reaches a threshold point where continued efforts to grow the economy lead to a decline in the quality of life (Neumayer, 1999). As a part of implementing sustainable development across communities, reliance on human impact, and what humans do to make this change happen, is crucial. Firstly, humans have traits that will impact the level of success to a change in sustainable development; humans are self-conscious, anticipatory, imaginative, creative beings. Therefore, the way in which humans act and choose to behave is non-measurable and cannot be confined to a set 'blueprint'. Humans are able to come up with new ideas and solutions for any given task. For example, creating a dam in the river to redirect water flow or with the use of a pump and hose to carry the water to the plantation which is in need. Both these solutions can work, but one would be more effective and efficient depending on the limitations of the land and education level of the human. Studies have shown that communities which are more educated are more likely to provide an open-cultural environment and have greater accessibility space. Other factors include human organisations, cultures, technology, and political systems. Human limitation includes human ethics and morals so the points at which a human will value one option over another and deem the actions as being ethical can vary. Time is a limitation which must be considered when wanting to realistically form a fully functional sustainably developed community. Equally as important is knowing how long a task will take, whether it is a long-term or short-term commitment, the viability of this method of practice and if it will in fact benefit the current community and future generations. As such, sustainable development must involve ongoing stakeholder involvement to ensure that development is responsive and reflexive (Hardy & Pearson, 2018). This leads us to the limitations of risk-taking, which involves the extent to which humans are willing to change for the community to become a sustainably developed community (Bossel, 1999). This also includes the money involved and the associated risks. If these risks and certainty are not kept into consideration and constantly measured, the effect on any sustainable developing could be detrimental. However, with a greater number of diverse individuals, more innovative options are possible for future sustainable development (Bossel, 1999).

Sustainable economic growth presents itself as a solution to retain the natural environment, resources, and employment opportunities for future generations (Santagata, 2002). The United Nations (UN) maintains a similar sentiment when it comes to the future of the world. The SDGs are a blueprint for achieving a better and more sustained future for all. They encompass the development that meets the needs of the present without compromising the needs of the future (United Nations, 2020). The seventeen goals address the global challenges, including those associated with climate change, environmental degradation, peace and justice (Robert, Parris & Leiserowitz, 2005). These goals are all interconnected, targeted for completion by the year 2030.

The tourism sector contributes to the progression of completing the SDGs, as the relationships' mechanism likewise promotes synergistic growth for the sector. A legislative position that supports sustained environmental development increases the probability of goal thirteen 'climate change' to be obtained in the stated timeline. Pennicott Wilderness Journeys is an example of a business that actively contributes to sustainable development. Their locally sourced business purchases (Pennicott Wilderness Journeys, 2020a), carbon offsetting (Pennicott Wilderness Journeys, 2020) and partnership with EarthCheck (Earth Check, 2020), an internationally recognised scientific benchmarking, certification and advisory group for tourism, provide evidence of tourism's ability to contribute to sustainable development and SDGs.

Authentic Leadership

The authentic leader appeared in literature as a direct response to a variety of societal shifts in the early twenty-first century. Events such as 9/11 called for strong and ethical leadership in the face of disaster. Essentially, effective and positive leaders were lacking. People desired a leader that could empathise with their followers and balance their decision-making between their own desires and the needs of the collective (Gardner et al., 2005). In contrast, authentic leadership theory emerged as in the face of ethical concerns in the contemporary business environment (Luthans & Avolio, 2003). This distinction solidifies the behavioural nuances of what makes an authentic leader who lacked in literature for several years (Crawford et al., 2020). A dyad of relationships comprises what makes an authentic leader, incorporating followership and authentic followers into the theory. This explains the effects behaviours have on the relationship between people, explicating the importance of each of these behaviours (see Crawford et al., 2019).

Authentic leaders exhibit behaviours of a leader who understands both themselves, as well as the people around them. Further, authentic leaders can enhance employee wellbeing by having a strong sense of positive ethics and being sincere (Avolio & Gardner, 2005). Through looking to maintain an awareness of others, authentic leaders provide aid to those around them, fostering a culture of positive organisational engagement, while considering contextual social values (see Table 2).

Table 2. Defining authentic leaders and their dimensions (Crawford et al., 2020)

Authentic leader	Influences and motivates followers to achieve goals through their sincerity and positive moral perspective, enabled through heightened awareness and balanced processing.
Awareness	Having insight into the behaviours of one's self, other individuals, and groups
Sincerity	Presenting one's true self to others, honestly and openly in all relationships with consideration for the context
Balanced processing	The tendency to consider all relevant information available and using this to make decisions that benefit the collective (e.g. followers).
Positive moral perspectives	The commitment to one's intrinsic ethical framework, and a willingness to subdue personal interests and ego to facilitate collective interests.
Informal influence	The ability to inspire and motivate individuals to accomplish goals of their own volition, regardless of rank or position.

While much research exists on authentic leadership, little attention has been given to its applicability to the tourism sector. Although, some instances of the application of leadership behaviours do exist. For example, authentic leaders have demonstrated a positive effect on enabling collective mindfulness, and thriving (Wu & Chen, 2019). Likewise, authentic leadership influences customer-centric organisational citizenship behaviours (Luu, 2019). Trust in leaders has also been demonstrated to moderate such a relationship (Qui et al., 2019). More recently, literature has emerged theorising a direct relationship between authentic leadership, organisational success, growth, and emergence in the tourism sector. Business tourists are theorised to develop stronger connections in their diverse interactions, as well as maintaining strong communication skills in the globalised world (Knox & Crawford, 2020). In terms of ecotourism and Tasmania, authentic leadership is likewise theorised to foster organisational success and growth (Knox et al., 2020).

The current leadership observed within the governing body of tourism in Tasmania, Tourism Tasmania, does not fully reflect that of effective authentic leadership practices. Given the nature of ecotourism in Tasmania and the vulnerability of the natural environment, one's ability to make ethical and morally sound decisions is paramount. Robert Pennicott, founder of the Tasmanian ecotourism business Pennicott Wilderness Journey, has been known for his inspiring leadership (Tasmanian Leaders, 2016). Given the nature of Pennicott's success, this chapter suggests that considered leadership behaviours will aid in fostering growth within the Tasmanian ecotourism industry as authentic leadership creates the foundations for positive social outcomes.

DISCUSSION

Authentic Leader Behavior

Tourism is reliant on quality leadership for the successful promotion and development of the industry (Wilson et al., 2001). The tourism sector, including ecotourism, reflects the increasing global pressure to see sustainable business development and practice (Björk, 2007; World Commission on Environment and Development, 1987). Poor leadership execution of tourism development can lead to negative impacts socio-culturally, economically, and environmentally (Beeton, 2006; Butler, 1974). Hence, the importance of authentic leaders, who can consider their context and environment (Avolio & Wernsing, 2008) while applying critical thinking skills (Waite et al., 2014). Authentic leaders are enabled by five behaviours: awareness, sincerity, positive moral perspective, balanced processing, and informal influence. Each behaviour has a distinct definition, yet authentic leaders can demonstrate a combination of these behaviours, sometimes emphasising certain behaviours over others. Therefore, it is valuable to separate each behaviour and examine its meaning and application independently.

It is important to note that an authentic leader must be an authentic individual; however, an individual who demonstrates authenticity is not necessarily a leader (Crawford et al., 2019). The importance of this distinction is relevant to the Tasmanian tourism industry. There are individuals in the industry that express themselves authentically but lack the leadership capabilities necessary to be an effective 'authentic leader'. For example, an authentic person exercising positional power in a managerial role does not give rise to a leader possessing leadership capabilities (Toor & Ofori, 2008; Zaleznik, 2004). The authors propose the five behaviours of authentic leaders as antecedents to the lack of sustainable practice evidence in the tourism industry. While Tasmania has proven a successful location for tourists worldwide, particularly in the ecotourism sector, there remain areas for improvement, and this will be addressed in the leadership gap in the following discussion.

Awareness

Self-awareness is the understanding of one's true self, the concept of themselves and how they make meaning from the world (Govern & Marsh, 2001; Avolio & Gardner, 2005). Based on Crawford et al.'s (2019) conceptualisation, self-awareness involves a leader's creation of meaning, the world around them and how they understand themselves. Authentic leaders develop a heightened awareness that is superior to an individual's level of self-awareness. An authentic leader possesses the awareness found within emotional intelligence research, such as the understanding of personal feelings and self-assessment as well as social awareness of others around them. Additionally, however, authentic leaders also have an awareness of the context in which other people are operating within. Not simply who others are and what they are doing, but an understanding of why. Authentic leaders are insightful, surrounding the behaviours of those around them (Crawford et al., 2019).

Heightened awareness in Tasmania's tourism leaders could lead towards more sustainable economic growth in the state. A greater understanding of employees would enable leaders to have a better grasp of the challenges they face. For example, a survey by Unions Tasmania and the Australia Institute Tasmania found that underemployment was a critical issue in Tasmania (Shine, 2017). It identified that sixty per cent of respondents wanted more working hours. Considering tourism is a major sector in Tasmanian employment, reducing underemployment could contribute to full and productive employment and decent work for all; a key factor in sustainable development goal eight. Yet, without authentic leaders that can exercise heightened awareness, tourism employees are unlikely to see a positive change in their circumstances. Bridestowe general manager James Ravens has been instrumental in the social sustainability of his business. As a regional Tasmanian business, engaging with the international tourist market can be challenging (Huang & McMurray, 2016). Bridestowe, however, has been effective at being innovative and understanding employee needs. They recognise there is a need for employment opportunities for young Tasmanians. The business capitalises on University students' need for casual summer work, which also meets the increased demand for Bridestowe experiences from international tourists during December and January (Piotrowski, 2020).

Sincerity

The nuances between authenticity and sincerity are difficult to distinguish and are often conflated (Ford & Harding, 2011; Trilling, 1971). Sincerity, as an authentic leadership behaviour, has implicit elements of sincerity, including open and honest relationships (Crawford et al., 2019). Authenticity is culminated as the internal relationship with self (Kernis & Goldman, 2006), whereas sincerity is the specific expressions of self, as actualised by a third party. If authenticity is the relationship with self, sincerity is the relationship with others. Therefore, authentic leaders who possess sincerity are true to themselves, their own standards and morals, yet equally builds trust with others through the expression of these, honesty and transparency. Sincerity enables an authentic leader to express their own core self within an interpersonal context (Crawford et al., 2019).

The prevalence of secretive, dismissive and excluding practices in the Tasmanian tourism industry demonstrates a lack of authentic leaders who enact sincerity. Tourism Tasmania's promotion body made internal changes after a 2016 report labelling the organisation negatively (Baker 2019). It detailed the lack of active listening and accountability towards staff members by the organisation's management. Tourism must have an inclusive, full and productive employment attitude that provides decent work for all (Adams, 2009). Sincerity offers an enabler for leaders of the industry to hold open, transparent, and honest relationships with their stakeholders. The inclusivity and decent work offered in the industry could result in sustained economic growth and better employment conditions for a greater demographic of tourism employees. For example, Pennicott Wilderness Journey's managing director Robert exemplifies sincerity and hosts a diverse workforce dynamic (Pennicott Wilderness Journeys, 2019). His businesses influence on the Bruny Island region is evident through the positive feedback from business owners, with Hotel Bruny's Tash Daniels claiming they now serve three hundred meals per day over summer and their quiet season has been reduced to two months over winter (Burgess, 2018).

Positive Moral Perspective

Leaders are facing decisions within organisations that have moral implications or outcomes. It is important to expand on the ethical and moral components of authentic leadership. An identified relationship exists between the ethicality and effectiveness of decision-making with followers' responses to their decisions (Iszatt-White et al., 2018). Ethical leaders were praised for virtuous decision-making but criticised for ineffective outcomes. Effectiveness was determined by obtaining organisational goals and virtuous actions through the avoidance of breaching ethical rules. Yet, Tasmanian tourism leaders have been at one end of the two extremes; organizationally effective *or* ethically correct. However, positive moral perspective is the authentic leadership behaviour that enables leaders to enact both ethical and effective decisions (Crawford et al., 2019).

An internalised moral perspective enables leaders to be guided by internal moral standards. Preventing them from giving into external pressures. However, Crawford et al. (2019) explore how this negates an important consideration of positive moral perspective. Expanding on moral identity theory (Zhu et al., 2011), a moral identity ensures that a leader's actions promote and protect the wellbeing of others too. Similarly, virtue is the contribution to the common good and the excellence of an individual's character, and can be developed (Newstead et al., 2019). Positive moral perspective incorporates not only the internal moral standards a leader has, but their moral identity also encompasses the consideration of others and the contribution to the common good. In the tourism industry, tourist visitation has a multiplier effect for many businesses within the region. Tasmanian tourism leaders that exhibit a positive moral perspective could operate within an ethical and effective industry that aims at its core, to contribute to broader economic, environmental, and social good. This ideal would generate an industry that seeks continuous improvement across all aspects of a tourists' experience in Tasmania. Where one offering would complement another in such a way that tourism businesses have a vested interest in each other. Not out of the selfish pursuit of financial gain but for the simple desire to be a part of the sustainable and 'good' movement being generated across the industry. For example, leaders of Tasmania's world-renowned Bridestowe Lavender Estate and Blue Derby Pods Ride, located within reasonable proximity, would see the value of pursuing the common good for the tourism industry. While these are two very different offerings, they are an example of highly successful tourism destinations in Tasmania that could benefit from the sustainability generated by authentic leaders.

Balanced Processing

Unbiased processing is the term to describe the process of analysing all relevant information before deciding (Kernis & Goldman, 2006). No person is completely objective, rational and unbiased; therefore, the concept has changed from its original meaning of unbiased processing (Gardner et al. 2011). Ideally, uncertainty should not influence rational decision making; however, findings from Hsu et al. (2005) demonstrate that it does, in fact, change decision making. Since human choice is based on the subjective value placed on alternatives, context is paramount to how authentic leaders process information and come to decisions. Authentic leaders are no exception, they too possess bias and subjective thought. Nonetheless, authentic leaders make decisions based on the potential positive outcomes that impact themselves and their collective (Crawford et al., 2019). "When making decisions, authentic leaders imperfectly consider their internal flaws, strengths, and biases (i.e. awareness)" (Crawford et al., 2019, p. 17).

Balanced processing has an important place in Tasmanian tourism because the impacts are effective and far-reaching. An authentic leader with balanced processing can assess the needs of the community and make decisions that benefit the collective. The collective benefits can include sustained economic growth, full and productive employment, and decent work for all. For example, James Raven from Bridestowe Lavender Estate makes informed decisions about his business that take into consideration the economic impact on the local region. This includes purchasing pork from the Scottsdale butcher and all café groceries from the Scottsdale Woolworths supermarket. Pennicott Wilderness Journeys (2020), similarly, injected AU$3.2 million into the regional economy with 78 per cent of purchases through local suppliers.

Informal Influence

Leadership influence is the ability to influence others to achieve shared goals. Crawford et al. (2019) propose the fifth authentic leader dimension; "informal influence is the ability to inspire and motivate individuals to accomplish goals of their own volition, regardless of rank or position" (p. 21). The informal influence behaviour is an important distinction between an authentic leader and an individual with authentic qualities. Authentic individuals behave in authentic ways, but authentic leaders influence their followers. Importantly, informal influence is the influence that occurs in the absence of recognised position, status or ranking. While authentic leaders in the Tasmanian tourism industry can be either formal or informal influencers, managers in the industry do not need authority to influence others if they are practising informal influence behaviour.

The authentic leader-follower relationship is such that authentic leaders empower and develop highly effective leaders with a multiplying effect. Thus, an informal influencing leader can facilitate growth in followers to the point that they are able to lead themselves. These followers become the new informal influencers within their own context. Therefore, authentic leaders with informal influence can build up the next generation of leaders within the tourism industry. For the tourism industry to be sustainable, socially, economically, and environmentally, the authentic leadership model must take a multiplying model to ensure longevity and success.

FUTURE RESEARCH DIRECTIONS AND PRACTICAL IMPLICATIONS

The argument for authentic leaders as facilitators of sustainable development in tourism, based on the Tasmanian context, provides practical implications for those seeking to improve and develop tourism practice. The application of authentic leader behaviours, in conjunction with sustainable development approaches (e.g. inclusive, full and productive employment, decent work for all, and sustained economic growth) provide the foundations necessary for leaders to facilitate sustainable development of tourism. The ecotourism emergence in Tasmania has stemmed from the desire to preserve Tasmania's relatively untouched natural environment. The extensive employment opportunities and economic growth, as a result, has encouraged continued efforts to preserve the environment which attracts tourism in the first place. Without authentic leaders in the tourism industry, the risk of exploitation of the environment due to increased monetary and economic gains has a detrimental effect on the SDG's tourism is contributing to.

This chapter aims to expand the literature on the role of tourism in achieving sustainable development goals by exemplifying the importance of authentic leadership. It has been demonstrated that several of the Tasmanian tourism successes, in generating inclusive, decent and productive employment along with increased economic growth, can be attributed to the authentic leadership found in the industry. Learning from the role of tourism in Tasmania, underdeveloped countries can progress towards eliminating unemployment but also contribute to a multitude of other SDG's that are interconnected, as previously mentioned. For example, the Indonesian island of Bali thrives off tourism, but could benefit from authentic leadership in enabling sustainable development in its tourism industry. In an announcement, Bali's Governor Pastika stated that "Bali will have to continue to stress strongly environmental conservation as nobody will [sic] like to spend their holidays on an island where the environment is destroyed" (Bali.com, 2015). Sustained economic growth and environmental conservation are possible for Bali with the appropriate use of authentic leadership. This chapter may provide further insight into contemporary issues and developments arising within the realm of authentic leadership, and the influence this specific leadership style can have on the tourism industry.

CONCLUSION

This chapter provided an explication on the Tasmanian tourism industry and how authentic leaders are facilitators of sustainable development. The critical review focused on Tasmanian tourism, authentic leaders and sustainable development practices. Firstly, this paper details the emergence of tourism in Tasmania, exemplifying stand-out tourism business throughout the state. Then the authors explore the leadership gap and what impacts poor leadership has had on the Tasmanian tourism sector. By referring to relevant tourism discourse and drawing upon current Tasmanian practice, sustainable development, sustainable development goal eight and the theory behind authentic leadership are discussed. It became evident through the synthesis that leadership is at the heart of sustainable development. Examples from Bridestowe Lavender Estate, Tasmanian Walking Company, Josef Chromy, Blue Derby Pods Ride and Pennicott Wilderness Journeys demonstrate the value of leaders, who can generate economic growth, longevity and employment opportunities that are inclusive, full, decent, and productive. This paper proposed that the five authentic leader behaviours explored provide the solution to the leadership gap currently experienced in Tasmania. Finally, the authors examined the implications and future research directions from the discussion. While recognising the areas of weakness in Tasmania, the emphasis is given to how tourism locations globally can learn from Tasmania, by seeking current and developing future authentic leaders, who facilitate sustainable development. Though research on authentic leadership is expanding, scholars are encouraged to continue to expand the theoretical work in shaping academic understanding and empirical application of leadership.

REFERENCES

Adams, D. (2009). *A social inclusion strategy for Tasmania*. Department of Premier and Cabinet.

Ashworth, J., & Thomas, B. (1999). Patterns of seasonality in employment in tourism in the UK. *Applied Economics Letters*, *6*(11), 735–739. doi:10.1080/135048599352312

Avolio, B. J., & Gardner, W. L. (2005). Authentic leadership development: Getting to the root of positive forms of leadership. *The Leadership Quarterly*, *16*(3), 315–338. doi:10.1016/j.leaqua.2005.03.001

Avolio, B. J., & Wernsing, T. S. (2008). Practicing authentic leadership. *Positive psychology: Exploring the best in people, 4*, 147-165.

Baker, E. (2019, October 11). Tourism Tasmania says it has made 'internal changes' since 2016 culture report labelled it a 'boys club'. *ABC News*. Retrieved from https://www.abc.net.au/news/2019-10-11/tourism-tasmania-defends-record-after-culture-report/11595532

Bali.com. (2015). Sustainable Tourism on Bali? *Bali.com*. Retrieved March, 12, 2020, from https://www.bali.com/news_Sustainable-Tourism-on-Bali-_161.html

Beeton, S. (2006). *Community development through tourism*. Landlinks Press. doi:10.1071/9780643093881

Bestic, R. (2019, December 18). Outdoor wonders and wild adventures in Australia's South. *Telegraph Newspaper*.

Björk, P. (2007). Definition paradoxes: from concept to definition. In J. Higham (Ed.), *Critical Issues in Ecotourism: Understanding a Complex Tourism Phenomenon* (pp. 23–45). Elsevier. doi:10.1016/B978-0-7506-6878-1.50006-2

Blue Derby Pods Ride. (2020). The Purpose. *Blue Derby Pods Ride*. Retrieved June, 8, 2020, from https://bluederbypodsride.com.au/the-purpose

Bossel, H. (1999). *Indicators for sustainable development: theory, method, applications*. International Institute for Sustainable Development.

Bradford Energy Solutions. (2013). *Case Study – Josef Chromy Wines*. Retrieved from https://www.bradfordenergy.com.au/-/media/bradford-solar/files/pdfs/case-study-josef-chromy-wines.pdf

Bramwell, B., & Lane, B. (2011). Critical research on the governance of tourism and sustainability. *Journal of Sustainable Tourism*, *19*(4-5), 411–421. doi:10.1080/09669582.2011.580586

Bridestowe Lavender Estate. (2020). Sustainability and efficiency matters. *Bridestowe Lavender Estate*. Retrieved June, 8, 2020, from https://bridestowelavender.com.au/explore-bridestowe-lavender-estate/the-bridestowe-farm/sustainability/

Buckley, R., & Sommer, M. (2001). *Tourism and protected areas: Partnerships in principle and practice*. CRC for Sustainable Tourism.

Burgess, G. (2018). Bruny Island tourist influx a double-edged sword for those calling it home. *ABC News*. Retrieved March 12, 2020, from https://www.abc.net.au/news/2018-12-20/bruny-island-off-tasmania-feels-growing-pains-of-tourism-boom/10582124

Butler, R. W. (1974). The social implications of tourist developments. *Annals of Tourism Research*, *2*(2), 100–111. doi:10.1016/0160-7383(74)90025-5

Cash, D., Clark, W., Alcock, F., Dickson, N., Eckley, N., Guston, D., & Mitchell, R. (2003). Knowledge systems for sustainable development. *Proceedings of the National Academy of Sciences of the United States of America*, *100*(14), 8086–8091. doi:10.1073/pnas.1231332100 PMID:12777623

Chalmers, S. (2020). Bushfires and coronavirus twin disasters hit Australia's tourism sector. *ABC News*. Retrieved March 12, 2020, from https://www.abc.net.au/news/2020-02-11/twin-disasters-coronavirus-bushfires-hit-tourism-sector/11949922

Cho, V. (2001). Tourism forecasting and its relationship with leading economic indicators. *Journal of Hospitality & Tourism Research (Washington, D.C.)*, *25*(4), 399–420. doi:10.1177/109634800102500404

Connor, L. (2016, March 11). 10 of the world's best eco-friendly luxury hotels. *Evening Standard*. Retrieved from https://www.standard.co.uk/lifestyle/travel/10-of-the-world-s-best-ecofriendly-luxury-hotels-a3201281.html

Crabb, A. (2019, October 8). Australia Talks National Survey reveals what Australians are most worried about. *ABC News*. Retrieved from https://www.abc.net.au/news/2019-10-08/annabel-crabb-australia-talks-what-australians-worry-about/11579644

Crawford, J., Dawkins, S., Martin, A., & Lewis, G. (2018). Conceptualising authentic followers and developing a future research agenda. In D. Cotter-Lockard (Ed.), *Authentic Leadership and Followership* (pp. 271–293). Palgrave Macmillan. doi:10.1007/978-3-319-65307-5_11

Crawford, J., Dawkins, S., Martin, A., & Lewis, G. (2019). Putting the leader back into authentic leadership: Reconceptualising and rethinking leaders. *Australian Journal of Management*, *45*(1), 114–133. doi:10.1177/0312896219836460

Crawford, J., Kelder, J., & Knox, M. (2020). What does it take to be a social entrepreneur?: Authentic leaders and their effect on innovation. In C. Dogru (Ed.), *Leadership Styles, Innovation, and Social Entrepreneurship in the Era of Digitalization* (pp. 282–310). IGI Global. doi:10.4018/978-1-7998-1108-4.ch011

Daly, H. (1990). Toward some operational principles of sustainable development. *Ecological Economics*, *2*(1), 1–6. doi:10.1016/0921-8009(90)90010-R

Delios, A., & Beamish, P. W. (1999). Geographic scope, product diversification, and the corporate performance of Japanese firms. *Strategic Management Journal*, *20*(8), 711–727. doi:10.1002/(SICI)1097-0266(199908)20:8<711::AID-SMJ41>3.0.CO;2-8

Denholm, M. (2019, October 11). Leaked report reveals 'unhealthy' workplace culture at tourism Tasmania. *The Australian*. Retrieved from https://www.theaustralian.com.au/travel/leaked-report-reveals-unhealthy-workplace-culture-at-tourism-tasmania/news-story/06e6fbdd33e5d06abaadccd21d2c0f86

Dowling, R. (2001). Oceania (Australia, New Zealand and South Pacific). In D. Weaver (Ed.), *The encyclopedia of ecotourism* (pp. 139–154). CABI Publishing.

EarthCheck. (2020). *Good for business. good for the planet*. EarthCheck.

East Coast Stockfeeds. (2020). *Grape Marc*. Retrieved from http://ecsf.com.au/grape-marc

Fathom. (2019). The top 15 places to travel in 2019. *Forbes*. Retrieved from <https://www.forbes.com/sites/fathom/2019/01/15/top-travel-destinations-2019/#44ef5012523e>

Ford, J., & Harding, N. (2011). The impossibility of the 'true self' of authentic leadership. *Leadership*, *7*(4), 463–479. doi:10.1177/1742715011416894

Ford, R., Williams, K., Bishop, I., & Webb, T. (2009). A value basis for the social acceptability of clearfelling in Tasmania, Australia. *Landscape and Urban Planning*, *90*(3-4), 196–206. doi:10.1016/j.landurbplan.2008.11.006

Friedman, B. M. (2017). The moral consequences of economic growth. In J. Imber (Ed.), *Markets, Morals, and Religion* (pp. 29–42). Routledge. doi:10.4324/9780203786222-2

Gardner, W., Avolio, B., Luthans, F., May, D., & Walumbwa, F. (2005). "Can you see the real me?" A self-based model of authentic leader and follower development. *The Leadership Quarterly*, *16*(3), 343–372. doi:10.1016/j.leaqua.2005.03.003

Gardner, W. L., Cogliser, C. C., Davis, K. M., & Dickens, M. P. (2011). Authentic leadership: A review of the literature and research agenda. *The Leadership Quarterly*, *22*(6), 1120–1145. doi:10.1016/j.leaqua.2011.09.007

Govern, J. M., & Marsch, L. A. (2001). Development and validation of the situational self-awareness scale. *Consciousness and Cognition*, *10*(3), 366–378. doi:10.1006/ccog.2001.0506 PMID:11697870

Grant, M., & Booth, A. (2009). A typology of reviews: An analysis of 14 review types and associated methodologies. *Health Information and Libraries Journal*, *26*(2), 91–108. doi:10.1111/j.1471-1842.2009.00848.x PMID:19490148

Hardy, A., & Pearson, L. J. (2018). Examining stakeholder group specificity: An innovative sustainable tourism approach. *Journal of Destination Marketing & Management*, *8*, 247–258. doi:10.1016/j.jdmm.2017.05.001

Hsu, M., Bhatt, M., Adolphs, R., Tranel, D., & Camerer, C. F. (2005). Neural systems responding to degrees of uncertainty in human decision-making. *Science*, *310*(5754), 1680–1683. doi:10.1126cience.1115327 PMID:16339445

Huang, X., & McMurray, A. (2016) *Regional Australia's engagement with China in the Asian century: Innovation, integration and impact*. RMIT University.

Iszatt-White, M., & Kempster, S. (2018). Authentic leadership: Getting back to the roots of the 'root construct'? *International Journal of Management Reviews*, *21*(3), 356–369. doi:10.1111/ijmr.12193

Jolliffe, L., & Farnsworth, R. (2003). Seasonality in tourism employment: Human resource challenges. *International Journal of Contemporary Hospitality Management*, *15*(6), 312–316. doi:10.1108/09596110310488140

Kernis, M., & Goldman, B. (2006). A multicomponent conceptualisation of authenticity: Theory and research. *Advances in Experimental Social Psychology*, *38*, 283–357. doi:10.1016/S0065-2601(06)38006-9

Knox, M., & Crawford, J. (2020). Authentic leaders and business tourism: An exploration of authentic relationships. In M. Bari, S. Shaheen, & M. Fanchen (Eds.), Accelerating Knowledge Sharing, Creativity, and Innovation Through Business Tourism. Hershey, PA: IGI Global.

Knox, M., Crawford, J., & Young, S. (2020). Ecotourism emergence in Tasmania through social entrepreneurs and authentic leaders. In D. Batabyal (Eds.), Global Entrepreneurial Trends in the Tourism and Hospitality Industry. Hershey, PA: IGI Global.

Lélé, S. (1991). Sustainable development: A critical review. *World Development*, *19*(6), 607–621. doi:10.1016/0305-750X(91)90197-P

Luthans, F. (2002). The need for and meaning of positive organisational behavior. *Journal of Organizational Behavior*, *23*(6), 695–706. doi:10.1002/job.165

Luthans, F., & Avolio, B. (2003). Authentic leadership development. In K. Cameron & R. Quinn (Eds.), *Positive Organizational Scholarship* (pp. 241–258). Berrett-Koehler Publishers.

MacNeill, J. (1989). Strategies for sustainable economic development. *Scientific American*, *261*(3), 154–165. doi:10.1038cientificamerican0989-154

Matysek, K., & Kriwoken, L. (2003). The natural state: Nature-based tourism and ecotourism accreditation in Tasmania, Australia. *Journal of Quality Assurance in Hospitality & Tourism*, *4*(1-2), 129–146. doi:10.1300/J162v04n01_07

Miller, G., & Twining-Ward, L. (2005). *Monitoring for a sustainable tourism transition: The challenge of developing and using indicators*. Cabi. doi:10.1079/9780851990514.0000

Morgan, R. (2019, July). Australians see the environment as the world's biggest problem but in Australia it's the economy. *Roy Morgan Research*. Retrieved from http://www.roymorgan.com/findings/8044-most-important-problems-facing-australia-world-july-2019-201907020950

Mosler, W. (1997). Full employment and price stability. *Journal of Post Keynesian Economics*, *20*(2), 167–182. doi:10.1080/01603477.1997.11490146

Neumayer, E. (1999). The ISEW: Not an index of sustainable economic welfare. *Social Indicators Research*, *48*(1), 77–101. doi:10.1023/A:1006914023227

Newstead, T., Dawkins, S., Macklin, R., & Martin, A. (2019). The virtues project: An approach to developing good leaders. *Journal of Business Ethics*. Advance online publication. doi:10.100710551-019-04163-2

Ogilvie, F. (2017). Wheels of progress: What happens when a rural town become the mountain bike capital of Australia? *ABC News*. Retrieved June, 8, 2020, from https://www.abc.net.au/news/2017-12-26/mountain-bike-trails-driving-major-change-in-derby/9276384?nw=0

Ooi, C. S., & Hardy, A. (2020). *Tourism in Tasmania*. Academic Press.

Pennicott Wilderness Journeys. (2019). Join our team. *Pennicott Wilderness Journeys*. Retrieved March, 12, 2020, from https://www.pennicottjourneys.com.au/journeys/jobs/

Pennicott Wilderness Journeys. (2020). Community sustainability. *Pennicott Wilderness Journeys*. Retrieved March, 12, 2020, from https://www.pennicottjourneys.com.au/about/community

Piotrowski, A. (2020). Aussie farm goes from brink of collapse to Chinese tourism hotspot. *Nine Now*. Retrieved March, 12, 2020, from https://9now.nine.com.au/a-current-affair/bridestow-lavender-estate-online-tourist-hotspot-tasmania/fe3ffb14-0d5b-4f7a-9667-ed66f7361d31

Robert, K., Parris, T., & Leiserowitz, A. (2005). What is sustainable development? Goals, indicators, values, and practice. *Environment*, *47*(3), 8–21. doi:10.1080/00139157.2005.10524444

Santagata, W. (2002). Cultural districts, property rights and sustainable economic growth. *International Journal of Urban and Regional Research*, *26*(1), 9–23. doi:10.1111/1468-2427.00360

Shine, R. (2017). Underemployment a struggle for Tasmanian worker with jobs but not enough hours, union survey finds. *ABC News*. Retrieved March, 12, 2020, from https://www.abc.net.au/news/2017-09-18/union-survey-reveals-extend-of-underemployment-in-tasmania/8957168

Stratford, E. (2006). Technologies of agency and performance: Tasmania together and the constitution of harmonious island identity. *Geoforum*, *37*(2), 273–286. doi:10.1016/j.geoforum.2005.03.001

Tasmanian Government, Tourism Tasmania, Tourism Council Tasmania. (2004). *Tourism 21: A New 10-year Vision*. Tasmanian Government, Tourism Tasmania and the Tourism Council of Australia.

Tasmanian Leaders. (2016). *Tasmanian Leaders Yearbook 2016*. Launceston, Australia: Tasmanian Leaders Inc. Retrieved January 29, 2020, from https://www.tasmanianleaders.org.au/wp-content/uploads/2019/09/TLP-Yearbook-2016_Web.pdf

Tasmanian Walking Company. (2020a). Community and environment. *Tasmanian Walking Company*. Retrieved June, 8, 2020 from https://www.taswalkingco.com.au/about-us/#accreditation

Tasmanian Walking Company. (2020b). Community and environment. *Tasmanian Walking Company*. Retrieved June, 8, 2020 from https://www.taswalkingco.com.au/community-and-environment/

The Wine Syndicate. (2020). *Josef Chromy Wines*. Retrieved from https://thewinesyndicate.ca/jchromy/

Toor, S. U. R., & Ofori, G. (2008). Leadership versus management: How they are different, and why. *Leadership and Management in Engineering*, *8*(2), 61–71. doi:10.1061/(ASCE)1532-6748(2008)8:2(61)

Tourism Research Australia. (2019a). *International visitors in Australia: Year ending March 2019*. Austrade.

Tourism Research Australia. (2019b). *State tourism satellite accounts: 2017-2018*. Austrade.

Tourism Tasmania. (2014). *Tasmanian Tourism Snapshot: Year ending September 2014*. Tourism Tasmania.

Tourism Tasmania. (2015). *Tasmanian Tourism Snapshot: Year ending September 2015*. Tourism Tasmania.

Tourism Tasmania. (2016). *Tasmanian Tourism Snapshot: Year ending September 2016*. Tourism Tasmania.

Tourism Tasmania. (2017). *Tasmanian Tourism Snapshot: Year ending September 2017*. Tourism Tasmania.

Tourism Tasmania. (2018). *Tourism Research: Tasmanian Tourism Snapshot*. Tourism Tasmania.

Tourism Tasmania. (2019a). *Tasmanian Tourism Snapshot: Year ending September 2019*. Tourism Tasmania.

Tourism Tasmania. (2019b). *Importance of Tourism*. Retrieved from https://www.tourismtasmania.com.au/archive/corporate-plan-2017-20/importance-of-tourism

Tovar, C., & Lockwood, M. (2008). Social Impacts of Tourism: An Australian regional case study. *International Journal of Tourism Research*, *10*(4), 365–378. doi:10.1002/jtr.667

Trilling, L. (1971). *Sincerity and authenticity*. Harvard University Press.

United Nations. (2020). *The Sustainable Development Agenda*. Retrieved from: <https://www.un.org/sustainabledevelopment/development-agenda/>

Waite, R., McKinney, N., Smith-Glasgow, M. E., & Meloy, F. A. (2014). The embodiment of authentic leadership. *Journal of Professional Nursing*, *30*(4), 282–291. doi:10.1016/j.profnurs.2013.11.004 PMID:25150413

Wall, G. (1993). International collaboration in the search for sustainable tourism in Bali, Indonesia. *Journal of Sustainable Tourism*, *1*(1), 38–47. doi:10.1080/09669589309450699

Walpole, M., & Goodwin, H. J. (2000). Local economic impacts of dragon tourism in Indonesia. *Annals of Tourism Research*, *27*(3), 559–576. doi:10.1016/S0160-7383(99)00088-2

WCED. (1987). World commission on environment and development. *Our Common Future, 17*, 1-91.

Whiley, M. (2020). Tourism in Australia is taking a hit with the 'double whammy' of bushfires and coronavirus travel bans, here's how the country is enticing tourists in an era of travel uncertainty. *Business Insider Australia*. Retrieved March, 12, 2020, from https://www.businessinsider.com.au/australia-extends-visas-young-backpacker-tourists-travel-bushfire-recovery-2020-2?r=US&IR=T

Whiteley, S. B. (1999). Calculating the sustainable yield of Tasmania's State forests. *Tasforests*, *11*(1), 23–34.

Wilkins, K. (2019, April 14). Bridestowe luxury accommodation development to be considered by Dorset Council. *The Examiner*.

Wilson, S., Fesenmaier, D., Fesenmaier, J., & Van Es, J. (2001). Factors for success in rural tourism development. *Journal of Travel Research*, *40*(2), 132–138. doi:10.1177/004728750104000203

Wine Tasmania. (2020). *Vin Zero*. Retrieved from http://winetasmania.com.au/vinzero

World Heritage Committee. (2007). The Tasmanian wilderness world heritage area. *Australian Government Department of the Environment and Heritage*. Retrieved June, 8, 2020, from https://www.environment.gov.au/system/files/resources/63307db8-6b28-44d8-807b-6979eeab9a41/files/tas-wilderness.pdf

Zaleznik, A. (2004). Managers and leaders. *Harvard Business Review*, 1. PMID:14723179

Zhu, W., Riggio, R. E., Avolio, B. J., & Sosik, J. J. (2011). The effect of leadership on follower moral identity: Does transformational/transactional style make a difference? *Journal of Leadership & Organizational Studies*, *18*(2), 150–163. doi:10.1177/1548051810396714

Chapter 17
Intercultural Dialogue as a Tool for Maintaining Sustainable Partnerships in Tourism

Teodora Kiryakova-Dineva
https://orcid.org/0000-0002-1374-9999
South-West University Neofit Rilski, Blagoevgrad, Bulgaria

Yana Chankova
South-West University Neofit Rilski, Blagoevgrad, Bulgaria

ABSTRACT

International partnerships have a reinforcing effect and lasting benefits for modern economy and social life. The principles of intercultural partnerships in tourism can be compared to other principles of sustainable development. This chapter recognises the ultimate role of sustainable development in the sphere of tourism, while applying the theory of intercultural dialogue as an approach to the SDG 17 and tracing back partnerships to the bridge where different cultures meet. Such meetings very often need a bridge to cross over cultural gaps. In view of the proposed research, one such bridge is presented by the International Tourism Fair Holiday and SPA EXPO 2020, and it is against the background of this event that the established partnerships mediated by intercultural dialogue are classified.

INTRODUCTION

With the advancement of the third millennium, the sustainable development of the countries and of the regions is increasingly dependent not so much on its envisaged goals but rather on its tools. Sustainability is nowadays understood as sustainability on a supranational level (Uyar, Karaman & Kilic, 2019; Rodríguez-Díaz & Pulido-Fernández, 2019; Lim, 2017; Anastasiadou, 2007). For this reason, sustainability is very often treated as an inherent goal of all nations and cannot be an isolated phenomenon. Researchers are inclined to think of sustainability only as a part of a larger international system and to

DOI: 10.4018/978-1-7998-5691-7.ch017

define it as a main feature of the global world. Much in this vein, the present authors have decided to have a closer look at the dialogue between cultures. Firstly, because it belongs to the scope of the international environment, and secondly, because, in its role of a tool, the cultural dialogue can contribute to achieving the global goals of sustainable development. One of its main functions is to maintain good partnerships and to manage international connections. In this regard, the intercultural dialogue has a close relation to Goal 17 and being not limited to any specific field (Ognjevic, 2019; Vicentijevic, 2016; Castro, Lundgren & Woodin, 2019), it can be investigated in view of the partnerships of tourism. This chapter considers one specific aspect of sustainable development (i.e. with respect to tourism), but the described interconnections can be transferred to other scopes of economic activities or human relations. The viewpoint presented here is that in the modern global tourism may be affected by many unpredictable factors, but more than ever, it desperately needs the support of good cooperation, partnering and skills for achieving dialogue between cultures.

Moreover, the achievement of best results, guest satisfaction and partnerships in the scope of international tourism requires not only the ability to communicate in a foreign language, but also awareness towards different types of cultures, such as the host culture, the guest culture and the culture of tourism itself. Many aspects of host and guest cultures have been studied so far (Tasci & Severt, 2016; Travis, 2011; Canavan, 2016; Mansfeld, 2019; Sonuç, 2020). Researchers and practitioners have long seen the benefits of optimising bilateral and multilateral relations (host and guest culture) through intercultural relations and intercultural dialogue (Truong & King, 2010; Castro, Lundgren, & Woodin, 2019). Educators have also acknowledged their significance and have included intercultural relations and intercultural dialogue in different disciplines of applied character in the scope of intercultural education (Aman, 2017; Gonçalves, 2011). The practical and theoretical achievements in the field still need optimisation, and there is only a limited number of studies devoted to the dialogue between cultures as a contributing tool for the development of partnerships and sustainable relations. For the reason that participants, especially in the sphere of tourism have two main roles – as service providers or as guests, and hence the above participants interact with each other on a daily basis, whereby mutual trust and respect are an indispensable part of the overall process, one can assume that the good skills for intercultural dialogue can be seen as a decisive advantage when choosing a destination to travel to and stay for a certain period of time.

In order to explore partnership relations in the tourism industry, the present chapter will discuss the role of partnerships and their management from the viewpoint of a chosen mega tourism event. It represents a stage where intensive intercultural dialogue takes place. Based on the authors' preliminary analysis, such a stage has been identified, when lots of tourism partnerships from different tourism branches are established, many roles are assumed, and the intercultural dialogue occurs in a natural way. Specifically, this is the stage of international tourism fairs and, consequently, the research focus comes on the partnerships taking place there.

The authors have chosen international tourism fairs over other possible objects of study, i.e. events with a similar purpose – international festivals, international conferences or other special events for attracting visitors (Lee & Arcodia, 2011; Brida, Disegna & Osti, 2012; Jamal & Getz 1999; Mariani & Giorgio, 2017). One reason for choosing exactly tourist fairs and exhibitions is the wide range character of the participants – institutional and governmental participants, travel agencies and tour operators, hotels, tourism organisations, associations, NGOs in tourism and stakeholders, individual private companies with supporting tourism activities, specialists and individuals with interest in tourism, tour guides, professionals in tourism and hospitality training, and interested potential tourists. As such mega event that provides the desired optimal scope for the intended research has been considered the Bulgarian most popular event in tourism, the International Holiday & SPA Expo, well-known in Bulgaria, on the Balkans and beyond.

BACKGROUND

The line of research exploring partnerships and collaborations in business has a long tradition. Partnership goals are usually associated with a division of responsibilities and an obligation to achieve and share common benefits. The dynamics of a relationship can vary based on the type of partnership's ultimate goal. Moreover, depending on factors such as the level of intervention, the involvement of the private sector (Yermachenko, Dekhtyar & Dorokhov, 2015), the cooperation of institutional partners (Bowen, 2004; Yeh, 2019), the inclusion of local communities (represented by other smaller municipal institutions) or non-governmental organisations, partnerships can differ in their sustainability. The type of relationship will also change if there is an international partner involved (Campos, 2014; Dyll-Myklebust, 2012).

The present paper is not aimed at considering the nature of one particular type of partnership, such as the most widely investigated public-private partnership (PPP). Despite being usually perceived as the most common form of partnerships, PPPs seldom reveal the other side of the coin. When searching for inner interconnections leading to sustainability in tourism, researchers should adopt a broader perspective. Furthermore, the involvement of the private sector as a party to a partnership, from a functional point of view, should take into account indicators such as investments and cash flow, management and impact, or, in the terms of Schubeler (1996), a comprehensive assessment of the public provision system. This study does not aim to make a financial assessment of the investments made in public-private partnerships, nor to evaluate the success of a partnership. Instead, the authors focus on treating the specifics of the Sustainable Development Goal 17, and in particular on investigating the opportunities for successful partnerships by applying the principles of the intercultural dialogue as a tool for maintaining a sustainable tourism industry.

The following chapter analyses the main types of partnerships in tourism established at the most important Bulgarian International Tourism Fair and consists of three parts. The first section deals with partnerships in general and with their role for sustainable development, the second one discusses intercultural dialogue as a tool for establishing and maintaining international relations and partnerships in tourism. The third section is more practically oriented and gives an outline of the Bulgarian International Fair Holiday & SPA Expo 2020, while discussing the role of the intercultural dialogue in the event.

Partnerships, Partnerships' Roles and the Sustainable Development in Tourism

The possible economic interpretations of the term partnership are revealed in different contexts and spheres: one possible perspective is attested in the scope of the industrial relationships (David, 2001; Marks, 2019); a second perspective is related to the importance of the socio-economic relationships (Bramwe & Lane, 2006); yet another perspective can be conceptualised when looking into the framework of projects implementation.

In the broad sense of the word, any relationship between the subjects of a market economy, the result of which is the signing of a contract, is a manifestation of cooperation between the companies (a form of intercompany relations). The policy of building up of intercompany relations, which are the bases of business partnerships (Ee, Halim, & Ramayah, 2012; Tuten & Urban, 2001), is one of the elements of the enterprise strategy and proves to be the foundation for successful competition (Hartman, & Dhanda, 2018; Innocent, 2016). Modern intercompany relations make part of a complex and purposeful process, including the mutual adaptation of activities. Such intercompany interactions, when executed between international companies, require the application of skills for intercultural communication and dialogue, because the partnership forged changes the business environment and expands its territorial boundaries. Under such conditions, the partnering with an international partner presupposes the knowledge of the principles of intercultural dialogue.

In recent times, and with no additional need for theoretical explanations, partnerships are being practised in almost every sphere of human interactions. In the sphere of entrepreneurial activity, finances and investments, partnership management can rely on suitable marketing tools and comply with the changing demands of consumers by considering the characteristics of the specific business activity.

Any partnership in the business environment can be defined as a type of economic relationship based on joint actions and efforts of the parties, united by a common interest (benefit for both parties), aimed at achieving specific goals and applying a combination of methods and forms for establishing a purposeful relationship. Partnerships provide companies with access to more diverse resources than a single company could have or could acquire. The goal of modern intercompany relations is to obtain the necessary access to any kind of resources (material and technical resources, finished products, financial environment), markets, technologies or distribution channels. As known, an enterprise is built on the interaction of market economy entities. Relations of collaboration between companies have always existed (in a formal or informal way), and the same holds true for the relations between a company and its counterparties (suppliers, customers, intermediaries, etc.). However, in recent years, and in view of the contractual relations in business, partnerships have changed significantly and have taken on new forms.

Most of the relationships between partners are regulated while they guarantee their own profit or while they apply common efforts to achieve commercial benefits. Many partnerships are documented as economic or legal in form, so that they accomplish the exchange of values.

When we discuss partnerships, we usually understand partnership management. Irrespective of the specific type of intervention, a partnership is marked by minimum bilateral efforts of effective beneficial forms of mutual action, is aimed at achieving common benefits, and makes use of different types of resources (labour, financial and material resources). The results accomplished by a partnership can be assessed in terms of the satisfaction of certain common goals and meeting the needs and requirements of partnering entities. To achieve these objectives, it is necessary for a partnership to rely on the common sense of coordinating, sharing, and acting. It can be assumed that partners need to communicate constructively as a prerequisite for building up a good partnership and, in this vein, dialogue and intercultural dialogue are indispensable for establishing valuable partnerships.

Partnerships and closer cooperations are applicable for the purposes of tourism development and bring benefit for the partnering organisations and for local communities as well. Some helpful directions in studying partnerships can be seen in the following main areas:

- destination management and governance,
- contribution to infrastructure,
- implementation of national and international projects,
- support of different types of tourism and contribution to sustainability.

As mentioned above, a much-debated form of a partnership in tourism is the public-private partnerships. The B2B partnerships are as numerous, but they are not discussed as often in the relevant literature because they involve investments of different kinds and the business in the field of tourism is always competitive.

The present study does not take an interest in the type of the partnerships studied (PPP, or B2B or NGO), but in the function of the partnering organisations as a part of a certain partnership. In this sense, the proposed analysis is carried out from a functionalist perspective. The authors have attested and considered three types of partners, which accomplish their roles on different economic levels: partners, representing the macroeconomy and national institutions – these are the institutional partners; partners, representing the business - and tourism companies and partners, represented by individuals or groups of individuals in their role of tourists[1].

The partnerships in tourism are like any other business or entrepreneurial relationships and can be developed in three directions: relations with the suppliers, relations with the consumers of the services and relations with partners connected with the provision/distribution of tourism products. Considering the diversity of the specific goals of a partnership, one can additionally distinguish between some characteristic types of partnerships: partnerships for hospitality, travel and tourism activities; partnerships for the use and construction of facilities; partnerships for transport; partnerships for supporting activities; partnerships with other tourism services providers. The various legal forms for such partnerships to occur may be a joint venture, licensing, industrial cooperation, contracting and subcontracting, management under a contract, etc.

Any tourism company (tour operator, travel agency, accommodation establishment, hotels, restaurants, etc.) needs effective partnerships - this is the main prerequisite for its successful functioning in the framework of the holistic process of hospitality, leisure and travel. Currently, the most important characteristic is the predisposition to cooperation and the constant search for the most effective partners, so that the organisation of activities can be carried out in accordance with the best market conditions. A successful and effective partnership in tourism helps a tourism company to achieve, maintain and strengthen its competitive advantages and to attract tourists.

There is still no unambiguous understanding of the essence of partnerships in business, though the terminology of partnerships is widely used in contemporary business practice (for example, the notions of strategic partner, production partner, financial partner, investment partner, etc.), and, by extension, partnerships can be regarded as a form of business organisations. In the studies on tourism and the economy of tourism, the issue of partnerships in tourism business is rarely tackled at present. The concept of partnerships in tourism for sustainable development of tourism is constantly understudied and often ignored as insignificant.

In view of our proposal, the partnerships in tourism can be classified in the four groups illustrated in figure 1.

We also believe that partnerships can be discussed in terms of their functionality in the world of intercultural dialogue. Intercultural dialogue can be successfully applied for achieving the best results in all the above types of partnerships and implemented as a valuable tool for increasing their effectiveness. What's more, the authors assume that intercultural dialogue can also be used as a tool that significantly contributes to the establishment of partnerships in tourism in a direct way since their very initial stage. It is out of doubt that cooperation and partnering can have a beneficial effect on the development of tourism activities and, finally, they can contribute to sustainability. What follows is a discussion of the two aspects of the dialogue in view of interculturality and partnership maintenance.

Figure 1. Types of partnerships in tourism (the authors)

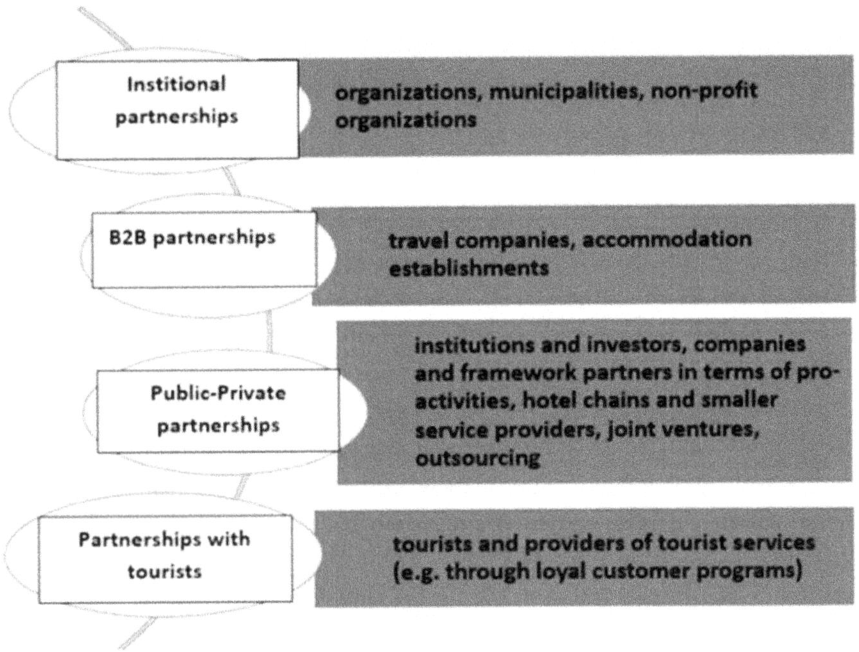

Two Aspects of the Dialogue – Interculturality and Partnership Maintenance

Modern research on the concept of dialogue studies it in the context of different roles and functions, but regardless of its specific application, the concept of dialogue inevitably affects the communicative side of communication. The concept of dialogue helps researchers understand many different aspects of the sciences that shape not only the space of communication but also the space of culture. The dialogue is also relevant to sensitivity in different contexts (i.e. religion, gender). In this sense, some authors feel justified to refer to this type of dialogue as a dialogue between cultures and the present authors share and support this definition.

The dialogue itself may involve many parties that interact as equal partners with each other. Equality here is not identified with their hierarchical function, equality is interpreted in terms of their equal behaviour in the exchange of messages, opinions, points of view. This feature has to do with the perception of dialogue as oriented towards the final result and towards a common understanding. Dialogue is, therefore, an open process. The outlined features of dialogue reveal that the dialogic partners are more than communicative partners; they accomplish special roles, e.g. business partners, services providers, users, tourists. The studied form of dialogue is process-oriented and leads to the achievement of common goals, and what's more it takes place over time. In the authors' view, a second essential characteristic of the dialogue is the establishment of partnership relations, which feature is not characteristic for the communication process. Along these lines, the present chapter focuses on the role of the dialogic process in cases when the dialogic partners exchange information during participation at common events.

Adding to the discussion on the global processes - social, economic, and cultural and taking into account the satisfaction of crises management needs, the contribution of dialogue can be more than expected. Furthermore, culture attributes can be conceived as a set of many sub-cultures and the latter is a good reason for identifying the sub-culture of tourism as a starting point in discussing sustainability and partnerships. Due to the fact that tourist activity is carried out under a certain set of cultural conditions (culture as travel and travel as a dialogue between cultures; tourism as partner relations between service providers, users of tourist services, guests, travellers), one can assume that tourism as a social phenomenon needs to avail of the benefits of cultural dialogue, especially in the scope of international tourism. Awareness of the rules of the dialogue between cultures in the forging of partnerships is the key to achieving satisfaction with tourism services to the greatest extent possible. The concept of dialogue can be extended in order to take up the establishment of relations for achieving sustainable partnerships in tourism, thus bring in intercultural dialogue into the analysis. Indeed, the intensity of partnerships in international tourism depends to a large extent on intercultural dialogue.

Following this line of analysis, the present chapter discusses the intercultural dialogue in the context of international tourism, and in particular, reflects on its implementation for partnerships maintenance. Notably, internationalisation, coming from all the borderline and tourism support areas, creates prerequisites for partnership and dialogue at the institutional, organisational (B2B and public-private) and individual levels. Provided that partnerships emerge in complex environments, and assuming that principles like mutual understanding, the achievement of common benefits, acceptance of otherness, etc. are of utmost importance, the discussion of the partnerships in tourism (and partnership maintenance) seems justified. One can consider, for example, the Mediterranean region which represents a conglomerate of many and diverse ethnic, cultural, religious and linguistic diversities. The partnerships in this region, as defined by Hadzhipetrova-Lachova (2017, p. 131) are aimed at establishing a zone of shared prosperity through economic and financial partnering and at building up a free trade area and attracting people through cooperation in the social, cultural and humanitarian spheres. Another goal of such partnerships is to promote the understanding between cultures and civil societies. Regardless of the various forms of international and intergovernmental interactions involved, it is necessary to emphasise that for the purpose of establishing future beneficial and effective partnerships in the tourism sector, many of the partnerships in the region envisage the intensification of the public-private sphere. This is not, however, to deny the fact that intercultural dialogue has always been an integral element in the exchange of socio-cultural practices in the region. Due to the newly emerged challenges in recent years, the appearance of socio-political instability in some of the Southeast Mediterranean countries, the presence of human flows and crises, close to the region, partnerships in the social, cultural and humanitarian sphere are needed more than ever.

The Intercultural Dialogue as a Mediator for Establishing and Maintaining Sustainable Partnerships in Tourism

Along with the growing importance of global interrelations and in view of the socio-cultural processes influencing travelling, hospitality and leisure, it is not surprising that the importance of cultural relations also increases, especially in the sphere of international tourism. Such kind of contacts not only stimulate the relations in tourism services but also have their effect on the intercultural relations (Elzarov, 2015; Leong, 2014) and consequently contribute to the sustainability of tourism destinations. In addition, such contacts positively affect the sustainability of local and regional communities (Kyurova, Madgerova, & Atanasova, 2016).

Tourism in the modern world makes one part of a complex set of socio-psychological, economic and geopolitical processes and makes itself manifest in various spheres of public life. It also initiates events, social happenings, meetings and different forms of partnerships. Thanks to the internationalisation and globalisation of tourist products, the latter can be promoted and maintained by means of communication and intercultural dialogue. Intercultural dialogue is a mediator and facilitator for building up sustainable partnerships in tourism, and it ensures the protection of cultural identity and acknowledges the cultural diversity of the tourism destination. The implementation of intercultural dialogue in tourism is firstly related to communication, existing international relations, contacts, policies and measures for the accomplishment of common goals and benefits. The intercultural dialogue takes place at the meeting point of cultures, and it is of crucial importance for the hospitality and tourism industry. International travel, provoked by different motives and seen as realised in different traditional and alternative tourism forms, inevitably involves direct and indirect contacts between two main parties - the providers of tourist services and the tourists. In great many cases, local and regional communities are also engaged.

Intercultural dialogue is an important tool for achieving cultural awareness at the personal and interpersonal level. The role of the cultural environment is confirmed in view of its contribution to sustainable development. For example, Tilbury and Mullà (2009) compare the role of intercultural dialogue with that of the cultural diversity, well-being, social justice and the health of the planet. All these aspects and some other related to tourism activity become of primary concern when partners from different cultures meet and establish partnerships for achieving common goals in tourism.

The reason for paying attention to the role of partnerships for the sustainable development of tourism derives from its connection with cross-border relations in whose scope many significant cross-cultural forms of dialogue emerge. Some further aspects can also be identified when investigating the contribution of intercultural dialogue for establishing and maintaining sustainable partnerships in tourism.

Culture, in its diversity of tangible, intangible and digital forms and modifications, plays an important role in ensuring mutual understanding between people of different nationalities, races, and ethnic groups, in protecting peace and human rights worldwide. Much in this spirit, every form of promoting intercultural and international partnerships, especially in problematic geopolitical regions (such as the Mediterranean region, the Black Sea region, the Middle East region, etc.) becomes of paramount concern. Intercultural dialogue is also extremely important in cases when partnerships are forged between people with highly divergent cultures.

The parameters of culture such as traditions, understandings, behaviours, customs, values and norms, social structure, etc. have a direct impact on tourism services, for example, they can influence the choice of a destination, the design of a tourism product, the marketing strategy for the destination, etc.

The practical implementation of the cultural parameters and the socio-cultural interactions has virtually no limits and can be found in studies on cultural anthropology, sociology, political science, marketing and consumer behaviour. The investigation of the various aspects of intercultural communication and dialogue feeds in and contributes to the theory and practice of tourism science.

In this sense, social and cultural dimensions, including intercultural communication, relationships and patterns of behaviour are extremely important in bridging the gaps between different cultures (i.e. via intercultural dialogue). Although the impact in such cases is not economically measurable, it may take priority over other factors. New implications emerge in searching for innovative means for mediating and facilitating dialogue and communication, and thus extra bonus benefits are achieved. For instance, such means of very effective character present themselves in view of the different forms of partnerships. Along these lines, the possibilities for intercultural partnerships can be conceived as part of the overall goals and principles of sustainable development. In such cases, the ensuing results and the advantages gained can be defined as consequent upon the beneficial and contributive application of intercultural dialogue.

Sustainability Through Intercultural Dialogue in the Sphere of Tourism

The present authors have made the proposal that sustainability can be attained through various forms of tourism partnerships. With regard to the specifics and particularities of partnerships, some relevant examples may be adduced as being crucial for the respective groups of tourists and providers of tourism services and achieved by establishing different types of institutional partnerships. Institutional partnerships will be discussed first.

Institutional partnerships are among the most important types of partnerships because they establish the integral connections of the individual sectors to different higher structures within and outside the framework of tourism. In the case of Bulgarian tourism, it can be said that the Ministry of Tourism is the largest governmental institution in charge of the largest network of institutional partnerships. These institutional partnerships are established in all major sectors of tourism, hospitality, travel and leisure. The forging of partnerships is justified by the need to provide extra transportation services for travelling to and covering all necessary destinations, e.g. places of culture, fitness and sports, wellness and spa, all kinds of events, medical aid and other tourism services. One of the most important issues in the implementation of institutional partnerships arises over the process of internationalisation. In order to exist and prosper, all sectors need to grow and attract investments, to participate in international projects, and to build up partnerships. Institutional partnerships in our chosen area often include those based on specialised partnerships in the field of tourism. For example, the Ministry of Tourism is directly responsible for all tourism associations and compiles and keeps a public National Tourism Register. Bulgarian tourist associations are of different types, e.g. regional, local, branch, product and professional, and they operate in accordance with their rules and goals. Partnerships among such organisations in Bulgaria and abroad can include partners from European and other international bodies and organisations in the field of tourism, which is necessarily declared in their statutes.

One specific feature of the institutional partnerships in the scope of tourism activities comes in the form of bilateral international agreements with countries in Europe, the Middle East and Africa. Important partnerships in this area are the bilateral partnerships with all neighbouring countries since these are the countries that generate Bulgaria's most intensive incoming tourism activity. Among the long-lasting agreements can be mentioned the Cooperation Agreement between the Government of the Republic of Bulgaria and the Government of the State of Israel, dated back to 1994.

Bilateral relations and good partnerships also emerge through the initiative for twin cities. For instance, the Bulgarian Black Sea coast city of Bourgas is a twin city to Rijeka (Croatia), Brasov (Romania), Rotterdam (Netherlands), Yalova, Sarayer city and Istanbul (Turkey), the city of Gomel (Belarus), the city of Miskolc (Hungary), the city of Krasnodar and the city of Vologda (Russia), the city of Alexandroupolis (Greece), the city of Yantai (People's Republic of China). The listed twin cities of Bourgas are also cities with developed tourism industry, and, in this sense, the twin cities initiative represents a suitable example for establishing purposeful partnerships and achieving common goals in tourism. Partnerships of this kind also testify to the successful and sustainable implementation of the intercultural dialogue model.

Another type of institutional partnerships arises in the process of negotiation and establishment of consortia aimed at the implementation of various international projects. Some of them target the development of partnerships for achieving sustainable results, consider for example the partnership forged in order to introduce the Common Assessment Framework between the Bulgarian Ministry of Tourism and the Bulgarian Institute of Public Administration. Another interesting example of institutional partnerships in the field of tourism is the partnership between higher education institutions and other international institutions and organisations. A case in point is the Project LMPT[2] implemented by the South-West University "Neofit Rilski" whose aims are to improve training in the field of sustainable tourism, to identify priority academic disciplines in the curricula of the three cycle system (BA-MA-PhD) at the universities of the 20 partnering countries and to ensure diplomas recognition. Another type of institutional partnership is realised through the participation of institutions in international organisations.

Further, the main type of partnership and indeed one of paramount importance is the public-private partnership. In terms of tourism development and the achievement of sustainable effects over the tourism economy and international relations, these partnerships can be defined as first and foremost partnerships that are usually established by larger institutions and by larger international companies. Such is the partnership between the Bulgarian Ministry of Tourism and ID Global Company and the partnership between the Ministry of Tourism and the Japan Southeast Europe Business Association (JSEEBA).

Another type of public-private partnership is realised through cooperation with international investors. Back to our field, such partnerships are established in the presence of specific investor intentions in the sector of tourism and destinations. They can also be developed for the sake of organising joint events and mega-events. One example comes from the International Conference on Sustainable Investment in Tourism, Bulgaria, 2019. The conference built on the usual formats but initiated the first-ever platform for tourism investment in the countries of Southeast Europe and the Mediterranean. The event was attended by about 300 delegates from home and abroad, including representatives of leading state institutions, leading hotel chains, investment funds and large-scale tour operators. Examples of successful partnerships can be found not only in the area of local government and business but also on an international scale in the sphere of tourism but what really matters is the fact that they result in competitive tourism products and activities. The sustainability of these partnerships is characterised by a network effect, and, in the long run, it stimulates the development of small businesses in the field of tourism services. In this vein, major efforts have been undertaken to develop sectors such as balneology and spa tourism, rural tourism, green tourism, sports, adventures, and extreme tourism.

The Contribution of Intercultural Dialogue for the Establishment of Partnerships: on the Example of the Bulgarian International Tourism Fair Holiday and SPA Expo

The Holiday & SPA Expo International Tourism Fair, Bulgaria[3] is a successful example for developing sustainable partnerships based on the implementation of intercultural dialogue as a tool for achieving better cooperation in the sphere of the Bulgarian tourism industry. This tourism fair is the most important tourism event, and exhibition in Bulgaria and with its 35 years of existence has established itself as the most successful sustainable tourism practice in the country. The participants in the Fair come from all national tourism branches. Additionally, the growing number of international participants in recent years indicates that the Fair is popular on the Balkans and in some remote countries as Japan, Israel and China as well. A historical look back reveals that the main exhibitors at this initial stage had a national background and represented the most famous Bulgarian seaside and mountain resorts. The main organisers were the Bulgarian tour operator Balkantourist and the Bulgarian Committee on Tourism. What the types and number of participants cannot reveal, however, is the fact that this event had been organised in view of the need for establishing dialogue and partnerships, the latter two being identified as crucial in this study. Firstly, this communicative event arose in a natural way and contributed to the development and establishment of stronger partnerships between tourism operators and tourism service providers.

Secondly, the enduring communication and meaningful dialogue between the participants in the Fair contributed to their future interactions for the purposes of tourism activities and ultimately resulted in the establishment of numerous partnerships, which proved to be sustainable and are still existent. As a long-established tradition, the current International Tourism Fair Holiday & SPA Expo 2020 brings together tourism companies from all areas of the hotel, travel and leisure sectors. Among them can be mentioned institutional and regional organisations such as tourism boards, industry organisations, private tourism companies, carrying out all types of tourism activities, hotels and accommodation facilities, spas and wellness centres, associations and companies for specialised types of tourism activities. Manufacturers of all types of tourist furnishing and suppliers of the hotel, spa and fitness equipment, specialised tourism software developers are also among the participants. Many advertising companies and media companies are also permanent participants at the Annual Holiday & SPA Expo event.

Except for being exhibitors and advertiser of their own destinations and tourist products, the participants have the opportunity to meet interested potential tourists as well. Any such exhibition involves the application of main marketing tools for the promotion of tourist products and services, and hence the total number of 360 exhibitors from 20 different countries is not surprising, however, the number of the established partnerships is more difficult to estimate. The organisers themselves are aware of their mediating role for building up partnerships, so, in addition to the exhibition, they have included a component envisaging the opportunity for direct business negotiations and direct meetings between stakeholders and all interested participants. Apart from sustaining a long-standing practice, the organisers provide participants with opportunities in various contexts to facilitate their contacts, to manage their direct meetings, and ultimately to establish partnerships. It is not by a chance that in their promotion strategy the organisers opted for a slogan which incorporates the perspective of having the "chance to bring business partners, suppliers and customers together" as a preferred tool. This tool the authors identify as intended dialogue for partnerships. A closer look at the framework of the above tourism fair reveals that several main groups of dialogue scenarios for building up partnerships can be identified, as listed here and presented in figure 2:

1. Dialogue by visiting the stands of exhibitors.
2. Dialogue by visiting the presentations of companies and organisations from the tourism industry.
3. Dialogue by joining B2B meetings.
4. Dialogue through participation at the annual meetings organised by the National Tourist Board.
5. Dialogue by participation in the discussion program of the Bulgarian Ministry of Tourism.
6. Dialogue through participation at the International Scientific Conference.
7. Dialogue by attending the informal accompanying program of the International Fair.;
8. Dialogue by participating in the traditional Holiday Hunting Initiative.;
9. Dialogue by attending the Annual Awards Holiday & Spa Expo 2020.

All mentioned forms of events, presentations and meetings can be regarded as occasions for establishing a dialogue for building up partnerships. Some of the dialogue forms are accomplished during the presentations and the various events for sharing of good practices, others - at the stands of the exhibitors or at the different meetings and smaller events. So, one can conclude that the whole program of the event International Fair Holiday & SPA Expo 2020 is accompanied by a plethora of dialoguing opportunities. What's more, the tourist destinations, the tourist products and the participants are overt carriers or inherently represent their respective cultures. This is a good reason to define the dialogue process studied

Figure 2. Dialogue forms for sustainable partnerships in tourism (the authors)

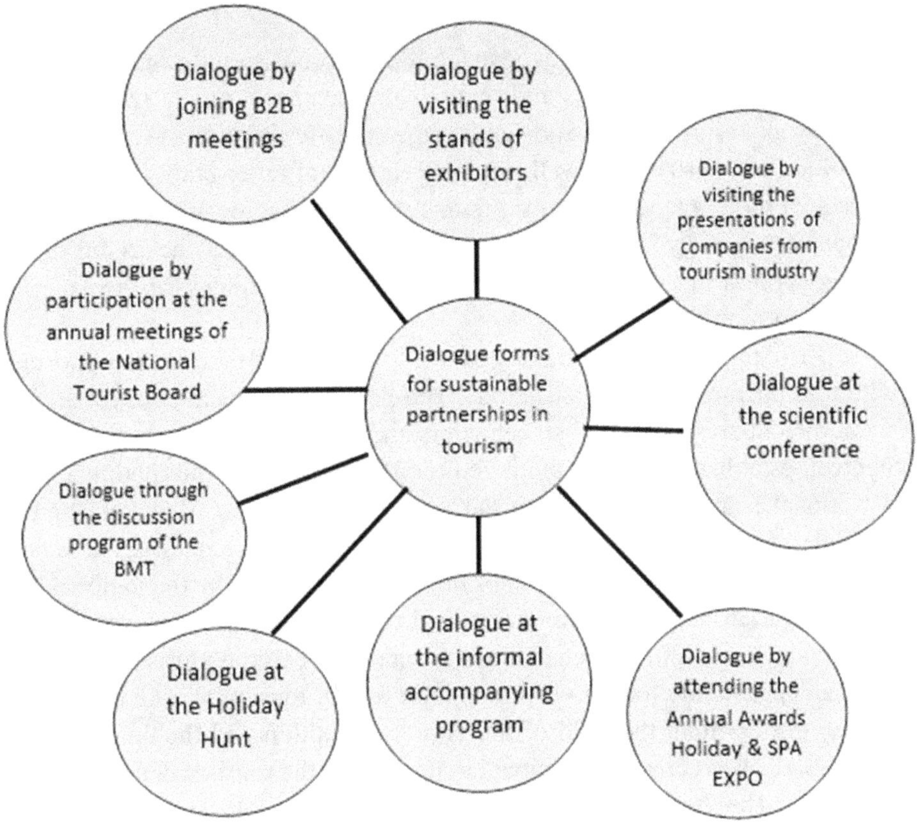

here from two viewpoints: as an instance of intercultural dialogue, on the one hand, and as a tool for maintaining sustainable partnerships, on the other. The establishment and promotion of intercultural dialogue is an important part of the program of the International Fair, without being explicitly announced. Such productive dialogue takes place, as previously mentioned, between the exhibitors and visitors, and it is a crucial element of the presentation, promotion and advertisement of each tourism product. Communicative and dialogue forms can also be found in some smaller events of the International Fair and do even become highlights in others. A case in point is the section *The Best of the Charter Programs in Turkey, Greece, Egypt, Morocco* which comprises tour operators, specialised in organising guided tours, cultural trips or individual holidays on the Balkans and all over the world as well as the sections *Global Trends 2020 and Summer 2020 Program Highlights – Greece, Turkey, Tunisia* which also avail of the role of intercultural dialogue in the pursuit of travel, vacations, adventures and leisure.

The role of dialogue in enhancing relations, cooperation and links in tourism is additionally reinforced by the initiation of specialised business to business meetings, which are not of interest to the general public but relate specifically to provisional partnerships in the field of tourism. Such opportunities are provided through a number of presentations: the presentation "Hotels point – Bulgarian B2B reservation system for hotels, transfers, car rentals and other travel services worldwide"; the presentation Growth and innovation between business to business tour operators; the presentation of new B2B booking platform for Spain – Spain in one click; the presentation "Top Destinations – Tunisia, Italy, Spain and special offers"; the presentation "Marketing Management for the Tourism Industry in 2020: Trends and Opportunities", etc.

In addition, the role of intercultural dialogue in facilitating professional communication is confirmed through some of the special presentations and sections. For instance, one of the special presentations, targeting both narrow and wide audiences discusses the specifics of offering airline tickets to all destinations and by all airlines partners, as well as of offering hotel reservations worldwide; another presentation is focused on the organisation of excursions and thematic tours in Bulgaria and abroad; still other presents a business strategy for the organisation and optimisation of successful sales. Intercultural dialogue is also highlighted in the presentation "Marketing Management for the Tourism Industry in 2020: Trends and Opportunities".

The famous super highlight in the program of the International Fair is the opportunity for direct negotiations with tour operators and organisations. The possibility for direct negotiations continues to arise interest in all direct participants and stakeholders and can be traced back to the very beginning of the Fair. Another well-established practice of intercultural dialogue at the International Fair is the Annual Meeting of the National Tourism Board, organised by the mentioned board and assisted by the Bulgarian Ministry of Tourism, the Ministry of Economy and the Sofia Municipality. These annual meetings are of utmost importance for the dissemination of the policies pursued by the national bodies monitoring the quality of Bulgarian national and international tourism. The discussions held at the meetings reaffirm the role of direct contact between institutions and between the tourism industry stakeholders. They also provide an opportunity for the visitors in their role of tourists to take an active part in policy making and hence in re-defining the field of tourism consumption and the quality of services in the tourism industry, as well as to actively participate in improving the tourism end-product and respecting host and guest cultures. The 7th Annual Meeting, in addition to its relevance to the challenging topics in tourism 2020, confirms the significance of possibilities for good communication between all actors in the tourism chain. The main highlights of this year's meeting come on issues related to the culture of native tourism, digitisation and internationalisation, as well as on the establishment of the Bulgarian national tourism destination brand.

Last but not least, a large number of presentations, meetings, promotions, discussions, and negotiations at the International Fair are carried out thanks to the active involvement and partnership of the Ministry of Tourism. The Bulgarian Ministry of Tourism is a key-role participant in the implementation of partnerships at all levels - inter-institutional partnerships, including municipalities and non-governmental organisations and public-private partnerships, comprising various types of organisations from the tourism industry.

Considering its successful cooperation policy, the Ministry of Tourism is able not only to publicise the opportunities for the development of a diversified tourist product, but also to promote to a high level the dialogue among all types of tourism partners on the national and international scale. At the same time, the main highlight of such discussions remains the promotion of Bulgaria as a sustainable tourism destination welcoming guests all through the year.

Along with the well-organised presentations of the profiled opportunities for tourism, travel and leisure, the International Tourism Fair purposefully encourages all kinds of networking activities and international partnerships. The program of the Holiday & SPA Expo 2020 bridges the cooperation between the tourism industry and science and creates a unique opportunity for stimulating dialogue of a different type. The first scientific conference, under the title 1st Congress of Universities of Southeast

Europe and Asia "Education in the Global World of New Technologies" was held at the time of the Fair and it marked the beginning of yet another kind of partnerships and endeavours for achieving sustainability of partnerships in the sphere of higher education.

With all of the above opportunities and benefits in mind, it is difficult to enumerate all positive sides and advantages of the small-scale events, meetings, presentations, etc. organised as part of the large-scale event International Tourism Fair in Bulgaria. Remarkably, the International Tourism Fair Holiday & SPA Expo 2020 has more to offer than just a large exhibition of tourism services providers and their tourist products. Resulting from the above analysis, attention can be drawn to some very important aspects of the Fair, namely the opportunities for establishing partnerships, and the specific contribution of intercultural dialogue as a tool for the effective implementation and maintenance of partnerships. The fact that 360 exhibitors from 20 countries participated in the 2020 International Fair and that it was visited by more than 20000 participants and visitors, professionals from the tourism industry and guests in the period from 12th to 14th of March reveals the importance of the sustainable partnerships that this large-scale event gave rise to. From the viewpoint of global tourism and in view of the establishment of future relations in international tourism, the studied event is all about maintaining sustainable partnerships in tourism and not about confirming leadership positions in the organisation of mega-events in the field of tourism, holidays, travel and leisure.

CONCLUSION

One of the main trends in the evolutionary development of the partnership system is its gradual but intensive movement from the category of a public-private aggregate to the category of an active tool for achieving the goals of sustainable development. We are now witnessing of how relations in all spheres are becoming more and more dependent on the reliability of partnerships and the cooperation among partners. Numerous cases are attested when effective and successful relations are established due to the skills for intercultural dialogue of the partners involved. It has been proved that, while maintaining some basic and generally accepted principles of economic relations, the cooperation system has a sufficient number of unique and exceptional features, which are not always explicitly mentioned but are revealed, made known and proven in due course of time and this is how partnerships work.

Tourism fairs, exhibitions and meetings must be thought of not just as places for setting up stands and distributing booklets but as platforms where real, contemporary and extremely fruitful dialogue takes place. This dialogue has the characteristics of a communication act between institutions, between organisations from the overall tourism industry, between stakeholders, the media, providers and consumers of the tourist product, and last but not least it has the characteristics of dialogue between cultures as tourism bridges cultures in a way. The fact that the profile range of participants is constantly expanding indicates that the number of tourism partners is increasing too. The establishment of numerous contacts at such an event further testifies to an essential feature of tourism, i.e. in addition to travel, recreation and leisure, tourism can be recognised as a crucial shared space for achieving contributive dialogue and for maintaining sustainable partnerships. Based on the complexity of contemporary socio-political contacts between different tourism branches from both geographically close and geographically distant regions, this study has analysed the sustainable development of partnerships, applying the principles of intercultural dialogue. Regarding the fact that the requirements for global tourism are increasingly connected to the Global Sustainable Development Goals, the context of the chosen topic has practical relevance.

Due to the complexity of the phenomena generated in cultural encounters and especially in view of the fact that tourism activities need the development of as many varied partnerships as possible (in the presence of institutions, tourism service providers, local communities and consumers of tourist products), the analysis presented here is based on a real event, whereby the established partnerships have a sustainable character. Based on concrete examples from the practice of the largest international tourist exhibition in Bulgaria, nine major types of partnerships have been identified. Future research, which opens against the background of the study being conducted, may investigate the intensity of these partnerships in due course of time. Future research may also apply a comparative approach in order to explore the types of partnerships forged at similar fairs, exhibitions and events in other countries and may study the extent to which intercultural dialogue can function as a universal tool for the maintenance of partnerships in tourism.

REFERENCES

Aman, R. (2017). Epistemological privilege in intercultural education. *Decolonising Intercultural Education*, 15–24. doi:10.4324/9781315616681-2

Anastasiadou, C. (2007). Group Politics and Tourism Interest Representation at the Supranational Level: Evidence from the European Union. *Tourism and Politics*, 59–69.

Bowen, H. (2004). Cultural tourism: The partnership between tourism and cultural heritage management. *Tourism Management*, *25*(3), 412–414. doi:10.1016/S0261-5177(03)00138-9

Bramwell, B., & Lane, B. (2006). *Tourism, collaboration, and partnership: politics, practice, and sustainability*. Viva Books Private Ltd.

Brida, J. G., Disegna, M., & Osti, L. (2012). Perceptions of Authenticity of Cultural Events: A Host–Tourist Analysis. *Tourism, Culture & Communication*, *12*(2), 85–96. doi:10.3727/109830413X13575858951121

Campos, M. J. Z. (2014). Partnerships, Tourism, and Community Impacts. *The Wiley Blackwell Companion to Tourism*, 567–577.

Canavan, B. (2016). Tourism culture: Nexus, characteristics, context and sustainability. *Tourism Management*, *53*, 229–243. doi:10.1016/j.tourman.2015.10.002

Castro, P., Lundgren, U., & Woodin, J. (2019). Intercultural dialogue. *Educational Approaches to Internationalization through Intercultural Dialogue*, 3–14. doi:10.4324/9780429444289-1

David, B. (2001). Partnership – participation – power: the meaning of empowerment in post-industrial society. *Partnership WorkingPolicy and Practice*, 243–259.

Dyll-Myklebust, L. (2012). *Public-Private-Community Partnership Model for Participatory Lodge (Tourism) Development*. Cultural Tourism and Identity., doi:10.1163/9789004234581_014

Ee, O., Halim, H. A., & Ramayah, T. (2012). The effects of partnership quality on business process outsourcing success in Malaysia: Key users perspective. *Service Business*, *7*(2), 227–253. doi:10.100711628-012-0152-z

Elzarov, Z. (2015). Community Stabilisation. *Journal of Security and Development*, *4*(1), 7–14.

Gonçalves, S. (2011). Intangible culture, cooperation and intercultural dialogue among university students. *Intercultural Education*, *22*(1), 83–95. doi:10.1080/14675986.2011.549647

Hadzhipetrova-Lachova, M. (2017). EU Policy Initiatives in the Mediterranean Region. *International Conference Knowledge-Based Organisation*, *23*(1), 130-134. 10.1515/kbo-2017-0020

Hartman, L. P., & Dhanda, K. K. (2018). Cross-Sector Partnerships: An Examination of Success Factors. *Business and Society Review*, *123*(1), 181–214. doi:10.1111/basr.12139

Jamal, T., & Getz, D. (1999). Community Roundtables for Tourism-related Conflicts: The Dialectics of Consensus and Process Structures. *Journal of Sustainable Tourism*, *7*(3-4), 290–313. doi:10.1080/09669589908667341

Kyurova, V., Madgerova, R., & Atanasova, A. (2016). Problems of the management and marketing of family business in tourism. *Horizon*, *20*. Advance online publication. doi:10.20544/HORIZONS.A.20.1.17.P01

Lee, I., & Arcodia, C. (2011). The Role of Regional Food Festivals for Destination Branding. *International Journal of Tourism Research*, *13*(4), 355–367. doi:10.1002/jtr.852

Lim, W. M. (2017). Exiting supranational unions and the corresponding impact on tourism: Some insights from a rejoinder to Brexit. *Current Issues in Tourism*, *21*(9), 970–974. doi:10.1080/13683500.2016.1272555

Mansfeld, Y. (2019). Pursued Cultural Sustainability and Sought Cultural Experiences: Host–Guest Perspectives. *Creating and Managing Experiences in Cultural Tourism*, 241–253. doi:10.1142/9789813233683_0015

Mariani, M. M., & Giorgio, L. (2017). The "Pink Night" festival revisited: Meta-events and the role of destination partnerships in staging event tourism. *Annals of Tourism Research*, *62*, 89–109. doi:10.1016/j.annals.2016.11.003

MarksJ. H. (2019). *The Perils of Partnership*. doi:10.1093/oso/9780190907082.001.0001

Ognjevic, T. (2019). Transformation of cultural identity, tourism and intercultural dialogue: medieval world heritage sites in Serbia between cultural monuments and shrines. *Tourism, Pilgrimage and Intercultural Dialogue: Interpreting Sacred Stories*, 121–129. doi:10.1079/9781789241129.0121

Schubeler, P. (1996). *Participation and Partnership in Urban Infrastructure Management*. Washington, DC: World Bank for the Urban Management Programme.

Sonuç, N. (2020). Culture, Tourism and Sustainability (Cultural Heritage and Sustainable Tourism, Social Sustainability of Tourism, Socio-Cultural Sustainability of Tourism). Encyclopedia of Sustainable Management, 1–7. doi:10.1596/0-8213-3650-9

Tasci, A. D. A., & Severt, D. (2016). A triple lens measurement of host–guest perceptions for sustainable gaze in tourism. *Journal of Sustainable Tourism*, *25*(6), 711–731. doi:10.1080/09669582.2016.1225746

Tilbury, D. & Mulà, I. (2009). *Review of Education for Sustainable Development Policies from a Cultural Diversity and Intercultural Dialogue Perspective: Identifying Opportunities for Future Action*. UNESCO Digital library.

Travis, A. S. (2011). Host cultures and tourism: is a culture sustainable? *Planning for Tourism, Leisure and Sustainability: International Case Studies*, 141–149. doi:10.1079/9781845937423.0141

Truong, T.-H., & King, B. (2010). Cultural Values and Service Quality: Host and Guest Perspectives. *Tourism, Culture & Communication*, *10*(1), 15–32. doi:10.3727/109830410X12629765735597

Tuten, T. L., & Ureban, D. J. (2001). An Expanded Model of Business-to-Business Partnership Formation and Success. *Industrial Marketing Management*, *30*(2), 149–164. doi:10.1016/S0019-8501(00)00140-1

Uyar, A., Karaman, A. S., & Kilic, M. (2019). Institutional drivers of sustainability reporting in the global tourism industry. *Tourism Economics*. doi:10.1177/1354816619886250

Vicentijevic, D. (2016). Strengthening intercultural dialogue by cross -border cooperation projects in the field of cultural tourism. *Kultura,* (152), 239–256. doi:10.5937/kultura1652239v

Yeh, C. M. (2019). The influence of foreign institutional investors, institutional directors, and the share pledge ratio of directors on financial performance of tourism firms. *Tourism Economics*, *26*(1), 179–201. doi:10.1177/1354816619835613

Yermachenko, V., Dekhtyar, N., & Dorokhov, O. (2015). Business Tourism Development on the Basis of Public–Private Partnership. *Tourism Analysis*, *20*(4), 433–439. doi:10.3727/108354215X14400815080640

ADDITIONAL READING

Freyer, W. (2009). *Tourismus: Einführung in die Fremdenverkehrsökonomie*. Walter de Gruyter.

Krasteva, R. (2016). Challenges faced by the tourism sector in the Greek islands as a result of the refugee crisis, Entrepreneurship (2), pp. 324-334.

Kyurova, V., & Koyundzhiyska-Davidkova, B. (2017). Current issues in family business management. *European Journal of Service Management*, *22*, 47–52. doi:10.18276/ejsm.2017.22-06

Leong, C. (2014). Social markers of acculturation. International Journal of Int. *Relations*, *38*, 120–132.

Nair-Venugopal, S. (2015). Issues of language and competence in intercultural business contexts. *Language and Intercultural Communication*, *15*(1), 29–45. doi:10.1080/14708477.2014.985304

Popova, N. & Mukova, L. (2018). On The Issue Of Russian Tourism In Bulgaria: Geopolitical, Economic And Legal Aspects. *Services in Russia and Abroad*, 17–31.

Reisinger, Y. (2009). *International tourism: Cultures and Behaviour*. Elsevier.

Rodríguez-Díaz, B., & Pulido-Fernández, J. I. (2019). Sustainability as a Key Factor in Tourism Competitiveness: A Global Analysis. *Sustainability*, *12*(1), 51. doi:10.3390u12010051

Roman, A., & Mauerhofer, V. (2019). Multilevel Coordination and Cooperation during Implementing Supranational Environmental Legislation: A Case Study on Invasive Alien Species. *Sustainability*, *11*(6), 1531. doi:10.3390u11061531

Stankova, M., Tsvetkov, T., & Ivanova, L. (2019). Tourist development between security and terrorism: Empirical evidence from Europe and the United States. *Oeconomia Copernicana*, *10*(2), 219–237. doi:10.24136/oc.2019.011

Stoykova, B. (2015). Authenticity of ethnic tourism (based on the example of the congregation of the Karakachans in Bulgaria). *International Review on Public and Nonprofit Marketing*, *12*(3), 297–313. doi:10.100712208-015-0138-3

Tapper, R. (2001). Tourism and socio-economic development. *Journal of Tourism Research*, *3*(5), 351–366. doi:10.1002/jtr.348

ENDNOTES

1. Hence, the current research pursues the functionalist perspective essentially.
2. Project "Licence, Master professionnels en formation ouverte et à distance pour le développement du tourisme durable en Chine, au Vietnam et au Kirghizstan » (Project LMPT), www.lpmt.eu
3. www.holidayfair-sofia.com

Chapter 18
Inclusivity in Higher Tourism Studies: A Necessary Reality

Almudena Barrientos-Báez
School of Tourism, Iriarte University, Spain

Eduardo Parra-López
Instituto de Investigación Social y Turismo (ISTUR), University of La Laguna, Spain

José Alberto Martínez-González
University of La Laguna, Spain

ABSTRACT

The university as an institution plays a key role in its role in improving the quality of life of people. On the other hand, education is shown as a fundamental pillar to achieving an ideal configuration of a heterogeneous, diversified, organized, and inclusive society. Teamwork and respect among the people who are part of the academic community are reflected in the development of the teaching units that make up the Teaching Guides adapted to the European Education Area. With this premise, the objective of this research is to establish the interrelationship between inclusive education in the university and the tourism industry. To this end, it will be analysed to what extent the Tourism Degree curriculum contains subjects related to inclusiveness, particularly with aspects such as disability and mobility. A thorough literature review on this field of study and the analysis of real cases of a sample of Spanish universities is carried out.

INTRODUCTION

This article is made in the hope of providing a little more visibility to a situation that is there and is likely to be ignored by young university students. The *White Book of Design for All at the university* (2006), was a before and after to favour substantial changes in the curricula of the different Spanish universities. On the other hand, different public institutions explicitly support this current of thought and

DOI: 10.4018/978-1-7998-5691-7.ch018

good work that support students and professors in supporting the University Training project in *Design for All People*. It is based on collaboration between different public administrations with private sector companies, both nationally and internationally (CRUE Spanish Universities, ONCE Foundation (National Organization of the Blind of Spain) and the Royal Disability Board). Also through the support provided to organisations, entities, specialists and promoters in studies, research and development, information, documentation and training (ONCE Fundation, CESyA, CENTAC).

Graduates in tourism during the exercise of their profession interact with people who may have mobility difficulties or pathologies that may make it difficult for them to approach and enjoy the tourist destination. Education, therefore, becomes the core of this research.

Currently, non-profit entities linked to specific groups and supported by state or supranational public organisations provide training courses on mobility and accessibility. In the case of this study, reference is made to the World Tourism Organization (UNWTO): an international institution whose main objective is to promote tourism in a sustainable, responsible and ACCESSIBLE way for the enjoyment of all. It is directly linked to the United Nations, and defends the application of the Global Code of Ethics for Tourism to minimise the impacts caused by the sector to the environment and society as a whole, giving additional educational importance to specific training in the making of ethical decisions, as Díaz del Campo defends (2012, p. 14). From here, it encourages teaching with active methodologies that motivate the student to train to increase and facilitate the inclusion and enjoyment of everyone in the tourism sector, both travellers and workers, in follow-up to the principles set by Munuera (2015). From here a conclusion is anticipated that refers to the importance of this knowledge acquired so that young people can be counted in the near future will be part of the tourism industry communicating daily with people who need specific and professional attention and taking advantage to this of its digital nativity (Alonso, et al., 2016) to compensate for the loss of relative importance of traditional media (Marta & Gabelas, 2013, p. 16; Bernal, 2010, p. 6) that would have played a more important role at the time of Sandoval (1998) or in the cases studied, by contrast, by Correyero (2003) and Infante (2014).

On the other hand, it is verified that in the Teaching Guides of the Degree in Tourism (2019-2020), issues concerning accessibility in the tourism field are hardly contextualised. Therefore it is understood that for students to be trained in this particular subject, the corresponding teacher training is essential so that they can program their subjects, including the present subject of study. The end goes beyond the purely academic because the graduate in Tourism must be able to manage, design and direct strategic plans for the improvement in accessibility of a hotel infrastructure of the beach, mountain, urban, a promenade, access to the cable car of a tourist resort, etc.

In 2014, the UNWTO published module I of the manual on Accessible Tourism for All, a work co-produced by the ONCE Foundation, and a year later the three remaining modules were published for professionals that make up the tourism sector. The manual details and bases the social inclusion of people with disabilities in Spain and in the European Accessible Tourism Network (ENAT). This is a manual in a digital version with easy access and recommended reading for students and teachers.

TOURISM AND ACCESSIBILITY

Improving the Accessibility of tourist facilities, cultural resources (from the postulated by Caldevilla et al., 2019) and natural resources among others, favours the destination and all those who live there, and serves as a stimulus to reinforce the premise of doing a "Tourism for All". On the other hand, increasing the levels of quality and accessibility has a direct impact on users and service providers because the result is beneficial for the best and greatest enjoyment of the free time of the former, and to increase the sales of the latter, generating income. According to the Observatory of Universal Accessibility of Tourism in Spain (2017, p. 57), people with disabilities spend 30% more than other types of tourists. Therefore, it is not far-fetched to think that the initial investment to convert a tourist infrastructure of this type, adapted and accessible, is compensated with subsequent gains from the sale of tourist services and/ or products to this same type of public. Technological innovation and proper professional performance come together to create solutions to problems that should not occur in the 21st century. Problems are related to mobility and the right to enjoy leisure time under the same conditions and guarantees as the rest of the population.

From the university institution (through academic awareness) and the Ministry of Industry, Commerce and Tourism (from the execution of actions) the adaptation of tourist infrastructure and the destinations themselves with certification programs in the field of accessibility the question is, are students aware of these programs, certifications and collaborations necessary to evolve positively towards the enjoyment of everyone's leisure and vacation time?

It is possible that inclusive education is not as present in training as it is believed or intended, but there is always time to improve and make changes by introducing brushstrokes of colour in the education field, or to implement action tables as proposed by Molina (2014) and Yazzo (2015). From these lines, the formative transversality is proposed for the optimal development of the teacher and student in this area. The article 148, 1, 18CE includes the powers on promotion and promotion of tourism by the Autonomous Communities. This fact entails that each Autonomous Community can treat accessibility differently and distantly from others. This fact is, therefore considered as a weakness in legal matters. On the other hand, the UNE 170001-1 standards (part 1: DALCO criteria – ambulation, apprehension, location and communication – to facilitate accessibility to the environment) and UNE 170001-2 (part 2: accessibility management system), are the basis on which to design, implement and maintain an accessibility management system (AENOR 2017) and the ISO 9001 quality certification, since it shares with this standard the tools of management systems (planning, objectives, corrective actions, preventive and universal accessibility vision). There is also the Q of Quality, distinctive that is granted to tourism infrastructures that meet the requirements of the Spanish Tourism Quality System (SCTE) regarding the provision of services.

This brand is promoted by the Ministry of Industry, Tourism and Commerce and is issued by the Institute for Spanish Tourism Quality (ICTE). In 2015, the ICTE awarded the *"Q for Tourism Quality"* award to the State Representative Platform for People with Physical Disabilities (PREDIF) for its trajectory and promotion of accessible tourism for the whole society as a whole. According to its president Francisco Sardón, the award recognised the work of all professionals of the Platform for promoting the principle of equal opportunities in the field of tourism and accessibility. Adding that applying the criteria of universal accessibility increases the possibility of having quality tourism. Training students on this topic every day more important is essential when dealing with segments of heterogeneous people (potential customers) whose purchase is, in most cases, intangible. Royal Decree 1/2013 stands out in its article 59: Obligation of public powers: Social awareness.

The public authorities will promote the implementation and maintenance of campaigns to raise awareness of society, accessible to people with disabilities, especially in the socio-sanitary, educational and professional fields, promoting recognition and respect for rights and the dignity of people with disabilities, so that as a whole, collaborate in their full inclusion in social life. The elimination of social barriers of exclusion in different contexts (personal, political, institutional, cultural, social and economic) and that affects the lives of these people is a concern of inclusion: transforming the culture, structure and educational practices in educational institutions to serve the diversity of educational needs of all kinds, which arise in society for personal characteristics (Boot, 2000).

In article 61 of the same one can read: The design of "*Design for All People*" will be included in the design of the Professional Training degrees and in the development of the corresponding curricula. Likewise, in the case of University Teachings, the Government will encourage universities to consider similar measures in the design of their degrees.

The Law on Equality, Non-Discrimination and Universal Accessibility (LIONDAU), approved on December 2, 2003, states that the Government will develop the training curriculum in *design for all,* in university programs – A provision that has not been carried out. In addition, in the tenth final provision of the same Law 51/2003, it was added that training on these principles and rights should be included in the university's studies plans.

After analysing the Teaching Guides of four Spanish Universities where the Degree in Tourism -Universidad Complutense de Madrid (UCM), University of Málaga (UMA), University of Valencia (UV) and Autonomous University of Barcelona (UAB) is taught - you can Note that subjects with compulsory character (OB) are not taught, nor in basic training (BA), nor in elective subjects (OP) subjects that correspond to the subject being analysed: inclusive tourism training from two different perspectives. On the one hand, the inclusion of students with disabilities in the University Degrees, and on the other, the inclusion of training that promotes recognition and respect for the rights and dignity of persons with disabilities.

According to Law 27/2005, of November 30, the Degree in Tourism aims to promote and encourage in its students the creation of their own values in the promotion of education and the culture of peace. In Law 51/2003, of December 2, respect for effective equality between men and women and equal opportunities, non-discrimination and universal accessibility of persons with disabilities are raised. However, the report Evaluation of the implementation of the principles of equal opportunities and universal accessibility in the curricula of the degrees of Spanish universities [IOAU] (2011), carried out by the Observatory University and Disability, reflects a low Implementation of the IOAU parameters: only three universities were registered in category C and the majority (85.7%), were in D, the lowest.

Below are 2 UNWTO initiatives for the benefit of this group: The ENAT (European Accessible Tourism Network) coordinates a training project, whose main objective is the learning and approach to Accessible Tourism and the subsequent obtaining of a certificate. The accessible road (The Accessible Road) is a guide that promotes and promotes the accessibility of travel in Quebec (Canada) informing about those tourist resources accessible to all.

The Access City Award of the European Commission recognises as many initiatives in Accessible Tourism as is the accessible leisure area on the Malvarrosa beach in Valencia, sailing activities for people with reduced mobility in Galicia and a large number of actions taken to conducted by the Ministry of Agriculture, Fisheries and Food that have adapted natural environments favouring access to them. From the inclusive approach, it is intended to respond to the educational needs of individuals, raised from multiple perspectives that relate to each other: culture, policies and educational practices (Booth & Ainscow, 2000, cited by Valenciano, 2009).

Echeita and Ainscow (2010) propose that inclusion should be seen as a permanent search for better ways of responding to the diversity of the student body. It is about seeing the difference as an opportunity to learn from it, taking advantage of it in the learning process. Yes, there is talk of inclusion as a process, for these authors, it is important to take into account the time, which implies that the transcendental changes do not happen from a moment to other.

MANAGEMENT OF PROGRAMS FOR ADAPTATION IN TOURISM INSTITUTIONS

Present your perspective on the issues, controversies, problems, etc., as they relate to theme and arguments supporting your position. Compare and contrast with what has been, or is currently being done as it relates to the chapter's specific topic and the main theme of the book. Public institutions in coordination with private organisations representing people with disabilities have developed and managed programs for the adaptation of tourist destinations and the training of the agents that make up the companies in the sector. Modifying regulations, implementing new challenges, objectives and collaborating in the dissemination of information to raise awareness and educate people according to the communication facilities mentioned by Tuñez et al. (2016, p. 249) and within the communication precepts for the development cited by Barranquero (2012, p. 74).

In the educational sector, the obligations of the state are, among others, to bet on a free, professional, innovative education with the capacity to adapt to the demands of society and to the changes that take place in it. The supervision and control on the part of the governance in the universities should have as main objective to pursue the quality of the educational service from the implantation of educational policies that will promote, among other elements, the company-university relationship facilitating the inclusion of the student in the working world while doing his studies and the integration of emotional education as a new shaping axis of the academic curriculum in the Degree in Tourism.

The agents that make up the tourism sector could be able to improve and promote accessibility and inclusive education to be able to integrate logically and professionally people who have some type of disability, taking advantage of ICT and its potential to achieve extraordinary results such as studied by Castleton (2017), correcting the deficiencies in its implementation observed by Guzmán et al. (2011, p. 444). In this way, diversity is recognised under an ethical framework that encompasses all the institutions that are part of this industry that provides entertainment, accommodation, gastronomy and transportation services, giving value to personal and collective leisure time. A recognition that happens to give the tourist 2.0 a 2.0 operator (Altamirano et al., 2016, p. 51).

In the traditional approach, each student is diagnosed to determine a category and solve the deficit that exists, however, in inclusive education the characteristics of each student are simply analysed to determine the supports that will be necessary. The traditional approach focuses on the student, while the inclusive approach does so in class. In short, in inclusive education, special programs are not used for certain students, but the needs of each student are met in the same classroom. In this sense, UNESCO understands that in inclusive education, students with special needs must have access to regular education universities, where they must be accommodated with pedagogical strategies focused on students so that their needs are met.

The objectives of inclusive education are:

- The social and academic inclusion of students since it is about including students not marginalising them from education.
- The single educational system, that is, there should be no special programs for students who have certain needs.

- Attention to all students according to their characteristics.
- The promotion of equal opportunities based on solidarity and the promotion of participation.
- The coordination of all the people who participate in the education sector.

It is currently agreed that it is not only excluded by people's economic situation (poverty), the exclusion is also related with the subject's difficulty in developing humanly, the lack of a project of life, the absence of social participation and problems of access to well-being and protection (Blanco, 2006).

Curricular Training

The university can be defined as the public or private institution of higher education, made up of different elements that interrelate with each other around an open, dynamic system capable of adapting to the changes taking place in society. From here, we can affirm that governance is the way of being managed by the university, indirectly affecting those services, companies, people and infrastructures that surround it. It is important to note that universities should be recognised for their skills and 7 or abilities to train professionals and carry out scientific research (Janeiro et al., 2013). Silva (2009, p. 71) indicates that they have to take into account the technological and behavioural changes of the population, the internationalisation of education and the incorporations or mergers to understand the need to transform these institutions.

The demands on the tourist labour market are increasing as qualified personnel are required who are not only trained in concepts and theories but in the social and cognitive capacities included in Emotional Intelligence. In this way, processes of frustration, erroneous responses to negative contingencies and other conflictive situations could be prevented that only with training and emotional control can be satisfactorily resolved.

Being competent in the professional sector of Tourism means knowing how to manage emotional skills and competences, theoretically acquired in classrooms that allow the worker (former student) to be competent and feel closer to the needs and demands of clients, which are the same as those of the society of each moment. This is how authors who investigate (Suberviola, 2011; Jiménez, 2009) emerge from this field of studies comparing academic performance and the role of university education (Ciarrochi et al., 2001; Extremera & Fernández-Berrocal, 2003; Sánchez et al., 2008; Trinidad & Johnson, 2002). They propose the figure of the teacher as a guide who also teaches about managing emotions, their own and those of others, thanks to more adequate training in their teaching careers.

The new university education policy must be adapted to the needs of the new tourist market. Likewise, it must direct its efforts and proposals initially to the area of teacher training so that it can transmit, through the proper use of teaching tools and learning styles, its knowledge and experiences to students who wish to learn and expand their competency horizons and that they may be at risk of social exclusion.

The Bachelor's degree in tourism guides student learning towards social competencies, omitting the use of diversity, ethical awareness and the drive for change. There are specific competencies that try to make the student aware of the importance of ICT and its use in tourism while trying to delve into the characteristics of the tourism sector. On the one hand, they try to address the management and exploitation of the main and/or potential tourist destinations through competencies. And on the other, they inquire about the principles that constitute it and the impacts that tourism produces at a socio-economic level. In this sense, the aim is to train the student in the knowledge and characteristics of the sector, away from emotional development.

The academic objective of the degree aims for the student to acquire competencies related to entrepreneurship, internationalisation, handling of economic and legal documentation in several languages, as well as negotiation in various cultural contexts. This is achieved through an approach that combines the theoretical rigour of the most prestigious and up-to-date sources and references with the study of real practical cases. Real practical cases that should also be directed towards situations with people with reduced mobility, lack of vision, deafness, or simply, as we mentioned previously, at risk of social exclusion. The degree in tourism aims to train multidisciplinary professionals, with global knowledge, and with the skills to carry out management activities related to the tourism sector. Adapted to the different qualifications that make up the European Higher Education Area (EHEA), the degree is divided into four academic courses that encompass a total of 38 four-month subjects (corresponding to subjects of Basic, Compulsory and Optional Training), External Practices and The Final Degree Project.

The Spanish educational policy must face the renewal that implies the need for a training offer in the Bachelor's and Postgraduate degrees in Tourism adapted to new needs (social skills, personalisation in the services, ethical conduct, dealing with people of diverse idiosyncrasies and culture, new technologies applied to personal treatment ...). This represents not only a challenge but an opportunity to adapt tourism training to the needs of the sector and to bring the university academic field closer to public tourism institutions and companies in the sector to establish synergies and improve human capital and research in tourism (Lillo, 2009). In addition to bringing the academic sphere closer to the reality of "diversity" and inclusiveness for all, in all areas of society.

According to the Curriculum Training in *Design for All People in Tourism* (2017) manual, the certification of universal accessibility in the tourism field is linked to the UNE 170001-1 standards (part 1: DALCO criteria - ambulation, apprehension, location and communication- to facilitate accessibility to the environment) and UNE 170001-2 (part 2: accessibility management system), AENOR and the ISO 9001 quality certification (planning, objectives, corrective and preventive actions) making clear the importance is given to universal accessibility to provide citizens with quality and solidarity with all.

According to Rubio (2015), there are no subjects in the curricula of the Degree in Tourism of Spanish universities whose content refers directly to Accessible Tourism or customer service with disabilities. He points out that there are two general competencies contained in the White Paper of the Degree in Tourism (2004) that could be related to customer service with disabilities: "Skills in interpersonal relationships" and "Recognition of diversity and multiculturalism".

In the current Teaching Guides, there is a greater presence of competencies related to ethical values and commitment to achieve universal accessibility of people with disabilities, democratic values and the firm recognition of diversity, along the lines defined by Castañeda (2017) regarding the city as a space for dialogue. However, there is hardly a real practice that attracts the attention and interest of the student, probably due to the lack of an approach according to emotional education as proposed by Benítez (2014).

Higher education as well as infant, primary or secondary education, needs scenarios in which the participation of the student in their process of human development in all its dimensions (comprehensiveness and participation), adapting to its changing realities (flexibility), in conditions of constant exchange with others (interculturality). On the other hand, university education should comply with the principles of integrality, participation, flexibility and interculturality so that it is of quality, fair and effective. Educational quality is described as the set of conditions that allow the permanent improvement of education at all levels. Thus, it can be said that quality is flexible because the circumstances and factors that surround it are also flexible. Thus, the revision of the Teaching Guides and their improvement is essential to achieve the proposed objectives.

The specific competencies must be internalised and focused on theatricalisation (interpretation and staging) of real situations that will help the student empathise with a large and extremely important group for the tourism sector. From here you can consider the need to train students in Emotional Intelligence and that the degree tries to address emotional skills to better understand consumer behaviour through their emotions in order to meet their desires and demands (Barrientos et al., 2019).

FUTURE RESEARCH DIRECTIONS

It is proposed as a future line of research to analyse the specific subjects and topics on disability that are taught in the degrees of tourism in Spain and subsequently carry out a comparison between Autonomous Communities and other community counties. Another line of work no less interesting is information and communication through social networks and social media on inclusive tourism. It would be more than interesting to continue in this line of research and deepen the weight that inclusive education has within the teaching guides of the degree in tourism.

Derived from the above, it is more than evident that the current training model must be analysed in order to discover whether or not it adapts to the current needs and challenges posed by the tourism sector, such as the competitive commercialisation of ethics (Barquero, 2017) or its adaptation to new teaching realities.

In addition, it is proposed to add specific and general competences as a transversal axis in the tourism degree to work on the inclusion of issues related to Universal Accessibility and *Design for All People.*

CONCLUSION

Accessible Tourism and Inclusive Tourism are contemplated as a professional opportunity for people with functional diversity who, with university training applied under certain guidelines, may work for others or be entrepreneurs, which could enrich the tourism industry, contributing to an increase in the volume of business dependent on the offer mentioned by Millán et al. (2016, p. 94) and avoiding the economic reductionism of interpretations such as César & Arnáiz (2016, p. 334).

In 2014 the *Declaration of San Marino* was published where they prayed seven appeals that meant the firm and safe start in the race – not free of obstacles –, in favour of the solidarity of the people:

1. Undertake awareness and training activities aimed at removing barriers due to the attitude of tourism and heritage professionals at all levels and persuading the private sector of the competitive advantages of investing in "tourism for all."
2. Ensure that global economic uncertainties do not undermine the right to the tourist experience and that the spirit of solidarity and international cooperation is further maintained and promoted to improve accessibility.
3. Observe the principles of universal design by creating new infrastructure, tourism products and services and improving existing ones, especially in the case of cultural heritage of tourist interest.
4. Streamline the development and application of "smart" technologies to provide objective information about all services (regardless of the degree of accessibility), in order to allow travellers to assess their own personal requirements regarding accessibility and, if necessary, and with the participa-

tion of civil society organisations, propose concrete solutions to improve accessibility and quality of tourism services.

5. Systematically incorporate universal accessibility in all components of the tourism value chain, including the physical environment, the transport system, information and communication channels and a full spectrum of public facilities and services.
6. Promote and disseminate good practices and real case studies in the field of universal accessibility in travel and tourism.
7. Strengthen cooperation with all competent entities working in the field of universal accessibility at international and regional levels, and in particular the European Union, and participate in public-private partnerships with the travel and tourism sector, Non-governmental entities and organisations of persons with disabilities for the formulation of new global policies in the field of universal accessibility.

In 2016, manuals on *Accessible Tourism* by the UNWTO, ENAT and ONCE Foundation, were published. They recognise profound and diverse changes in society and include that the university must collaborate directly in the training of professionals who know the special needs and are able to design activities for all people.

As published in Curriculum Training in *Design for All People in Tourism* (2017) it is found that: The University of Almería in its Tourism Degree introduces the optional subject "Disability Care", corresponding to the Psychology Study Plan (Plan 2010) and is offered in 25 degrees. The Complutense University of Madrid addresses Universal Accessibility in Tourism within the topics "Challenges of tourism today" and "Universal Accessibility in tourism", in the subject "Sociology of Tourism and Leisure". The optional "Unit: Aging and Disability" (6 ECTS), taught by the Faculty of Social Work of the same University, includes among its descriptors "Leisure and free time and Accessible Tourism". The European University Miguel de Cervantes includes in the Degree in Physical Activity and Sports Sciences the subject "Physical Exercise and Disability", the Degree in Architecture at the University of Alcalá de Henares incorporates the subject "Universal accessibility in urban environments". And the Degree in Fundamentals of Architecture of the Polytechnic University of Cartagena has in its curriculum with the optional subject of Universal Accessibility.

They are clear examples of willingness and collaboration with the citizen and, therefore, with society. Based on Organic Law 4/2007, of April 12, which modifies the Organic Law of Universities (LOMLOU), and stresses that: all curricula proposed by universities must take into account that training in any professional activity It must be done from the respect and promotion of Human Rights and the principles of Universal Accessibility and Design for All.

After analysing the legal framework and the reality that occurs in the curricula of the Degrees in Tourism in Spain, the adaptation of the Teaching Guides towards the promotion and theoretical and practical knowledge of Accessible Tourism and Inclusive Tourism is suggested. In the same way that the advancement of technologies is causing new challenges, subjects, methodologies and knowledge tests to be promoted, it is also necessary to train professionals for customer service with various needs and for the management and planning of products and services Tourist for everyone. Globalisation, the ease of being able to travel, and technological progress, have diversified the supply and demand of tourism products and services, and, therefore, the type of customer that demands them has diversified. Training students taking these factors into account could guarantee everyone's solidarity and cooperation with everyone.

As previously mentioned, there are certifications for accessible destinations, products and services with the aim of classifying them and facilitating their recognition through social networks, websites, training schools and the university itself. Private companies act under the mandatory regulations for physical disability, leaving the rest of the sensory or intellectual disabilities in oblivion. The availability of adapted facilities in tourist establishments is essential for the feedback received from customers, their loyalty and the resulting competitive and economic benefit.

Training students in values, emotional intelligence, emotion management and attending to the needs of people with disabilities is a premise to normalise the characteristics and abilities of each one (Barrientos, 2019). The article Economic profitability of accessible tourism for the tourism sector in Spain (Domínguez, et al., 2017) points out that customer service must be professionalised in specific markets, and normalise the heterogeneous capabilities found. Updating and completing the Teaching Guides of the Degree in Tourism innovating in practical exercises, with tools that facilitate teamwork, would provide students with a solidary and civic attitude towards others.

During the study period, the inclusion, awareness and training of future graduates must be emphasised so that they are able to create and adapt tourism resources for all. Otherwise, ethics would be lacking indirectly infringing on discrimination between equals. It is about guaranteeing equal opportunities (Ortiz, 2010). There is a clear absence of training applied to the practice of inclusive tourism. In turn, professional experts have appeared in: manager of accessible tourism and inclusive tourism; manager of basic conditions of accessibility and non-discrimination in the means of transport and manager of accessible destinations among others.

As a limitation in the present study is the lack of research before that carried out on this specific topic. Societies today are betting (some more than others) on inclusion not only academic but also to be present in all moments and circumstances of people´s lives.

REFERENCES

Alonso, M. H., Gonzálvez, J. E., & Bartolomé, A. (2016). Ventajas e inconvenientes del uso de dispositivos electrónicos en el aula: Percepción de los estudiantes de grados en comunicación. *Revista de Comunicación de la SEECI*, *20*(41), 136–154. doi:10.15198eeci.2016.41.136-154

Altamirano, V. P., Túñez, J. M., & Valarezo, K. P. (2016). Promoción turística 2.0. Análisis de las redes sociales de los gobiernos iberoamericanos. *Revista Opción*, *32*(9), 32–53.

Barquero, M. (2017). La ética en el sector del alojamiento turístico como Fuente de ventaja competitiva. *Revista de Comunicación de la SEECI*, *42*(42), 119–130. doi:10.15198eeci.2017.42.119-130

Barranquero, A. (2012). De la comunicación para el desarrollo a la justicia ecosocial y el buen vivir. CIC. *Cuadernos De Información Y Comunicación, 17*, 63-78. doi:10.5209/rev_CIYC.2012.v17.39258

Barrientos-Báez, A. (2019). Teoría del Triple Rol: Inteligencia Emocional, Turismo y Educación. *Cuadernos Artesanos de Comunicación, 158.*

Barrientos-Báez, A., Barquero, M., & Rodríguez, J. (2019). La educación emocional como contenido transversal para una nueva política educativa: El caso del grado de turismo. *Revista Utopía y Praxis Latinoamericana*, *24*(4), 147–165.

Benítez, L. (2014). Educación emocional para la comunicación transcultural. *Perspectivas de la Comunicación, 7*(1), 7–24.

Bernal, A. I. (2010). Definición conceptual de los medios de comunicación por un grupo de jóvenes españoles. El valor de Internet. *Vivat Academia. Revista de Comunicación, 112*, 22–32. doi:10.15178/va.2010.112.22-32

Blanco, R. (2006). La Equidad y la Inclusión Social: Uno de los Desafíos de la Educación y la Escuela Hoy. REICE. *Revista Electrónica Iberoamericana sobre Calidad, Eficacia y Cambio en Educación, 4*, 1–15.

Booth, T. (2000). *Progreso en la educación inclusiva. Estudio Temático para la evaluación de educación para todos*. UNESCO.

Booth, T., & Ainscow, M. (2004). *Índice de Inclusión: Desarrollando el aprendizaje y la participación en las escuelas.* UNESCO/OREALC.

Caldevilla Domínguez, D., García García, E., & Barrientos-Báez, A. (2019). La importancia del turismo cultural como medio de dignificación del turista y de la industria. *Mediaciones Sociales, 18*, 59–69. doi:10.5209/meso.65117

Castañeda, M. (2017). La Ciudad Como Escenario De Diálogo. Apuntes Desde La Comunicación Intercultural. *Perspectivas de la Comunicación, 10*(2), 119–142.

Castleton, A. (2017). Tecnología e identidad: El caso de los Inuit y Facebook. CIC. *Cuadernos de Información y Comunicación, 22*(0), 107–125. doi:10.5209/CIYC.55970

César, F., & Arnáiz, S. M. (2016). Capitalismo, sociedad y turismo. *Revista Opción, 32*(13), 319–338.

Ciarrochi, J., Chan, A., & Bajgar, J. (2001). Personalidad y diferencias individuales. *Elsevier, 31*(7), 1105-1119.

Correyero, B. (2003). La propaganda turística española en los años del aislamiento internacional. *Historia y Comunicación Social, 8*, 47-61. https://revistas.ucm.es/index.php/HICS/article/view/HICS0303110047A

Díaz del Campo, J. (2012). Objetivos pedagógicos básicos en la enseñanza ética de la comunicación. *Vivat Academia. Revista de Comunicación, 15*(121), 1–16. doi:10.15178/va.2012.121.1-16

Domínguez, T., Fraiz, J., & Elisa, A. (2017). Economic profitability of accessible tourism for the tourism sector in Spain. *Tourism Economics, 19*(6), 1385–1399. doi:10.5367/te.2013.0246

Echeita, G., & Ainscow, M. (2010). *La Educación inclusiva como derecho. Marco de referencia y pautas de acción para el desarrollo de una revolución pendiente.* En II Congreso Iberoamericano de Síndrome de Down. https://www.webcitation.org/6vTVhQziA

Extremera, N., & Fernández-Berrocal, P. (2003). La Inteligencia Emocional en el contexto educativo: Hallazgos científicos de sus efectos en el aula. *Review of Education, 332*, 97–116.

Formación Curricular en Diseño para Todas las Personas en Turismo. (2017). *Crue Universidades Españolas.*

Guzmán, T., Borges, O., & Castillo, A. M. (2011). Desarrollo económico local y turismo comunitario en países en vías de desarrollo. Un estudio de caso. *Revista de Ciencias Sociales, 17*(3), 432–446.

Infante, E. (2014). Guerra, historia, turismo y prensa: bases de la puesta en valor del patrimonio cultural de Niebla (Huelva) durante el franquismo. *Historia Y Comunicación Social, 19,* 153-170. doi:10.5209/rev_HICS.2014.v19.47290

Janeiro, P., Proença, I., & Gonçalves, V. C. (2013). Open Innovation: Factors explaining universities as service firm innovation services. *Journal of Business Research, 66*(10), 2017–2023. doi:10.1016/j.jbusres.2013.02.027

Jiménez, I. (2009). *Una aproximación psicosocial a la relación entre Inteligencia Emocional y rendimiento académico*. Universidad de Jaén.

Marta, C., & Gabelas, J. A. (2013). Hábitos de consumo televisivo de ficción entre los universitarios que estudian Comunicación. *Revista de Comunicación de la SEECI, 17*(31), 14–33. doi:10.15198eeci.2013.31.14-33

Millán, G., Pérez, L. M., & Martínez, R. (2016). Factores que determinan el crecimiento del turismo en destinos religiosos. *Revista de Ciencias Sociales, 22*(1), 85–97.

Molina, R. (2014). En tránsito hacia la educación inclusiva en las Instituciones de Educación Superior de Colombia: Buenas prácticas. Revista Inclusiones, 12-22.

Munuera, M. P. (2015). Inclusión sociolaboral de las personas con discapacidad. Mediación laboral. *Mediaciones Sociales, 14*(14), 161–179. doi:10.5209/rev_MESO.2015.n14.51564

Ortiz, M. (2010). Ineficacia de las medidas para garantizar la igualdad de oportunidades. *Revista Iberoamericana de Educación, 51*(3), 1022–6508. doi:10.35362/rie5131829

Premio de "Q de Calidad Turística" por su labor en el turismo accesible. (n.d.). https://www.predif.org/index.php?q=predif-premio-q-de-calidad-tur%C3%ADstica-por-su-labor-en-el-turismo-accesible

Rubio, M. J. (2015). Formación en materia de accesibilidad en el sector turístico. *Estudios turísticos,* 203-204.

Sánchez, M., Fernández-Berrocal, P., Montañés, J., & Latorre, J. (2008). Does emotional intelligence depend on gender? The socialisation of emotional competencies in men and woman and its implications. *Electronic Journal of Research in Educational Psychology, 15,* 455–474.

Sandoval Martín, M. T. (1998). Promoción turística a través del sector audiovisual. El caso de Canarias. *Revista Latina de Comunicación Social, 9.* https://mdc.ulpgc.es/utils/getfile/collection/rldcs/id/158/filename/330.pdf

Silva, R. (2009). Balanced scorecard – BSC: gestão de esnino superior, gestão profissionalizada e qualidade de ensino para instituições de ensino superior privado. *Juruá, 186.*

Suberviola, I. (2011). Auto-percepción del profesorado sobre su formación en educación emocional. Vivat Academia. *Revista de Comunicación, 14*(117), 1154–1167.

Trinidad, D., & Johnson, C. (2002). The association between emocional intelligence and early adolescent tobacco and alcohol use. *Personality and Individual Differences*, *32*(1), 95–105. doi:10.1016/S0191-8869(01)00008-3

Túñez López, M., Altamirano, V., & Valarezo, K. P. (2016). Comunicación turística colaborativa 2.0: Promoción, difusión e interactividad en las webs gubernamentales de Iberoamérica. *Revista Latina de Comunicación Social*, *71*, 249–271.

UNESCO. (2008). *La educación inclusiva: el camino hacia el futuro.* https://www.webcitation.org/6vTY885Ay

UNWTO. (2014). Manual sobre Turismo Accesible para Todos: principios, herramientas y buenas prácticas. *Módulo I: turismo accesible - definición y contexto.* https://www.e-unwto.org/doi/book/10.18111/9789284416486

Valenciano, G. (2009). *Construyendo un concepto de educación inclusiva: una experiencia compartida.* Instituto Universitario de Integración en la Comunidad/ Colección Investigación.

Yazzo, M. A. (2015). Educación inclusiva superior. Una mirada desde la acreditación de calidad en Colombia. *Revista Inclusiones*, *2*(4), 317–326.

KEY TERMS AND DEFINITIONS

Accessibility: Accessibility is the set of characteristics and factors that facilitate communication, access and enjoyment for all citizens in the world.

High Education: It is the education that is taught at the university to complete the academic stage of the student including professional and formal aspects with the aim of labour inclusion.

Inclusive Education: Is that education based on principles of solidarity and teaching for all. Especially for those who could suffer some type of social exclusion due to their disability, culture or social condition.

Inclusive Tourism: Inclusive tourism is the total of activities, services and tourist products that allow their enjoyment by and for all regardless of the person´s disability.

Study Plans: A study plan is a template that summarises everything that will be taught in certain areas or subjects.

Teaching Guides: A didactic or teaching guide is a document that admits different possibilities of a specific subject for the correct learning, development and understanding by the student.

Tourism Degree: The degree in tourism enables the student to enter the job market in the tourism sector by learning specifically about the creation, management, direction, insertion and distribution of tourism products and/or services. In Europe, it is a university career structured in 4 academic courses made up of 60 credits (ECTS) each.

University: The University is that public or private institution of higher education made up of different specialisations or faculties that grant degrees based on the learning and development of subjects taught over 4 academic years.

Chapter 19
Managing Partnerships Between Tourism and Hospitality Stakeholders:
The Case of Health and Wellness Products in Portugal

Ricardo Jorge da Costa Guerra
Polytechnic Institute of Guarda, Portugal

Eduardo Cordeiro Gonçalves
University Institute of Maia, Portugal

Vítor Figueiredo
https://orcid.org/0000-0003-2965-0050
University of Beira Interior, Portugal

ABSTRACT

This study presents a reflection on a new and resized (re)interpretation of the thermal spa sector in Portugal. Taking a holistic view of the strategic health and wellness tourism product, with the example of a case study (thermal spas in Portugal), it materialises a new proposal for a strategic management model. The study analyses the perspective of networking, which characterises a good part of the sector. Special attention is given to the training of organisations and their connection to integrated strategies to make the best use of resources, to the distinguishing orientation of preparation and placement of the tourism product on the market. The aim was to listen to thermal spa managers in Portugal regarding complementary services associated with their core offer, the partnership relationships they have with local stakeholders and what type, as well as the motivations for networking and development perspectives of the thermal spa sector in Portugal. The results presented are incorporated in this guideline to establish a competitive offer.

DOI: 10.4018/978-1-7998-5691-7.ch019

INTRODUCTION

Networks of businesses for competitiveness are a current reality and can be found in all sectors of activity (Lui & Ngo, 2005; Street & Cameron, 2007; Mazzarol et al., 2013; Marques & Franco, 2020). Alliances or networks occur when two or more organisations decide to join their resources and efforts to achieve common goals (Mendonça et al., 2015). Tourism is no exception, particularly because of the considerable amount of complementarity involved and the atomisation of supply (Chim-Miki & Batista-Canino, 2018; Islam et al., 2018). Setting up collaborative networks or even coopetitive networks is seen as fundamental to building consistent images of tourist sites, to sell products, to ensure a unique consumption experience or even to increase the competitiveness of companies and destinations, thereby creating added value for visitors and making it more likely that they will stay (Cortese et al., 2018; Karampela et al., 2019; Kvasnova et al., 2019).

Furthermore, another advantage of partnerships and networks among the actors involved in the supply side is that they enable destinations to sell service packages and experiences in an integrated and aggregated manner, which makes them more attractive to visitors and thus enhances competitiveness and encourages the building of a positive image for destinations (Marcoz et al., 2016).

Hospitality and restaurants/cafés, as core components of the tourism offer, play a fundamental role in driving and managing networks, particularly when health and wellness products are at hand since they normally involve longer stays in comparison with other tourism products.

This study aims to use interviews with thermal spa managers to understand how they view the importance of partnerships, specifically with hotels and restaurants. In order to do this, a theoretical context is provided, followed by the method used. Subsequently, the results are analysed and discussed. The study ends by presenting conclusions, limitations, implications, and suggestions for future research.

THEORETICAL CONTEXTUALISATION

Building a network-based offer has numerous advantages, despite the complexity of this task, as Islam, Ruhanen and Ritchie (2018) point out, primarily because there are often tense relationships between governments, local authorities, rich elites and local residents, which are exacerbated when the destinations are protected areas. This complexity and tension are explained, at least in part, by the studies of van der Zee et al. (2017), when they state that tourist destinations are complex systems of interrelated entities, where there is no chain of command. Despite this, the entities (or stakeholders) operating at destinations acknowledge the importance of networks and cooperation and mention the need for them to obtain training in techniques for cooperation, identification of benefits and trust-building because, despite sharing common objectives, difficulties and barriers to creating networks persist (Karampela et al., 2019; Cortese et al., 2018). It is, therefore hoped that stakeholders are able to reach a balance between individual competition and cooperation, for everyone's benefit (Gajdosik et al., 2017).

Authors such as Chim-Miki and Batista-Canino (2018), Karampela et al. (2019), and Kvasnova et al. (2019) suggest that successful networks and cooperation increase the quality of tourists' experiences and contribute to destinations' sustainability and performance. Moreover, they encourage development in tourism and the participation of private entities in the process. Networks and cooperation are seen as the bases of developing tourism and sources of competitive advantages. These authors make destinations' competitiveness depend on partnerships and networks between the public, private, and non-profit sectors and communities. Tourist destinations should, therefore, be managed as an integrated co-production network, in other words, as a system that combines individual resources, products, and services to be offered to tourists (Cortese et al., 2018).

Even when discussing coopetition models in tourism, cooperation emerges as one of seven factors to be taken into consideration (Chim-Miki & Batista-Canino, 2018). Cortese et al. (2018), on the other hand, point out innovation as a crucial tool to meet tourists' aspirations and needs and ensure destinations' economic sustainability, giving networking as a fundamental factor, since it allows actors and their resources to be organised to capitalise on them with added benefits for all stakeholders.

For example, in the case of agrotourism, effective performance by organisations and the achievement of economies of scale and revenue is impossible without the existence of networks and trust among actors involved in the offer. To be successful, networks should have a range of critical features: stable funding; involvement of the public, private and non-profit sectors; awareness of the independence of each actor; trust, communication, ability to think long-term, expertise, knowledge, and entrepreneurial direction and professionalism (Cortese et al., 2018, Kvasnova et al., 2019).

This networked management of tourist destinations makes it possible to share information, knowledge, technology, and other relevant resources to find new solutions to leverage tourists' experience. At health and wellness destinations, hospitality and restaurants/cafés play a fundamental role in stimulating and implementing networks and in achieving competitive advantages over competing destinations (Marcoz et al. 2016). As we know, there are currently several interpretations of concepts connected to this theme, often according to geographic context, particularly North American and/or European (Goodrich, 1994). The health and wellness tourism offer has seen significant development and diversification, something which makes it possible to integrate a range of public and solidify that corresponding variation in concepts. Additionally, with the expansion and privatisation of fields of health, there has been a broad diversification in services and spaces where leisure and health come together, the importance of which has led to the adoption of the term health tourism and/or health and wellness tourism (Gustavo, 2010; Guerra, 2019). The PENT report – *Plano Estratégico Nacional do Turismo, Revisão e Objetivos 2013-2015* (National Strategic Plan for Tourism, Revision and Goals 2013-2015) – within the ten strategic products for Portugal, defines health tourism as a feature bringing together medical, thermal, spa and thalassotherapy components (Turismo de Portugal, 2013). A similar slant is applied in the planning of *Estratégia Turismo 2027* (Tourism Strategy 2027) (Turismo de Portugal, 2017), which presents it as an emerging asset. Health and wellness tourism, therefore, has the profile of a composite and complex tourism product encompassing multiple subproducts, specifically medical tourism, beauty tourism, thalassotherapy, spas, climate-related health tourism, health and wellness resorts, residences and, in particular, the resizing of thermal spas as a whole.

With the exponential growth of this market and its corresponding specialisation, activity has revealed itself to be multifaceted (Gustavo, 2010) and is sited on the boundaries of areas of medicine, health, and leisure, revealing a sense of interdependence in order to meet the motivations of customers who seek out these spaces (Guerra, 2018). The concept has been evolving since the World Health Organization's classification in 1948 that defines health as a complete state of physical, mental, and social wellbeing, and not simply the absence of disease or infirmity (WHO, 1948). In turn, in 1972, the International Union of Official Travel Organisations (today's UNWTO – World Tourism Organisation) advanced the first definition of health tourism, a concept entwined with the use of health facilities that make use of natural, climatic and thermal resources. In 1981, the same organisation updated the concept and recognised that health tourism programs should provide tourists with an improvement in their health level in a primarily preventative sense. This means that places are included where natural resources are considered tourist destinations and their users tourists, encompassing both cure and prevention.

We still have a large range of concepts concerning the nomenclature used, and this has been increasingly clarified by several authors, such as Voigt (2010), who reviews the different terms used, such as *wellness tourism* (e.g., Nahrstedt, 2004; Smith & Puczkó, 2008; Steiner & Reisinger, 2006), *health tourism* (Douglas, 2001; Hall, 2003), *health-care tourism* (Henderson, 2003), *wellbeing tourism* (Inside Story, 2007), *holistic tourism* (Kelly & Smith, 2009; Smith, 2003), *spa tourism* (Puczkó & Bacharov, 2006), *thermal tourism* (Erfurt-Cooper & Cooper, 2009; Radnic et al., 2009) and *medical tourism* (Connell, 2006). The emergence of new concepts connected to the subjects' responsibility for the management of their health can be observed (Nossa & Caldeira, 2014). Alongside the doctrine of self-accountability, concepts such as *wellness* emerged, which Cunha (2006) notes as the result of combining the concepts of *wellbeing* and *fitness* (Erfurt-Cooper & Cooper, 2009). As a rule, definitions of health tourism diverge considerably, with some focusing on the supply side and some on the demand side. In a recent document, the UNWTO established the concept of health and wellness tourism as including types of tourism that have the primary aim of contributing to physical, mental and/or spiritual health following medical and wellbeing action that develop people's ability to satisfy their own needs and better fit into their social environment (UNWTO, 2019).

The importance of health and wellness tourism for tourist destinations, society, the economy, culture, and experiences marks it out as a tourism product with the potential for development (Ramos, 2005). Health and wellness tourism also helps lessen regional asymmetries and generate more investment, whether directly in the core activity or indirectly in support activities, creating a multiplying effect that benefits local economic agents (Ramos, 2008; Fazenda et al., 2009).

The concept of cross-selling can also be observed, along with a diversification of the offer and other products complementing it. The development model for each unit should be integrated with the different products found at the destination in a global common offer. Support services for the main activity, such as accommodation, restaurants/cafés, transport, and other services, which are not primary aspects in a choice of destination, set high-quality standards. At these types of destinations, health and wellness tourism units are the core product and other local agents should adapt to the needs of this type of visitor (Guerra, 2018).

Portugal has a wealth of water resources defined by a network of mineral waters that have numerous therapeutic properties and are richly mineralised, as in the case of natural mineral springs (Acciaiuoli, 1944; Cantista, 2010). The territories where thermal springs are located have seen, throughout history, examples of the use of these spaces for health and leisure (Ramos, 2005).

Thermal spas in Portugal have seen several high and low points, which have led in many cases to the disappearance and closure of various thermal spa resorts. Today, the concept has a broader scope with the inclusion of wellness recognised by Decree-Law no. 142/2004 of 11 June 2004. The recent (re)attribution of Social Security copayment for National Health Service (SNS of Portugal) users has breathed new life into the sector, as has the recognition of the treatments' therapeutic goals. In recent years, many thermal parks have been renovated in Portugal, while products and services have been adapted to new market trends.

According to recent data presented by the Portuguese Thermal Spa Association (ATP), Portuguese thermal spas saw 40,000 customers for therapy programs in 2018, representing more than 420,000 days of treatment. About wellness programs, more than 87,000 customers were recorded, representing more than 106,000 days of thermal practices. According to ATP, the sector's current priorities in Portugal are to create value, promote health and a healthy lifestyle, quality and innovation, and also research and development.

METHOD

According to Quivy and Campenhoudt (1998), the data collection process should have the following characteristics: accurate, properly defined, able to be applied again under the same conditions, and adapted to the type of problem and phenomena in question. A mixed methodology was used for the collection of quantitative and qualitative data. This appeared to us to be the most suitable choice to achieve the goals in mind, with a thorough document analysis also being carried out in accordance with the themes defined for the study and complemented with primary data collection on the Portuguese thermal spa offer. The combination of these two methods was essential for defining the relevance and reach of the study.

The use of the method applied was underpinned by the results of the questionnaires addressed to Portuguese thermal spa managers. In 2014, there were a total of 38 thermal spas in operation (DGEG, 2019). Application of the data collection tool covered all of that universe, and a questionnaire was drawn up aiming at measuring the variables based on the questions contained in it, as well as using a scale to measure answers to test operational hypotheses (Hill & Hill, 2005).

Dissemination of the questionnaire involved the cooperation of the Portuguese Thermal Spa Association and was sent to the relevant spa managers' email addresses. Twenty-four answers were received from spa managers, corresponding to 63% of those who were sent questionnaires. The aim was, therefore, to hear from spa managers in Portugal about services complementing the core offer and about partnership relationships established with local stakeholders in a coopetition rationale. The goal was also to find out about the motivations for networking and future perspectives for the development of the thermal spa sector in Portugal. The results of the data collection were analysed using SPSS statistical tests and content analysis techniques in order to ensure validation of the problem and categorisation of the data, making it comprehensible and interpretable.

This questionnaire survey is a useful instrument for identifying and understanding the organisation, management, and promotion models at Portuguese thermal spas and their perception about the thermal spa sector, particularly the management of partnerships with tourism and hospitality stakeholders. The results presented are within this guideline to achieve common goals, following a rationale of complementarity in the use of resources, but also to solidify a competitive offer.

RESULTS

Twenty-four Portuguese thermal spa managers were heard in this research, who replied to several questions relating to the need for good management of partnerships with different stakeholders that have interests in tourist development in their areas. The greatest number of replies were from the Center Region (12), followed by the North Region (9), with two thermal spa resorts in the Alentejo (2) and a single resort located in the Algarve.

It is important, therefore, to understand management models at each thermal spa resort, with a factor of analysis relating to possible development perspectives at each resort and the dynamics embedded in terms of establishing partnerships. Table 1 shows that more than half (14) of the thermal spa resorts surveyed have a private management model, seven have a public management model and three have a mixed public-private model.

Table 1. Distribution of surveyed thermal spa resorts by management model (the authors, 2020)

Management model	N	%
Public	7	29.2%
Private	14	58.3%
Public/private	3	12.5%
Total	24	100.0%

Most of the surveyed thermal spa resorts (71%) are responsible for managing more tourist services connected to thermal spas, particularly accommodation, restaurants/cafés, tourist entertainment, among others. However, we can highlight the fact that seven thermal spa resorts only manage the baths themselves, although there may be services offered by other private or public initiatives available at those locations.

Accommodation and restaurant/café services are the most common ones to be connected to thermal spa baths, whether under integrated management or through partnerships and networks between separate companies. Other services include connected tourist entertainment and museum spaces (Table 2).

Table 2. Distribution of surveyed thermal spa resorts by type of connected tourist services (the authors, 2020)

		N	%
Connected accommodation	No	4	23.5%
	Yes	13	76.5%
	Total	17	100.0%
Connected restaurants/cafés	No	3	17.6%
	Yes	14	82.4%
	Total	17	100.0%
Connected tourist entertainment	No	12	70.6%
	Yes	5	29.4%
	Total	17	100.0%
Connected museum	No	15	88.2%
	Yes	2	11.8%
	Total	17	100.0%

Figure 1. Local stakeholders who have partnership relationships with the surveyed thermal spa resorts (the authors, 2020)

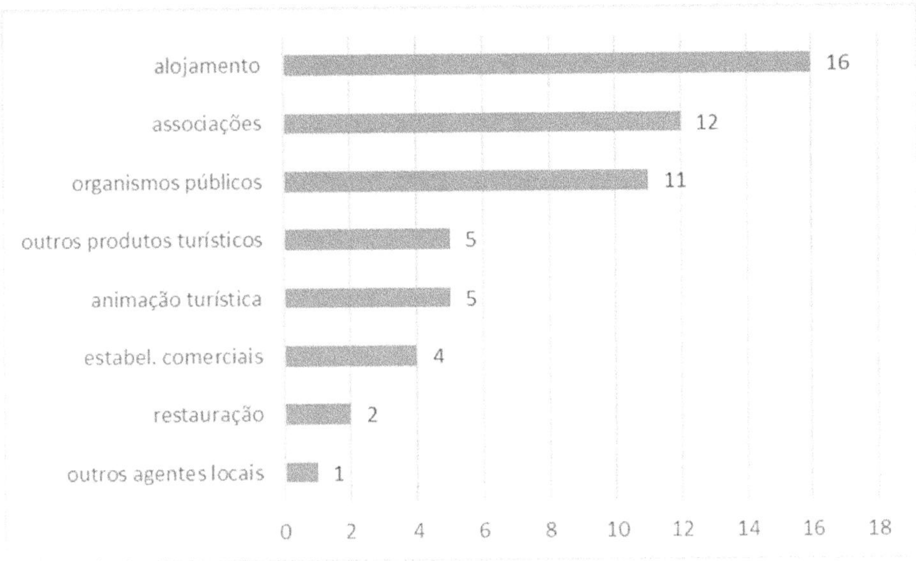

The vast majority of thermal spa resorts surveyed (82.6%) stated that they have partnership relationships with other local stakeholders, with whom they develop complementarity agreements. Figure 1 shows the types of partnerships and the number of each type.

The accommodation sector is the type of local stakeholder that is most represented, bearing in mind that almost half the thermal spa resorts do not manage this type of offer. Associations and public bodies are important driving forces in local development and are natural partners for thermal spa resorts. There is a clear trend towards cross-selling with other tourist products that complement the existing offer at thermal spa resorts. The small number of partnerships with local restaurants/cafés is also notable.

When questioned about the types of partnerships that exist between the thermal spa resorts surveyed and local stakeholders, almost half the partnerships (44%) are connected to agreements to provide commercial discounts, 31% to services complementing the existing offer, and 22% connected to advertising/dissemination.

The term thermal spa cluster corresponds to a structure of several thermal spa resorts that are geographically close to one another and create a common, attractive and competitive offer to complement their offers and what is offered by the others. According to the perception of Portuguese thermal spa managers/directors, the vast majority (74%) believe that investing in structuring a thermal spa cluster is feasible and desirable, but on the condition that it is accompanied and monitored by a strategic plan, while around 22% of those surveyed are also in no doubt that it would be an advantage for destinations and tourist agents, although they do not mention the need for a common strategic plan. This data is justified by the fact that an increase in the size of the offer is considered to be very important for it to be solidified, to make use of synergies, to attempt to achieve complementarity with different types of treatments and therapeutic instructions and also to enhance the sector's credibility. Another important point to mention is the fact that through investment in an attractive thermal spa cluster, it is possible to be better prepared to attract the international market and therefore receive customers of other nationalities.

Figure 2. Perception of managers/directors of the surveyed thermal spa resorts about whether or not the thermal spas should create a common offer when present at the same destination (the authors, 2020)

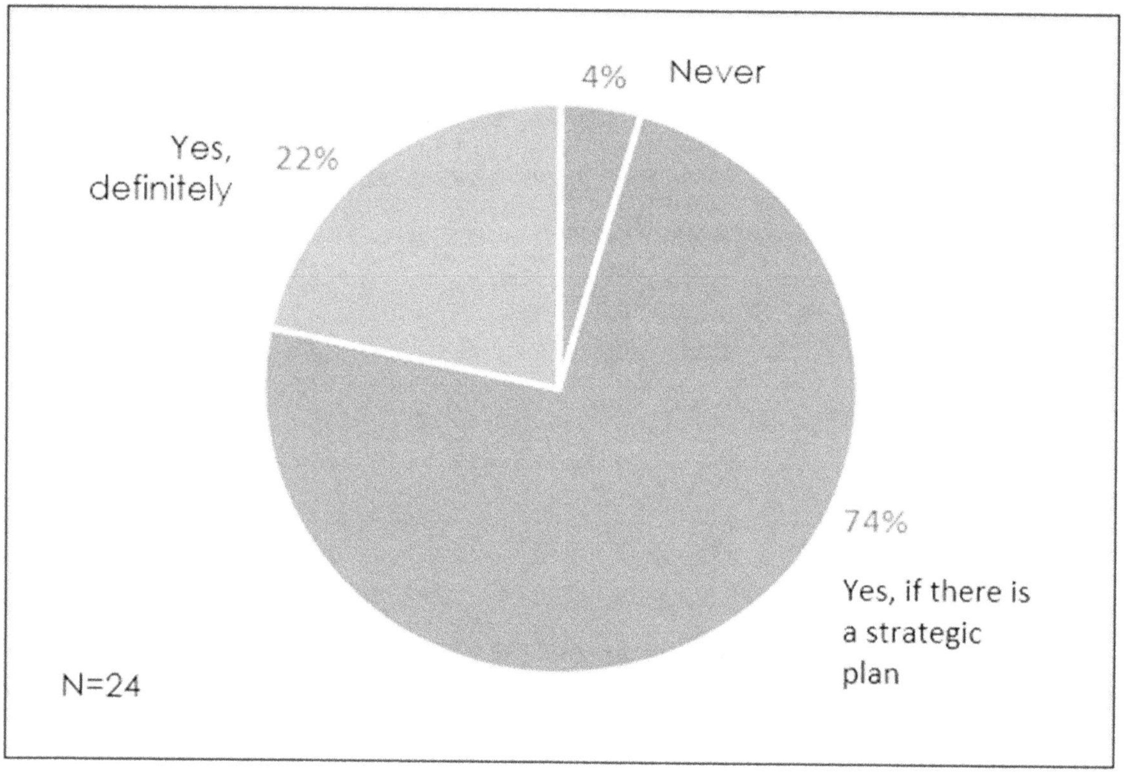

The managers/directors of Portuguese thermal spas were also questioned about their perspective on the future of the thermal spa resorts they manage (on a scale of 1–Very Poor to 10–Excellent), demonstrating clear dispersion with answers ranging between 3 and 8, although with most answers going to 7 and 8. The most optimistic respondents emphasised the benefits of treatments using thermal spa waters, the ability to innovate and establish partnerships and work in networks and complementarity as features that aim to develop their potential. When pointing out the main competitiveness factors to be explored or improved, they again highlight the ability to stimulate networking, alongside better promotion and dissemination of the sector, improvements in the service provided, human resource training, and even innovation and diversification in the offer.

CONCLUSION

Health and wellness tourism can be an important local driver and creates investment and jobs. As a rule, these destinations are found in non-urban areas and help encourage people to stay in the area, primarily contributing to reducing regional asymmetries and encouraging the emergence of more direct and indirect investment in support activities, creating a multiplying effect that benefits agents in the area. Cross-selling with other tourism products can be an important complement for health and wellness tourism to provide an increase in the quality of destinations' performance and tourist experiences, as mentioned in the literature review (Cortese et al., 2018). The respondents acknowledge that partnerships favourable leverage competitiveness, the image, and choice of health and wellness tourist destinations, as defended by Chim-Miki & Batista-Canino (2018), by making it possible to integrate core resources and attractions, support resources and factors, policies, planning, and management, which have a determining impact on demand, taking into consideration the sustainability and competitiveness of companies and destinations.

Along with the exploratory component of this research, the empirical study revealed that respondents believed that the development of Portuguese thermal spas should follow a national and regional strategic development plan, which makes it possible to foresee the role that the state and local authorities can play in drawing up those plans, as well as in the management and/or sponsorship of their implementation itself. As reported in the literature (Cortese et al., 2018), the managers recognise that local synergies should be used in a common strategy to increase the competitiveness of the respective thermal spa units and the surrounding area.

We can, therefore, suggest that Portuguese thermal spa resort managers/directors believe in the added value that comes from establishing partnerships with different local tourism and hospitality stakeholders, which improve the success of their businesses and enhance their scale to increase competitiveness and visibility in the market.

Despite acknowledging partnerships as essential and strategic to the success of all involved, the reality shows that, with certain stakeholders such as restaurants, they are still in their early stages, which may mean a lack of culture and training in cooperation or the existence of tensions among the different organisations at the destination which make cooperative work and networking more difficult. These situations have already been found in other studies (Zee et al., 2017; Karampela et al., 2019; Cortese et al., 2018). Nonetheless, we must bear in mind that 71% of the thermal spa managers surveyed manage connected services, with the greatest number involving accommodation and restaurants/cafés. This situation may be explained by the existence of tensions with other local actors that block the development of partnerships with external companies operating in the same sectors.

In reality, unlike what Chim-Miki & Batista-Canino (2018) tell us when presenting coopetition in tourism as a factor for success, stakeholders in thermal spa tourist destinations in Portugal appear to have difficulties cooperating with competing organisations, with the study indicating greater ease in establishing partnerships when the services and products to be offered are complementary rather than competing.

In theoretical terms, this study helps expand literature in Portugal on the current state of partnerships in health and wellness tourism. Furthermore, it reveals a set of restrictions mentioned by the respondents concerning setting up partnerships.

In practice, this research makes it possible for authorities with responsibilities for defining policy for the sector to make decisions and act, since the respondents recognise the need for integrated planning, which is currently non-existent at local, regional and national level, thereby hindering the sector's operations. Another practical implication is for companies in the sector who need to organise themselves, cooperate and establish partnerships to achieve success for the destinations as a whole and their business in particular. To do so, partnerships in health and wellness tourism, specifically thermal spa tourism, are a reality that should be enhanced and amplified, investing in coopetition and not only partnerships that group together complementary, non-competing services.

Finally, this study points out some paths to follow to improve the quality of services and products offered through investment in scientific studies about the therapeutic quality of mineral waters, defining high-quality services/certification, the improvement of communication strategies, diversification of the offer and even the modernisation of facilities.

With regard to limitations and suggestions for future research, we recognise that the questionnaire was too broad, and so it is suggested that a study be carried out on thermal spa resort managers that are more focused on the issues of coopetition and cooperation.

It would be similarly important to understand what tourists' perceptions are regarding partnerships between organisations in the sector and whether or not these contribute to their satisfaction and the quality of the tourist experience. Another path of research involves understanding why the number of thermal spas open to the public varies from year to year or why destinations with high-quality mineral waters do not have thermal spa activity. Lastly, another path is how the pandemic has changed the health and wellness tourism sector, setting out the main difficulties and the strategies used by operators to overcome them.

REFERENCES

Acciaiuoli, L. (1944). *Águas de Portugal, Minerais e de Mesa* (IV Volume). Lisboa: Ministério da Economia, DGMSG.

Alpoim, M. (2010). *Análise à procura termal*. Aveiro: UA.

Cantista, A. (2010). O termalismo em Portugal. *Anales de Hidrología Medica 2008-2010, 3*(1), 79–107.

Chim-Miki, A. F., & Batista-Canino, R. M. (2018). Development of a tourism coopetition model: A preliminary Delphi Study. *Journal of Hospitality and Tourism Management, 37*, 78–88. doi:10.1016/j.jhtm.2018.10.004

Cortese, D., Giacosa, E., & Cantino, V. (2018). Knowledge sharing for coopetition in tourist destinations: The difficult path to the network. *Review of Managerial Science*. Advance online publication. doi:10.100711846-018-0322-z

DGEG. (2019). *Mapa das Águas Minerais – October of 2019*. http://www.dgeg.gov.pt/

Erfurt-Cooper, P., & Cooper, M. (2009). Health and Wellness Tourism – Spas and hot springs. Bristol.

Fazenda, N.; Ladeiras, A.; Vieira, T.; Barbosa, D. (2009). Programa de Acção de Enfoque Temático "Turismo de Saúde e Bem-Estar" para o Porto e Norte de Portugal. In *Agenda Regional de Turismo*. Lisbon: TP.

Gajdosik, T., Gajdosiková, Z., Maráková, V., & Flagestad, A. (2017). Destination structure revisited in view of the community and corporate model. *Tourism Management Perspectives, 24*, 54–63. doi:10.1016/j.tmp.2017.07.009

Guerra, R. (2018). Turismo de Saúde e Bem-Estar: para uma estratégia de desenvolvimento local. O caso das Caldas da Cavaca. Maia: Ed. ISMAI e CEDTUR.

Guerra, R., & Gonçalves, E. (2019). "O turismo de saúde e bem-estar como fator de desenvolvimento local: Uma análise à oferta termal portuguesa". *PASOS. Revista de Turismo y Patrimonio Cultural, 17*(2), 453–472. doi:10.25145/j.pasos.2019.17.030

Gustavo, N. (2010). *Os novos espaços de lazer, turismo e saúde em Portugal: o caso dos SPAS*. Tese de Doutoramento. Universidade de Coimbra.

Hill, M. M., & Hill, A. (2005). *Investigação por questionário* (2nd ed.). Edições Sílabo.

Islam, M. W., Ruhanen, L., & Ritchie, B. W. (2018). Adaptive co-management: A novel approach to tourism destination governance? *Journal of Hospitality and Tourism Management*, *37*, 97–106. doi:10.1016/j.jhtm.2017.10.009

Karampela, S., Papapanos, G., & Kizos, T. (2019). Perceptions of Agritourism and Cooperation: Comparisons between an Island and a Mountain Region in Greece. *Sustainability*, *11*(680), 1–18. doi:10.3390u11030680

Kvasnová, D., Gajdosik, T., & Maráková, V. (2019). Are partnership enhancing tourism destination competitiveness? *Acta Universitatis Agriculturae et Silviculturae Mendelianae Brunensis*, *67*(3), 811–821. doi:10.11118/actaun201967030811

Lui, S., & Ngo, H. (2005). The influence of structural and process factors on partnership satisfaction in interfirm cooperation. *Group & Organization Management*, *30*(4), 378–397. doi:10.1177/1059601103259113

Marcoz, E. M., Mauri, C., Maggioni, I., & Cantù, C. (2016). Benefits from Service Bundling in Destination Branding: The Role of Trust in Enhancing Cooperation among Operators in the Hospitality Industry. *International Journal of Tourism Research*, *18*(3), 260–268. doi:10.1002/jtr.2002

Marques, I. C. P., & Franco, M. (2020). Cooperation networks in the area of health: Systematic literature review. *Scientometrics*, *122*(3), 1727–1750. doi:10.100711192-019-03341-3

Mazzarol, T., Limnios, E., & Reboud, S. (2013). Cooperatives as a strategic network of small firms: Case studies from Australian and French cooperatives. *Journal of Co-operative Organization and Management*, *1*(2), 27–40. doi:10.1016/j.jcom.2013.06.004

Mendonça, V., Varajão, J., & Oliveira, P. (2015). Cooperation Networks in the Tourism Sector: Multiplication of Business Opportunities. *Procedia Computer Science*, *64*, 1172–1181. doi:10.1016/j.procs.2015.08.552

Mueller, H., & Kaufmann, E. (2001). Wellness tourism: Market analysis of a special health tourism segment and implications for the hotel industry. *Journal of Vacation Marketing*, *7*(1), 5–17. doi:10.1177/135676670100700101

Nahrstedt. (2004). Wellness. In *The tourism and leisure industry* (pp. 181-198). New York: Haworth Hospitality Press.

Nossa & Caldeira. (2014). Turismo medicalizado. In Trabalho em saúde, desigualdade e políticas públicas (pp. 1-31). Braga: Univ. Minho.

Quivy, R., & Campenhoudt, L. (1998). *Manual de Investigação em Ciências Sociais*. Gradiva.

Ramos, A. (2005). *O Termalismo em Portugal : Dos factores de obstrução à revitalização pela dimensão turística*. Tese de Doutoramento. Universidade de Aveiro.

Ramos, A. (2008). Lazer, Turismo e Termalismo. Inovação e Complementaridade no Turismo de Saúde. *Revista Turismo & Desenvolvimento, 9*, 7-18.

Smith, M., & Puczkó, L. (2008). Health and Wellness Tourism. Elsevier. doi:10.4324/9780080942032

Street, C. T., & Cameron, A. (2007). External relationships and the small business: A review of small business alliances and network research. *Journal of Small Business Management, 45*(2), 239–266. doi:10.1111/j.1540-627X.2007.00211.x

Turismo de Portugal. (2013). *Plano Estratégico Nacional do Turismo - Revisão e Objetivos 2013-2015.* Lisbon: TP.

Turismo de Portugal. (2017). *Estratégia Turismo 2027.* Lisbon: TP.

UNWTO. (2019). *UNWTO Tourism Definitions*. UNWTO. doi:10.18111/9789284420858

Van der Zee, E., Gerrets, A., & Vanneste, D. (2017). Complexity in the governance of tourism networks: Balancing between external pressure and internal expectations. *Journal of Destination Marketing & Management, 6*(4), 296–308. Advance online publication. doi:10.1016/j.jdmm.2017.07.003

Voit, C. (2010). The concept of Health Tourism. An Overview. In C. Voigt & C. Pforr (Eds.), *Welness tourism. A destination perspective*. Routledge.

WHO. (1948). *Officials Records of the World Health Organization*. United Nations.

Compilation of References

Aall, C. (2014). Sustainable tourism in practice: Promoting or perverting the quest for a sustainable development? *Sustainability, 6*(5), 2562–2583. doi:10.3390u6052562

Abreu, A. (2012). Príncipe Island, UNESCO Biosphere Reserve, a tool for the sustainable development. *2nd Meeting of the World Network of Island and Coastal Biosphere Reserves.*

Acciaiuoli, L. (1944). *Águas de Portugal, Minerais e de Mesa* (IV Volume). Lisboa: Ministério da Economia, DGMSG.

Adams, W. (2008). *Green Development, Environment and sustainability in the Third World.* Academic Press.

Adams, D. (2009). *A social inclusion strategy for Tasmania.* Department of Premier and Cabinet.

Adams, S. M., Gupta, A., Haughton, D. M., & Leeth, J. D. (2007). Gender differences in CEO compensation: Evidence from the USA. *Women in Management Review, 76*(6), 830–846. doi:10.1108/09649420710743662

Adams, W. (2004). Biodiversity Conservation and the Eradication of Poverty. *Science, 306*(5599), 1146–1149. doi:10.1126cience.1097920 PMID:15539593

ADB. (2012). *Democratic Republic of São Tomé and Príncipe, Country Strategic Paper, 2012-2016.* Academic Press.

AGE. (2018). *Programas Educativos Estrela Geopark (Ensino Básico e Secundário).* http://www.geoparkestrela.pt/aplication/views/assets/documents/doc_conteudos/Programas_educativos_Ensino_Basico_18_19.pdf

Aguiar, C., Rodrigues, O., Azevedo, J., & Domingos, T. (2016). Montanha. In Ecossistemas e Bem-Estar Humano. Avaliação para Portugal do Millenium Ecosystem Assessment. Lisboa, Escolar Editora.

AICEP. (2017). Ficha de Mercado. Academic Press.

Akama, J., & Kieti, D. (2007). Tourism and socio-economic development in developing countries: A case study of Mombasa resort in Kenya. *Journal of Sustainable Tourism, 15*(6), 733–748. doi:10.2167/jost543.0

Akinboade, O., & Braimoh, L. (2010). International tourism and economic development in South Africa: A granger causality test. *International Journal of Tourism Research, 12*(2), 149–163. doi:10.1002/jtr.743

Alcock, P. (1993). *Understanding Poverty.* Macmillan.

Aldırmaz Akkaya, F. (2018). *Türkiye'de kadın kooperatiflerinin sürdürülebilir kırsal turizmdeki önemi ve finansal performanslarının analizi.* Akademisyen Kitabevi.

Alegre, J., & Pou, L. (2003). El turismo de sol y playa desde la perspectiva de los consumidores Europeos. Un nuevo desafío en el desarrollo turístico de Baleares. *Serie Estudios Regionales Islas Baleares*, 149-176.

Alén, E., Luis Nicolau, J., Losada, N., & Domínguez, T. (2014). Determinant factors of senior tourist length of stay. *Annals of Tourism Research*, *49*, 19–32. doi:10.1016/j.annals.2014.08.002

Alles, M., T., F. (2009). Turismo Accesible: Importancia de la Accesibilidad para el sector turístico. *Entelequia*, (9), 211-224.

Alonso, M. H., Gonzálvez, J. E., & Bartolomé, A. (2016). Ventajas e inconvenientes del uso de dispositivos electrónicos en el aula: Percepción de los estudiantes de grados en comunicación. *Revista de Comunicación de la SEECI*, *20*(41), 136–154. doi:10.15198eeci.2016.41.136-154

Alpoim, M. (2010). *Análise à procura termal*. Aveiro: UA.

Altamirano, V. P., Túñez, J. M., & Valarezo, K. P. (2016). Promoción turística 2.0. Análisis de las redes sociales de los gobiernos iberoamericanos. *Revista Opción*, *32*(9), 32–53.

Aluttis, C., Bishaw, T., & Frank, M. W. (2014). The workforce for health in a globalised context – global shortages and international migration. *Global Health Action*, *7*(1), 23611. Advance online publication. doi:10.3402/gha.v7.23611 PMID:24560265

Alves, G. M., Sousa, B. M., & Machado, A. (2020). The Role of Digital Marketing and Online Relationship Quality in Social Tourism. In J. Santos & Ó. Silva (Eds.), *Digital Marketing Strategies for Tourism, Hospitality, and Airline Industries* (pp. 49–70). IGI Global. doi:10.4018/978-1-5225-9783-4.ch003

Aman, R. (2017). Epistemological privilege in intercultural education. *Decolonising Intercultural Education*, 15–24. doi:10.4324/9781315616681-2

Aminudin, N., Yahya, W. S. Z., & Sumarjan, N. (2015). Supply analysis in the continuity of community-based tourism. In S. M. Radzi, N. Sumarjan, C. T. Chick, M. S. M. Zahari, Z. Mohi, M. F. S. Bakhtiar, & F. I. Anuar (Eds.), *Theory and practice in hospitality and tourism research* (pp. 247–251). CRC Press.

Ampumuza, C., van der Heijden, F., & Hendriks, N. (2008). *Women Empowerment through tourism.* Retrieved from http://videa.ca/wp-content/uploads/2015/08/Women empowerment-through-Ecotourism.pdf

Anastasiadou, C. (2007). Group Politics and Tourism Interest Representation at the Supranational Level: Evidence from the European Union. *Tourism and Politics*, 59–69.

Anderson, W. (2015). Cultural tourism and poverty alleviation in rural Kilimanjaro, Tanzania. *Journal of Tourism and Cultural Change*, *13*(3), 208–224. doi:10.1080/14766825.2014.935387

Anup, K. C., & Rijal, K., & Sapkota. (2014). Role of ecotourism in environmental conservation and socioeconomic development in Annapurna conservation area, Nepal. *International Journal of Sustainable Development and World Ecology*, *22*(3), 251–258.

Ardianto, E. (2012). Dimensions of community-based tourism. A case study of surf community in Bali, Indonesia. *Journal of Tourism Research*, *5*(B), 38-46.

Ariza, C., Maselli, D., & Kohler, T. (2013). Mountains: Our Life, Our Future. Progress and Perspectives on Sustainable Mountain Development from Rio 1992 to Rio 2012 and Beyond. Bern: Swiss Agency for Development and Cooperation (SDC), Centre for Development and Environment (CDE).

Arnstein, S. R. (1969). A ladder of citizen participation. *Journal of the American Planning Association*, *35*(4), 216–224.

Ashley, C. (2006). *How Can Governments Boost the Local Economic Impacts of Tourism?* Options and Tools. SNV and ODI.

Ashley, C., & Roe, D. (2002). Making tourism work for the poor: Strategies and challenges in South Africa. *Development Southern Africa, 19*(1), 61–82. doi:10.1080/03768350220123855

Ashworth, J., & Thomas, B. (1999). Patterns of seasonality in employment in tourism in the UK. *Applied Economics Letters, 6*(11), 735–739. doi:10.1080/135048599352312

Asker, S., Boronyak, L., Carrard, N., & Paddon, M. (2010). *Effective community-based tourism: A best practice manual.* Sustainable Tourism Cooperative Research Centre.

Ataberk, E., & Baykal, F. (2011). Utilisation of natural and cultural resources of Dikili (Izmir) for tourism. *Procedia: Social and Behavioral Sciences, 19*, 173–180. doi:10.1016/j.sbspro.2011.05.121

Aurora Project. (2018). *Overview| The Aurora Project.* Retrieved from https://auroraproject.com.au/overview

Ausland, A. (2010a, August 27). Poverty taxonomy 2.0. *Huffington Post.* Retrieved from https://www.huffingtonpost.com/aaron-ausland/poverty-tourism-taxonomy-_b_697399.html

Ausland, A. (2010b). *Poverty tourism: A debate in need of typological nuance. Staying for tea.* Retrieved from http://stayingfortea.org/2010/08/17/poverty-tourism-a-debate-in-need-of-typological-nuance/

Avolio, B. J., & Wernsing, T. S. (2008). Practicing authentic leadership. *Positive psychology: Exploring the best in people, 4*, 147-165.

Avolio, B. J., & Gardner, W. L. (2005). Authentic leadership development: Getting to the root of positive forms of leadership. *The Leadership Quarterly, 16*(3), 315–338. doi:10.1016/j.leaqua.2005.03.001

Azevedo, J. C., Cadavez, V., Arrobas, M., & Pires, J. M. (2016). Sustentabilidade da montanha portuguesa: realidades. In J. C. Azevedo, V. Cadavez, M. Arrobas, & J. M. Pires (Eds.), *Sustentabilidade da Montanha Portuguesa: Realidades e Desafios* (pp. 9–38). IPB.

Aziz, A., Barzeker, G., Ajuhar, Z., & Idris, N. H. (2015). Criterion and Indicators for Monitoring Ecotourism Sustainability in a Protected Watershed Sustainable Tourism: A Delphi Consensus. *IOSR Journal of Environment Science. Toxicology and Food Technology, 9*(3), 1–9.

Azman, N., Halim, S. A., Liu, O. P., Saidin, S., & Komoo, I. (2010). Public Education in Heritage Conservation for Geopark Community. *Procedia: Social and Behavioral Sciences, 7*(C), 504–511. doi:10.1016/j.sbspro.2010.10.068

Bagri, S. C., & Kala, D. (2016). Residents' attitudes toward tourism development and impacts in Koti-Kanasar, Indroli, Pattyur tourism circuit of Uttarakhand, India. *PASOS Journal of Tourism and Cultural Heritage, 14*(1), 23–39. doi:10.25145/j.pasos.2016.14.002

Baker, E. (2019, October 11). Tourism Tasmania says it has made 'internal changes' since 2016 culture report labelled it a 'boys club'. *ABC News.* Retrieved from https://www.abc.net.au/news/2019-10-11/tourism-tasmania-defends-record-after-culture-report/11595532

Bali.com. (2015). Sustainable Tourism on Bali? *Bali.com.* Retrieved March, 12, 2020, from https://www.bali.com/news_Sustainable-Tourism-on-Bali-_161.html

Balli, F., Balli, H. O., & Jean Louis, R. (2016). The impacts of immigrants and institutions on bilateral tourism flows. *Tourism Management, 52*, 221–229. doi:10.1016/j.tourman.2015.06.021

Banco Portugal (BP). (2019). *Evolução das Economias dos PALOPS e de Timor-Leste 2018-2019.* Banco Portugal.

Bansal, S. P., & Kumar, J. (2011). Ecotourism for community development: A stakeholder's perspective in Great Himalayan National Park. *International Journal of Social Ecology and Sustainable Development*, *2*(2), 31–40. doi:10.4018/jsesd.2011040103

Baptista, J. (2011). The tourists of developmentourism– representations 'from below'. *Current Issues in Tourism*, *14*(7), 651–667. doi:10.1080/13683500.2010.540314

Barbacena, H. (2012). *A globalização do turismo e a competitividade dos destinos turísticos* (Doctoral dissertation).

Barney, J. (1991). Firm resources and sustained competitive advantage. *Journal of Management*, *17*(1), 99–120. doi:10.1177/014920639101700108

Barquero, M. (2017). La ética en el sector del alojamiento turístico como Fuente de ventaja competitiva. *Revista de Comunicación de la SEECI*, *42*(42), 119–130. doi:10.15198eeci.2017.42.119-130

Barranquero, A. (2012). De la comunicación para el desarrollo a la justicia ecosocial y el buen vivir. CIC. *Cuadernos De Información Y Comunicación*, *17*, 63-78. doi:10.5209/rev_CIYC.2012.v17.39258

Barrett, C. B., & Arcese, P. (1995). Are Integrated Conservation-Development Projects (ICDPs) Sustainable? On the conservation of large mammals in sub-Saharan Africa. *World Development*, *23*(7), 1073–1084. doi:10.1016/0305-750X(95)00031-7

Barrientos-Báez, A. (2019). Teoría del Triple Rol: Inteligencia Emocional, Turismo y Educación. *Cuadernos Artesanos de Comunicación, 158.*

Barrientos-Báez, A., Barquero, M., & Rodríguez, J. (2019). La educación emocional como contenido transversal para una nueva política educativa: El caso del grado de turismo. *Revista Utopía y Praxis Latinoamericana*, *24*(4), 147–165.

Bartram, D. (2014). Migration and quality of life in the global context. In W. Glatzer, L. Camfield, V. Møller, & M. Rojas (Eds.), *Global Handbook of Quality of Life - Exploration of Well-Being of Nations and Continents* (pp. 491–503). Springer.

Barzeker, G., Aziz, A., Mariapan, M., Ismail, M. H., & Hoseni, S. M. (2011). Delphi Technique for Generating Criterion and Indicators in Monitoring Ecotourism Sustainability in Northern Forest of Iran: Case Study on Dohezar and SehezarWatersheds, Folia. *Forestalia Pronica*, *53*(2), 130-141.

Bashar, M. A. (2018). Vision on biodiversity: Ecotourism and biodiversity conservation in Bangladesh. *Journal of Biodiversity Conservation & Bio-resource Management*, *4*(1), 1–10. doi:10.3329/jbcbm.v4i1.37871

Batı Akdeniz Kalkınma Ajansı, B. A. K. A. (2012). Ekoturizm sektör raporu. Academic Press.

Batterbury, S., & Toscano, M. (2018). Seeking justice through interdisciplinary environmental education at postgraduate level: Lessons from Melbourne, Australia. *Revista Internacional de Educación para la Justicia Social*, *7*(1), 141–156. doi:10.15366/riejs2018.7.1.007

Baum, T. (2013). *International perspectives on women and work in hotels, catering and tourism.* Retrieved from https://www.ilo.org/wcmsp5/groups/public/---dgreports/---gender/documents/publication/wcms_209867.pdf

Bazazo, I., Nasseef, M., Mukattesh, B., Kastero, D., & Al-Hallaq, M. (2017). Assessing the Glass Ceiling Effect for Women in Tourism and Hospitality. *Journal of Management and Strategy*, *8*(3), 51. doi:10.5430/jms.v8n3p51

Bechtel, R. B., Corral Verdugo, V., & de Queiroz Pinheiro, J. (1999). Environmental belief systems: United States, Brazil, and Mexico. *Journal of Cross-Cultural Psychology*, *30*(1), 122–128. doi:10.1177/0022022199030001008

Beeton, S. (2006). *Community Development through Tourism.* Landlink Press. doi:10.1071/9780643093881

Beham, B., Drobnič, S., Präg, P., Baierl, A., & Eckner, J. (2019). Part-time work and gender inequality in Europe: A comparative analysis of satisfaction with work–life balance. *European Societies*, *21*(3), 378–402. doi:10.1080/14616696.2018.1473627

Benítez, L. (2014). Educación emocional para la comunicación transcultural. *Perspectivas de la Comunicación*, *7*(1), 7–24.

Ben, M. (2005). *Sustainability: Definition and five core principles a new framework*. Sustainability Laboratory Publication.

Bennett, O., Ashley, C., Roe, D., & Britain, G. (1999). *Sustainable tourism and poverty elimination study: a report to the department for international development*. Deloitte & Touche.

Benson, M., & O'Reilly, K. (2016). From lifestyle migration to lifestyle in migration: Categories, concepts and ways of thinking. *Migration Studies*, *4*(1), 20–37. doi:10.1093/migration/mnv015

Bernal, A. I. (2010). Definición conceptual de los medios de comunicación por un grupo de jóvenes españoles. El valor de Internet. *Vivat Academia. Revista de Comunicación*, *112*, 22–32. doi:10.15178/va.2010.112.22-32

Bertram, G. (2006). Introduction: The MIRAB model in the twenty-first century. *Asia Pacific Viewpoint*, *47*(1), 1–13. doi:10.1111/j.1467-8373.2006.00296.x

Bestic, R. (2019, December 18). Outdoor wonders and wild adventures in Australia's South. *Telegraph Newspaper*.

Best, J. W. (1977). *Research in education* (3rd ed.). Englewood Cliffs, NJ: Prentice Hall, Inc.

Bickford, N., & Glancy, S. (2010). *The case for change: Women working in hospitality, leisure, travel and tourism.* United Kingdom: People 1st. Available at: http://www.women1st.co.uk/system/assets/files/000/000/004/origi-nal/Women_1st_Case_For_Change_Executive_Summary_Novem-ber_2010.pdf?1342707875

Biddulph, R. (2018). Social enterprise and inclusive tourism. Five cases in Siem Reap, Cambodia. *Tourism Geographies*, *20*(4), 610–629. doi:10.1080/14616688.2017.1417471

Biosphere Tourism. (2015). Retrieved from: https://www.biospheretourism.com/en/blog/17-sustainable-tourism-examples-for-this-2017/2

Björk, P. (2007). Definition paradoxes: from concept to definition. In J. Higham (Ed.), *Critical Issues in Ecotourism: Understanding a Complex Tourism Phenomenon* (pp. 23–45). Elsevier. doi:10.1016/B978-0-7506-6878-1.50006-2

Blackstock, K. (2005). A critical look at community based tourism. *Community Development Journal: An International Forum*, *40*(1), 39–49. doi:10.1093/cdj/bsi005

Blake, A., Arbache, J. S., Sinclair, M. T., & Teles, V. (2008). Tourism and Poverty Relief. *Annals of Tourism Research*, *35*(1), 107–126. doi:10.1016/j.annals.2007.06.013

Blanco, R. (2006). La Equidad y la Inclusión Social: Uno de los Desafíos de la Educación y la Escuela Hoy. REICE. *Revista Electrónica Iberoamericana sobre Calidad, Eficacia y Cambio en Educación*, *4*, 1–15.

Blau, F. D., & Kahn, L. M. (1994). Rising wage inequality and the US gender gap. *The American Economic Review*, *84*(2), 23–28.

Blazey, M. A. (1992). Travel and retirement status. *Annals of Tourism Research*, *19*(4), 771–783. doi:10.1016/0160-7383(92)90066-X

Blose, J., Mack, R. W., & Pitts, R. E. (2015). The Influence of Message Framing on Hotel Guests' Linen-Reuse Intentions. *Cornell Hospitality Quarterly*, *56*(2), 145–154. doi:10.1177/1938965514556480

Blue Derby Pods Ride. (2020). The Purpose. *Blue Derby Pods Ride*. Retrieved June, 8, 2020, from https://bluederby-podsride.com.au/the-purpose

Bohdanowicz, P., & Martinac, I. (2003). Attitudes towards sustainability in chain Hotels – results of a European survey. *Proceedings of the CIB International Conference on Smart and Sustainable Built Environment (SASBE2003).*

Bohnet, I. (2016). *What works: Gender Equality by Design*. The Belknap Press of Harvard University Press. doi:10.4159/9780674545991

Boluk, K., Cavaliere, C., & Higgins-Desbiolles, F. (2019). A critical framework for interrogating the United Nations Sustainable Development Goals 2030 Agenda in tourism. *Journal of Sustainable Tourism*, *27*(7), 847–864. doi:10.1080/09669582.2019.1619748

Boo, E. (1990). *Ecotourism: the potentials and the pitfalls* (Vol. 1). World Wildlife Fund.

Bookbinder, M. P., Dinerstein, E., Rijal, A., Cauley, H., & Rajouria, A. (1998). Ecotourism's Support of Biodiversity Conservation. *Conservation Biology*, *12*(6), 1399–1404.

Booth, T., & Ainscow, M. (2004). *Índice de Inclusión: Desarrollando el aprendizaje y la participación en las escuelas.* UNESCO/OREALC.

Booth, T. (2000). *Progreso en la educación inclusiva. Estudio Temático para la evaluación de educación para todos.* UNESCO.

Bossel, H. (1999). *Indicators for sustainable development: theory, method, applications*. International Institute for Sustainable Development.

Bottrill, C. G., & Pearce, D. G. (1995). Ecotourism: Towards a key elements approach to operationalising the concept. *Journal of Sustainable Tourism*, *3*(1), 45–54. doi:10.1080/09669589509510707

Bowdin, G., Johnny, A., William, O., Rob, H., & Ian, M. (2012). *Events Management*. Academic Press.

Bowen, H. (2004). Cultural tourism: The partnership between tourism and cultural heritage management. *Tourism Management*, *25*(3), 412–414. doi:10.1016/S0261-5177(03)00138-9

Boyce, C., & Neale, P. (2006). *Conducting in-depth interviews: a guide for Designing and Conducting In-Depth Interviews for Evaluation Input*. Pathfinder International.

Bradford Energy Solutions. (2013). *Case Study – Josef Chromy Wines*. Retrieved from https://www.bradfordenergy.com.au/-/media/bradford-solar/files/pdfs/case-study-josef-chromy-wines.pdf

Braga, C. A. P., & Dömeland, D. (2009). *Debt Relief Beyond Lessons Learned and Challenges Ahead*. Academic Press.

Bramwell, B., & Lane, B. (2006). *Tourism, collaboration, and partnership: politics, practice, and sustainability*. Viva Books Private Ltd.

Bramwell, B., & Lane, B. (2008). Priorities in Sustainable Tourism Research. *Journal of Sustainable Tourism*, *16*(1), 1–4. doi:10.2167/09669580803489612

Bramwell, B., & Lane, B. (2011). Critical research on the governance of tourism and sustainability. *Journal of Sustainable Tourism*, *19*(4-5), 411–421. doi:10.1080/09669582.2011.580586

Brandão, F., Costa, C., & Buhalis, D. (2018). Tourism innovation networks: A regional approach. *European Journal of Tourism Research*, *18*, 33–56.

Brida, J. G., Disegna, M., & Osti, L. (2012). Perceptions of Authenticity of Cultural Events: A Host–Tourist Analysis. *Tourism, Culture & Communication, 12*(2), 85–96. doi:10.3727/109830413X13575858951121

Bridestowe Lavender Estate. (2020). Sustainability and efficiency matters. *Bridestowe Lavender Estate.* Retrieved June, 8, 2020, from https://bridestowelavender.com.au/explore-bridestowe-lavender-estate/the-bridestowe-farm/sustainability/

Briedenhann, J., & Wickens, E. (2004). Tourism routes as a tool for the economic development of rural areas—Vibrant hope or impossible dream? *Tourism Management, 25*(1), 71–79. doi:10.1016/S0261-5177(03)00063-3

Brilha, J. B. R. (2005). *Património Geológico e Geoconservação: a conservação da Natureza na sua vertente geológica.* Palimage Editores http://www.dct.uminho.pt/docentes/pdfs/jb_livro.pdf

Brilha, J.B.R. (2009). A Importância dos Geoparques no Ensino e Divulgação das Geociências. *Geologia USP, 5*, 27-33.

Brockhaus, S., Amos, C., Fawcett, A. M., Knemeyer, M., & Fawcett, S. E. (2017). Please clap! How customer service quality perception affects the authenticity of sustainability initiatives. *Journal of Marketing Theory and Practice, 25*(4), 396–420. doi:10.1080/10696679.2017.1345594

Brockington, D., Igoe, J., & Schmidt-Soltau, K. (2006). Conservation, Human Rights, and Poverty Reduction. *Conservation Biology, 20*(1), 250–252. doi:10.1111/j.1523-1739.2006.00335.x PMID:16909680

Brohman, J. (1996). New directions in tourism for the Third World. *Annals of Tourism Research, 23*(1), 48–70. doi:10.1016/0160-7383(95)00043-7

Brundtland, G. H. (1987). Our common future—Call for action. *Environmental Conservation, 14*(4), 291–294. doi:10.1017/S0376892900016805

Bruner, A., Hanks, J., & Hannah, L. (2003). *How much will effective protected area systems cost?(updated version).* http://www.citeulike.org/group/342/article/432012

Bruner, A., Gullison, R., Rice, R., & da Fonseca, G. (2001). Effectiveness of Parks in Protecting Tropical Biodiversity. *Science, 291*(5501), 125–128. doi:10.1126cience.291.5501.125 PMID:11141563

Buckley, R. (1994). A framework for ecotourism. *Annals of Tourism Research, 21*(3), 661–665. doi:10.1016/0160-7383(94)90126-0

Buckley, R., & Sommer, M. (2001). *Tourism and protected areas: Partnerships in principle and practice.* CRC for Sustainable Tourism.

Büge, M., Meijer, K., Wittmer, H., Bordon, I., Hein, J., Lindenberg, N., Rodriguez De Francisco, J. C., Richerzhagen, C., Willem Te Velde, D., & Wolff, P. (2015). *International financial instruments for biodiversity conservation in developing countries – financial mechanisms and enabling policies for forest biodiversity Background paper for the European Report on Development 2015.* Academic Press.

Bunruamkaew, K., & Murayama, Y. (2012). Land Use and Natural Resources Planning for Sustainable Ecotourism Using GIS in Surat Thani, Thailand. *Sustainability, 4*(1), 412–429. doi:10.3390u4030412

Burek, C. V., & Prosser, C. D. (2008). The history of geoconservation: An introduction. Geological Society, London, Special Publications. *Geological Society of London, Special Publications, 300*, 1–5. doi:10.1144/SP300.1

Burgess, G. (2018). Bruny Island tourist influx a double-edged sword for those calling it home. *ABC News.* Retrieved March 12, 2020, from https://www.abc.net.au/news/2018-12-20/bruny-island-off-tasmania-feels-growing-pains-of-tourism-boom/10582124

Burns, P. M. (2004). Tourism planning: A third way? *Annals of Tourism Research, 31*(1), 24–43. doi:10.1016/j.annals.2003.08.001

Bursztyn, I., Bartholo, R., & Delamero, M. (2009). Turismo para quem? Sobre caminhos de desenvolvimento e alternativas para o turismo no Brasil. In R. Bartholo, D. G. Sansolo, & I. Bursztyn (Eds.), *Turismo de base comunitária: Diversidade de olhares e experiências brasileiras* (pp. 76–91). Letra e Imagem.

Butcher, J. (2017). Citizenship, global citizenship and volunteer tourism: A critical analysis. *Tourism Recreation Research, 42*(2), 129–138. doi:10.1080/02508281.2017.1295172

Butler, R. (1998). Sustainable tourism-looking backwards in order to progress? *Sustainable Tourism: A Geographical Perspective*, 25-34.

Butler, R. W. (1974). The social implications of tourist developments. *Annals of Tourism Research, 2*(2), 100–111. doi:10.1016/0160-7383(74)90025-5

Calabria.jblasa.com. (2015). Available at: https://calabria.jblasa.com/wp-content/uploads/2015/10/Gizzeria-lido.jpg

Caldeira, M., Silva, N. & Nunes, F. (2016). Turismo acessível em Guimarães. *Oportunidade e Desafio para uma cidade inclusiva Brasil & Norte Holos, 4*, 341-357.

Caldevilla Domínguez, D., García García, E., & Barrientos-Báez, A. (2019). La importancia del turismo cultural como medio de dignificación del turista y de la industria. *Mediaciones Sociales, 18*, 59–69. doi:10.5209/meso.65117

Camacho-Gómez, M. (2016). Chocolate Business in the Hospitality Industry. *ECORFAN Journal ECORFAN Journal-Mexico Journal, 7717*(17), 7–17.

Camilleri, M. (2017). The Tourism Industry: An Overview. *Tourism, Hospitality & Event Management*, 3–27. doi:10.1007/978-3-319-49849-2_1

Campos, M. J. Z. (2014). Partnerships, Tourism, and Community Impacts. *The Wiley Blackwell Companion to Tourism*, 567–577.

Canadian Tourism Commission. (2007). *Guidelines: Survey Procedures for Tourism Economic Impact Assessments of Gated Events and Festivals*. Alberta Tourism, Parks, Recreation & Culture. Retrieved from https://www.destinationbc.ca/getattachment/Research/Resources-for-esearchers/How-To-Guides/Gated_Events_Festivals-economic-impact.pdf.aspx

Canavan, B. (2016). Tourism culture: Nexus, characteristics, context and sustainability. *Tourism Management, 53*, 229–243. doi:10.1016/j.tourman.2015.10.002

Cantafio, G. U. (2017). *Merging Economic Development with Clusters Theory and Environmental Sustainability* (Unpublished PhD thesis). Reggio Calabria, Università degli Studi di Reggio Calabria Mediterranea.

Cantista, A. (2010). O termalismo em Portugal. *Anales de Hidrología Medica 2008-2010, 3*(1), 79–107.

Canziani, B. F. (2014). Values-based Internships: Combining TEFI Values, Career Internships, and Community Engagement. *Journal of Teaching in Travel & Tourism, 14*(2), 129–148. doi:10.1080/15313220.2014.907957

Cape Verde Government Website. (n.d.). Available at https://www.governo.cv/

Capriello, A., & Rotherham, I. D. (2011). Building a preliminary model of event management for rural communities. *Journal of Hospitality Marketing & Management, 20*(3–4), 246–264. doi:10.1080/19368623.2011.562411

Cardoso, M. (2015). Migrações e Estratégias de Desenvolvimento dos Pequenos Estados Insulares em Desenvolvimento. Estudos de caso: Cabo Verde e São Tomé e Príncipe. *Cadernos de Estudos Africanos*, *29*(29), 103–128. doi:10.4000/cea.1804

Carrigan, M., & Attalla, A. (2001). The myth of the ethical consumer – do ethics matter in purchase behaviour? *Journal of Consumer Marketing*, *18*(7), 560–578. doi:10.1108/07363760110410263

Carrilho, P., & Perista, H. (2016). *Portugal: High and rising emigration in a context of high, but decreasing, unemployment*. Eurofound. Accessed 31 May 2020, from https://www.eurofound.europa.eu/publications/article/2016/portugal-high-and-rising-emigration-in-a-context-of-high-but-decreasing-unemployment

Carrillo-Hermosilla, J., & Könnola, T. (2009). *Eco-innovation. When sustainability and competitiveness shake hands*. Palgrave.

Carter, A. (2013). *The political theory of global citizenship* (2nd ed.). Routledge. doi:10.4324/9780203183397

Carvalho, I., Costa, C., Lykke, N., & Torres, A. (2018). What Could be Done? Tourism,Gender (In)Equality and Women Managers' Views of What Should Change. *Proceedings of the International Conference on Gender Research*.

Cash, D., Clark, W., Alcock, F., Dickson, N., Eckley, N., Guston, D., & Mitchell, R. (2003). Knowledge systems for sustainable development. *Proceedings of the National Academy of Sciences of the United States of America*, *100*(14), 8086–8091. doi:10.1073/pnas.1231332100 PMID:12777623

Castañeda, M. (2017). La Ciudad Como Escenario De Diálogo. Apuntes Desde La Comunicación Intercultural. *Perspectivas de la Comunicación*, *10*(2), 119–142.

Castleton, A. (2017). Tecnología e identidad: El caso de los Inuit y Facebook. CIC. *Cuadernos de Información y Comunicación*, *22*(0), 107–125. doi:10.5209/CIYC.55970

Castro, P., Lundgren, U., & Woodin, J. (2019). Intercultural dialogue. *Educational Approaches to Internationalization through Intercultural Dialogue*, 3–14. doi:10.4324/9780429444289-1

Caton, K., Schott, C., & Daniele, R. (2014). Tourism's Imperative for Global Citizenship. *Journal of Teaching in Travel & Tourism*, *14*(2), 123–128. doi:10.1080/15313220.2014.907955

Cavagnaro, E., & Curiel, G. (2012). *The three levels of sustainability*. Greenleaf.

Cavagnaro, E., Düweke, A., & Melissen, F. (2018). The host-guest relationship is the key to sustainable hospitality: Lessons learned from a Dutch case study. *Hospitality & Society*, *8*(1), 23–44. doi:10.1386/hosp.8.1.23_1

CBI. (2015). *Tailored intelligence study: Community-Based Tourism (CBT) in Bolivia*. Ministry of Foreign Affairs of the Netherlands CBI - Centre for the Promotion of Imports from developing countries.

CBI. (2020). *What are the opportunities for community-based tourism from Europe?* Retrieved March 20, 2020, from https://www.cbi.eu/market-information/tourism/community-based-tourism/community-based-tourism-europe

CE. (2004). *Mountain Areas in Europe: Analysis of mountain areas in EU member states, acceding and other European countries*. https://ec.europa.eu/regional_policy/sources/docgener/studies/pdf/montagne/mount1.pdf

Ceballos-Lascurain, H. (1993). Ecotourism as a worldwide phenomenon. *Ecotourism as a Worldwide Phenomenon*, 12-14.

Ceballos-Lescuráin H. (1987). The future of ecotourism. *Mexico Journal*, 13–14.

Cella, D. F. (1994). Quality of life: Concepts and definition. *Journal of Pain and Symptom Management*, *9*(3), 186–192. doi:10.1016/0885-3924(94)90129-5 PMID:8014530

Censis. (2018). *Rapporto sulla situazione sociale del paese*. Angeli.

César, F., & Arnáiz, S. M. (2016). Capitalismo, sociedad y turismo. *Revista Opción, 32*(13), 319–338.

Ceylan, Y. (2019). Sürdürülebilir Turizm Kapsamında Turizmde Eko Eko Etiketler. *Turizm Ekonomi ve İşletme Dergisi, 1*(1), 65–80.

Chalmers, S. (2020). Bushfires and coronavirus twin disasters hit Australia's tourism sector. *ABC News*. Retrieved March 12, 2020, from https://www.abc.net.au/news/2020-02-11/twin-disasters-coronavirus-bushfires-hit-tourism-sector/11949922

Chambers, R. (2013). *Ideas for development* (2nd ed.). Earthscan. doi:10.4324/9781849771665

Chambers, R., & Conway, G. (1992). *Sustainable rural livelihoods: practical concepts for the 21st century*. Institute of Development Studies.

Chandel, A., & Mishra, S. (2016). Ecotourism revisited: Last twenty-five years. *Czech Journal of Tourism, 5*(2), 135–154. doi:10.1515/cjot-2016-0008

Chatterjee, R. (2018). *A new survey finds 81 percent of women have experienced sexual harassment.* Retrieved from https://www.npr.org/sections/thetwo-way/2018/02/21/587671849/a-new-survey-finds-eighty-percent-of-women-have-experienced-sexual-harassment

Cheung, C., Baum, T., & Hsueh, A. (2017). Workplace sexual harassment: Exploring the experience of tour leaders in an Asian context. *Current Issues in Tourism, 21*(13), 1468–1485. doi:10.1080/13683500.2017.1281235

Chim-Miki, A. F., & Batista-Canino, R. M. (2018). Development of a tourism coopetition model: A preliminary Delphi Study. *Journal of Hospitality and Tourism Management, 37*, 78–88. doi:10.1016/j.jhtm.2018.10.004

Cho, V. (2001). Tourism forecasting and its relationship with leading economic indicators. *Journal of Hospitality & Tourism Research (Washington, D.C.), 25*(4), 399–420. doi:10.1177/109634800102500404

Cho, Y. (2014). *Different shades of green consciousness: The interplay of sustainability labeling and environmental impact on product evaluations*. Springer Science and Business Media Dordrecht.

Ciarrochi, J., Chan, A., & Bajgar, J. (2001). Personalidad y diferencias individuales. *Elsevier, 31*(7), 1105-1119.

Ciulli, F., Kolk, A., & Boe-Lillegraven, S. (2019). Circularity Brokers: Digital Platform Organisations and Waste Recovery in Food Supply Chains. *Journal of Business Ethics*, 1–33. Retrieved October 16, 2019, from. doi:10.100710551-019-04160-5

Cláudio Corallo. (2018). *As 13 melhores imagens em Chocolate from Claudio Corallo no Pinterest | Chocolates, Coisas engraçadas e Palavras*. https://www.pinterest.pt/culinarious/chocolate-from-claudio-corallo/

Cobbinah, P., Black, R., & Thwaites, R. (2013). Dynamics of Poverty in Developing Countries: Review of Poverty Reduction Approaches. *Journal of Sustainable Development, 6*(9), 25–35. doi:10.5539/jsd.v6n9p25

Coelho, C. (2014). *Turismo acessível para todos, um paradigma emergente e um desafio para a oferta turística – O caso dos espaços museológicos e empreendimentos turísticos de Cascais* (Master's thesis). Escola Superior de Hotelaria e Turismo do Estoril.

Cohn, S. (2017). The Determinants of the Division of Labor between Men and Women in Paid Employment in the Global North and South: How Occupational Sex-typing Informs the Study of Gender and Development. *Sociology of Development (Oakland, Calif.), 3*(1), 1–23. doi:10.1525od.2017.3.1.1

Concu, N., & Atzeni, G. (2012). Conflicting preferences among tourists and residents. *Tourism Management, 33*(6), 1293–1300. doi:10.1016/j.tourman.2011.12.009

Cone, E. B., Westerman, M. E., Nguyen, D.-D., Stern, K. L., DesLoges, J., & Koo, K. (2020). Gender-Based Differences in Career Plans, Salary Expectations, and Business Preparedness Among Urology Residents. *Urology*. Advance online publication. doi:10.1016/j.urology.2020.04.123 PMID:32540301

Connor, L. (2016, March 11). 10 of the world's best eco-friendly luxury hotels. *Evening Standard.* Retrieved from https://www.standard.co.uk/lifestyle/travel/10-of-the-world-s-best-ecofriendly-luxury-hotels-a3201281.html

Conrad, J., Ferraro, P., Alm, J., Cummings, R., Morrison P., C. J., Taylor, L., Uchida, T., & Weitzman, M. (2001). *Habitat Conservation: the dynamics of direct and indirect payments The Dynamics of Direct and Indirect Payments.* Academic Press.

Cooper, C., & Hall, M. (2008). *Contemporary tourism: An international approach.* Elsevier.

Coriolano, L. N., Araújo, A. M., & Vasconcelos, F. P. (2009). *Arranjos produtivos locais do turismo comunitário: Atores e cenários em mudanças*. EDUECE.

Correll, S. J. (2004). Constraints into preferences: Gender, status, and emerging career aspirations. *American Sociological Review, 69*(1), 93–113. doi:10.1177/000312240406900106

Correyero, B. (2003). La propaganda turística española en los años del aislamiento internacional. *Historia y Comunicación Social, 8,* 47-61. https://revistas.ucm.es/index.php/HICS/article/view/HICS0303110047A

Cortese, A. D., & Rowe, D. (2017). Sustainability strategic plan. In *Sustainable Haverford.* Haverford College, Council on Sustainability and Social Responsibility Final Draft.

Cortese, D., Giacosa, E., & Cantino, V. (2018). Knowledge sharing for coopetition in tourist destinations: The difficult path to the network. *Review of Managerial Science*. Advance online publication. doi:10.100711846-018-0322-z

Costa, C., Breda, Z., & Carvalho, I. (2011). Gender inequalities in tourism employment: The Portuguese case. *Revista Turismo & Desenvolvimento, 15*(1), 37–52.

Crabb, A. (2019, October 8). Australia Talks National Survey reveals what Australians are most worried about. *ABC News*. Retrieved from https://www.abc.net.au/news/2019-10-08/annabel-crabb-australia-talks-what-australians-worry-about/11579644

Crawford, J., Dawkins, S., Martin, A., & Lewis, G. (2018). Conceptualising authentic followers and developing a future research agenda. In D. Cotter-Lockard (Ed.), *Authentic Leadership and Followership* (pp. 271–293). Palgrave Macmillan. doi:10.1007/978-3-319-65307-5_11

Crawford, J., Dawkins, S., Martin, A., & Lewis, G. (2019). Putting the leader back into authentic leadership: Reconceptualising and rethinking leaders. *Australian Journal of Management, 45*(1), 114–133. doi:10.1177/0312896219836460

Crawford, J., Kelder, J., & Knox, M. (2020). What does it take to be a social entrepreneur?: Authentic leaders and their effect on innovation. In C. Dogru (Ed.), *Leadership Styles, Innovation, and Social Entrepreneurship in the Era of Digitalization* (pp. 282–310). IGI Global. doi:10.4018/978-1-7998-1108-4.ch011

Croes, R., & Vanegas, M. (2008). Tourism and Poverty Alleviation: A Co-integration analysis. *Journal of Travel Research, 47*(1), 94–103. doi:10.1177/0047287507312429

Crossley, É. (2012). Poor but happy: Volunteer tourists' encounters with poverty. *Tourism Geographies, 14*(2), 235–253. doi:10.1080/14616688.2011.611165

Csedő, K. (2008). Negotiating skills in the global city: Hungarian and Romanian professionals and graduates in London. *Journal of Ethnic and Migration Studies, 34*(5), 803–823. doi:10.1080/13691830802106093

Cunha, L., & Cordeiro, A. M. (2000). A Montanha em Portugal – Uma perspectiva geográfica. Reunión de Expertos en Desarrollo Rural: Regiones Periféricas Europeas, las Áreas de Montaña.

Cunha, L. (2003). A montanha do centro português: espaço de refúgio, território marginal e recurso para o desenvolvimento local. In *Território, Ambiente e Trajectórias de Desenvolvimento* (pp. 72–85). Centro de Estudos Geográficos, Universidade de Coimbra.

Curtis, K. R., & Slocum, S. L. (2016). Role of Sustainability Certification Programs in Reducing Food Waste in Tourism. *Journal of Developments in Sustainable Agriculture, 11*, 1–7. doi:10.11178/jdsa.11.1

Cusack, D., & Dixon, L. (2006). Community-based ecotourism and sustainability: Cases in Bocas del Toro Province, Panama and Talamanca, Costa Rica. *Journal of Sustainable Forestry, 22*(1-2), 157–182. doi:10.1300/J091v22n01_09

Daghri, T., Jorio, H., & Kasmi, S. (2018). Communicating a new vision of tourism: The case of Volubilis, Meknès, Morocco. *International Scientific Conference on Economic and Social Development Proceedings*, 297-306.

Daily Sabah Turkey. (2019). *Hotels fighting food waste to be recognised.* Retrieved October 7, 2019, from https://www.dailysabah.com/turkey/2019/04/15/hotels-fighting-food-waste-to-be-recognized

Dallimer, M., & King, T. (2008). Habitat preferences of the forest birds on the island of Príncipe, Gulf of Guinea. *African Journal of Ecology, 46*(3), 258–266. doi:10.1111/j.1365-2028.2007.00891.x

Daly, H. (1990). Toward some operational principles of sustainable development. *Ecological Economics, 2*(1), 1–6. doi:10.1016/0921-8009(90)90010-R

Darcy, S. (2006). *Setting a research agenda for accessible tourism.* Sustainable Tourism CRC.

Darcy, S., Cameron, B., & Pegg, S. (2010). Accessible tourism and sustainability: Adiscussion and case study. *Journal of Sustainable Tourism, 18*(4), 515–537. doi:10.1080/09669581003690668

Darcy, S., & Dickson, T. (2009). A Whole-of-Life Approach to Tourism: The Case for Accessible Tourism Experiences. *Journal of Hospitality and Tourism Management, 16*(1), 32–44. doi:10.1375/jhtm.16.1.32

David, B. (2001). Partnership – participation – power: the meaning of empowerment in post-industrial society. *Partnership WorkingPolicy and Practice*, 243–259.

Dax, T. (2008). *The role of mountain regions in territorial cohesion: a contribution to the discussion on the Green Paper on territorial cohesion.* Euromontana. https://www.euromontana.org/wpcontent/uploads/2014/08/final_dax_report_april2008_enlight.pdf

Dax, T. (2019). Development of mountainous regions: smart specialisation approaches as a means to overcoming peripheralisation. In I. Kristensen, A. Alexandre Dubois, & J. Teräs (Eds.), *Strategic Approaches to Regional Development. Smart Experimentationin Less-Favoured* (pp. 52–67). Routledge.

de Kadt, E. (1979). Social planning for tourism in the developing countries. *Annals of Tourism Research, 6*(1), 36–48. doi:10.1016/0160-7383(79)90093-8

Debarbieux, B., Oiry Varacca, M., Rudaz, G., Maselli, D., Kohler, T., & Jurek, M. (Eds.). (2014). Tourism in Mountain Regions: Hopes, Fears and Realities. In Sustainable Mountain Development Series. Geneva, Switzerland: UNIGE, CDE, SDC.

Debarbieux, B., & Rudaz, G. (2015). *The Mountain: A Political History from the Enlightenment to the Present*. University of Chicago Press. doi:10.7208/chicago/9780226031255.001.0001

Dejonckheere, M., & Vuaghn, L. (2019). Semistructured interviewing in primary care research: A balance of relationship and rigour. *BMJ Journals*, *7*(2), 1–8. doi:10.1136/fmch-2018-000057 PMID:32148704

Delgado, R. (2018). Community inclusiveness for tourism governance and globalisation: The case of Kanagawa prefecture, Japan. *International Scientific Conference on Economic and Social Development Book of Proceedings*, 642-647.

Delios, A., & Beamish, P. W. (1999). Geographic scope, product diversification, and the corporate performance of Japanese firms. *Strategic Management Journal*, *20*(8), 711–727. doi:10.1002/(SICI)1097-0266(199908)20:8<711::AID-SMJ41>3.0.CO;2-8

Denholm, M. (2019, October 11). Leaked report reveals 'unhealthy' workplace culture at tourism Tasmania. *The Australian*. Retrieved from https://www.theaustralian.com.au/travel/leaked-report-reveals-unhealthy-workplace-culture-at-tourism-tasmania/news-story/06e6fbdd33e5d06abaadccd21d2c0f86

Devereux, P. (2008). International Volunteering for Development and Sustainability: Outdated paternalism or a radical response to globalisation? *Development in Practice*, *18*(3), 357–370. doi:10.1080/09614520802030409

Devile, E. L., Jesus, F., & Cruz, A. I. (2011). O desenvolvimento de turismo acessível na lousã: Perspectivas dos agentes locais do sector do turismo. *International Conference on Tourism & Management Studies, 1*, 529-539.

Devile, E. (2009). O desenvolvimento do Turismo Acessível: Dos argumentos sociais aos argumentos de mercado. *Revista Turismo & Desenvolvimento, 11*, 39–46.

DGADR. (n.d.). *Fator de desenvolvimento rural*. Retrieved March 14, 2020, from https://www.dgadr.gov.pt/diversificacao/turismo-rural/fator-de-desenvolvimento-rural

DGEG. (2019). *Mapa das Águas Minerais – October of 2019*. http://www.dgeg.gov.pt/

Diakonidze, M. (2018). Tourism industry and prospects of development of the Georgian labor market. *Managerial Issues in Modern Business,* 98-103.

Dias, C. (2017). São Tomé e Príncipe | Na ilha do Príncipe, como em muitas outras, já se ouve a orquestra do Titanic. *Público*.

Díaz del Campo, J. (2012). Objetivos pedagógicos básicos en la enseñanza ética de la comunicación. *Vivat Academia. Revista de Comunicación, 15*(121), 1–16. doi:10.15178/va.2012.121.1-16

Dien, N. X., Hoa, H. T., Huong, N. T. T., Thuy, L. T., Nga, N. T. Q., & Nguyen, P. T. (2019). Sustainable development of Vietnam's Marine Tourism. *Environmental Management and Sustainable Development, 8*(4).

Dinç, A., & Kocan, N. (2012). *Ecotourism Planning of Murat Mountain (Usak, Kutahya/Turkey). International Conference on Applied Life Sciences (ICALS2012), Turkey, September 10- 12, 2012*. Retrieved from http://creativecommons.org/licenses/ by/3.0

Dini, A., & Lippit, V. (2009). *Poverty, from Orthodox to Heterodox Approaches: a Methodological Comparison Survey*. University of California- Riverside.

do Ambiente, A. (n.d.). *Sistema Nacional de Informação de Ambiente*. https://sniamb.apambiente.pt/

Do Le, T. T. (2017). *The reasons and solutions for the under-representation of women in the hotels' senior management* (Thesis). Submitted to Helsinki Metropolia University of Applied Sciences. Available at: https://www.theseus.fi/bitstream-handle/10024/129302/Do

do Valle, P. O., Mendes, J., & Guerreiro, M. (2012). Residents' Participation in Events, Events Image, and Destination Image: A Correspondence Analysis. *Journal of Travel & Tourism Marketing*, *29*(7), 647–664. doi:10.1080/10548408.2012.720151

Dolezal, C. (2011). Community-based tourism in Thailand: (Dis-)illusions of authenticity and the necessity for dynamic concepts of culture and power. *ASEAS – Austrian Journal of Southeast Asian Studies*, *4*(1), 129–138.

Domínguez, T., Fraiz, J., & Elisa, A. (2017). Economic profitability of accessible tourism for the tourism sector in Spain. *Tourism Economics*, *19*(6), 1385–1399. doi:10.5367/te.2013.0246

Dowling, R. (2001). Oceania (Australia, New Zealand and South Pacific). In D. Weaver (Ed.), *The encyclopedia of ecotourism* (pp. 139–154). CABI Publishing.

Dowling, R. (2009). Geotourism's contribution to Local and Regional Development. In C. Neto de Carvalho & J. C. Rodrigues (Eds.), *Geoturismo & Desenvolvimento Local* (pp. 15–37). Idanha-a-Nova.

Dowling, R. (2011). Geotourism's global growth. *Geoheritage*, *3*(1), 1–13. doi:10.100712371-010-0024-7

Dowling, R. (2015). Geotourism's contribution to sustainable tourism. In M. Hughes, D. Weaver, & C. Pforr (Eds.), *The Practice of Sustainable Tourism: Resolving the paradox London* (pp. 207–227). Routledge.

Drexhage, J., & Murphy, D. (2010). Sustainable development: From Brundtland to Rio 2012. Background paper for the high-level panel on global sustainability. New York: United Nations.

Drexler, C., Braun, V., Christie, D., Claramunt, B., Dax, T., Jelen, I., Kanka, R., Katsoulakos, N., Le Roux, G., Price, M., Scheurer, T., & Weingartner, R. (2016). *Mountains for Europe's Future—A Strategic Research Agenda.* Bern, Switzerland, and Innsbruck, Austria: Mountain Research Inititative (MRI) and Institut für Interdisziplinäre Gebirgsforschung (IGF). http://mri.scnatweb.ch/en/resources/publications/mri-publications/2573-mountains-for-europe-s-future-a-strategic-research-agenda/file

Dritsakis, N. (2004). Tourism as a long-run economic growth factor: An empirical investigation for Greece using causality analysis. *Tourism Economics*, *10*(3), 305–316. doi:10.5367/0000000041895094

Dwyer, L., Seetaram, N., Forsyth, P., & King, B. (2014). Is the migration-tourism relationship only about VFR? *Annals of Tourism Research*, *46*, 130–143. doi:10.1016/j.annals.2014.03.009

Dyer, P., Aberdeen, L., & Schuler, S. (2007). Tourism impacts on an Australian indigenous community. *Tourism Management*, *24*(1), 83–95. doi:10.1016/S0261-5177(02)00049-3

Dyll-Myklebust, L. (2012). *Public-Private-Community Partnership Model for Participatory Lodge (Tourism) Development.* Cultural Tourism and Identity., doi:10.1163/9789004234581_014

EarthCheck. (2020). *Good for business. good for the planet.* EarthCheck.

East Coast Stockfeeds. (2020). *Grape Marc.* Retrieved from http://ecsf.com.au/grape-marc

EC. (2004). *Mountain areas in Europe: Analysis of mountain areas in the EU Member States, acceding and other European countries.* Stockholm: Nordregio. https://ec.europa.eu/regional_policy/sources/docgener/studies/pdf/montagne/mount1.pdf

Echeita, G., & Ainscow, M. (2010). *La Educación inclusiva como derecho. Marco de referencia y pautas de acción para el desarrollo de una revolución pendiente.* En II Congreso Iberoamericano de Síndrome de Down. https://www.webcitation.org/6vTVhQziA

Eder, W., & Patzak, M. (2004). Geoparks – geological attractions: A tool for public education, recreation, and sustainable economic development. *AUC Geographica*, *51*(2), 235–246. doi:10.18814/epiiugs/2017/v40i4/017036

Edwards, M., & Gaventa, J. (2014). *Global citizen action*. Routledge. doi:10.4324/9781315071916

EEA. (2010). *Europe's ecological backbone: recognising the true value of our mountains. Copenhagen*. European Environment Agency. https://www.eea.europa.eu/publications/europes-ecological-backbone

Ee, O., Halim, H. A., & Ramayah, T. (2012). The effects of partnership quality on business process outsourcing success in Malaysia: Key users perspective. *Service Business*, *7*(2), 227–253. doi:10.100711628-012-0152-z

Ehsan, S., Leman, M., & Begum, R. (2012). Geotourism: A Tool for Sustainable Development of Geoheritage Resources. *Advanced Materials Research*, *622-623*, 1711–1715. doi:10.4028/www.scientific.net/AMR.622-623.1711

Elkington, J. (1994). Towards the Sustainable Corporation: Win-Win-Win Business Strategies for Sustainable Development. *California Management Review*, *36*(2), 90–100. doi:10.2307/41165746

Ellis, F. (1999). *Rural livelihood diversity in developing countries: evidence and policy implications*. Overseas Development Institute London.

Elzarov, Z. (2015). Community Stabilisation. *Journal of Security and Development*, *4*(1), 7–14.

Emmanuel, A., Nsor, C. A., & Boadi, S. (2017). The socio-cultural impact of ecotourism on park-adjacent communities in Ghana. *African Journal of Hospitality, Tourism and Leisure*, *6*(2), 1–14.

Envia. (2018). *Fundacion En Via*. Retrieved from https://www.envia.org/

Erfurt-Cooper, P., & Cooper, M. (2009). Health and Wellness Tourism – Spas and hot springs. Bristol.

Erskine, L. M., & Meyer, D. (2012). Influenced and influential: The role of tour operators and development organisations in tourism and poverty reduction in Ecuador. *Journal of Sustainable Tourism*, *20*(3), 339–357. doi:10.1080/09669582.2011.630470

Eshetu, A. (2014). Ecotourism as a viable strategy for livelihood diversification and sustainable natural resource management in Ethiopia (from eco-development paradigm point of view). *Journal of Environmental Science and Water Resources*, *3*(2), 40–52.

ETC/WTO. (2005). *City tourism and culture - The European Experience*. European Travel Commision. Retrieved from www.etc-corporate.org

EU. (2008). *Green Paper on Territorial Cohesion*. European Commission.

European Commission. (2018). *ERASMUS+ 2017 in numbers Portugal*. Accessed 30 November 2019, from https://ec.europa.eu/programmes/erasmus-plus/about/statistics_en

European Food Banks Federation. (2019). *Towards the Next Decade Together: FEBA Annual Report 2018*. Retrieved October 16, 2019, from https://www.eurofoodbank.org/images/cont/feba-annual-report-2018-final_file.pdf

European Institute for Gender Equality. (2020). *Tourism: Relevance of gender in the policy area*. Retrieved from https://eige.europa.eu/gender-mainstreaming/policy-areas/tourism

Eurostat. (2019). *Employment in services*. Accessed 15 October 2019, from https://ec.europa.eu/

Eurostat. (2019). *Purchasing power adjusted GDP per capita*. http://https://ec.europa.eu/eurostat/databrowser/view/sdg_10_10/default/table?lang=en

Eusébio, C., & Figueiredo, E. (2014). Turismo e desenvolvimento sustentável de destinos rurais. In E. Kastenholz, C. Eusébio, M. J. Carneiro, & J. Lima (Eds.), *Reinventar o turismo rural em Portugal: Cocriação de experiências turísticas sustentáveis* (pp. 51–58). UA Editora.

Extremera, N., & Fernández-Berrocal, P. (2003). La Inteligencia Emocional en el contexto educativo: Hallazgos científicos de sus efectos en el aula. *Review of Education, 332*, 97–116.

Fadahunsi, J. T. (2011). Application of Geographical Information System (GIS) Technology to Tourism Management in Ile-Ife, Osun State, Nigeria. *Pacific Journal of Science and Technology, 12*(2), 274–283.

Falls, V. (2013). *Turismo Acessível para Todos - Recomendações OMT*. Academic Press.

Farsani, N. T., Coelho, C., & Costa, C. (2012). Geotourism and geoparks as gateways to socio-cultural sustainability in Qeshm rural areas, Iran. *Asia Pacific Journal of Tourism Research, 17*(1), 30–48. doi:10.1080/10941665.2011.610145

Fathom. (2019). The top 15 places to travel in 2019. *Forbes*. Retrieved from <https://www.forbes.com/sites/fathom/2019/01/15/top-travel-destinations-2019/#44ef5012523e>

Fazenda, N.; Ladeiras, A.; Vieira, T.; Barbosa, D. (2009). Programa de Acção de Enfoque Temático "Turismo de Saúde e Bem-Estar" para o Porto e Norte de Portugal. In *Agenda Regional de Turismo*. Lisbon: TP.

Fennell, D. A., & Dowling, R. K. (2003). Ecotourism Policy and Planning. CABI Publishing.

Ferguson, J. (1990). *The anti-politics machine:" development," depoliticisation, and bureaucratic power in Lesotho*. CUP Archive.

Ferguson, L. (2011). Promoting gender equality and empowering women? Tourism and the third Millennium Development Goal. *Current Issues in Tourism, 14*(3), 235–249. doi:10.1080/13683500.2011.555522

Fernandes, G. (2012). Políticas públicas e equidade territorial em zonas de montanha: O exemplo da Cordilheira Central. In Respuestas de la Geografía Ibérica a la crisis actual, XIII Coloquio Ibérico de Geografía. AGE e APG, Santiago de Compostela.

Fernandes, G. (2015). *Geografia, cultura e desenvolvimento turístico em territórios de montanha: Das perspetivas às logicas de atuação e governação territorial*. Coleção Instituto Politécnico da Guarda.

Fernandes, G. (2017). Tourism, ecocultural heritage, development strategies, and territorial valorisation in the border mountains of the Iberian central mountain range. *Revista Egitania Scientia, 17*, 67–81.

Fernandes, G. (2018). Resilience and territorial dynamics in the mountains in the extreme western areas of the Iberian Peninsula Range: Challenges of governance, territorial cohesion, heritage and tourism. In E. Gomez-Pellón (Ed.), *Rural worlds, social sustainability, and local landscapes in the globalisation era. Case studies in southern Europe* (pp. 265–288). Thomson Reuters Editorial ARANZADI.

Fernandes, G. (2020). Dynamics of tourism development and resilience in the medium mountains of Portugal: Challenges for the sustainable management in mountains of the Central Mountain Range. In D. Kala & S. Bagri (Eds.), *Global Opportunities and Challenges for Rural and Mountain Tourism* (pp. 263–282). IGI Global. doi:10.4018/978-1-7998-1302-6.ch015

Ferraro, J., & Kiss, A. (2002). Direct Payments to Conserve Biodiversity. *Science*, 298. PMID:12459569

Ferreira, J., & Sousa, B. (2020) Experiential Marketing as Leverage for Growth of Creative Tourism: A Co-creative Process. In Advances in Tourism, Technology and Smart Systems. Smart Innovation, Systems and Technologies, (vol. 171. pp 567-577). Springer, Singapore Systems, Smart Innovation, Systems and Technologies 171. doi:10.1007/978-981-15-2024-2_49

Fontes, A., & Monteiro, I. (2009). O Projecto "Lousã, Destino de Turismo Acessível": um estudo de caso da aplicação de uma abordagem sistémica ao Turismo Acessível. *Revista Turismo & Desenvolvimento*, (11), 61-72.

Font, X., & McCabe, S. (2017). Sustainability and marketing in tourism: Its contexts, paradoxes, approaches, challenges and potential. *Journal of Sustainable Tourism*, *25*(7), 869–883. doi:10.1080/09669582.2017.1301721

Food and Agriculture Organization of the United Nations. (2019). *FAO in Egypt*. Retrieved October 5, 2019, from http://www.fao.org/egypt/programmes-and-projects/food-loss-waste-reduction/en/?fbclid=IwAR0i8NJyHSTnJKy0vdfeMq7PhbmkFvWc7rAX4-mC1yyEkKLV_oMvzNGldU8

Food Recovery Network. (2019). *2018 Annual Report*. Retrieved October 16, 2019, from https://static1.squarespace.com/static/555b5cf1e4b0864ccf1a0156/t/5c7041d141920266f5dce0ed/1550860780285/FY18+Annual+Report+for+Website.pdf

Ford, J., & Harding, N. (2011). The impossibility of the 'true self' of authentic leadership. *Leadership*, *7*(4), 463–479. doi:10.1177/1742715011416894

Ford, R., Williams, K., Bishop, I., & Webb, T. (2009). A value basis for the social acceptability of clearfelling in Tasmania, Australia. *Landscape and Urban Planning*, *90*(3-4), 196–206. doi:10.1016/j.landurbplan.2008.11.006

Formación Curricular en Diseño para Todas las Personas en Turismo. (2017). *Crue Universidades Españolas*.

Freestone, O. M., & McGoldrick, P. J. (2008). Motivations of the ethical consumer. *Journal of Business Ethics*, *79*(4), 445–467. doi:10.100710551-007-9409-1

Frenzel, F. (2013). Slum tourism in the context of the tourism and poverty (relief) debate. *Journal of the Geographical Society of Berlin*, *144*(2).

Frenzel, F., & Koens, K. (2012). Slum tourism: Developments in a young field of interdisciplinary tourism research. *Tourism Geographies*, *14*(2), 195–212. doi:10.1080/14616688.2012.633222

Frey, M. L., Schafer, K., Buchel, G., & Patzak, M. (2016). *Geoparks: A Regional European and Global Policy*. Geotourism. doi:10.1016/B978-0-7506-6215-4.50014-2

Friedman, B. M. (2017). The moral consequences of economic growth. In J. Imber (Ed.), *Markets, Morals, and Religion* (pp. 29–42). Routledge. doi:10.4324/9780203786222-2

Frueh, S. (1988). *Report to WWF on Tourism to Protected Areas*. Academic Press.

Gajdosik, T., Gajdosiková, Z., Maráková, V., & Flagestad, A. (2017). Destination structure revisited in view of the community and corporate model. *Tourism Management Perspectives*, *24*, 54–63. doi:10.1016/j.tmp.2017.07.009

Gama, I., & Sanoussi, F. (2017). African Economic Outlook. Academic Press.

Garcia-Garcia, G., Woolley, E., Rahimifard, S., Colwill, J., White, R., & Needham, L. (2017). A Methodology for Sustainable Management of Food Waste. *Waste and Biomass Valorization*, *8*(6), 2209–2227. doi:10.100712649-016-9720-0 PMID:32010409

Gardner, W. L., Cogliser, C. C., Davis, K. M., & Dickens, M. P. (2011). Authentic leadership: A review of the literature and research agenda. *The Leadership Quarterly*, *22*(6), 1120–1145. doi:10.1016/j.leaqua.2011.09.007

Gardner, W., Avolio, B., Luthans, F., May, D., & Walumbwa, F. (2005). "Can you see the real me?" A self-based model of authentic leader and follower development. *The Leadership Quarterly*, *16*(3), 343–372. doi:10.1016/j.leaqua.2005.03.003

Garrod, B., Wornell, R., & Youell, R. (2006). Re-conceptualising rural resources as countryside capital: The case of rural tourism. *Journal of Rural Studies*, *22*(1), 117–128. doi:10.1016/j.jrurstud.2005.08.001

Gartner, C. M. (2008). Tourism, Development, and Poverty Reduction: A Case Study from Nkhata Bay, Malawi (Master of Environmental Studies in Geography - Tourism Policy and Planning). University of Waterloo.

Gascon, J. (2015). Pro-poor tourism as a strategy to fight rural poverty: A critique. *Journal of Agrarian Change*, *15*(4), 499–518. doi:10.1111/joac.12087

Gaucher, D., Friesen, J., & Kay, A. C. (2011). Evidence that gendered wording in job advertisements exists and sustains gender inequality. *Journal of Personality and Social Psychology*, *101*(1), 109–128. doi:10.1037/a0022530 PMID:21381851

GEF. (2018). *Sao Tome and Principe | Global Environment Facility*. https://www.thegef.org/country/sao-tome-and-principe

Gerdt, S. O., Wagner, E., & Schewe, G. (2019). The relationship between sustainability and customer satisfaction in hospitality: An explorative investigation using eWOM as a data source. *Tourism Management*, *74*, 155–172. doi:10.1016/j.tourman.2019.02.010

Getz, & Page, S. J. (2014). Progress and prospects for event tourism research. *Tourism Management, 52*, 593–631. doi:10.1016/j.tourman.2015.03.007

Getz, D. (2014). *The forms and functions of planned events: Past and future. The Future of Events and Festivals*. Academic Press.

Getz, D., & Page, S. J. (2007). Event Studies Theory, research and policy for planned events. In Development (3rd ed., vol. 134). Academic Press.

Getz, D. (2008). Event tourism: Definition, evolution, and research. *Tourism Management*, *29*(3), 403–428. doi:10.1016/j.tourman.2007.07.017

Getz, D., & Frisby, W. (1988). Evaluating Management Effectiveness In Community-Run Festivals. *Journal of Travel Research*, 21–27. doi:10.1177/004728758802700105

Gheorghe, C. (2013). Ecotourism: Definition and Concepts. *Journal of Tourism Studies and Research in Tourism*, *15*, 56–60.

Giampiccoli, A., & Kalis, J. H. (2012). Community-based tourism and local culture: The case of the Mamapondo. *Pasos Revista de Turismo y Patrimonio Cultural*, *10*(1), 173–188. doi:10.25145/j.pasos.2012.10.017

Giampiccoli, A., & Saayman, M. (2017). Community-based tourism, responsible tourism, and infrastructures development and poverty. *African Journal of Hospitality, Tourism and Leisure*, *6*(2), 1–28.

Giebelhausen, M., Chun, H. E. H., Cronin, J. J. Jr, & Hult, T. M. (2016). Adjusting the warm-glow thermostat: How incentivizing participation in voluntary green programs moderates their impact on service satisfaction. *Journal of Marketing*, *80*(4), 56–71. doi:10.1509/jm.14.0497

Gil, J. D. B., Reidsma, P., Giller, K., Todman, L., Whitmore, A., & van Ittersum, M. (2019). Sustainable development goal 2: Improved targets and indicators for agriculture and food security. *Ambio*, *48*(7), 685–698. doi:10.100713280-018-1101-4 PMID:30267284

Gil-Soto, E., Armas-Cruz, I., Morini-Marrero, S., & Ramos-Henriquez, J. M. (2019). Hotel guests' perceptions of environmental friendly practices in social media. *International Journal of Hospitality Management*, *78*, 59–67. doi:10.1016/j.ijhm.2018.11.016

Giovannoni, E., & Fabietti, G. (2010). *What is sustainability? A review of the concept and its applications. Springer International Publishing Switzerland*. Department of Business and Law, University of Siena.

Glenuig Inn. (2019). *Being Green*. Retrieved October 17, 2019, from https://glenuig.com/being-green/

Gløersen, E., Price, M., Borec, A., Dax, T., & Giordano, B. (2016). Cohesion in Mountainous Regions of the EU—Research for REGI Committee, European Parliament, Directorate-General for Internal Policies Policies Department B. In *Structural and Cohesion Policies, Regional Development, IP/B/REGI/IC/2015_175*. Brussels, Belgium: European Parliament. https://www.europarl.europa.eu/RegData/etudes/STUD/2016/573420/IPOL_STU%282016%29573420_EN.pdf

Goldin, C., Kerr, S. P., Olivetti, C., & Barth, E. (2017). The expanding gender earnings gap: Evidence from the LEHD-2000 Census. *The American Economic Review*, *107*(5), 110–114. doi:10.1257/aer.p20171065

Goldstein, N. J., Cialdini, R. B., & Griskevicius, V. (2008). A room with a viewpoint: Using social norms to motivate environmental conservation in hotel. *The Journal of Consumer Research*, *35*(3), 35. doi:10.1086/586910

Gomes, H., Fernandes, M., Castro, E., & Vieira, G. &. (2017). The (Serra da) Estrela Aspiring Geopark (Portugal): Preserving geoheritage, while promoting science and its links to local communities. *Geophysical Research Abstracts* Vol. 19. EGU General Assembly.

Gomiero, T. (2016). Soil degradation, land scarcity and food security: Reviewing a complex challenge. *Sustainability*, *8*(3), 281. doi:10.3390u8030281

Gonçalves, S. (2011). Intangible culture, cooperation and intercultural dialogue among university students. *Intercultural Education*, *22*(1), 83–95. doi:10.1080/14675986.2011.549647

Goodwin, H., & Santilli, R. (2009). *Community-based tourism: A success?* ICRT Occasional paper 11, 1-37.

Goodwin, H. (1996). In pursuit of ecotourism. *Biodiversity and Conservation*, *5*(3), 277–291. doi:10.1007/BF00051774

Govern, J. M., & Marsch, L. A. (2001). Development and validation of the situational self-awareness scale. *Consciousness and Cognition*, *10*(3), 366–378. doi:10.1006/ccog.2001.0506 PMID:11697870

Government of Kenya. (2011). *Tourism Act 2010*. Government Printer.

Government of Kenya. (2013). Sector plan for tourism 2013-2017. Nairobi, Kenya: Author.

Government of Kenya. (2014). Kenya National Bureau of Statistics. Economic Survey. Nairobi, Kenya: Author.

Governo Cabo Verde (GCV). (2016). *Programa do Governo – IX Legislatura*. Cabo Verde.

Governo Cabo Verde (GCV). (2018). *PEDS: Plano Estratégico de Desenvolvimento Sustentável 2017/2021*. Cabo Verde: Ministério Finanças – Direção Nacional Planeamento.

Governo Cabo Verde (GCV). (2019). Grandes Opções do Plano *Estratégico de Desenvolvimento Sustentável do Turismo para 2018-2030*. Cabo Verde: Boletim Oficial I Série, N.º 2.

Graeber, D. (2007). Army of Altruists: on the alienated right to do good; Essay. *Harper's Magazine*, *314*(1880), 31.

Grant, M., & Booth, A. (2009). A typology of reviews: An analysis of 14 review types and associated methodologies. *Health Information and Libraries Journal*, *26*(2), 91–108. doi:10.1111/j.1471-1842.2009.00848.x PMID:19490148

Gray, M. C., & Shankar, S. (2018). *The role of informal wage negotiations and formal promotion processes in explaining the gender wage gap: Centre for Social research & Methods*. Australian National University.

Green Lodging News and Greenview. (2017). *Green Lodging Trends Report 2017*. Retrieved October 2, 2019, from https://greenview.sg/wp-content/uploads/2017/09/green-lodging-trends-report-2017.pdf

Griffin, T., & Dimanche, F. (2017). Urban tourism: The growing role of VFR and immigration. *Journal of Tourism Futures*, *3*(2), 103–113. doi:10.1108/JTF-10-2016-0036

Guerra, R. (2018). Turismo de Saúde e Bem-Estar: para uma estratégia de desenvolvimento local. O caso das Caldas da Cavaca. Maia: Ed. ISMAI e CEDTUR.

Guerra, R., & Gonçalves, E. (2019). "O turismo de saúde e bem-estar como fator de desenvolvimento local: Uma análise à oferta termal portuguesa". *PASOS. Revista de Turismo y Patrimonio Cultural*, *17*(2), 453–472. doi:10.25145/j.pasos.2019.17.030

Gullison, R., Rice, R., & Blundell, A. (2000). 'Marketing'species conservation. *Nature*, *404*(923). PMID:10801103

Güngör, S., & Bozyigit, R. (2011). Tourism potential of Gazipasa (Antalya-Turkey). *Procedia: Social and Behavioral Sciences*, *19*, 231–239. doi:10.1016/j.sbspro.2011.05.128

Gursoy, D., Kim, K., & Uysal, M. (2004). Perceived impacts of festivals and special events by organisers: An extension and validation. *Tourism Management*, *25*(2), 171–181. doi:10.1016/S0261-5177(03)00092-X

Gustafson, P., & Cardozo, E. L. (2017). Language use and social inclusion in international retirement migration. *Social Inclusion (Lisboa)*, *5*(4), 69–77. doi:10.17645i.v5i4.1133

Gustavo, N. (2010). *Os novos espaços de lazer, turismo e saúde em Portugal: o caso dos SPAS*. Tese de Doutoramento. Universidade de Coimbra.

Guzmán, T., Borges, O., & Castillo, A. M. (2011). Desarrollo económico local y turismo comunitario en países en vías de desarrollo. Un estudio de caso. *Revista de Ciencias Sociales*, *17*(3), 432–446.

Hadzhipetrova-Lachova, M. (2017). EU Policy Initiatives in the Mediterranean Region. *International Conference Knowledge-Based Organisation*, *23*(1), 130-134. 10.1515/kbo-2017-0020

Hall, C. (2019). Constructing sustainable tourism development: The 2030 Agenda and the managerial ecology of sustainable tourism. *Journal of Sustainable Tourism*, *27*(7), 1044–1060. doi:10.1080/09669582.2018.1560456

Hall, C. M. (1991). *Introduction to tourism in Australia: Impacts, planning and development*. Wesley and Longman.

Hall, C. M. (2010). Changing paradigms and global change: From sustainable to steady-state tourism. *Tourism Recreation Research*, *35*(2), 131–145. doi:10.1080/02508281.2010.11081629

Han, H., Hsu, L. T., & Lee, J.-S. (2009). Empirical investigation of the roles of attitudes toward green behaviors, overall image, gender, and age in hotel customers' eco-friendly decision-making process. *International Journal of Hospitality Management*, *28*(4), 519–528. doi:10.1016/j.ijhm.2009.02.004

Hardeman, G., Font, X., & Nawijn, J. (2017). The power of persuasive communication to influence sustainable holiday choices: Appealing to self-benefits and norms. *Tourism Management*, *59*, 484–493. doi:10.1016/j.tourman.2016.09.011

Harding, J. (2013). *Qualitative data analysis from start to finish*. Sage Publication.

Hardy, A., & Pearson, L. J. (2018). Examining stakeholder group specificity: An innovative sustainable tourism approach. *Journal of Destination Marketing & Management*, *8*, 247–258. doi:10.1016/j.jdmm.2017.05.001

Harrison, D. (2008). Pro-poor tourism: A critique. *Third World Quarterly*, *29*(5), 851–868. doi:10.1080/01436590802105983

Hartman, L. P., & Dhanda, K. K. (2018). Cross-Sector Partnerships: An Examination of Success Factors. *Business and Society Review*, *123*(1), 181–214. doi:10.1111/basr.12139

Hassan, A., & Rahimi, R. (2016). Consuming "Innovation" in Tourism: Augmented Reality as an Innovation Tool in Digital Tourism Marketing. In Global dynamics in travel, tourism, and hospitality (pp. 130-147). IGI Global.

Haughton, J. H., & Khandker, S. R. (2009). *Handbook on poverty and inequality*. World Bank Publications.

Hawkins, D. E., & Lamoureux, K. (2001). Global growth and magnitude of ecotourism. In D. B. Weaver (Ed.), *The encyclopedia of ecotourism*. CABI International.

HBD. (2020). *The Island of Príncipe*. https://press.hbdprincipe.com/pdf/press_release/Beyond_Green-Principes_sustainability_story.pdf

Higgins-Desbiolles, F. (2006). More than an "industry": The forgotten power of tourism as a social force. *Tourism Management*, *27*(6), 1192–1208. doi:10.1016/j.tourman.2005.05.020

Higgins-Desbiolles, F. (2008). Justice tourism and alternative globalisation. *Journal of Sustainable Tourism*, *16*(3), 345–364. doi:10.1080/09669580802154132

Hill, M. M., & Hill, A. (2005). *Investigação por questionário* (2nd ed.). Edições Sílabo.

Hoey, B. A. (2005). From pi to pie: Moral narratives of noneconomic migration and starting over in the postindustrial Midwest. *Journal of Contemporary Ethnography*, *34*(5), 586–624. doi:10.1177/0891241605279016

Holden, A., Sonne, J., & Novelli, M. (2011). Tourism and poverty reduction: An interpretation by the poor of Elmina, Ghana. *Tourism Planning & Development*, *8*(3), 317–334. doi:10.1080/21568316.2011.591160

Holdnak, A., & Holland, S. M. (1996). Edu-tourism: Vacationing to learn. *Parks & Recreation*, *31*(9), 72–75.

Hoobler, J. M., Wayne, S. J., & Lemmon, G. (2009). Bosses' perceptions of family-work conflict and women's promotability: Glass ceiling effects. *Academy of Management Journal*, *52*(5), 939–957. doi:10.5465/amj.2009.44633700

Hopp, K. J. (2015). How Effective Are Gender Quotas in Achieving Meaningful Change for Women? A Case Study of Argentina. *Interstate - Journal of International Affairs*, *2015/2016*(1). http://www.inquiriesjournal.com/a?id=1238

Hose, T. (2012). 3gs for modern geotourism. *Geoheritage*, *4*(1-2), 7. doi:10.100712371-011-0052-y

Hose, T. (2005). Geoturismo europeo. Interpretacíon geológica y promoción de la conservación geológica para turistas. In D. Barretino, W. A. P. Winbledon, & E. Gallego (Eds.), *Patrimonio geológico: conservación y gestión*. Istituto Tecnológico Geomineiro de España.

Hose, T. (2011). The English origins of geotourism (as a vehicle for geo-conservation) and their relevance to current studies. *Acta Geographica Slovenica*, *51*(3), 343–359. doi:10.3986/AGS51302

Hose, T., & Vasiljevi, D. A. (2012). Defining the nature and purpose of modern geotourism with particular reference to the United Kingdom and South-East Europe. *Geoeheritage*, *4*(1-2), 25–43. doi:10.100712371-011-0050-0

Hsu, M., Bhatt, M., Adolphs, R., Tranel, D., & Camerer, C. F. (2005). Neural systems responding to degrees of uncertainty in human decision-making. *Science*, *310*(5754), 1680–1683. doi:10.1126cience.1115327 PMID:16339445

Huang, X., & McMurray, A. (2016). *Regional Australia's engagement with China in the Asian century: Innovation, integration and impact*. RMIT University.

Hudson, S., & Hudson, L. (2020). *A Worldwide Guide to Retirement Destinations*. Cambridge Scholars Publishing.

Huffman, M. L., & Velasco, S. C. (1997). When more is less: Sex composition, organisations, and earnings in US firms. *Work and Occupations*, *24*(2), 214–244. doi:10.1177/0730888497024002005

Hulme, D., & Murphree, M. (2001). African wildlife and livelihoods: the promise and performance of community conservation. Academic Press.

Hunter, C. J. (1995). On the need to re-conceptualise sustainable tourism development. *Journal of Sustainable Tourism*, *3*(3), 155–165. doi:10.1080/09669589509510720

Hurley, G. (2015). *Financing for development and small island developing states: A snapshot and ways forward. UNDP and UN-OHRLLS*. https://sustainabledevelopment.un.org/content/documents/2181(UNDP & OHRLLS 2015) Financing for development and SIDS A snapshot and ways forward.pdf

ILO. (2017). *Rapid market assessment of responsible tourism ın Vietnam*. International Labour Organization.

IMF. (2013). *IMF Country Report nº14/2, Democratic Republic of São Tomé and Príncipe*. Academic Press.

INE. (2019). *Estatísticas Demográficas - 2018*. Lisboa: Instituto Nacional de Estatística, IP.

INE-STP. (2019). *Boletim Estátistico da Direção Geral do Turismo e Hotelaria*. https://www.ine.st/index.php/component/phocadownload/file/467-boletim-estatistico-da-direcao-geral-do-turismo-e-hotelaria

Infante, E. (2014). Guerra, historia, turismo y prensa: bases de la puesta en valor del patrimonio cultural de Niebla (Huelva) durante el franquismo. *Historia Y Comunicación Social, 19,* 153-170. doi:10.5209/rev_HICS.2014.v19.47290

Instituto Nacional Estatística (INE). (2015). *III Inquérito às Despesas e Receitas Familiares (IDRF)*. Cabo Verde: INE.

International Labour Organization (ILO). (2017). *Reducing gender gaps would significantly benefit women, society and the economy*. Retrieved 15 February 2020, from https://www.ilo.org/global/about-the-ilo/newsroom/news/WCMS_557266/lang--en/index.htm

International Labour Organization (ILO). (2018). *What lies behind gender pay gaps. Global Wage Report 2018/19*. Retrieved 7 June 2020, https://www.ilo.org/wcmsp5/groups/public/---dgreports/---dcomm/---publ/documents/publication/wcms_650553.pdf

International Labour Organization (ILO). (2020). *Labour statistics on women*. Retrieved 7 June 2020, from https://ilostat.ilo.org/topics/women/

International Monetary Fund (IMF). (2010). *Kenya: poverty reduction strategy paper*. IMF Country Report No. 10/224.

International Vision University. (2017). Yaratıcı Turizm Kapsamında Şirince Köyü İçin Bir Uygulama Modeli Önerisi ve Köyün Yaratıcı Turizm Kaynakları. In *Mediterranean International Congress on Social Sciences MECAS II*. International Vision University Ohrid.

Inui, Y., Wheeler, D., & Lankford, S. (2006). Rethinking tourism education: What should schools teach. *Journal of Hospitality, Leisure, Sport and Tourism Education*, *5*(2), 25–35. doi:10.3794/johlste.52.122

Ioannides, D., & Petersen, T. (2003). Tourism "non-entrepreneurship" in peripheral destinations: A case study of small and medium tourism enterprises in Bornholm. *Tourism Geographies*, *5*(4), 408–435. doi:10.1080/1461668032000129146

IOM. (2010). *Mainstreaming Migration into Development Planning. A Handbook for Policy-Makers and Practitioners*. Geneva: Global Migration Group. Accessed 1 June 2020, from https://publications.iom.int/es/books/mainstreaming-migration-development-planning-handbook-policy-makers-and-practitioners

IOM. (2019). *World Migration Report 2020*. International Organization for Migration.

Iorio, M., & Corsale, A. (2010). Rural tourism and livelihood strategies in Romania. *Journal of Rural Studies*, *26*(2), 152–162. doi:10.1016/j.jrurstud.2009.10.006

Islam, M. W., Ruhanen, L., & Ritchie, B. W. (2018). Adaptive co-management: A novel approach to tourism destination governance? *Journal of Hospitality and Tourism Management*, *37*, 97–106. doi:10.1016/j.jhtm.2017.10.009

Ismail, M. N. I., Hanafiah, M. H., Aminuddin, N., & Mustafa, N. (2016). Community-based homestay service quality, visitor satisfaction, and behavioral intention. *Procedia: Social and Behavioral Sciences*, *222*, 398–405. doi:10.1016/j.sbspro.2016.05.192

Iszatt-White, M., & Kempster, S. (2018). Authentic leadership: Getting back to the roots of the 'root construct'? *International Journal of Management Reviews*, *21*(3), 356–369. doi:10.1111/ijmr.12193

Jakubiak, C., & Iordache-Bryant, I. (2017). Volunteer tourism in Romania as/for global citizenship. *Tourism Recreation Research*, *42*(2), 212–222. doi:10.1080/02508281.2017.1299344

Jamal, T., & Getz, D. (1999). Community Roundtables for Tourism-related Conflicts: The Dialectics of Consensus and Process Structures. *Journal of Sustainable Tourism*, *7*(3-4), 290–313. doi:10.1080/09669589908667341

Jamal, T., & Stronza, A. (2009). Collaboration theory and tourism practice in protected areas: Stakeholders, structuring and sustainability. *Journal of Sustainable Tourism*, *17*(2), 169–189. doi:10.1080/09669580802495741

Janeiro, P., Proença, I., & Gonçalves, V. C. (2013). Open Innovation: Factors explaining universities as service firm innovation services. *Journal of Business Research*, *66*(10), 2017–2023. doi:10.1016/j.jbusres.2013.02.027

Jarvis, M., Lange, G. M., Hamilton, K., Desai, D., Fraumeni, B., Edens, B., ... Ruta, G. (2011). *The changing wealth of nations: measuring sustainable development in the new millennium*. Academic Press.

Jiménez, I. (2009). *Una aproximación psicosocial a la relación entre Inteligencia Emocional y rendimiento académico*. Universidad de Jaén.

Johnson, C. (2013). *Social Entrepreneurship - Call for papers*. https://www.emeraldgrouppublishing.com/products/journals/call_for_papers.htm?id=4978

Jolliffe, L., & Farnsworth, R. (2003). Seasonality in tourism employment: Human resource challenges. *International Journal of Contemporary Hospitality Management*, *15*(6), 312–316. doi:10.1108/09596110310488140

Juma, L., & Khademi-Vidra, A. (2019). Community-based tourism and sustainable development of rural regions in Kenya; perceptions of the citizenry. *Sustainability*, *11*(17), 4733. doi:10.3390u11174733

Kahveci, C., Karacan, E., & Kosnick, K. (2020). Tactical mobility: navigating mobile ageing and transnational retirement between Turkey and Germany. A comparison between Turkish-German and German retirees. *Journal of Ethnic and Migration Studies*. Doi:10.1080/1369183X.2020.1724532

Kala, D., & Bagri, S. C. (2018). Barriers to local community participation in tourism development: Evidence from mountainous state Uttarakhand, India. *Tourism (Zagreb)*, *66*(3), 318–333.

Kalev, A., & Deutsch, G. (2018). Gender inequality and workplace organisations: Understanding reproduction and change. In *Handbook of the Sociology of Gender* (pp. 257–269). Springer. doi:10.1007/978-3-319-76333-0_19

Kambur, O., Petryshchenko, N., & Sehorina, N. (2018). Factors affecting the development of tourism industry. *Conference on Economic and Social Development Proceedings*, 125-133.

Kapos, V. (2000). Developing a map of the world's mountain forests. In Forests in sustainable mountain development: a state-of-knowledge report of 2000, (pp. 4-9). IUFRO Series N. 5 Wallingford, Reino Unido, Cab International.

Karampela, S., Papapanos, G., & Kizos, T. (2019). Perceptions of Agritourism and Cooperation: Comparisons between an Island and a Mountain Region in Greece. *Sustainability*, *11*(680), 1–18. doi:10.3390u11030680

Kasalak, M. A. (2015). Dünya'da Ekoturizm Pazarı ve Ekoturizm'in Ülke Gelirlerine Katkıları. *Journal of Recreation and Tourism Research*, *2*(2), 20–26.

Kastenholz, E., Eusébio, C., Figueiredo, E., & Lima, J. (2011). Accessibility as a competitive advantage of a tourist destination - the case of Lousã. In Field Guide for Case Study Research in Tourism, Hospitality, and Leisure, Advances in Culture, Tourism & Hospitality Research book series. Emerald Publishers.

Kayat, K. (2014). Community-based rural tourism: A proposed sustainability framework. *SHS Web of Conferences, 12.*

Kc, S. (2013). Roles and Challenges of Women in Tourism Sector of Western Nepal: A Micro- Ethnographic Study. *Nepal Tourism And Development Review*, *2*(1), 32–59. doi:10.3126/ntdr.v2i1.7379

Kensbock, S., Bailey, J., Jennings, G., & Patiar, A. (2014). Sexual Harassment of Women Working as Room Attendants within 5-Star Hotels. *Gender, Work and Organization*, *22*(1), 36–50. doi:10.1111/gwao.12064

Kenya National Bureau of Statistics (KNBS). (2019). *2019 Kenya population and housing census results.* Available Online: https://www.knbs.or.ke/?p=5621

Kepe, T., Saruchera, M., & Whande, W. (2004). Poverty alleviation and biodiversity conservation: A South African perspective. *Oryx*, *38*(382), 143–145. doi:10.1017/S0030605304000262

Kernis, M., & Goldman, B. (2006). A multicomponent conceptualisation of authenticity: Theory and research. *Advances in Experimental Social Psychology*, *38*, 283–357. doi:10.1016/S0065-2601(06)38006-9

Khader, B. F. Y., Yigezu, Y. A., Duwayri, M. A., Niane, A. A., & Shideed, K. (2019). Where in the value chain are we losing the most food? The case of wheat in Jordan. *Food Security*, *11*(5), 1009–1027. doi:10.100712571-019-00962-7 PMID:31976025

Kibara, O., Odhiambo, N., & Njuguna, J. (2012). Tourism and economic growth in Kenya: An empirical investigation. *International Business and Economics Research Journal*, *11*(5), 512–528. doi:10.19030/iber.v11i5.6970

Kihima, B. O. (2010). *Towards Unlocking Tourism Potential in Less Visited Destinations in Kenya: The Case of Western Kenya Region* (PhD Dissertation). University of Lyon, France.

Kihima, B. (2015). Domestic tourism in Kenya: Trends, initiatives and practices. *The East African Review*, *50*, 22–39.

Kim, J. (2016). Neoliberalism and the Politics of Social Enterprises in South Korea: The Dynamics of Neoliberal Governmentality and Hegemony (PhD). Georgia State University.

Kimenyi, M., Mwega, F., & Ndung'u, N. (2019). *The African lions: Kenya country case study.* Available Online: https://www.brookings.edu/wp-content/uploads/2016/07/kenya-country-case.pdf

Kim, H. J., Chen, M. H., & Jang, S. S. (2006). Tourism expansion and economic development: The case of Taiwan. *Tourism Management*, *27*(5), 925–933. doi:10.1016/j.tourman.2005.05.011 PMID:32287716

Kim, M. (2000). Women paid low wages: Who they are and where they work. *Monthly Labor Review*, *123*, 26.

Kim, N., & Lee, C. K. (2012). Granger causality among tourism development, foreign direct investment and economic growth: The case of Korea. *Korean Journal of Tourism Research*, *27*(3), 43–57.

Kim, S. B., & Kim, D. Y. (2014). The effects of message framing and source credibility on green messages in hotels. *Cornell Hospitality Quarterly*, *55*(1), 64–75. doi:10.1177/1938965513503400

Kiss, A. (2002). *"Making biodiversity conservation a land use priority" in Getting Biodiversity Projects to Work Towards More Effective Conservation and Development.* Academic Press.

Kiss, A. (2004). Is community-based ecotourism a good use of biodiversity conservation funds? *Trends in Ecology & Evolution*, *19*(5), 231–237. doi:10.1016/j.tree.2004.03.010 PMID:16701261

Klarin, T. (2018). The concept of sustainable development: From its beginning to the contemporary issues. *Zagreb International Review of Economics & Business*, *21*(1), 67–94. doi:10.2478/zireb-2018-0005

Knight-Grofe, J., & Rauh, K. (2016). *A World of Learning: Canada's Performance and Potential in International Education 2016*. Canadian Bureau for International Education.

Knox, M., & Crawford, J. (2020). Authentic leaders and business tourism: An exploration of authentic relationships. In M. Bari, S. Shaheen, & M. Fanchen (Eds.), Accelerating Knowledge Sharing, Creativity, and Innovation Through Business Tourism. Hershey, PA: IGI Global.

Knox, M., Crawford, J., & Young, S. (2020). Ecotourism emergence in Tasmania through social entrepreneurs and authentic leaders. In D. Batabyal (Eds.), Global Entrepreneurial Trends in the Tourism and Hospitality Industry. Hershey, PA: IGI Global.

Koca, G. (2019). Evaluation of traditional Şirince Houses According to sustainable construction principles. *International Journal of Architecture & Planning*, *7*(1), 30–49. doi:10.15320/ICONARP.2019.65

Koens, K., & Thomas, R. (2016). "You know that's a rip-off": Policies and practices surrounding micro-enterprises and poverty alleviation in South African township tourism. *Journal of Sustainable Tourism*, *24*(12), 1641–1654. doi:10.1080/09669582.2016.1145230

Kogovsek, M., & Kogovsek, M. (2015). *Hospitality and tourism gender issues remain unsolved: A call for research.* Retrieved from http://www.quaestus.ro/en/wp-content/uploads/2012/02/kogovsek.metka-1.pdf

Koh, H. (2017). *The unseen scandal of hotel food waste. Eco-business*. Retrieved on October 5, 2019, from https://www.eco-business.com/news/the-unseen-scandal-of-hotel-food-waste/

Koikkalainen, S. (2011). Highly skilled Finns in the European labor market: Why do they move abroad? In E. Heikkilä & S. Koikkalainen (Eds.), *Finns Abroad: New Forms of Mobility and Migration* (pp. 46–61). Institute of Migration.

Korner, C., & Ohsawa, M. (2005). Mountain systems. In R. Hassan, R. Scholes, & N. Ash (Eds.), *Ecosystems and Human Well-being: Current State and Trends* (Vol. 1, pp. 2–23). Island Press.

Kotob, F. (2000). *What is sustainability?* University of Wollongong, Faculty of Business.

Kubo, A. (2004). *The sustainability of coastal tourism. Research into the Kenya portfolio of TUI and its environmental performance.* Breda.

Kvasnová, D., Gajdosik, T., & Maráková, V. (2019). Are partnership enhancing tourism destination competitiveness? *Acta Universitatis Agriculturae et Silviculturae Mendelianae Brunensis*, *67*(3), 811–821. doi:10.11118/actaun201967030811

Kyurova, V., Madgerova, R., & Atanasova, A. (2016). Problems of the management and marketing of family business in tourism. *Horizon*, *20*. Advance online publication. doi:10.20544/HORIZONS.A.20.1.17.P01

Laderchi, C. R., Saith, R., & Stewart, F. (2003). *Does it matter that we don't agree on the definition of poverty? A comparison of four approaches*. University of Oxford.

Laeis, G. C. M., & Lemke, S. (2016). Social entrepreneurship in tourism: Applying sustainable livelihoods approaches. *International Journal of Contemporary Hospitality Management*, *28*(6), 1076–1093. doi:10.1108/IJCHM-05-2014-0235

Lai, P. H., & Nepal, S. K. (2005). Local perspectives of ecotourism development in Tawushan Nature Reserve, Taiwan. *Journal of Tourism Management*, *27*(6), 1117–1129. doi:10.1016/j.tourman.2005.11.010

Lane, B. (1994). What is rural tourism? *Journal of Sustainable Tourism*, *2*(1–2), 7–21. doi:10.1080/09669589409510680

Lapeyre, R. (2010). Community-based tourism as a sustainable solution to maximise impacts locally? The Tsiseb Conservancy case, Namibia. *Development Southern Africa, 27*(5), 757–772. doi:10.1080/0376835X.2010.522837

Lee, I., & Arcodia, C. (2011). The Role of Regional Food Festivals for Destination Branding. *International Journal of Tourism Research, 13*(4), 355–367. doi:10.1002/jtr.852

Lee, S., & Oh, H. (2014). Effective Communication Strategies for Hotel Guests' Green Behavior. *Cornell Hospitality Quarterly, 55*(1), 52–63. doi:10.1177/1938965513504029

Legazpe, N., & Davia, M. A. (2019). Women's Employment and Childcare Choices in Spain Through The Great Recession. *Feminist Economics, 25*(2), 173–198. doi:10.1080/13545701.2019.1566754

Legrand, W., Sloan, P., & Chen, J. S. (2016). *Sustainability in the hospitality industry: principles of sustainable operations* (3rd ed.). Routledge. doi:10.9774/gleaf.9781315690261

Lélé, S. (1991). Sustainable development: A critical review. *World Development, 19*(6), 607–621. doi:10.1016/0305-750X(91)90197-P

Lerner, M., & Haber, S. (2001). Performance factors of small tourism ventures: The interface of tourism entrepreneurship and the environment. *Journal of Business Venturing, 16*(1), 77–100. doi:10.1016/S0883-9026(99)00038-5

Lim, W. M. (2017). Exiting supranational unions and the corresponding impact on tourism: Some insights from a rejoinder to Brexit. *Current Issues in Tourism, 21*(9), 970–974. doi:10.1080/13683500.2016.1272555

Li, W. (2004). Environmental management indicators for ecotourism in China's nature reserve: A case study in Tianmushan Nature Reserve. *Tourism Management, 25*(5), 559–564. doi:10.1016/j.tourman.2003.06.001

Llorca-Rodríguez, C., García-Fernández, R., & Casas-Jurado, C. (2018). Domestic versus inbound tourism in poverty reduction: Evidence from panel data. *Current Issues in Tourism*, 1–20. doi:10.1080/13683500.2018.1494701

Loach, K., Rowley, J., & Griffiths, J. (2017). Cultural sustainability as strategy for the survival of museums and libraries. *International Journal of Cultural Policy, 23*(2), 186–198. doi:10.1080/10286632.2016.1184657

Lofstorm, M., & Wang, Ch. (2019). Immigrants and entrepreneurship. *IZA World of Labor, 85*(2). Advance online publication. doi:10.15185/izawol.85.v2

Longhurst, R. (2010). Semi-structured interviews and focus groups. In N. R. Clifford, S. French, & G. Valentine (Eds.), *Key methods in geography* (2nd ed.). Sage Publications.

Lopes, A. S., & Almeida, P. S. (2018). Socioenvironmental sustainability actions in hotel management. *Revista Brasileira de Ecoturismo, 11*(3), 378–398. doi:10.34024/rbecotur.2018.v11.6689

Lorber, J. (2001). *Gender inequality*. Roxbury.

Lui, S., & Ngo, H. (2005). The influence of structural and process factors on partnership satisfaction in interfirm cooperation. *Group & Organization Management, 30*(4), 378–397. doi:10.1177/1059601103259113

Lu, J., & Nepal, S. (2009). Sustainable tourism research: An analysis of papers published in the Journal of Sustainable Tourism. *Journal of Sustainable Tourism, 17*(1), 5–16. doi:10.1080/09669580802582480

Luo, X., & Bao, J. (2019). Exploring the impacts of tourism on the livelihoods of local poor: The role of local government and major investors. *Journal of Sustainable Tourism, 27*(3), 344–359. doi:10.1080/09669582.2019.1578362

Luthans, F. (2002). The need for and meaning of positive organisational behavior. *Journal of Organizational Behavior, 23*(6), 695–706. doi:10.1002/job.165

Luthans, F., & Avolio, B. (2003). Authentic leadership development. In K. Cameron & R. Quinn (Eds.), *Positive Organizational Scholarship* (pp. 241–258). Berrett-Koehler Publishers.

Luthra, R., Platt, L., & Salamonska, J. (2016). Types of migration: The motivations, composition, and early integration patterns of "new migrants" in Europe. *The International Migration Review*, *52*(2), 368–403. doi:10.1111/imre.12293

Lv, Z. (2019). Deepening or lessening? The effects of tourism on regional inequality. *Tourism Management*, *72*, 23–26. doi:10.1016/j.tourman.2018.11.009

Lyons, K., Hanley, J., Wearing, S., & Neil, J. (2012). Gap year volunteer tourism: Myths of Global Citizenship? *Annals of Tourism Research*, *39*(1), 361–378. doi:10.1016/j.annals.2011.04.016

Ma, B. (2010). *A Trip into the Controversy: A Study of Slum Tourism Travel Motivations*. Paper presented at the 2009-2010 Penn Humanities Forum on Connections.

Macedo, C., & Sousa, B. (2019, July-September). A acessibilidade no etourism: Um estudo na ótica das pessoas portadoras de necessidades especiais. *PASOS. Revista de Turismo y Patrimonio Cultural*, *17*(4), 709–723. doi:10.25145/j.pasos.2019.17.050

Machado, M. (2010). Evolução Urbana e Turistificação dos lugares: Um olhar sobre o Rio de Janeiro. *Revista Turismo & Desenvolvimento*, *2*(13), 817–825.

Machado, M. (2014). Turismo e mobilidade reduzida. *Revista Turismo & Desenvolvimento*, *4*(21), 365–376.

Machado, P. (2009). Turismo Acessível para Todos. *Revista Turismo & Desenvolvimento*, *11*, 24.

Macias-Alonso, I., Alkhaled, S., Paunova, M., & Villeséche, F. (2018). Women, Gender Stereotypes, & Leadership: Overcoming the Negatives or Accentuating the Positives? *Proceedings of the Seventy-eighth Annual Meeting of the Academy of Management*. 10.5465/AMBPP.2018.16184symposium

MacNall, S. G., Lang, D. Q., & Sobieszcyk, T. (2016). Ecotourism in Costa Rica and Vietnam: Is ıt sustainable? *Sustainability: The Journal of Record*, *9*(3), 144–154. doi:10.1089/SUS.2016.29058.sgm

MacNeill, J. (1989). Strategies for sustainable economic development. *Scientific American*, *261*(3), 154–165. doi:10.1038cientificamerican0989-154

Mahadevan, R., Amir, H., & Nugroho, A. (2016). Regional Impacts of Tourism-Led Growth on Poverty and Income: Inequality: A Dynamic General Equilibrium Analysis for Indonesia. *Tourism Economics*, 1–18.

Mahadevan, R., & Suardi, S. (2019). Panel evidence on the impact of tourism growth on poverty, poverty gap and income inequality. *Current Issues in Tourism*, *22*(3), 253–264. doi:10.1080/13683500.2017.1375901

Mair, J., & Martí, I. (2006). Social entrepreneurship research: A source of explanation, prediction, and delight. *Journal of World Business*, *41*(1), 36–44. doi:10.1016/j.jwb.2005.09.002

Mair, J., Robinson, J., & Hockerts, K. (Eds.). (2006). *Social entrepreneurship*. Palgrave Macmillan. doi:10.1057/9780230625655

Maleka, M., Ntanjana, A., Tshipala, N., & du Plessi, L. (2018). Employment condition differences based on gender: A case of adventure tourism employees in Gauteng. Academic Press.

Malek, A., & Costa, C. (2015). Integrating communities into tourism planning through social innovation. *Tourism Planning & Development*, *12*(3), 281–299. doi:10.1080/21568316.2014.951125

Malhotra, N., & Birks, D. (2006). *Marketing Research: An Applied Approach*. Pearson. doi:10.1108/S1548-6435(2006)2

Mansfeld, Y. (2019). Pursued Cultural Sustainability and Sought Cultural Experiences: Host–Guest Perspectives. *Creating and Managing Experiences in Cultural Tourism*, 241–253. doi:10.1142/9789813233683_0015

Manwa, H., & Manwa, F. (2014). Poverty alleviation through pro-poor tourism: The role of Botswana forest Reserves. *Sustainability*, *6*(9), 5697–5713. doi:10.3390u6095697

Manyara, G., & Jones, E. (2007). Best practice model for community capacity-building: A case study of community-based tourism enterprises in Kenya. *Preliminary Communication*, *55*(4), 403–415.

Manyara, G., & Jones, E. (2007). Community-based tourism enterprises development in Kenya: An exploration of their potential as avenues of poverty reduction. *Journal of Sustainable Tourism*, *15*(6), 628–644. doi:10.2167/jost723.0

Manyara, G., Jones, E., & Botterill, D. (2006). Tourism and poverty alleviation: The case for indigenous enterprise development in Kenya. *Tourism, Culture & Communication*, *7*(1), 19–37. doi:10.3727/109830406778493579

Marcoz, E. M., Mauri, C., Maggioni, I., & Cantù, C. (2016). Benefits from Service Bundling in Destination Branding: The Role of Trust in Enhancing Cooperation among Operators in the Hospitality Industry. *International Journal of Tourism Research*, *18*(3), 260–268. doi:10.1002/jtr.2002

Mariani, M. M., & Giorgio, L. (2017). The "Pink Night" festival revisited: Meta-events and the role of destination partnerships in staging event tourism. *Annals of Tourism Research*, *62*, 89–109. doi:10.1016/j.annals.2016.11.003

Market Report Cape Verde. (2018). Available at https://www.hospitalitynet.org/file/152008345.pdf

Marko, P., & Jelena, D. (2014). Systems thinking and alternative business model for responsible tourist destination. *Kybernetes*, *43*(3/4), 480–496. doi:10.1108/K-07-2013-0132

MarksJ. H. (2019). *The Perils of Partnership*. doi:10.1093/oso/9780190907082.001.0001

Marques, I. C. P., & Franco, M. (2020). Cooperation networks in the area of health: Systematic literature review. *Scientometrics*, *122*(3), 1727–1750. doi:10.100711192-019-03341-3

Marta, C., & Gabelas, J. A. (2013). Hábitos de consumo televisivo de ficción entre los universitarios que estudian Comunicación. *Revista de Comunicación de la SEECI*, *17*(31), 14–33. doi:10.15198eeci.2013.31.14-33

Martinez, J. M., Martin, J. M., Fernandez, J. A., & Mogorron-Guerrero, H. (2019). An analysis of the stability of rural tourism as a desired condition for sustainable tourism. *Journal of Business Research*, *100*, 165–174. doi:10.1016/j.jbusres.2019.03.033

Martínez-Roget, F., Moutela, J. A. T., & Núñez, J. C. E. (2015). Chaves do êxito do turismo em espaço rural (TER): Evidências a partir da lealdade turística na Rede das Aldeias do Xisto. *Revista Portuguesa de Estudos Regionais*, *40*, 65–81.

Marujo, M. N., & Carvalho, P. (2010). Turismo, planeamento e desenvolvimento sustentável. *Turismo e Sociedade*, *3*(2), 147–161. doi:10.5380/tes.v3i2.19635

Matthews, S. (2006). Responding to poverty in the light of the post-development debate: Some insights from the NGO Enda Graf Sahel. *Africa Development. Afrique et Developpement*, *31*(4), 52–72.

Matysek, K., & Kriwoken, L. (2003). The natural state: Nature-based tourism and ecotourism accreditation in Tasmania, Australia. *Journal of Quality Assurance in Hospitality & Tourism*, *4*(1-2), 129–146. doi:10.1300/J162v04n01_07

Mayo, M. (2005). *Global citizens: Social movements and the challenge of globalisation*. Zed Books.

Mazzarol, T., Limnios, E., & Reboud, S. (2013). Cooperatives as a strategic network of small firms: Case studies from Australian and French cooperatives. *Journal of Co-operative Organization and Management*, *1*(2), 27–40. doi:10.1016/j.jcom.2013.06.004

Mbaiwa, J. (2015). Ecotourism in Botswana: 30 years later. *Journal of Ecotourism*, *14*(2–3), 204–222. doi:10.1080/14724049.2015.1071378

McCaston, M. K., & Rewald, M. (2005). Unifying Framework for Poverty Eradication & Social Justice: The Evolution of CARE's Development Approach: Technical report. Atlanta, GA: Cooperative for Assistance and Relief Everywhere (CARE).

McGehee, N. (2006). *Volunteer Tourism: Sustainable Innovation in Tourism, or just "Pettin' the Critters"?* Best Education Network Think Tank VII: Innovations for Sustainable Tourism.

McKinsey & Company. (2013). *Women matter 2013: Gender diversity in top manage-ment: Moving corporate culture, moving boundaries.* Author.

McKinsey & Company. (2016). *Women matter 2016: Reinventing the workplace to un-lock the potential of gender diversity.* Author.

Medina-Muñoz, D. R., Medina-Muñoz, R. D., & Gutiérrez-Pérez, F. J. (2016). The impacts of tourism on poverty alleviation: An integrated research framework. *Journal of Sustainable Tourism*, *24*(2), 270–298. doi:10.1080/09669582.2015.1049611

Meekes, J., & Hassink, W. H. (2020). *Fired and pregnant: Gender differences in job flexibility outcomes after job loss.* Academic Press.

Mekawy, M. A. (2012). Responsible slum tourism: Egyptian experience. *Annals of Tourism Research*, *39*(4), 2092–2113. doi:10.1016/j.annals.2012.07.006

Meléndez, G., Fermeli, G., Moreira, J. C., Escorihuela, J., & Basso, A. (2001). What do we mean when we say Geotourism International Congress Geotourism in Action, 1, 2011, Arouca. Proceedings Arouca: Geopark Arouca, 97–100.

Mendonça, V., Varajão, J., & Oliveira, P. (2015). Cooperation Networks in the Tourism Sector: Multiplication of Business Opportunities. *Procedia Computer Science*, *64*, 1172–1181. doi:10.1016/j.procs.2015.08.552

Menezes, V. O., & Cunha, S. K. (2017). Ecoinovação em redes hoteleiras globais: designs, barreiras, estímulos e motivações. In V. O. Menezes, M. B. Maçaneiro, & S. K. Cunha (Eds.), *Observatório de ecoinovação: aspectos teóricos e casos ilustrativos* (pp. 167–184). CRV.

Mercer, J., Kurvits, T., Kelman, I., & Mavrogenis, S. (2014). Ecosystem-Based Adaptation for Food Security in the AIMS SIDS: Integrating External and Local Knowledge. *Sustainability*, *6*(9), 5566–5597. doi:10.3390u6095566

Messerli, B. (2012). Global change and the world's mountains. *Mountain Research and Development*, *32*(S1), 55–63. doi:10.1659/MRD-JOURNAL-D-11-00118.S1

Meyer-Cech, K., & Pröbstl, U. (2013). *Sustainable tourism in mountainous regions. In Tourism as a Challenge WIT Transactions on State of the Art in Science and Engineering* (Vol. 73). WIT Series on Tourism Today.

Meyer, J. L., & Helfman, G. S. (1993). The ecological basis of sustainability. *Ecological Applications*, 569–571. PMID:27759309

Mgonja, J. T., Sirima, A., Backman, K. F., & Backman, S. J. (2015). Cultural community-based tourism in Tanzania: Lessons learned and way forward. *Development Southern Africa*, *32*(3), 377–391. doi:10.1080/0376835X.2015.1010710

Michopoulou, E., & Buhalis, D. (2011). *Technology Platforms and Challenges. Accessible tourism: Concepts and issues.* Channel View Publications.

Migon, P. (2018). Geoheritage and geotourism. A European perspective. The Heritage Matters Series. Doi:10.2478/logos-2018-0017

Millán, G., Pérez, L. M., & Martínez, R. (2016). Factores que determinan el crecimiento del turismo en destinos religiosos. *Revista de Ciencias Sociales, 22*(1), 85–97.

Millar, M., & Baloglu, S. (2011). Guests' preferences for green guest room attributes. *Cornell Hospitality Quarterly, 52*(3), 302–311. doi:10.1177/1938965511409031

Miller, G., & Twining-Ward, L. (2005). *Monitoring for a sustainable tourism transition: The challenge of developing and using indicators.* Cabi. doi:10.1079/9780851990514.0000

Ministério de Economia. (2019). *Principais Indicadores Económicos de Portugal. Gabinete de Estratégia e Estudos.* Accessed 17 November 2019, from https://www.gee.gov.pt/pt/publicacoes/indicadores-e-estatisticas/principais-indicadores-economicos-de-portugal

Ministry of Tourism and Wildlife. (2016). Kenya's tourism agenda 2018-2022. Nairobi, Kenya: Author.

Miśkiewicz, K. (2016). Promoting geoheritage in geoparks as an element of educational tourism. In A. Szponar & S. Toczek-Werner (Eds.), *Geotourism: organisation of the tourism and education in the geoparks in the Middle-Europe Mountains.* University of Business in Wrocław.

Misturelli, F., & Heffernan, C. (2008). What is poverty? A diachronic exploration of the discourse on poverty from the 1970s to the 2000s. *European Journal of Development Research, 20*(4), 666–684. doi:10.1080/09578810802464888

Mitchell, J., & Ashley, C. (2010). *Tourism and Poverty Reduction: Pathways and Prosperity.* Earthscan.

Mitchell, J., & Li, S. (2017). Autonomy found: estimating the local benefit from tourism in SIDS: The case of Cape Verde. *Journal of Policy Research in Tourism, Leisure & Events, 9*(2), 182–200. doi:10.1080/19407963.2016.1261145

Mitchell, M. A., & Mitchell, S. J. (2001). Consumer Experience Tourism. *Journal of Food Products Marketing, 6*(3), 1–16. doi:10.1300/J038v06n03_01

Mokhtari, D., Roostaei, S., Khodadadi, M., Ahmadi, M., Ebrahimi, O., & Shahabi, H. (2019). Evaluation of the Role of Environmental Education in Manesht and Ghelarang Geotourism Destination, Iran. *Journal of Quality Assurance in Hospitality & Tourism,* 203–218. doi:10.1080/1528008X.2019.1616039

Molina, R. (2014). En tránsito hacia la educación inclusiva en las Instituciones de Educación Superior de Colombia: Buenas prácticas. Revista Inclusiones, 12-22.

Moniz, A. (2006). *Sustentabilidade do Turismo em Ilhas de Pequena Dimensão: O Caso dos Açores* (Master's thesis). Universidade dos Açores.

Montanari, M. G., & Giraldi, J. M. E. (2013). Competitividade no turismo: Uma comparação entre Brasil e Suíça. *Revista Brasileira de Pesquisa em Turismo, 7*(1), 92–113. doi:10.7784/rbtur.v7i1.564

Monteiro, A. (2017). *Evolução do Desenvolvimento da Oferta Turística do Turismo em Espaço Rural em Monsaraz.* Universidade Europeia.

Montes, R. & Aragall, F. (2009). Turismo Acessível ou Turismo para Todos? Sustentabilidade do Negócio. *Revista Turismo & Desenvolvimento,* (11), 141-145.

Moraes, A. (2006). Competitividade e sustentabilidade: É possível as cidades turísticas serem sustentáveis e competitivas ao mesmo tempo? *PASOS Revista de Turismo y Patrimonio Cultural*, *4*(3), 443–446. doi:10.25145/j.pasos.2006.04.031

Moreira, H. (2014). Estrangeiros e nascidos no estrangeiro residentes em Portugal, CPLP em destaque. *Revista de Estudos Demográficos*, *53*, 91–116.

Morgan, N., Pritchard, A., & Roger, P. (2004). Destination Branding : Creating the unique destination propostition. *Book*, 335.

Morgan, R. (2019, July). Australians see the environment as the world's biggest problem but in Australia it's the economy. *Roy Morgan Research.* Retrieved from http://www.roymorgan.com/findings/8044-most-important-problems-facing-australia-world-july-2019-201907020950

Moric, I. (2013). The role and challenges of rural tourism development in transition countries: Montenegro experiences. *Turizam*, *17*(2), 84–95. doi:10.5937/Turizam1302084M

Morris, A. (2015). *A practical introduction to in-depth interviewing*. SAGE Productions., doi:10.4135/9781473921344

Mosler, W. (1997). Full employment and price stability. *Journal of Post Keynesian Economics*, *20*(2), 167–182. doi:10.1080/01603477.1997.11490146

Mottiar, Z., Boluk, K., & Kline, C. (2018). The roles of social entrepreneurs in rural destination development. *Annals of Tourism Research*, *68*, 77–88. doi:10.1016/j.annals.2017.12.001

Moufakkir, O., & Kelly, I. (2015). Developing a tourism poverty reduction strategy. In Y. Reisinger (Ed.), *Transformational Tourism: Host Perspectives*. CABI. doi:10.1079/9781780643922.0180

Mtapuri, O., Giampiccoli, A., & Jugmohan, S. (2015). Community-based tourism affinity index: A visitor"s approach. *African Journal of Hospitality, Tourism and Leisure*, *4*(2), 1–13.

Mueller, H., & Kaufmann, E. (2001). Wellness tourism: Market analysis of a special health tourism segment and implications for the hotel industry. *Journal of Vacation Marketing*, *7*(1), 5–17. doi:10.1177/135676670100700101

Muganda, M., Sahli, M., & Smith, K. (2010). Tourism's contribution to poverty alleviation: A community perspective from Tanzania. *Development Southern Africa*, *27*(5), 629–646. doi:10.1080/0376835X.2010.522826

Munuera, M. P. (2015). Inclusión sociolaboral de las personas con discapacidad. Mediación laboral. *Mediaciones Sociales*, *14*(14), 161–179. doi:10.5209/rev_MESO.2015.n14.51564

Murphy, P. E. (1985). *Tourism: A community approach*. Methuen.

Musavengane, R., Siakwah, P., & Leonard, L. (2019). The nexus between tourism and urban risk: Towards inclusive, safe, resilient and sustainable outdoor tourism in African cities. *Journal of Outdoor Recreation and Tourism*, *29*, 100254. doi:10.1016/j.jort.2019.100254

Nadda, V. (2009). *Exploring adventure tourism potential in central Himalayas*. VDM publishers.

Nahrstedt. (2004). Wellness. In *The tourism and leisure industry* (pp. 181-198). New York: Haworth Hospitality Press.

Nail, J., & Wearing, S. (1999). *Reed Educational and Professional Publishing Ltd.* Linacre House.

Nelson, F. (2007). *East Africa: Trends in coastal tourism and strategies for promoting sustainable development. Centre on Ecotourism and Sustainable Development*. Stanford University.

Neto, F. (2003). A new approach to sustainable tourism development: Moving beyond environmental protection. *Natural Resources Forum*, *27*(3), 212–222. doi:10.1111/1477-8947.00056

Neumayer, E. (1999). The ISEW: Not an index of sustainable economic welfare. *Social Indicators Research*, *48*(1), 77–101. doi:10.1023/A:1006914023227

Newsome, D., & Dowling, R. (2018). Geoheritage and Geotourism. In Geoheritage: Assessment, Protection, and Managemen Book. Elsevier Inc. doi:10.1016/B978-0-12-809531-7.00017-4

Newstead, T., Dawkins, S., Macklin, R., & Martin, A. (2019). The virtues project: An approach to developing good leaders. *Journal of Business Ethics*. Advance online publication. doi:10.100710551-019-04163-2

Newton, M. (2012). *TripAdvisor and Kiva Launch Partnership*. Retrieved from http://ir.tripadvisor.com/news-releases/news-release-details/tripadvisor-and-kiva-launch-partnership

Ng, T. W., Eby, L. T., Sorensen, K. L., & Feldman, D. C. (2005). Predictors of objective and subjective success: A meta_analysis. *Personnel Psychology*, *58*(2), 367–408. doi:10.1111/j.1744-6570.2005.00515.x

Nicholson, R. E., & Pearce, D. G. (2001). *Why Do People Attend Events : A Comparative Analysis of Visitor Motivations at Four South Island Events*. Academic Press.

Nicolau, J. L., & Mas, F. J. (2006). The influence of distance and prices on the choice of tourist destinations: The moderating role of motivations. *Tourism Management*, *27*(5), 982–996. doi:10.1016/j.tourman.2005.09.009

Nielsen. (2014). *Global consumers are willing to put their money where their heart is when it comes to goods and services from companies committed to social responsibility*. Retrieved at: https://www.nielsen.com/us/en/press-room/2014/global-consumers-are-willing-toput-their-money-where-their-heart-is.html

Niñerola, A., Sánchez-Rebull, M. V., & Hernández-Lara, A. B. (2019). Tourism research on sustainability: A bibliometric analysis. *Sustainability*, *11*(1), 1377–1387. doi:10.3390u11051377

Nisthar, S., Mustafa, A. M. M., & Nufile, A. A. M. (2017). An Analysis of Socioeconomic Issues In Relation To the Role of Women in Travel and Tourism Sector: A Study on Global Perspectives. *The International Journal for Economics and Business Management*, *7*(1), 282–293.

Njoya, T., & Seetaram, N. (2017). Tourism contribution to poverty alleviation in Kenya: A dynamic computable general equilibrium analysis. *Journal of Travel Research*, *57*(4), 513–524. doi:10.1177/0047287517700317 PMID:29595836

Nocca, F. (2017). The Role of Cultural Heritage in Sustainable Development: Multidimensional Indicators as Decision-Making Tool. *Sustainability*, *9*(10), 1882. doi:10.3390u9101882

Nossa & Caldeira. (2014). Turismo medicalizado. In Trabalho em saúde, desigualdade e políticas públicas (pp. 1-31). Braga: Univ. Minho.

Novelli, M., Morgan, N., Mitchell, G., & Ivanov, K. (2016). Travel philanthropy and sustainable development: The case of the Plymouth–Banjul Challenge. *Journal of Sustainable Tourism*, *24*(6), 824–845. doi:10.1080/09669582.2015.1088858

Nunkoo, R., Hall, C. M., Rughoobur-Seetah, S., & Teeroovengadum, V. (2019). Citation practices in tourism research: Toward a gender conscientious engagement. *Annals of Tourism Research*, *79*, 102755. doi:10.1016/j.annals.2019.102755

Nunnally, J. C., & Bernstein, I. H. (1994). The Assessment of Reliability. *Psychometric Theory*, *3*, 248–292.

Nussbaum, M. (2002). Education for citizenship in an era of global connection. *Studies in Philosophy and Education*, *21*(4-5), 289–303. doi:10.1023/A:1019837105053

Oates, J. F. (1999). *Myth and reality in the rain forest: how conservation strategies are failing in West Africa*. University of California Press.

OCDE. (2018). *Geographical Distribution of Financial Flows to Developing Countries 2018*. OECD Publishing.

OECD. (2011). *Education at a Glance 2011: OECD Indicators*. OECD Publishing.

OECD. (2017). *Education at a Glance 2017: OECD Indicators*. Paris: OECD Publishing. Accessed 29 November 2019, from https://www.oecd-ilibrary.org/education/education-at-a-glance-2017_eag-2017-en

OECD. (2018). *OECD Tourism Trends and Policies, 2018*. Advance online publication. doi:10.1787/tour-2018-en

OECD. (2019). *Policies to enhance Sustainable Development*. Available at: https://www.oecd.org/greengrowth/1869800.pdf

OECD. (2019a). *2019 International Migration and Displacement Trends and Policies Report to the G20*. OECD, IOM, UNHCR. Accessed 1 December 2019, from https://www.oecd.org/migration/mig/G20-migration-and-displacement-trends-and-policies-report-2019.pdf

OECD. (2019b). *Education at a Glance 2019: OECD Indicators*. Paris: OECD Publishing. Accessed 30 November 2019, from https://www.oecd-ilibrary.org/education/education-at-a-glance-2019_f8d7880d-en

OECD. (2019c). *International Migration Outlook 2019*. Paris: OECD Publishing. doi:10.1787/c3e35eec-en

OECD/European Union. (2015). *Indicators of Immigrant Integration 2015: Settling*. Paris: OECD Publishing; doi:10.1787/9789264234024-

OECD/European Union. (2018). *Inclusive Entrepreneurship Policies: Country Assessment Notes*. Portugal 2018.

Ogilvie, F. (2017). Wheels of progress: What happens when a rural town become the mountain bike capital of Australia? *ABC News*. Retrieved June, 8, 2020, from https://www.abc.net.au/news/2017-12-26/mountain-bike-trails-driving-major-change-in-derby/9276384?nw=0

Ognjevic, T. (2019). Transformation of cultural identity, tourism and intercultural dialogue: medieval world heritage sites in Serbia between cultural monuments and shrines. *Tourism, Pilgrimage and Intercultural Dialogue: Interpreting Sacred Stories*, 121–129. doi:10.1079/9781789241129.0121

Oh, C. O. (2005). The contribution of tourism development to economic growth in the Korean economy. *Tourism Management*, *26*(1), 39–44. doi:10.1016/j.tourman.2003.09.014

Oishi, N., & Hamada, I. (2020). Quality of life in Japan and emigration: The perspectives of Japanese skilled immigrants in Australia. In M.-Ch. Tsai & N. Iwai (Eds.), *Quality of Life in Japan: Contemporary Perspectives on Happiness* (pp. 193–214). Springer. doi:10.1007/978-981-13-8910-8_9

Okazaki, E. (2008). A community-based tourism model: Its conception and use. *Journal of Sustainable Tourism*, *16*(5), 511–529. doi:10.1080/09669580802159594

Okech, R., Haghiri, M., & George, B. P. (2012). Rural tourism as a sustainable development alternative: An analysis with special reference to Luanda, Kenya. *Revista de Cultura e Turismo*, *6*(3), 36–54.

Oliveira, A. (2008). Planejamento participativo como instrumento de desenvolvimento turístico responsável. *Red de Revistas Científicas de América Latina, 8*(3), 22-28.

Ondicho, T. (2010). *Tourism, Power and Politics: The Challenges of Maasai Involvement in Tourism Development* (Doctor's Thesis). Massey University, Palmerston North, New Zealand.

Ondicho, T. (2016). Poverty alleviation in Kenya: A review of why anti-poverty strategies have failed. In A. A. Bard, J. B. Ndohvu, & B. Tiberius (Eds.), *Poverty and Human Rights: East African Experiences* (pp. 82–94). Focus Publishers.

Ondicho, T. (2018). Indigenous ecotourism as a poverty eradication strategy: A case study of the Maasai people in the Amboseli region of Kenya. *African Study Monographs*, *56*, 87–109.

Ooi, C. S., & Hardy, A. (2020). *Tourism in Tasmania*. Academic Press.

Orrenius, P. M., & Zavodny, M. (2013). Immigrants in risky occupations. In A. F. Constant & K. F. Zimmermann (Eds.), International Handbook on the Economics of Migration (pp. 214-226). Cheltenham: Edward Elgar Publishing Limited. doi:10.4337/9781782546078.00019

Ortiz, M. (2010). Ineficacia de las medidas para garantizar la igualdad de oportunidades. *Revista Iberoamericana de Educación*, *51*(3), 1022–6508. doi:10.35362/rie5131829

Osbaldiston, N., Picken, F., & Denny, L. (2020). Exploring emotional reflexivity in British lifestyle migration to Australia. *Population Space and Place*, *2328*(5). Advance online publication. doi:10.1002/psp.2328

Paese24.it. (2015). *Regata del Pollino. Rafting sul fiume Lao per una sfida all'ultimo… gommone | Paese24.it*. Available at: http://www.paese24.it/28384/pollino/regata-del-pollino-rafting-sul-fiume-lao-per-una-sfida-allultimo-gommone.html

Pakdil, F., & Kurtulmuşoğlu, F. B. (2017). Using quality function deployment for environmentally sustainable hotels: A combined analysis of customer and manager point of view. *European Journal of European Research*, *16*, 252–275.

Panapress. (n.d.). *São Tomé e Príncipe: sul-africano investe 12 milhões em resort na ilha do Príncipe*. Retrieved March 30, 2020, from https://hotelcontinentalluanda.com/index.php/pt/noticias/541-sao-tome-e-principe-sul-africano-investe-12-milhoes-de-euros-em-resort-na-ilha-do-principe

Panfiluk, E. (2015). Impact of a Tourist Event of a Regional Range on the Development of Tourism. *Procedia: Social and Behavioral Sciences*, *213*, 1020–1027. doi:10.1016/j.sbspro.2015.11.520

Pansiri, J. (2008). The effects of characteristics of partners on strategic alliance performance in the SME dominated travel sector. *Tourism Management*, *29*(1), 101–115. doi:10.1016/j.tourman.2007.03.023

Parent, S., Klein, J. L., & Jolin, L. (2009). Le développement communautaire local et le tourisme communautaire: Une analyse conceptuelle comparative. *Journal for Communication Studies*, *4*(1), 73–89.

Pawson, S., D'Arcy, P., & Richardson, S. (2017). The value of community-based tourism in Banteay Chhmar, Cambodia. *Tourism Geographies: An International Journal of Tourism Space, Place and Environment*, *19*(3), 378–397.

Pawson, S., D'Arcy, P., & Richardson, S. (2017). The value of community-based tourism in Banteay Chhmar, Cambodia. *Tourism Geographies: An International Journal of Tourism Space. Place and Environment*, *19*(3), 378–397.

Peixoto, J., & Sabino, C. (2009). *Portugal: Immigration, the labour market and policy in Portugal: Trends and prospects*. IDEA Working Papers, nº 6 April 2009. Citizens and Governance in a Knowledge-based Society.

Pennicott Wilderness Journeys. (2019). Join our team. *Pennicott Wilderness Journeys*. Retrieved March, 12, 2020, from https://www.pennicottjourneys.com.au/journeys/jobs/

Pennicott Wilderness Journeys. (2020). Community sustainability. *Pennicott Wilderness Journeys*. Retrieved March, 12, 2020, from https://www.pennicottjourneys.com.au/about/community

Petersen, T., & Saporta, I. (2004). The opportunity structure for discrimination. *American Journal of Sociology*, *109*(4), 852–901. doi:10.1086/378536

Pfeffer, J., & Davis-Blake, A. (1987). The effect of the proportion of women on salaries: The case of college administrators. *Administrative Science Quarterly*, *32*(1), 1–24. doi:10.2307/2392740

Phi, G. (2017). *A Critical Analysis of Microfinance Tourism and Poverty Alleviation: Characteristics, Opportunities and Constraints* (PhD thesis). Griffith University, Brisbane, Australia.

Phi, G., Dredge, D., & Whitford, M. (2013). *Fostering Global Citizenship through the Microfinance-tourism Nexus*. Paper presented at the 7th Annual Tourism Education Futures Initiative Conference 2013 (TEFI7): Tourism Education for Global Citizenship: Educating for Lives of Consequence, Oxford, UK.

Phi, G. T., & Whitford, M. (2017). Social entrepreneurship in Vietnam. In *Managing Growth and Sustainable Tourism Governance in Asia and the Pacific* (pp. 132–139). UNWTO.

Phi, G. T., Whitford, M., & Dredge, D. (2017). Knowledge dynamics in the tourism-social entrepreneurship nexus. In *Social Entrepreneurship and Tourism* (pp. 155–172). Springer. doi:10.1007/978-3-319-46518-0_9

Phi, G. T., Whitford, M., Dredge, D., & Reid, S. (2017). Educating tourists for global citizenship: A microfinance tourism providers' perspective. *Tourism Recreation Research*, *42*(2), 235–247. doi:10.1080/02508281.2017.1290738

Phi, G. T., Whitford, M., & Reid, S. (2016). What's in the black box? Evaluating anti-poverty tourism interventions utilising theory of change. *Current Issues in Tourism*, 1–16.

Phommavong, S., & Sorensson, E. (2014). Ethnic tourism in Lao PDR: Gendered divisions of labour in community-based tourism for poverty reduction. *Current Issues in Tourism*, *17*(4), 350–362. doi:10.1080/13683500.2012.721758

Pinto da Silva, F., Brandão, F., & Sousa, B. (2019). Towards socially sustainable tourism in cities: Local community perceptions and development guidelines. *Enlightening Tourism. A Pathmaking Journal*, *9*(2), 168-198.

Piotrowski, A. (2020). Aussie farm goes from brink of collapse to Chinese tourism hotspot. *Nine Now*. Retrieved March, 12, 2020, from https://9now.nine.com.au/a-current-affair/bridestow-lavender-estate-online-tourist-hotspot-tasmania/fe3ffb14-0d5b-4f7a-9667-ed66f7361d31

Pleumarom, A. (2012). *The Politics of Tourism, Poverty Reduction and Sustainable Development*. Third World Network.

Plummer, R., & Fennell, A. (2009). Managing protected areas for suatainable tourism:prospects for adaptative co-management. *Journal of Sustainable Tourism*, *17*(2), 149–168. doi:10.1080/09669580802359301

Poças, F. (2016). *Sustainable Development in Small Islands Developing States: Public/Private Cooperation in the case of Príncipe Island*. ISCTE.

Polak, P. (2009). *Out of Poverty: What Works When Traditional Approaches Fail*. Berrett-Koehler Publishers.

Pordata. (2019). *Retrato de Portugal na Europa*. Lisboa: Fundação Francisco Manuel dos Santos. Accessed 30 November 2019, from https://www.pordata.pt/ebooks/PT_EU2019v20191020/mobile/index.html

Portalecalabria.com. (2015). Available at: http://www.portalecalabria.com/site/servizi/trasporti_calabria.asp

Porter, M., & Van der Linde, C. (1995). Toward a new conception of the environmental-competitiveness relationship. *The Journal of Economic Perspectives*, *9*(4), 1995. doi:10.1257/jep.9.4.97

PR Newswire. (2017). *World wildlife fund, American Hotel & Lodging Association and the Rockefeller Foundation bring hotel brands together to prevent hotel food waste: Hotel properties kick-off 12-week pilot programs to reduce food waste in hotels*. Retrieved October 7, 2019, from https://search.proquest.com/docview/1879099480?accountid=17116

Pratt, S., Gibson, D., & Mavono, A. (2013). Tribal tourism in Fiji: An application and extension of Smith's 4Hs of indigenous tourism. *Asia Pacific Journal of Tourism Research*, *18*(8), 894–912. doi:10.1080/10941665.2012.717957

Premio de "Q de Calidad Turística" por su labor en el turismo accesible. (n.d.). https://www.predif.org/index.php?q=predif-premio-q-de-calidad-tur%C3%ADstica-por-su-labor-en-el-turismo-accesible

Priskin, J. (2001). Assessment of natural resources for nature-based tourism: The case of the Central Coast Region of Western Australia. *Tourism Management*, *22*(6), 637–648. doi:10.1016/S0261-5177(01)00039-5

Pritchard, A. (2018). Predicting the next decade of tourism gender research. *Tourism Management Perspectives*, *25*, 144–146. doi:10.1016/j.tmp.2017.11.014

Público. (2019). *Nunca houve tantos estrangeiros a viver em Portugal: são mais de 480 mil*. 28 de junho de 2019. Accessed 25 November 2019, from https://www.publico.pt/2019/06/28/sociedade/noticia/ha-480-mil-imigrantes-viver-portugal-maior-numero-ha-registo-1878007

Quinn, B. (2009). *Festivals, events, and tourism*. The SAGE Handbook of Tourism Studies. doi:10.4135/9780857021076.n27

Quivy, R., & Campenhoudt, L. (1998). *Manual de Investigação em Ciências Sociais*. Gradiva.

Raajpoot, N. A., Sharma, A., & Chebat, J.-C. (2008). The role of gender and work status in shopping center patronage. *Journal of Business Research*, *61*(8), 825–833. doi:10.1016/j.jbusres.2007.09.009

Racz, A., Marchesi, V. V., & Ivana Crnković, I. (2018). Economical, Environmental and Ethical Impact of Food Wastage in Hospitality and Other Global Industries. *European Journal of Bioethics*, *9*(1), 25–42. doi:10.21860/j.9.1.1

Ramchander, P. (2003). Towards the responsible management of the socio-cultural impact of township tourism. *Tourism and politics: Global frameworks and local realities*, 149-173.

Ramos, A. (2008). Lazer, Turismo e Termalismo. Inovação e Complementaridade no Turismo de Saúde. *Revista Turismo & Desenvolvimento, 9*, 7-18.

Ramos, A. (2005). *O Termalismo em Portugal : Dos factores de obstrução à revitalização pela dimensão turística*. Tese de Doutoramento. Universidade de Aveiro.

Rauwald, K. S., & Moore, C. F. (2002). Environmental attitudes as predictors of policy support across three countries. *Environment and Behavior*, *34*(6), 709–739. doi:10.1177/001391602237243

Rayalwar, A. P. (2020). Role of ICT in Women's Entrepreneurship Development. *Studies in Indian Place Names*, *40*(55), 32–37.

Rebutin, J. (2009). *Social Enterprise & Tourism: The key to a better Integration of Indigenous populations*. Université de Toulon et du Var.

Redford, H. K., Levy, M. A., Sanderson, E. W., & Sherbinin, A. (2008). What is the role for conservation organisations in poverty alleviation in the world's wild places? *Oryx*, *42*(04), 516. doi:10.1017/S0030605308001889

ReFED. (2016). *A Roadmap to Reduce U.S. Food Waste by 20 Percent: Executive Summary*. Retrieved October 10, 2019, from https://www.refed.com/downloads/Executive-Summary.pdf

Relego, D. (2017). *São Tomé e Príncipe. Academic Press.*

Remington, J., & Kitterlin-Lynch, M. (2017). Still pounding on the glass ceiling: A study of female leaders in hospitality, travel, and tourism management. *Journal of Human Resources in Hospitality & Tourism*, *17*(1), 22–37. doi:10.1080/15332845.2017.1328259

Republic of Kenya. (2018). *Kenya Tourism Agenda 2018–2022*. Government of Kenya Ministry of Tourism & Wildlife.

Republic of Kenya. (2019). Draft Community Based Tourism Framework. Republic of Kenya.

Responsible travel to Morocco and Peru. (2017). Retrieved from: https://www.wanderlust.co.uk/content/sustainable-tourist-projects-maldives-morocco-thailand/

Reynolds. (2001). *Conservation of Exploited Species*. Academic Press.

Rieder, P., & Wyder, J. (1997). Economic and political framework for the sustainability of mountain áreas. In Mountains of the world: A global priority. Carnforth, Reino Unido, Parthenon.

Rivas, R., & Gibson-Light, M. (2016). Exploring culture through in-depth interviews: Is it useful to ask people about what they think, mean, and do? *Cinta de Moebio*, *57*(57), 316–329. doi:10.4067/S0717-554X2016000300007

Robert, K., Parris, T., & Leiserowitz, A. (2005). What is sustainable development? Goals, indicators, values, and practice. *Environment*, *47*(3), 8–21. doi:10.1080/00139157.2005.10524444

Rodrigues, J. C., & Neto de Carvalho, C. (2010). Património geológico no Geopark Naturtejo: Base para uma estratégia de geoturismo. *E-Terra*, *18*(11), 1–4.

Rodríguez-Pose, A. (2010). *Do institutions matter for regional development?* http://personal.lse.ac.uk/RODRIGU1

Roe, D. (2008). The origins and evolution of the conservation-poverty debate: A review of key literature, events and policy processes. *Oryx*, *42*(04), 491. doi:10.1017/S0030605308002032

Rolfes, M. (2010). Poverty tourism: Theoretical reflections and empirical findings regarding an extraordinary form of tourism. *GeoJournal*, *75*(5), 421–442. doi:10.100710708-009-9311-8

Rubio, M. J. (2015). Formación en materia de accesibilidad en el sector turístico. *Estudios turísticos*, 203-204.

Ruhanen, L., Weiler, B., Moyle, B., & McLennan, C. (2015). Trends and patterns in sustainable tourism research: A 25-year bibliometric analysis. *Journal of Sustainable Tourism*, *23*(4), 517–535. doi:10.1080/09669582.2014.978790

Russell, P. (2000). Community-based tourism. *Travel & Tourism Analyst*, (5), 89–116.

Ryan, G. W., & Bernard, H. R. (2000). Data management and analysis methods. In *Handbook of Qualitative Research* (2nd ed., pp. 769–802). Sage Publications, Inc.

Saayman, M., Rossouw, R., & Krugell, W. (2012). The impact of tourism on poverty in South Africa. *Development Southern Africa*, *29*(3), 462–487. doi:10.1080/0376835X.2012.706041

Sabina, J., & Nicolae, J. (2013). Gender Trends in Tourism Destination. *Procedia: Social and Behavioral Sciences*, *92*, 437–444. doi:10.1016/j.sbspro.2013.08.698

Sahli, M., & Nowak, J. J. (2007). Does Inbound Tourism Benefit Developing Countries? A Trade Theoretic Approach. *Journal of Travel Research*, *45*(4), 426–434. doi:10.1177/0047287506295948

Salazar, N. B. (2004). Developmental tourists vs. development tourism: A case study. In A. Raj (Ed.), *Tourist behaviour: A psychological perspective* (pp. 85–107). Kanishka Publishers.

Salazar, N. B. (2012). Community-based cultural tourism: Issues, threats and opportunities. *Journal of Sustainable Tourism*, *20*(1), 9–22. doi:10.1080/09669582.2011.596279

Salda Gölü Rehberi- Türkiye'nin Maldivleri. (2020). Retrieved from: http://www.bizevdeyokuz.com./SaldaGölüRehberi/Türkiye'ninMaldivleri

Sampson, D. (2019). *Portugal the country of choice for immigrants*. The Portugal News. https://www.theportugalnews.com/news/portugal-the-country-of-choice-for-immigrants/51452

Sánchez-Hernández, M. I., González-López, Ó. R., Buenadicha-Mateos, M., & Tato-Jiménez, J. L. (2019). Work-life balance in great companies and pending issues for engaging new generations at work. *International Journal of Environmental Research and Public Health, 16*(24), 5122. doi:10.3390/ijerph16245122 PMID:31847460

Sánchez, M., Fernández-Berrocal, P., Montañés, J., & Latorre, J. (2008). Does emotional intelligence depend on gender? The socialisation of emotional competencies in men and woman and its implications. *Electronic Journal of Research in Educational Psychology, 15*, 455–474.

Sandoval Martín, M. T. (1998). Promoción turística a través del sector audiovisual. El caso de Canarias. *Revista Latina de Comunicación Social, 9.* https://mdc.ulpgc.es/utils/getfile/collection/rldcs/id/158/filename/330.pdf

Santagata, W. (2002). Cultural districts, property rights and sustainable economic growth. *International Journal of Urban and Regional Research, 26*(1), 9–23. doi:10.1111/1468-2427.00360

Sardinha, J. (2014). Lifestyle migrants in Central Portugal. Strategies of settlement and socialization. In M. Janoschka & H. Haas (Eds.), *Contemporary Geographies of Leisure, Tourism and Mobility: Contested Spatialities Lifestyle Migration and Residential Tourism* (pp. 174–189). Routledge.

Scheyvens, R. (2002). *Tourism for development: Empowering communities*. Prentice Hall.

Scheyvens, R. (2007). Exploring the Tourism-Poverty Nexus. *Current Issues in Tourism, 10*(2-3), 231–254. doi:10.2167/cit318.0

Scheyvens, R. (2012). *Tourism and Poverty*. Routledge. doi:10.4324/9780203831595

Scheyvens, R., & Biddulph, R. (2018). Inclusive tourism development. *Tourism Geographies, 20*(4), 589–609. doi:10.1080/14616688.2017.1381985

Scheyvens, R., & Hughes, E. (2019). Can tourism help to "end poverty in all its forms everywhere"? The challenge of tourism addressing SDG1. *Journal of Sustainable Tourism, 27*(7), 1061–1079. doi:10.1080/09669582.2018.1551404

Scheyvens, R., & Russell, M. (2012). Tourism, land tenure and poverty alleviation in Fiji. *Tourism Geographies, 14*(1), 1–25. doi:10.1080/14616688.2011.593188

Schmetzke, A. (2012). Guest editorial: Accessibility of Web-based information resources for people with disabilities. *Library Hi Tech, 20*(2), 135–136. doi:10.1108/07378830210432534

Schmidt-Soltau, K., & Brockington, D. (2004a). The social and environmental impacts of wilderness and development. *Oryx, 38*(382), 140–142.

Schubeler, P. (1996). *Participation and Partnership in Urban Infrastructure Management*. Washington, DC: World Bank for the Urban Management Programme.

Schultz, P. W., & Zelezny, L. (1999). Values as predictors of environmental attitudes: Evidence for consistency across 14 countries. *Journal of Environmental Psychology, 19*(3), 255–265. doi:10.1006/jevp.1999.0129

Scott Dunn. (2018). *Here Be Dragons – Bom Bom Island*. Academic Press.

Sebele, S. (2010). Community-based tourism ventures, benefits and challenges: Khama rhino sanctuary trust, central district, Botswana. *Tourism Management, 31*(1), 136–146. doi:10.1016/j.tourman.2009.01.005

Seibert, G. (2016). São Tomé and Príncipe 1975-2015: Politics and economy in a former plantation colony. *Estudos Ibero-Americanos, 42*(3), 987–1012. doi:10.15448/1980-864X.2016.3.22842

Selin, S., & Chavez, D. (1995). Developing an evolutionary tourism partnership model. *Annals of Tourism Research, 22*(4), 844–856. doi:10.1016/0160-7383(95)00017-X

Sen, A. (2000). *Social exclusion: Concept, application, and scrutiny. Social Development Papers No. 1*. Office of Environment and Social Development: Asian Development Bank.

Serafimova, M., & Petrevska, B. (2018). Female Entrepreneurship in Tourism: A Strategic Management Perspective in Macedonia. *Journal of Applied Economics and Business, 6*(1), 21–32.

Serageldin, I., & Steer, A. (Eds.). (1994). *Making development sustainable: from concepts to action*. The World Bank. doi:10.1596/0-8213-3042-X

Shaffer, P. (2008). *New Thinking on Poverty: Implications for Globalisation and Poverty Reduction Strategies*. DESA Working Paper No. 65.

Sharpley, R., & Harrison, D. (2019). *A research agenda for tourism and development*. Edward Elgar. doi:10.4337/9781788112413

Sharpley, R., & Naidoo, P. (2010). Tourism and Poverty Reduction: The case of Mauritius. *Tourism and Hospitality Planning and Development, 7*(2), 145–162. doi:10.1080/14790531003737169

Sharpley, R., & Telfer, D. (Eds.). (2008). *Tourism and Development: Concepts and Issues*. Channel View Publications.

Shaw, D., Grehan, E., Shiu, E., Hassan, L., & Thomson, J. (2005). An exploration of values in ethical consumer decision making. *Journal of Consumer Behaviour, 4*(3), 185–200. doi:10.1002/cb.3

Shedenov, U., Litvishko, O., Kazbekov, B., Suyunchaliyeva, M., & Kazbekova, K. (2019). Improvement of ecological tourism on the principles of sustainable economic development. *E3S Web of Conferences, 135*, 04047. Doi:10.1051/e3sconf/201913504047

Shekhar, S., Kumar, P., Chauhan, G., & Thakkar, M. (2019). *Conservation and Sustainable Development of Geoheritage, Geopark, and Geotourism: a Case Study of Cenozoic Successions of Western Kutch*. Geoheritage. doi:10.100712371-019-00362-5

Sheldon, P. J., & Daniele, R. (Eds.). (2017). *Social Entrepreneurship and Tourism*. Springer. doi:10.1007/978-3-319-46518-0

Shen, F., Hughey, K. F., & Simmons, D. G. (2008). Connecting the sustainable livelihoods approach and tourism: A review of the literature. *Journal of Hospitality and Tourism Management, 15*(01), 19–31. doi:10.1375/jhtm.15.1.19

Shen, S., Sotiriadis, M., & Zhou, Q. (2020). We must be aware that poverty is a multidimensional concept and because of that, it is extremely difficult to manage. *Sustainability, 12*, 1–21. doi:10.3390u12041470

Shet, R. R., & Verma, P. (2019). Spiritual Intelligence and Mental Health: Study of Working Mother. *Global Health & Medical Tourism (GloHMT)*.

Shine, R. (2017). Underemployment a struggle for Tasmanian worker with jobs but not enough hours, union survey finds. *ABC News*. Retrieved March, 12, 2020, from https://www.abc.net.au/news/2017-09-18/union-survey-reveals-extend-of-underemployment-in-tasmania/8957168

Shultz, L. (2007). Educating for Global Citizenship: Conflicting Agendas and Understandings. *The Alberta Journal of Educational Research, 53*(3), 248–258.

Sigüenza, M. C., Brotons, M., & Huete, R. (2013). *The evolution of gender inequality in tourism employment in Spain*. Academic Press.

Silva, A. (2014). *A importância dos georrecursos para a divulgação e ensino das Geociências* (Dissertação de Mestrado). Universidade Nova de Lisboa.

Silva, C. (2017). *Turismo Acessível – Estudo de caso da cidade de Coimbra* (Master's thesis). Instituto Politécnico de Coimbra.

Silva, R. (2009). Balanced scorecard – BSC: gestão de esnino superior, gestão profissionalizada e qualidade de ensino para instituições de ensino superior privado. *Juruá, 186.*

Silva, L., & Leal, J. (2015). Rural tourism and national identity building in contemporary Europe: Evidence from Portugal. *Journal of Rural Studies*, *38*, 109–119. doi:10.1016/j.jrurstud.2015.02.005

Simmons, D. G. (1994). Community participation in tourism planning. *Tourism Management*, *15*(2), 98–108. doi:10.1016/0261-5177(94)90003-5

Simpson, F., Chapman, M., & Mahne, L. (1998). Partnership Approaches to Tourism and Rural Development in Post Socialist Europe, the Experience of Notranjski Kras, Slovenia. *Rural Tourism Management: Sustainable Options International Conference.*

Simpson, M. C. (2009). An integrated approach to assess the impacts of tourism on community development and sustainable livelihood. *Community Development Journal: An International Forum*, *44*(2), 186–208. doi:10.1093/cdj/bsm048

Situmorang, D. B. M., & Mirzanti, I. R. (2012). Social Entrepreneurship to Develop Ecotourism. *Procedia Economics and Finance*, *4*(0), 398–405. doi:10.1016/S2212-5671(12)00354-1

Sloan, P., Legrand, W., & Simons-Kaufmann, C. (2014). A survey of social entrepreneurial community-based hospitality and tourism initiatives in developing economies: A new business approach for industry. *Worldwide Hospitality and Tourism Themes*, *6*(1), 51–61. doi:10.1108/WHATT-11-2013-0045

Smith, M., & Puczkó, L. (2008). Health and Wellness Tourism. Elsevier. doi:10.4324/9780080942032

Smith, N. (2018). *Gender quotas on boards of directors.* Retrieved from http://womensboardaward.dk/wp-content/uploads/2019/02/2018-12-06-Nina-Smith-Gender-quotas-on-boards-of-directors.pdf

SNV Netherlands Development Organisation. (2007). Tourism and development: Agendas for action. Nairobi, Kenya: Author.

Snyman, S. (2014). The impact of ecotourism employment on rural household incomes and social welfare in six southern African countries. *Tourism and Hospitality Research*, *14*(1-2), 37–52. doi:10.1177/1467358414529435

Soares, A., & Breda, Z. (2014). Destinos turísticos Acessíveis e Pessoas com Mobilidade Reduzida – Uma realidade? O caso da cidade de Aveiro. *Revista Turismo & Desenvolvimento*, *3*(21/22), 383–396.

Sofield, T. (2003). *Empowerment for sustainable tourism development.* Pergamon.

Sohail, M. S. (2017). Green marketing strategies: How do they influence consumer-based brand equity? J. *Global Business Advancement*, *10*(2), 229–243. doi:10.1504/JGBA.2017.084607

Solmaz, S. A. (2019). Ekim). Sürdürülebilir Turizm ve Türkiye'de Çevre Politikaları. *Uluslararası Sosyal Araştırmalar Dergisi*, *12*(66), 1176–1186.

Sonuç, N. (2020). Culture, Tourism and Sustainability (Cultural Heritage and Sustainable Tourism, Social Sustainability of Tourism, Socio-Cultural Sustainability of Tourism). Encyclopedia of Sustainable Management, 1–7. doi:10.1596/0-8213-3650-9

Soumaré, I. (2015). Does FDI improve economic development in North African countries? *Applied Economics*, *47*(51), 5510–5533. doi:10.1080/00036846.2015.1051655

Sousa, B., & Rodrigues, S. (2019). The role of personal brand on consumer behaviour in tourism contexts: the case of Madeira. *Enlightening Tourism. A Pathmaking Journal, 9*(1), 38-62. doi:10.33776/et.v9i1.3597

Sousa, B., & Silva, M. (2019). Creative tourism and destination marketing as a safeguard of the cultural heritage of regions: The case of Sabugueiro Village. *Revista Brasileira de Gestão e Desenvolvimento Regional, 15*(5), 78-92.

Spenceley, A. (2005). Nature-based tourism and environmental sustainability in South Africa. *Journal of Sustainable Tourism*, *13*(2), 136–168. doi:10.1080/09669580508668483

Spenceley, A., Ashley, C., & Kock, M. (2009). *Tourism-led poverty reduction programme: Core training module*. Centro Comércio Internacional.

Stamarski, C., & Son Hing, L. (2015). Gender inequalities in the workplace: The effects of organisational structures, processes, practices, and decision makers' sexism. *Frontiers in Psychology*, *6*(5), 34–54. doi:10.3389/fpsyg.2015.01400 PMID:26441775

Steinicke, E., & Neuburger, M. (2012). The impact of community-based afro-alpine tourism on regional development. *Mountain Research and Development*, *32*(4), 420–430. doi:10.1659/MRD-JOURNAL-D-11-00102.1

Stilwell, B., Diallo, K., Zurn, P., Dal Poz, M. R., Adams, O., & Buchan, J. (2003). Developing evidence-based ethical policies on the migration of health workers: Conceptual and practical challenges. *Human Resources for Health*, *1*(8), 8. Advance online publication. doi:10.1186/1478-4491-1-8 PMID:14613524

Stilwell, B., Diallo, K., Zurn, P., Vujicic, M., Adams, O., & Dal Poz, M. (2004). Migration of health-care workers from developing countries: Strategic approaches to its management. *Bulletin of the World Health Organization*, *82*, 595–600. PMID:15375449

STP. M. F. (2018). *Plano Estratégico e de Marketing para o Turismo de São Tomé e Príncipe*. https://interlusofona.info/wp-content/uploads/2018/10/Plano-Estratégico-e-de-Marketing-para-o-Turismo-de-São-Tomé-e-Príncipe.pdf

Stratford, E. (2006). Technologies of agency and performance: Tasmania together and the constitution of harmonious island identity. *Geoforum*, *37*(2), 273–286. doi:10.1016/j.geoforum.2005.03.001

Street, C. T., & Cameron, A. (2007). External relationships and the small business: A review of small business alliances and network research. *Journal of Small Business Management*, *45*(2), 239–266. doi:10.1111/j.1540-627X.2007.00211.x

Strickland-Munro, J. K., Allison, H. E., & Moore, S. A. (2010). Using resilience concepts to investigate the impacts of protected area tourism on communities. *Annals of Tourism Research*, *37*(2), 499–519. doi:10.1016/j.annals.2009.11.001

Stueve, A. M., Cooks, S., & Drew, D. (2002). The Geotourism Study: Phase I. Executive Summary. Washington, DC: Travel Industry Association of America.

Suberviola, I. (2011). Auto-percepción del profesorado sobre su formación en educación emocional. Vivat Academia. *Revista de Comunicación*, *14*(117), 1154–1167.

Sunarminto, T., Mijiarto, J., & Prabowo, E. D. (2019). Socioeconomic and cultural impacts of oil palm plantation development in Indonesia. *IOP Conf. Series: Earth and Environmental Science, 336*. 10.1088/1755-1315/336/1/012008

Supriharjoa, R. D., Rahmawatia, D., Santoso, E. B., Setiawana, R. P., & Pradinie, K. (2016). Factors influencing community-based heritage sustainability in Kampung Kemasan, Gresik. *Procedia: Social and Behavioral Sciences*, *227*, 498–502. doi:10.1016/j.sbspro.2016.06.106

Susskind, A. (2014). Guests' reactions to in-room sustainability initiatives: An experimental look at product performance and guest satisfaction. *Cornell Hospitality Quarterly*, *3*(55), 228–238. doi:10.1177/1938965514533744

Sutawa, G. K. (2012). Issues on Bali tourism development and community empowerment to support sustainable tourism development. *Procedia Economics and Finance*, *4*, 413–422. doi:10.1016/S2212-5671(12)00356-5

Świercz, A., & Smorzewska, E. (2015). Selected examples of interactive teaching methods in the Centre of Geoeducation in the city of Kielce (Poland). *Procedia: Social and Behavioral Sciences*, *174*, 680–686. doi:10.1016/j.sbspro.2015.01.601

Szczygiel, N. (2018). Integrating international experience in business education. In M. Zakrzewska, & R. Gruenwald (Eds.), *Proceedings of 11th ICEBE & 7th ICIE & PEESA III International Conference on Engineering and Business Education, Innovation and Entrepreneurship, and Capacity Building in Higher Education* (pp. 347-352). University of Szczecin.

Szczygiel, N., & Santana, S. (2018). 'Call me if you need': Social support experiences under economic and social change. *Social Work in Health Care*, *57*(9), 794–810. doi:10.1080/00981389.2018.1508112 PMID:30118651

Szymanska, E. (2018). Sustainable tourism: Theory versus practice. *International Scientific Conference on Economic and Social Development Book of Proceedings*, 17-25.

Taplin, J., Dredge, D., & Scherrer, P. (2014). Monitoring and evaluating volunteer tourism: A review and analytical framework. *Journal of Sustainable Tourism*, *22*(6), 874–897. doi:10.1080/09669582.2013.871022

Tasci, A. D. A., & Severt, D. (2016). A triple lens measurement of host–guest perceptions for sustainable gaze in tourism. *Journal of Sustainable Tourism*, *25*(6), 711–731. doi:10.1080/09669582.2016.1225746

Tasmanian Government, Tourism Tasmania, Tourism Council Tasmania. (2004). *Tourism 21: A New 10-year Vision*. Tasmanian Government, Tourism Tasmania and the Tourism Council of Australia.

Tasmanian Leaders. (2016). *Tasmanian Leaders Yearbook 2016*. Launceston, Australia: Tasmanian Leaders Inc. Retrieved January 29, 2020, from https://www.tasmanianleaders.org.au/wp-content/uploads/2019/09/TLP-Yearbook-2016_Web.pdf

Tasmanian Walking Company. (2020a). Community and environment. *Tasmanian Walking Company*. Retrieved June, 8, 2020 from https://www.taswalkingco.com.au/about-us/#accreditation

Tasmanian Walking Company. (2020b). Community and environment. *Tasmanian Walking Company*. Retrieved June, 8, 2020 from https://www.taswalkingco.com.au/community-and-environment/

Teece, D. J., Pisano, G., & Shuen, A. (1997). Dynamic capabilities and strategic management. *Strategic Management Journal*, *18*(7), 509–533. doi:10.1002/(SICI)1097-0266(199708)18:7<509::AID-SMJ882>3.0.CO;2-Z

Terborgh, J. (1999). *Requiem for nature*. Island Press.

The Global Environment Facility (GEF). (2018). *Funding | Global Environment Facility*. Academic Press.

The WHOQOL Group. (1995). The World Health Organization quality of life assessment (WHOQOL): Position paper from the World Health Organization. *Social Science & Medicine*, *41*(10), 1403–1409. doi:10.1016/0277-9536(95)00112-K PMID:8560308

The Wine Syndicate. (2020). *Josef Chromy Wines*. Retrieved from https://thewinesyndicate.ca/jchromy/

The World Bank. (2005). *São Tomé and Príncipe Joint IDA-IMF Staff and Advisory Note on the Poverty Reduction Strategy Paper*. https://documents.worldbank.org/curated/en/824571468299160559/pdf/31726.pdf

The World Bank. (2017). *Women and tourism: Designing for inclusion.* Retrieved from https://documents.worldbank.org/curated/en/401321508245393514/pdf/120477-WP-PUBLIC-Weds-oct-18-9am-ADD-SERIES-36p-IFCWomenandTourismfinal.pdf

The World Bank. (2018). *São Tomé e Príncipe: aspectos gerais.* https://www.worldbank.org/pt/country/saotome/overview

The World Bank. (2019). *Country profile.* Accessed 25 November 2019, from https://data.worldbank.org/country/portugal

The World Bank. (2019). *São Tomé and Príncipe Recent Developments.* http://pubdocs.worldbank.org/en/766771492188172104/mpo-stp.pdf

Theingthae, S. (2017). Sustainability of Community Based Ecotourism Development after the Impact of Tsunami Disasters: Comparison between Buddhism Community and Muslim Community in Phuket Province, Thailand. *Journal of Tourism Research & Hospitality*, *6*(4), 1–10. doi:10.4172/2324-8807.1000175

Thomas, K. D., William, N. B., & Trotz, M. A. (2014). The Sustainability of Ecotourism Activities: Development of An Accessible, Applicable and Efficient Tool for Assessment in the Caribbean Region. *The West Indian Journal of Engineering*, *36*(2), 51–62.

Thomlinson, E., & Getz, D. (1996). The Question of Scale in Ecotourism: Case Study of Two Small Ecotour Operators in the Mundo Maya Region of Central America. *Journal of Sustainable Tourism*, *4*(4), 183–188. doi:10.1080/09669589608667267

Tilbury, D. & Mulà, I. (2009). *Review of Education for Sustainable Development Policies from a Cultural Diversity and Intercultural Dialogue Perspective: Identifying Opportunities for Future Action.* UNESCO Digital library.

Tisdell, C. (2014). *The MIRAB Model of Small Island Economies in the Pacific and their Security Issues: Revised Version.* Social Economics, Policy and Development Working Papers.

Tisdell, C. A. (2016). Models of the international economic dependence of Pacific Microstates: A critical review with important implications for international policies and relations. *Journal of Self-Governance and Management Economics*, *4*(2), 7–27. doi:10.22381/JSME4220161

Todaro, M. P. (1995). *Reflections on economic development.* Books.

Tomé, M. (2014). Turismo e Mobilidade. Reduzida. *Revista Turismo & Desenvolvimento*, *4*(21/22), 365–376.

Tomé, M. (2018). Restrictive factors for the tourism. The perception of real and potential demand about the public security in Rio de Janeiro (Brazil). *Estudios y Perspectivas en Turismo*, *27*(4), 968–984.

Toor, S. U. R., & Ofori, G. (2008). Leadership versus management: How they are different, and why. *Leadership and Management in Engineering*, *8*(2), 61–71. doi:10.1061/(ASCE)1532-6748(2008)8:2(61)

Torkington, K. (2010). Defining lifestyle migration. *Dos Algarves*, *19*, 99–111.

Torres, R., & Momsen, J. M. (2004). Challenges and potential for linking tourism and agriculture to achieve pro-poor tourism objectives. *Progress in Development Studies*, *4*(4), 294–318. doi:10.1191/1464993404ps092oa

Tosun, C. (2001). Challenges of sustainable tourism development in the developing world: The case of Turkey. *Tourism Management*, *22*(3), 289–303. doi:10.1016/S0261-5177(00)00060-1

Tourism Research Australia. (2019a). *International visitors in Australia: Year ending March 2019.* Austrade.

Tourism Research Australia. (2019b). *State tourism satellite accounts: 2017-2018.* Austrade.

Tourism Research Institute (TRI). (2018). Tourism sector performance report, 2018. Nairobi, Kenya: Author.

Tourism Tasmania. (2014). *Tasmanian Tourism Snapshot: Year ending September 2014*. Tourism Tasmania.

Tourism Tasmania. (2015). *Tasmanian Tourism Snapshot: Year ending September 2015*. Tourism Tasmania.

Tourism Tasmania. (2016). *Tasmanian Tourism Snapshot: Year ending September 2016*. Tourism Tasmania.

Tourism Tasmania. (2017). *Tasmanian Tourism Snapshot: Year ending September 2017*. Tourism Tasmania.

Tourism Tasmania. (2018). *Tourism Research: Tasmanian Tourism Snapshot*. Tourism Tasmania.

Tourism Tasmania. (2019a). *Tasmanian Tourism Snapshot: Year ending September 2019*. Tourism Tasmania.

Tourism Tasmania. (2019b). *Importance of Tourism*. Retrieved from https://www.tourismtasmania.com.au/archive/corporate-plan-2017-20/importance-of-tourism

Tovar, C., & Lockwood, M. (2008). Social Impacts of Tourism: An Australian regional case study. *International Journal of Tourism Research*, *10*(4), 365–378. doi:10.1002/jtr.667

Travis, A. S. (2011). Host cultures and tourism: is a culture sustainable? *Planning for Tourism, Leisure and Sustainability: International Case Studies*, 141–149. doi:10.1079/9781845937423.0141

Trejos, B., & Chiang, L. N. (2009). Local economic linkages to community-based tourism in rural Costa Rica. *Singapore Journal of Tropical Geography*, *30*(3), 373–387. doi:10.1111/j.1467-9493.2009.00375.x

Trilling, L. (1971). *Sincerity and authenticity*. Harvard University Press.

Trinidad, D., & Johnson, C. (2002). The association between emocional intelligence and early adolescent tobacco and alcohol use. *Personality and Individual Differences*, *32*(1), 95–105. doi:10.1016/S0191-8869(01)00008-3

Tripadvisor. (2017). *TripAdvisor Network Effect and the Benefits of Total Engagement*. Retrieved from https://www.tripadvisor.com/TripAdvisorInsights/w828

Truong, T.-H., & King, B. (2010). Cultural Values and Service Quality: Host and Guest Perspectives. *Tourism, Culture & Communication*, *10*(1), 15–32. doi:10.3727/109830410X12629765735597

Truong, V. (2014). Pro-poor tourism: Looking backward as we move forward. *Tourism Planning & Development*, *11*(2), 228–242. doi:10.1080/21568316.2013.864996

Tseng, S. C., & Hung, S. W. (2013). A framework identifying the gaps between customers' expectations and their perceptions in green products. *Journal of Cleaner Production*, *59*, 174–184. doi:10.1016/j.jclepro.2013.06.050

Tucker, H., & Boonabaana, B. (2012). A critical analysis of tourism, gender and poverty reduction. *Journal of Sustainable Tourism*, *20*(3), 437–455. doi:10.1080/09669582.2011.622769

Túñez López, M., Altamirano, V., & Valarezo, K. P. (2016). Comunicación turística colaborativa 2.0: Promoción, difusión e interactividad en las webs gubernamentales de Iberoamérica. *Revista Latina de Comunicación Social*, *71*, 249–271.

Turismo de Portugal. (2013). *Plano Estratégico Nacional do Turismo - Revisão e Objetivos 2013-2015*. Lisbon: TP.

Turismo de Portugal. (2017). *Estratégia Turismo 2027*. Lisbon: TP.

Turkish National Network. (2020). *Cittaslow Turkey*. Retrieved from: https://www.cittaslow.org/network/turkish-national-network

Turkish Statistical Institute. (2020). *Tourism Statistics*. Retrieved from: http://www.tuik.gov.tr/PreTablo.do?altid=1072

Tuten, T. L., & Ureban, D. J. (2001). An Expanded Model of Business-to-Business Partnership Formation and Success. *Industrial Marketing Management*, *30*(2), 149–164. doi:10.1016/S0019-8501(00)00140-1

Twining-Ward, L. (2007). *A toolkit for monitoring and managing community-based tourism*. School of Travel Industry Management, University of Hawai.

Tyler, D., & Dangerfield, J. M. (1999). Ecosystem Tourism: A Resource-based Philosophy for Ecotourism. *Journal of Sustainable Tourism*, *7*(2), 146–158. doi:10.1080/09669589908667332

Tyson, L. (2019). *An economist explains why women are paid less*. Retrieved from https://www.weforum.org/agenda/2019/03/an-economist-explains-why-women-get-paid-less/

UN. (1992). *Earth Summit UN Conference on Environment and Development (1992)*. Academic Press.

UNCTAD. (2012). *The Least Developed Countries Report 2012*. Academic Press.

UNEP & WTO. (1993). *Sustainable Development Goals*. Retrieved from: https://www.unwto.org/sustainable-development

UNEP & WTO. (2005). *Making tourism more sustainable: A guide for policymakers*. Retrieved from: https://www.unwto.org/sustainable-development

UNEP&WTO. (2012). *Tourism in the green economy: Background report*. Retrieved from: https://www.unwto.org/sustainable-development

UNESCO. (2008). *La educación inclusiva: el camino hacia el futuro*. https://www.webcitation.org/6vTY885Ay

UNESCO. (2012). *The Island of Príncipe*. Academic Press.

UNESCO. (2018). *Biosphere Reserves*. http://www.unesco.org/new/en/natural-sciences/environment/ecological-sciences/biosphere-reserves/

UNESCO. (2020). *World Heritage Sustainable Tourism Toolkit*. Retrieved from: http://whc.unesco.org/sustainabletourismtoolkit/guide-2-—-case-study-ichkeul-national-park-tunisia

UNHCR. (2015). *Handbook for Emergencies* (4th ed.). Office of the United Nations High Commissioner for Refugees.

United Nations (UN). (2015). *The Millennium Development Goals Report*. New York: UN.

United Nations Development Programme. (2018). *Human Development Indices and Indicators: 2018 Statistical Update*. New York: UNDP. Accessed 26 November 2019, from http://www.hdr.undp.org/sites/default/files/2018_human_development_statistical_update.pdf

United Nations General Assembly. (2015). Transforming our world: The 2030 Agenda for sustainable development. Resolution 15-16301. United Nations.

United Nations World Tourism Organisation (UNWTO). (2014). Available online: http://step.unwto.org/en/content/background-and-objectives

United Nations. (2003). *Human Development Report, Millennium Development Goals: A Compact among Nations to End Human Poverty*. Oxford University Press.

United Nations. (2016). *Transforming our world: the 2030 Agenda for Sustainable Development*. Retrieved from https://sustainabledevelopment.un.org/post2015/transformingourworld

United Nations. (2018). *Measuring International Labour Mobility*. UNECE.

United Nations. (2019). *About the Sustainable Development Goals*. Retrieved from https://www.un.org/sustainabledevelopment/

United Nations. (2020). Retrieved from: https://www.un.org/wcm/webdav/site/climatechange/shared/gsp/docs/GSP1-6_Background%20on%20Sustainable%20Devt.pdf

United Nations. (2020). *The Sustainable Development Agenda*. Retrieved from: <https://www.un.org/sustainabledevelopment/development-agenda/>

United Nations. (2020). *United Nations: Gender equality and women's empowerment*. Retrieved from https://www.un.org/sustainabledevelopment/gender-equality/

UN-OHRLLS. (2017). *Small Island Developing States, Small Islands Big(ger) Stakes*. http://www.sustainablesids.org/wp-content/uploads/2017/08/UN_SIDS_booklet_5x6-5_062811_web.pdf

UNWTO & UN Global Compact Network Spain. (2016). *The tourism sector and the Sustainable Development Goals: Responsible tourism – A global commitment*. UNWTO & UN Global Compact Network Spain.

UNWTO. (2014). Manual sobre Turismo Accesible para Todos: principios, herramientas y buenas prácticas. *Módulo I: turismo accesible - definición y contexto*. https://www.e-unwto.org/doi/book/10.18111/9789284416486

UNWTO. (2014). *Tourism in Small Island Developing States*. SIDS.

UNWTO. (2015). *Tourism and the Sustainable Development Goals*. UNWTO.

UNWTO. (2015, September 28). *UNWTO welcomes the adoption of the Sustainable Development Goals*. UNWTO. Retrieved from http://media.unwto.org/press-release/2015-09-28/unwto-welcomes-adoption-sustainable-development-goals

UNWTO. (2017). UNWTO tourism highlights 2015 edition. UNWTO.

UNWTO. (2018). *Tourism and the Sustainable Development Goals – Good Practices in the Americas*. UNWTO.

UNWTO. (2019). *UNWTO Tourism Definitions*. UNWTO. doi:10.18111/9789284420858

UNWTO. (2019a). *UNWTO world tourism barometer* (nov 2019). *UNWTO World Tourism Barometer, 17*(8), 1–44.

UNWTO. (2019b). *International tourism highlights* (2019 edition). World Tourism Organization.

UNWTO. (2020). *Small Islands Developing States (SIDS)*. https://www.unwto.org/sustainable-development/small-islands-developing-states

Uusimäki, L., & Johansson, K. (2020). The Vulnerability of Pregnancy and the Motherhood Myth. In L. Henderson, A. Black, & S. Garvis (Eds.), *Re)birthing the Feminine in Academe. Palgrave Studies in Gender and Education. Palgrave Macmillan*. doi:10.1007/978-3-030-38211-7_8

Uyar, A., Karaman, A. S., & Kilic, M. (2019). Institutional drivers of sustainability reporting in the global tourism industry. *Tourism Economics*. doi:10.1177/1354816619886250

Valenciano, G. (2009). *Construyendo un concepto de educación inclusiva: una experiencia compartida*. Instituto Universitario de Integración en la Comunidad/ Colección Investigación.

Valentine, P. S. (1993). Ecotourism and nature conservation: A definition with some recent developments in Micronesia. *Tourism Management, 14*(2), 107–115. doi:10.1016/0261-5177(93)90043-K

Valle-García, S. E. (2014). Ecotourism: Sustainable indigenous policies and its effects in Mayan communities in Southern Mexico. *WIT Transactions on Ecology and The Environment*, 187.

van Beers, T. (2011). *Discourse in Tourism Development: A Narrative of Inclusive Business* (MSc Thesis Leisure, Tourism and Environment). Wageningen University.

Van der Zee, E., Gerrets, A., & Vanneste, D. (2017). Complexity in the governance of tourism networks: Balancing between external pressure and internal expectations. *Journal of Destination Marketing & Management*, *6*(4), 296–308. Advance online publication. doi:10.1016/j.jdmm.2017.07.003

Van Teijlingen, E., & Hundley, V. (2002). The importance of pilot studies. *Nursing Standard*, *16*(40), 33–36. doi:10.7748/ns2002.06.16.40.33.c3214 PMID:12216297

Vanegas, M. (2015). The triangle of poverty, economic growth, and inequality in Central America: Does tourism matter? *Worldwide Hospitality and Tourism Themes*, *6*(3), 277–292. doi:10.1108/WHATT-03-2014-0014

Velázquez, G., Valderrama, J., Ruiz, J., Martínez, O., Morales, L., Verján, R., & Flores, J. (2010). Community-based tourism and the development of local small businesses in rural Baja California, Mexico. *WIT Transactions on Ecology and the Environment*, *142*, 653–663. doi:10.2495/SW100591

Veltmeyer, H., & Petras, J. (2011). *Beyond Pragmatic Neoliberalism: From Social Inclusion and Poverty Reduction to Equality and Social Change*. Paper presented at the UNESCO "Rethinking Development: Ethics and Social Inclusion", Mexico City.

Verniers, C., & Vala, J. (2018). Justifying gender discrimination in the workplace: The mediating role of motherhood myths. *PLoS One*, *13*(1), e0190657. doi:10.1371/journal.pone.0190657 PMID:29315326

Vicentijevic, D. (2016). Strengthening intercultural dialogue by cross -border cooperation projects in the field of cultural tourism. *Kultura*, (152), 239–256. doi:10.5937/kultura1652239v

Vikan, A., Camino, C., Biaggio, A., & Nordvik, H. (2007). Endorsement of the new ecological paradigm: A comparison of two Brazilian samples and one Norwegian sample. *Environment and Behavior*, *39*(2), 217–228. doi:10.1177/0013916506286946

Vilas-Boas, V., & Sousa, B. (2019). The Role of Service Quality in Predisposition for Portuguese Online Commerce. *Quality - Access to Success, 20*(173), 12-17.

Viviparchi.eu. (2015). *Parco avventura orme nel parco-divertimento-zagarise-catanzaro*. Available at: https://www.viviparchi.eu/strutture-dettaglio.asp?catalogueProdId=356&catalogueCatId=3&catalogueClassId=

Voit, C. (2010). The concept of Health Tourism. An Overview. In C. Voigt & C. Pforr (Eds.), *Welness tourism. A destination perspective*. Routledge.

Volkmann, C. K., Tokarski, K. O., & Ernst, K. (2012). *Background, Characteristics and Context of Social Entrepreneurship Social entrepreneurship and social business*. Springer. doi:10.1007/978-3-8349-7093-0

Waite, R., McKinney, N., Smith-Glasgow, M. E., & Meloy, F. A. (2014). The embodiment of authentic leadership. *Journal of Professional Nursing*, *30*(4), 282–291. doi:10.1016/j.profnurs.2013.11.004 PMID:25150413

Wallace, G. N., & Pierce, S. M. (1996). An evaluation of ecotourism in Amazonas, Brazil. *Annals of Tourism Research*, *23*(4), 843–873. doi:10.1016/0160-7383(96)00009-6

Wall, G. (1993). International collaboration in the search for sustainable tourism in Bali, Indonesia. *Journal of Sustainable Tourism*, *1*(1), 38–47. doi:10.1080/09669589309450699

Wall, G. (1997). Is Ecotourism Sustainable? *Environmental Management*, *21*(4), 483–491. doi:10.1007002679900044 PMID:9175538

Walpole, M., & Goodwin, H. J. (2000). Local economic impacts of dragon tourism in Indonesia. *Annals of Tourism Research, 27*(3), 559–576. doi:10.1016/S0160-7383(99)00088-2

Walpole, M., Goodwin, H., & Ward, K. (2001). Pricing Policy for Tourism in Protected Areas: Lessons from Komodo National Park, Indonesia. *Conservation Biology, 15*(1), 218–227. doi:10.1111/j.1523-1739.2001.99231.x

Wattanakuljarus, A., & Coxhead, I. (2008). Is tourism-based development good for the poor? A general equilibrium analysis for Thailand. *Journal of Policy Modeling, 30*(6), 929–955. doi:10.1016/j.jpolmod.2008.02.006

Waudby, B., & Poulston, J. (2017). Sexualisation and harassment in hospitality workplaces: Who is responsible? *International Journal of Culture, Tourism and Hospitality Research, 11*(4), 483–499. doi:10.1108/IJCTHR-10-2016-0102

WCED. (1987). World commission on environment and development. *Our Common Future, 17*, 1-91.

Wearing, S., & Darcy, S. (2011). Inclusiveness of the 'Othered' in Tourism. *Cosmopolitan Civil Societies: An Interdisciplinary Journal, 3*(2), 18–34.

Weeden, K. A., Newhart, M., & Gelbgiser, D. (2018). *Occupational segregation. Stanford Center on Poverty and Inequality*. Retrieved from https://inequality.stanford.edu/sites/default/files/Pathways_SOTU_2018_occupational-segregation.pdf

Wehril, R., Priskin, J., Demarmels, S., Kolberg, S., Schaff, D., Schwarz, J., Truniger, F., & Stettlerand, J. (2013). *How to communicate sustainable tourism products effectively to customers: An explorative study presented at the World Tourism Forum Lucerne*. Lucerne University of Applied Sciences and Arts.

Whiley, M. (2020). Tourism in Australia is taking a hit with the 'double whammy' of bushfires and coronavirus travel bans, here's how the country is enticing tourists in an era of travel uncertainty. *Business Insider Australia*. Retrieved March, 12, 2020, from https://www.businessinsider.com.au/australia-extends-visas-young-backpacker-tourists-travel-bushfire-recovery-2020-2?r=US&IR=T

Whiteley, S. B. (1999). Calculating the sustainable yield of Tasmania's State forests. *Tasforests, 11*(1), 23–34.

WHO. (1948). *Officials Records of the World Health Organization*. United Nations.

Widawski, K., Jary, Z., Oleśniewicz, P., Owczarek, P., Markiewicz-Patkowska, J., & Zaręba, A. (2018). Attractiveness of protected areas for geotourism purposes from the perspective of visitors: The example of Babiogórski National Park (Poland). *Open Geosciences, 10*(1), 358–366. doi:10.1515/geo-2018-0028

Wiernik, M., Ones, S. D., & Dilchert, S. (2013). Age and environmental sustainability: A meta-analysis. *Journal of Managerial Psychology, 28*(7/8), 826–856. doi:10.1108/JMP-07-2013-0221

Wilkins, K. (2019, April 14). Bridestowe luxury accommodation development to be considered by Dorset Council. *The Examiner.*

Williams, A. M., & Hall, M. C. (2002). Tourism, migration, circulation and mobility: The contingencies of time and place. In A. M. Williams & M. C. Hall (Eds.), *Tourism and Migration: New Relationships Between Production and Consumption* (pp. 1–52). Kluwer Academic Publishers. doi:10.1007/978-94-017-3554-4_1

Wilson, S., Fesenmaier, D., Fesenmaier, J., & Van Es, J. (2001). Factors for success in rural tourism development. *Journal of Travel Research, 40*(2), 132–138. doi:10.1177/004728750104000203

Wiltshier, P., & Robinson, P. (2011). Community Tourism. *Research Themes for Tourism,* 97-99.

Wine Tasmania. (2020). *Vin Zero*. Retrieved from http://winetasmania.com.au/vinzero

Wishitemi, B., Momanyi, S., Ombati, B., & Okello, M. (2015). The link between poverty, environment and ecotourism development in areas adjacent to Maasai Mara and Amboseli protected areas, Kenya. *Tourism Management Perspectives*, *16*, 306–317. doi:10.1016/j.tmp.2015.07.003

Wood, S., Jones, K., & Brauman K. (2018). Distilling the role of ecosystem services in the Sustainable Development Goals. *Ecosystem Services, 29*(Part A), 70-82. Doi:10.1016/j.ecoser.2017.10.010

Wood-Dauphinée, S., Exner, G., Bostanci, B., Exner, G., Glass, C., & Jochheim, K. A. (2002). Quality of life in patients with spinal cord injury-basic issues, assessment, and recommendations. *Restorative Neurology and Neuroscience*, *20*(3-4), 135–149. PMID:12454362

Woodley, A. (1992). Tourism and Sustainable Development: the Community Perspective. In J. G. Nelson, R. Butler, & G. Wall (Eds.), *Tourism and Sustainable Development: Monitoring* (pp. 135–147). Planning and Managing.

Workplace Gender Equality Agency. (2020). *About workplace gender equality*. Australian Government. Retrieved from https://www.wgea.gov.au/topics/about-workplace-gender-equality

World Bank. (2020). *Fragility and conflict: On the front lines of the fight against poverty*. World Bank.

World Heritage Committee. (2007). The Tasmanian wilderness world heritage area. *Australian Government Department of the Environment and Heritage*. Retrieved June, 8, 2020, from https://www.environment.gov.au/system/files/resources/63307db8-6b28-44d8-807b-6979eeab9a41/files/tas-wilderness.pdf

World Trade Organization. (2002). *Tourism and poverty alleviation*. Author.

World Travel & Tourism Council (WTTC). (2018). *Travel & Tourism economic impact 2018 in Cape Verde*. WTTC.

World Travel & Tourism Council. (2018). *Travel & Tourism Economic Impact. 2018 Portugal*. Author.

World Travel & Tourism Council. (2019). *Economic Impact 2019 World*. Retrieved from https://www.wttc.org/-/media/files/reports/economic-impact-research/regions-2019/world2019.pdf

World Travel & Tourism Council. (2019). *Travel and Tourism: Driving Women's Success*. Retrieved from https://www.wttc.org/-/media/files/reports/2019/driving-womens-success-2019.pdf

Wrelton, E. (2006). *'Reality Tours' to Chiapas, Mexico: The Role of Justice Tourism in Development*. Academic Press.

WTO. (1998). *Guide for local authorities on developing sustainable tourism*. https://www.e-unwto.org/doi/pdf/10.18111/9789284402809

WWF Internacional. (2001). *Guidelines for community-based ecotourism development*. Retrieved March 14, 2020, from https://wwf.panda.org/wwf_news/?12002/Guidelines-for-Community-based-Ecotourism-Development

Xiang, L., & Petrick, J. F. (2006). A review of festival and event motivation studies. *Event Management*, *9*(4), 239–245. doi:10.3727/152599506776771526

Yadav, R., Balaji, M. S., & Jebarajakirthy, C. (2019). How psychological and contextual factors contribute to travelers' propensity to choose green hotels? *International Journal of Hospitality Management*, *77*, 385–395. doi:10.1016/j.ijhm.2018.08.002

Yalınkılıç, M. K., & Arpa NY. (2005). *Türkiye'deki Korunan Alanlar ve Ekoturizm*. Korunan Doğal Alanlar Sempozyumu Sözlü Bildiriler Kitabı, 3-13, Isparta. Yeşil P. 2009. Kelkit Havzası Tokat Kesiminin Biyosfer Rezervi.

Yazzo, M. A. (2015). Educación inclusiva superior. Una mirada desde la acreditación de calidad en Colombia. *Revista Inclusiones*, *2*(4), 317–326.

Yeh, C. M. (2019). The influence of foreign institutional investors, institutional directors, and the share pledge ratio of directors on financial performance of tourism firms. *Tourism Economics, 26*(1), 179–201. doi:10.1177/1354816619835613

Yenil, Ü., & Akyazı, N. (2019). Sürdürülebilir Turizm Kapsamında Kültürel İz Koridoru Değerlendirmesi. *Niğde Ömer Halis Demir Üniversitesi Mühendislik Bilimleri Dergisi, 8*(2), 1179–1192.

Yergeau, M. (2020). Tourism and local welfare: A multilevel analysis in Nepal's protected areas. *World Development, 127*, 1–19. doi:10.1016/j.worlddev.2019.104744

Yermachenko, V., Dekhtyar, N., & Dorokhov, O. (2015). Business Tourism Development on the Basis of Public–Private Partnership. *Tourism Analysis, 20*(4), 433–439. doi:10.3727/108354215X14400815080640

Yu, L., Wang, G., & Marcouiller, D. (2019). A scientometric review of pro-poor tourism research: Visualisation and analysis. *Tourism Management Perspectives, 30*, 75–88. doi:10.1016/j.tmp.2019.02.005

Yu, Q., & Yin, J. (2018). Evaluating urban eco-tourism resources and environment: A case study in Shanghai Chenshan Botanical Garden, China. *E&ES, 121*(3), 032008. doi:10.1088/1755-1315/121/3/032008

Zaleznik, A. (2004). Managers and leaders. *Harvard Business Review*, 1. PMID:14723179

Zamora, N., & Obando, V. (2001). *Biodiversity and tourism in Costa Rica.* Instituto Costarricense de Turismo, ICT; Sistema Nacional de Áreas de Conservación, SINAC; Ministerio del Ambiente y Energía, MINAE; Instituto Nacional de Biodiversidad, INBio.

Zarandian, N., Shalbafian, A., Ryan, C., & Bidokhti, A. A. (2016). Islamic pro-poor and volunteer tourism—The impacts on tourists: A case study of Shabake Talayedaran Jihad, Teheran—A research note. *Tourism Management Perspectives, 19*, 165–169. doi:10.1016/j.tmp.2015.12.005

Zebrowski, R.L. (2009). *Social Entrepreneurship as a New Paradigm: Rethinking Development Intervention Ideologies.* Academic Press.

Zeppel, H. (2006). Indigenous Ecotourism: Sustainable Development and Management. Ecotourism Series No.3. CABI. doi:10.1079/9781845931247.0000

Zhao, W., & Ritchie, B. (2007). Tourism and poverty alleviation: An integrative research framework. *Current Issues in Tourism, 10*(2-3), 119–143. doi:10.2167/cit296.0

Zhong, L. Z., & Liu, L. (2017). Ecotourism development in China: Achievements, problems and strategies. *Journal of Resources and Ecology, 8*(5), 441–448. doi:10.5814/j.issn.1674-764x.2017.05.001

Zhu, W., Riggio, R. E., Avolio, B. J., & Sosik, J. J. (2011). The effect of leadership on follower moral identity: Does transformational/transactional style make a difference? *Journal of Leadership & Organizational Studies, 18*(2), 150–163. doi:10.1177/1548051810396714

Zouros, N., & Mc Keever, P. (2008). The European Geoparks Network. *Proceedings. 3rd International UNESCO Conference on Geoparks,* 11-12.

About the Contributors

Filipa Brandão holds a Ph.D. in Tourism and a MSc in Tourism Management and Development from the University of Aveiro. She is Assistant Professor at the at the Department of Economics, Management, Industrial Engineering and Tourism (DEGEIT) of the University of Aveiro. She is a full member of the Research Unit in 'Governance, Competitiveness and Public Policies' of the University of Aveiro (GOVCOPP-UA), Vice- Director of the Master in Tourism Management and Planning, and Associate Editor of Tourism & Development Journal. She authored and co-authored books, book chapters and papers published in international journals. She is also involved in applied research and consultancy projects in the field of Tourism and Leisure and is a member of the organizing and scientific committees of international tourism conferences. Her research and scientific production focus on tourism innovation, regional development, social network analysis, and tourism destinations' planning and management.

Zelia Breda holds a PhD in Tourism, a MA in Chinese Studies (Business and International Relations) and a BSc in Tourism Management and Planning from the University of Aveiro, where the editor is an Assistant Professor, at the Department of Economics, Management, and Industrial Engineering, holding the position of Director of the Master in Tourism Management and Planning. The editor is a member of the Research Unit 'Governance, Competitiveness and Public Policies' of the University of Aveiro; a founding member and vice-president of the Observatory of China, and a senior researcher of the tourism R&D spin-off company Idtour. The editor is also a member of the editorial and scientific boards of a few academic national and international journals, as well a member of the organizing and scientific committees of international tourism conferences. The editor has authored and co-authored several national and international papers and communications on tourism. The editor has also been taking part in several research projects in the tourism field, both as member of the team and as a consultant.

* * *

Manuel Au-Yong-Oliveira did his post-doctoral studies at the University of Aveiro (2016-2019) with the title "Elements that contribute to organisational competitiveness". He has a PhD in Industrial Engineering and Management, from the Faculty of Engineering of the University of Porto (2012; distinction awarded to the thesis), an MBA from Cardiff Business School (1993; distinction awarded to the dissertation). He is an Assistant Professor at the University of Aveiro (DEGEIT), where he lectures at the undergraduate, master's and doctoral levels on marketing, innovation, strategy and research methods. He has over 200 research publications. Manuel is the Director of the Master's in Management and a member of the Executive Committee at DEGEIT - the Department of Economics, Management, Industrial Engineering and Tourism at the University of Aveiro.

Almudena Barrientos-Báez holds a PhD in Education and is a Professor at the Iriarte University School of Tourism, attached to the University of La Laguna (Tenerife-Spain) and the European University; a Master in Management of Protocol, Production, Organization and Design of Events (Communication area) from the University Camilo José Cela, and a Master in Management of Tourist Accommodation from the University of Girona; and a Degree in Tourism (EUTI-ULL) and Teaching (University of Valencia). She is part of the DEBATv project "Televised Electoral Debates in Spain: Models, Process, Diagnosis and Proposal" (CSO2017-83159-R), R+D+I project financed by the Ministry of Science, Innovation and Universities and the State Research Agency (AEI) of the Government of Spain, with the support of the European Regional Development Fund (ERDF) of the European Union (EU). She is also part of the project "New values, governance, financing and public audiovisual services for the Internet society: European and Spanish contrasts" (RTI2018-096065-B-I00) of the R+D+I State Research Program oriented to the Challenges of the Society of the Ministry of Science, Innovation and Universities (MCIU), State Research Agency (AEI) and the European Regional Development Fund (FEDER). Her main research interests are tourism education, communication, social media, business management and intermediation, emotional intelligence, and gender.

Tessa Campbell is a third-year Business student, majoring in Business Management and International Logistics at The University of Tasmania. She is interested is organisational culture, behaviour and leadership styles, along with logistics and strategic development within an organisation. She is fascinated with the intricacies of leadership styles within gender dominated industries, and may pursue further research in this area.

Giuseppe Umberto Cantafio is a Lecturer at the University of Sunderland in London, in the Post-graduation in Business and Management, teaching Global Strategy, Managing Strategy, Entrepreneurship and Innovation, Innovation Management, Operations, Global Strategy, Project Management, e-Logistics and Supply Chain Management, and Marketing. His interests are GIS software and Data Analysis, Urban Planning, Innovation, Environmental Sustainability.

Emanuel Castro holds a Bachelor's degree in Geography, specialisation in Environmental Studies, a Master's Degree in Geography and Spatial Planning, and he is pursuing a PhD in Geography at the University of Coimbra. Between January 2003 and September 2017, he lectured at the Polytechnic Institute of Guarda and has participated in different research projects in the areas of Tourism, Landscape Analysis, Territorial Development and Geographic Information Systems. Currently, he is the Executive Coordinator of the Estrela Geopark Association, and manager of the Estrela UNESCO Global Geopark brand.

Elena Cavagnaro is a Professor of Sustainability in Hospitality and Tourism at Stenden University of Applied Sciences in the Netherlands. She is also a Visiting Professor at the University of Derby (UK) and the University of Bergamo (Italy). She has consulted several organisations in sectors such as hospitality, retail, and health care on sustainability strategy and implementation. In line with her understanding of sustainability as a multi-dimensional and multi-layered concept, her research focuses on issues that run across and connect the social, organisational and individual layers of sustainability. The book "Three Levels of Sustainability" that she co-authored with G. Curiel is a bestseller.

Yana Chankova is an Associate Professor at the Department of Foreign Languages at the South-West University "Neofit Rilski" of Blagoevgrad. Her major line of inquiry is based on multilingualism, cultural studies, strategies of intercultural communication and historical linguistics. She is a co-author of the Bulgarian translations of three poems from "The Poetic Edda".

Monika Chodilíková graduated in English Language for Business Administration from the Tomas Bata University in Zlín. Currently, she is a student of a Master's Degree in Marketing Communications at the same university. This study was conducted while she was an Erasmus exchange Master's student at the University of Aveiro.

Anna Cui is a current undergraduate completing a Bachelor of Business Management at the University of Tasmania. She is interested in organizational leadership and its impact on various stakeholders, including the wider community. She believes that good leaders are foundational within organisations and for the health and well-being of society. Based on her own experience and skills in organisational management and leadership, she sees the need for continued research in the area of promoting ethical leadership in Australia.

Vanessa de Oliveira Menezes graduated in Tourism from the UNIOESTE, Brazil, with a specialisation in Human Resources Management from the UFPR, Brazil. Vanessa has a Master in Tourism and Hotel Management from the University of Extremadura, Spain, a PhD in Business Management from the Universidade Positivo, Brazil, and a post-doctorate from NHL Stenden University of Applied Sciences (2020), the Netherlands. She is a Professor at the Department of Tourism at Unicentro in Brazil. Her academic experience as a lecturer was devoted to technical, undergraduate, and postgraduate courses, dedicating her studies on the following topics: tourism, tourism and hospitality management, sustainability in hospitality, tourism and hospitality innovation, human resources in tourism and hotels.

Luke DeJesus is in his final year of a double Bachelor of Business and Economics at the Tasmanian School of Business and Economics at the University of Tasmania. Luke's career progression through his current career in sales has shaped his studies majoring in analytical analysis and marketing. Luke seeks to further this knowledge in postgraduate study, with a focus on macroeconomics and its relationship with human agency and behaviour. Luke's ambition to contribute to the field and research of economic theory has led to his investigation in areas such Fintech and the Tasmanian Tourism industry.

Gonçalo Poeta Fernandes is a Coordinator Professor at the Polytechnic Institute of Guarda – ESTH/ Scientific-Technical Unit of Tourism and Hospitality (Portugal). He holds a PhD in Geography and Regional Planning, from the Faculty of Social Sciences and Humanities of the New University of Lisbon. His research areas include Tourism and Development, Planning and Territorial Management, Public Policy and Border Areas, Rural Entrepreneurship, Territory and Tourism Destinations, Mountains and Geotourism. He is a Researcher at the Centre for Tourism Research, Development and Innovation (CI-TUR), at the Interdisciplinary Centre of Social Sciences (CICS.NOVA) and at the Inland Development Research Unit from the Polytechnic Institute of Guarda.

Vítor Figueiredo holds a PhD in Management from the University of Beira Interior, where he is an Invited Assistant Professor and a Researcher at NECE/UBI.

Eduardo Cordeiro Gonçalves holds a PhD from the University of Porto. He is Professor at the University Institute of Maia – ISMAI, and Coordinator of RG1 – Tourism at CETRAD, CEDTUR.

Ricardo Guerra holds a PhD in Tourism, Leisure and Culture from the University of Coimbra. He is Professor at the School of Tourism and Hospitality Management of the Polytechnic Institute of Guarda, and Researcher at CETRAD, CITUR and UDI/IPG.

Nadine Heleno holds an MSc in Tourism Management and Planning from the University of Aveiro, having as topics of interest Sustainable Development Goals and Community-Based Tourism, and how the tourism can contribute positively for a local and regional sustainable development. Nadine also holds a BSc in Tourism from the University of Aveiro.

Teodora Kiryakova-Dineva is a scientist with an interdisciplinary profile. Her research interests lie in tourism studies, intercultural studies and all types of intangible assets, including cultural management and theoretical and practical approaches to contemporary phenomena. She is an Associate Professor and academic member of the South-West University "Neofit Rilski", Blagoevgrad, Bulgaria.

Matthew Knox is a recent graduate of the University of Tasmania with a passion for understanding how to create a better world. His current work is focused on understanding authentic followers and their role in enabling ethical leader behaviours.

Iveta Košárková earned a Bachelor's Degree in English for Business Administration from Tomas Bata University (TBU) in Zlín. Her bachelor's thesis, titled "Masaryk and Wilson: Founding Fathers of Czechoslovakia" was awarded a prize within the university by a committee during the diploma exams, in 2018. She is studying Marketing Communications at the Faculty of Multimedia Communications of TBU. The study was conducted while Iveta was an Erasmus exchange Master student at the University of Aveiro.

Patryk Lewandowski is a student of logistics at the University of Lodz, with extensive experience and management skills. His passion is travelling and sustainable development. Environmental issues are very important to him, which is why, in his works, he raises the problem of hunger, waste and exploitation at work. This article was written while the author was on an Erasmus exchange as a Master's student at the University of Aveiro.

Cecília Lobo graduated in Languages and Business Relations from the University of Aveiro in 2017. In 2019, on that same educational institution, she completed her Master's Degree in Tourism Management and Planning, from the same university, presenting a dissertation entitled "The importance of events on a destination's image: The case of the city of Viseu". Her research interests cover several topics, such as event tourism, as well as sustainability, addressing social and environmental perspectives.

Mariana Marques is a higher education teacher, with a PhD in Communication Sciences, a master's degree in Tourism and a graduate in Marketing and Communication. She is also a PhD candidate in Tourism at the University of Aveiro and a researcher at CICANT, where she has been developing research in the tourism field.

José Alberto Martínez González is a Professor of Business Organization, Digital Economy and Tourism at the Department of Business Management and Economic History of the University of La Laguna. He holds a Degree in Economic Sciences and in Psychopedagogy, and a PhD in Psychology and in Economy, Business and Tourism from the University of La Laguna.

Isabel Silva Martins is a PhD Student in Tourism at the University of Aveiro and has a Master in Management from the ISEG Lisbon School of Economics & Management. She is an invited Assistant at the Polytechnic Institute of Cávado and Ave, where she lectures in the areas of Management, Tourism and Marketing (Analysis of Tourism Markets, Planning and Management of Tourism Destinations and Accessible Tourism). She is also a member of GOVCOPP Research Unit, being part of the Tourism and Development Research Group. Her research interests are souvenirs, shopping, authenticity, identity and geographical representations.

Alda Matos holds a PhD from the University of Law, Economics and Social Sciences (Paris II), a Master's in Economics (Paris II) and a Degree in Economics from the University of Coimbra. Since 2018, she teaches at the Infante D. Henrique Nautical School, having been previously affiliated with the European University (2007-2013) and the Autonomous University of Lisbon (1990-2003).

Catarina Meira is a graduate in Tourism Management from the Polytechnic Institute of Cávado and Ave, and manager-administrative at Fundação AMA Autismo. She is also a volunteer at Refood in Portugal.

Vipin Nadda is currently serving as a Lecturer and Programme Manager (Postgraduate - Tourism & Hospitality) at the University of Sunderland in London. He is a science graduate with an MBA in Marketing, MTA in Tourism and a PhD in Tourism Management. He has more than fifteen years of experience in academia and industry, having published two books and presented papers in various international seminars and global conferences. He also freelances as 'Education development consultant' for the University of East London, Glyndwr University, and Anglia Ruskin University. He is the Director of studies for PhD with Cardiff Metropolitan University, 'Member of Academic Council' and Chief Examiner for Confederation of Tourism and Hospitality, London. He has been lecturing a variety of courses ranging from Marketing, HR in 'Business and Management' to 'International Tourism and Hospitality Management' in the UK as well as overseas.

Joan Mwihaki Nyika is an Environmental Science PhD student at the University of South Africa and a Tutorial Fellow at the Technical University of Kenya, Geoscience and the Environment Department. She obtained MSc and BSc degrees in Land and Water Management and Biochemistry from the University of Nairobi and Jomo Kenyatta University of Agriculture and Technology, Kenya in 2017 and 2011, respectively. Her PhD work revolves around solid waste management and the effects of landfill leachate on soil and groundwater. She has interests in sustainable environmental management, water management and ecosystem management. Joan has published more than 15 articles and 5 book chapters in peer-reviewed journals and books, respectively. She has and is mentoring both undergraduate and diploma students and has more than 5 years of experience teaching and training in environmental science.

Georgia Papadopoulou is an Assistant Professor at Al Ghurair University in Dubai, UAE. She has previously served as a Lecturer at the Dublin City University on secondment to the Princess Nourah University in Riyadh, Saudi Arabia. She teaches courses in management and economics in both undergraduate and graduate programs. Her research interests include cruise tourism, tourism economics, tourists' satisfaction and motivations, residents' well-being and happiness.

Eduardo Parra López is a Professor of Business Organisation, Digital Economy and Tourism at the University of La Laguna. Is Visiting Scholar of Universidad del Valle (Guatemala). He was the President of the Spanish Association of Scientific Experts in Tourism (AECIT, 2015-2020). He is graduated in Economic and Business Sciences, holds a Diploma in Business Studies, a Master in International Business Program, and a PhD from the University of La Laguna. Researcher at Institute of Social Science and Tourism (ISTUR).

Giang Phi is an Assistant Professor in Tourism Analysis, Innovation and Change at the Faculty of Humanities of Aalborg University, Denmark. She holds a PhD in Tourism Management from Griffith University, Australia and researches in the areas of event and tourism management, social entrepreneurship, design thinking and sustainable development. Giang is also interested in addressing tourism methodological issues with innovative research methods, such as Q-methodology and Theory of Change.

Kristýna Pšeničková is a Master student in Enterprise in the International Environment at the Faculty of Economics at the Technical University of Liberec. In 2018, she obtained a Bachelor's degree in Economics and Management of International Trade at the same university. She participated in the Erasmus study programme at the University of Aveiro in Portugal in 2019. She cooperated in the Project SGS "Employee Value Proposition Identification (EVP): Research into factors affecting employees in choosing a job". This study was conducted while she was on an Erasmus exchange mobility as a Master's student at the University of Aveiro.

Josh Rowlands is an ambitious and entrepreneurial individual, he firmly believes that sound leadership, particularly Authentic leadership, is of upmost importance in the development of business, including tourism. He is interested in the tourism industry and Tasmania, with a particular focus on the Eastern Market. He has lived in China for several years, and his experience overseas has helped him to understand how Authentic Leadership can enable tourism operators in Tasmania to effectively engage with this in Tasmania can effectively engage with this increasingly important tourism market.

Eduardo Moraes Sarmento is an international consultant in projects for the European Union and several countries, such as Cape Verde, Angola, and Portugal. He holds a PhD in Economics from ISEG - Lisbon School of Economics & Management (University of Lisbon) with public examinations in Management/Tourism at the University of Aveiro. He is Full Professor at Lusófona University, with teaching experience in several private and public universities. He is also a Researcher at Center for African and Development Studies (CEsA, University of Lisbon) and president of CIPES (Lusófona University), being author and co-author of many papers published in international peer-reviewed journals (Q1, Q2, and others). He is a reviewer for many journals, member of scientific commissions, and coordinator of several master dissertations and doctoral thesis, as well as jury member, either as President or Discussant.

Lakhvinder Singh is working as an Assistant Professor of Tourism in Government College, Kaithal, affiliated to Kurukshetra University, Kurukshetra, Haryana, India. He holds a PhD Degree, with UGC-NET qualification, in Tourism Administration & Management. He has more than three years of teaching experience and has more than 30 research publications in his credit. He is in the Editorial/Advisory Board of different journals of international and national repute. His areas of interest include tourist behaviour, tourism marketing, tourism impact, culture & religious tourism, e-tourism and ICT in tourism.

Bruno Sousa is a Professor at the Polytechnic Institute of Cavado and Ave (IPCA, Portugal) and Head of the Master Programme in Tourism Management. He holds a PhD in Marketing and Strategy and was Market Analyst at Sonae Distribuição – Modelo e Continente, S.A. (2006 to 2009) and Marketing Assistant at the newspaper "O Jogo" at Controlinveste (2005). He has received several awards: Teaching Award of the School of Economics and Management of the University of Minho 2015/2016; Best Thesis in Tourism Award; ICIEMC 2015; Management Graduation, University of Minho Award; Best performance (2006); Merit Scholarship for Students in Public Higher Education Awards of Merit Scholarship by the University of Minho in 2001/02 - 2002/03 - 2003/04. He is a researcher at the Centre for Tourism Research, Development and Innovation (CiTUR) and Applied Management Research Unit (UNIAG). He is author or co-author of several papers, and his research interests include tourism management, marketing and strategy. He serves in the Editorial Board of several peer-reviewed scientific journals and is member of the scientific committee of several national and international congresses and conferences.

Iwona Staniec works as an Assistant Professor at the Department of Management of Lodz University of Technology, Poland. Her research interests are data analysis, statistics, econometrics, economic forecasting, risk management, innovation risk, risk analysis and assessment, technological entrepreneurship, creating models and forecasting risk in public and private pension systems and social security. She has benefited from several research grants to carry out her research activity.

Nina Szczygiel is an Assistant Professor at the Department of Economics, Management, Industrial Engineering and Tourism of the University of Aveiro, Portugal, and a research fellow at the Research Unit on Governance, Competitiveness and Public Policies (GOVCOPP). She holds a PhD in Industrial Management from the University of Aveiro with a grant from the Portuguese Foundation for Science and Technology (FCT). She is particularly interested in quality of life and well-being in societies, organisations and individuals. She has been working on intersectoral, inter-organisational and social networks, integration and mobility. She has authored and/or co-authored papers in journals, conference proceedings, book chapters and a book.

Rosa Branca Tracana is a Professor at the School of Education, Sports and Communication of the Polytechnic Institute of Guarda, Portugal, in the area of science education. She is a member of the Scientific Council of Estrela Geopark Association. Her areas of expertise are science education, environmental education and biology. She has supervised several masters dissertations and published 25 international articles and nine chapters in international books. She also has participated in several international projects and participated in several scientific commissions of international conferences. She is a member of the Editorial Board of four international scientific journals and also serving as a referee of several ones.

Gül Yılmaz graduated from Istanbul University, Institute of Social Sciences, Department of Human and Economic Geography in 1995. She lectured at Trakya University and Çanakkale Onsekiz Mart University on General Tourism, Tourism Management, Food & Beverage Management, Housekeeping, and Tourism Planning etc. She involved in various practices in the field of Food & Beverage Management. After being retired from Çanakkale Onsekiz Mart University, she established the Gastronomy Department at Applied Sciences Academy of Okan University in between years 2008-2011. By opening the Cookery Certificate Program at Continuing Education Center of Okan University, she ensured the training of certified cooks for the sector. She participated as an arbitrator at International Cookery Contests. She lectured World Cuisines course along with the expert chefs of the sector. She is still serving as Academic Member at Arel University, School of Economics and Administrative Sciences, Department of Tourism and Hospitality Management.

Index

H

I

J

K

L

M

N

O

P

Q

R

S

T

U

V

W

www.ingramcontent.com/pod-product-compliance
Lightning Source LLC
LaVergne TN
LVHW060404300926
849162LV00006B/1338